Chronobiology & Chronomedicine
Basic Research and Applications
Proceedings of the 7th Annual Meeting of the European
Society for Chronobiology, Marburg 1991

CHRONOBIOLOGY & CHRONOMEDICINE

BASIC RESEARCH AND APPLICATIONS

PROCEEDINGS OF THE 7TH ANNUAL MEETING
OF THE EUROPEAN SOCIETY FOR CHRONOBIOLOGY,
MARBURG 1991

EDITED BY
CHRISTOPH GUTENBRUNNER,
GUNTHER HILDEBRANDT
AND RUDOLF MOOG

PETER LANG
Frankfurt am Main · Berlin · Bern · New York · Paris · Wien

Die Deutsche Bibliothek - CIP-Einheitsaufnahme

Chronobiology & chronomedicine : basic research and
applications ; proceedings of the ... annual meeting of the
European Society for Chronobiology ... - Frankfurt am Main ;
Berlin ; Bern ; New York ; Paris ; Wien : Lang, 1993
   ISSN 0939-2351

NE: European Society for Chronobiology; Chronobiology and
   chronomedicine

7. Marburg, 1991. - 1993
   ISBN 3-631-45239-X

ISBN 3-631-45239-X
© Verlag Peter Lang GmbH, Frankfurt am Main 1993
Alle Rechte vorbehalten.

Das Werk einschließlich aller seiner Teile ist urheberrechtlich
geschützt. Jede Verwertung außerhalb der engen Grenzen des
Urheberrechtsgesetzes ist ohne Zustimmung des Verlages
unzulässig und strafbar. Das gilt insbesondere für
Vervielfältigungen, Übersetzungen, Mikroverfilmungen und die
Einspeicherung und Verarbeitung in elektronischen Systemen.

Printed in Germany 1 2 3 5 6 7

PREFACE

Since the first research activities in the field of chronobiology more than 50 years ago, its importance in biology and medicine dramatically increased. On the one hand basic research about the mechanisms and properties of biological rhythms made great progress, on the other hand the applications in medicine extended. The latter concerns both diagnosis and therapy.

In order to intensify both scientific exchange and discussion between scientists working in diffenent disciplines and living in different European countries in 1985 the European Society for Chronobiology was founded in Leiden (The Netherlands). One of the most important aim of the founding assembly was to promote the exchange between Eastern and Western European countries. However, during the first six meetings of this society, there were many difficulties to realize this purpose. The new political situation in Europe just before the start of the 7th Annual Meeting of the European Society for Chronobiology (held in Marburg, May 30 - June 2, 1991) opened borders for the international scientific exchange too. Hence 70 scientists from "Eastern European" countries were able to take part and to give insights in their work. All in all 257 scientist from 26 countries attended the meeting.

The VIIth Annual Meeting of the European Society for Chronobiology under the presidency of L. RENSING (Bremen) was organized by G. HILDEBRANDT (Marburg) and J. SCHUH (Halle). Corresponding to the increasing interest in Chronobiology, the number of scientific contributions was higher than in the preceding meetings of this society. They concerned the following themes: coordination and interaction, entrainment, photoperiod, and of light effects, clinical aspects, development and aging, basic and cellular aspects, monitoring, data analysis, occupational aspects, chronopharmacology and -toxicology. For special topics review lectures werde held.

Additionally to the presentation of new experimental results, the meeting included a special workshop on "Coupled Oscillators" and a satellite symposion on "Chronobiological Aspects of Physical Medicine and Cure Treatment". The latter was held in Bad Wildungen after the end of the Marburg meeting. Because of the special significance of these two topics, both symposia are documented in this volume.

Starting with the 2nd Annual Meeting, the contributions to the annual meetings of the European Society for Chronobiology were documented in the series "Chronobiology and Chronomedicine, basic research and applications". This series is continued with this volume.

The editors are grateful to Mrs. E. Gonnermann for the graphical assistance, and the Peter Lang-Verlag for the cooperation in the production of this volume.

Chr. Gutenbrunner
G. Hildebrandt
R. Moog

CONTENTS

BASIC AND CELLULAR ASPECTS   1

Cyclic variations of amplitude in cultures of neonatal rat heart cells continuously measured by a laser scattering technique
W. LINKE, W. BOLDT, J. SCHUH & M. WUSSLING   3

Chronobiological aspects of computerized cultures of *Chlorella Vulgaris*
H. BÖHM, H. ARNDT & H.-G. HIEKEL   8

Circadian rhythms in ecdysteroid titers in insect hemolymph
B. CYMBOROWSKI   12

Daily changes of ß-n-acetyl-d-glucosaminidase (ec 3.2.1.30.) from rat submandibular gland assayed by rocket immunoelectrophoresis
A. LITYŃSKA   16

Regulation of pineal gland function in rats - the role of nonapeptides
R. RIEMANN & S. REUSS   21

Is there a circadian rhythm of cytochrome P-450 IIE1 activity?
D. PANKOW, P. WOLNA & P. HOFFMANN   26

Motilar activity of intestine of white rats and its 24-hour rhythm
B. HARBACH & I. MLETZKO   30

Rat liver structure: Circadian rhythm of some morphometric parameters
C. DOLCI, F. CARANDENTE, L. VIZZOTTO, A. MONTARULI & A. MIANI   35

Time intervals and amplitudes of ultradian oscillations of carbon dioxide emission ($\dot{V}CO_2$) in several endothermic species (mice, rats, guinea-pigs, quail, chicks, monkeys and premature infants)
V. GOURLET, M. STUPFEL, A. PERRAMON, P. MERAT, G. PUTET & L. COURT   41

Seasonal differences in qualitative and quantitative indices of myocardial ultrastructure in rabbits
E.S. MATYEV, V.A. FROLOV, S.M. CHIBISOV & V. MOGYLEVSKY   47

Are there seasonal differences in rates of regeneration in red-spotted newts, *Notophthalmus Viridescens*?
M.F. BENNETT   53

Seasonal changes of thermoregulation in species of different ecological specialization
M.P. MOSHKIN & E.I. ZHIGULINA   57

Long-term variations in solar and geomagnetic activities
J. STRESTIK & I. CHARVATOVA   63

II

DEVELOPMENT AND AGING 69

Development of cardiovascular and temperature rhythms in neonates
U. SITKA, D. WEINERT, F. HALBERG, J. SCHUH &
W. RUMLER 70

Chronobiometry of rhythmogenesis in newborn infants
R. SAMMECK & U. STEPHANI 77

Dynamic variation of circadian amplitudes and acrophases depending on age
J. SCHUH, D. WEINERT, G. D. GUBIN & I. P. KOMAROV 82

The effects of thyroid state on rhythm coherence: Interactions with aging
D.L. McEACHRON & N.T. ADLER 88

Ontogeny and 24-h rhythms
H.-G. MLETZKO & I. MLETZKO 93

Spectral analysis of heart rate variability in premature newborns
N. HONZÍKOVÁ, B. FIŠER & E. KONVIČKOVÁ 98

Evaluation of a neonatal cardiovascular risk score from blood pressure and heart rate characteristics
J.R. FERNÁNDEZ, R.C. HERMIDA, D.E. AYALA, A. MOJÓN, A. REY, J. RODRÍGUEZ-CERVILLA, J.R. FERNÁNDEZ-LORENZO & J.M. FRAGA 103

ENTRAINMENT, PHOTOPERIOD, LIGHT EFFECTS 111

A hypothesis on the evolutionary origins of photoperiodism based on circadian rhythmicity of melatonin in phylogenetically distant organisms
R. HARDELAND, B. PÖGGELER, I. BALZER &
G. BEHRMANN 113

Neuropeptides in the hypothalamic suprachiasmatic nuclei: Seasonal changes in a mammalian circadian oscillator
S. REUSS 121

On the mechanism of cyst induction in *Gonyaulax Polyedra*: Roles of photoperiodism, low temperature, 5-methoxylated indoleamines, and proton release
I. BALZER & R. HARDELAND 128

Are neurosecretory cells involved in the circadian control of locomotor activity of *Drosophila Melanogaster*?
C. HELFRICH-FÖRSTER 134

On the circadian rhythms in microphthalmic mice
H. HILBIG 141

Consequences of modified zeitgeber-conditions during juvenile phase for the circadian system of adult mice
D. WEINERT, H. EIMERT & J. SCHUH   145

The valence of photoperiod and feeding as 'zeitgebers' of the kidney rhythms of peptide-cleaving enzymes
D. BALSCHUN, S. JANKE & J. SCHUH   151

Differential effects of exogenous melatonin on correlates of oestrus in domestic rabbits
R. HUDSON   157

Weather-induced interdiurnal variations of L/D-proportion and of brightness affects man in same manner as flights in direction of latitudes
L. KLINKER & H. PIAZENA   163

Serum β-glucuronidase activity after light exposure
A. FALKENBACH, U. UNKELBACH, J. CRONIBUS & U. SEIFFERT   168

Neither 0-hour nor 12-hour temporal relations of 5-htp and L-dopa inhibit hibernation in the thirteen-lined ground squirrel, *Spermophilus Tridecemlineatus*
J.T. BURNS & K.A. ROTHENBERGER   172

An endogenous circadian rhythm in the swimming activity of the blind Mexican Cave Fish
S. CORDINER & E. MORGAN   177

Nonlinear dynamics caused by phase shifts
G. TSCHUCH & T. SÜSSMUTH   182

Circadian and circannual activity rhythms of *Microtus Brandti*4 (radde, 1861)
A. STUBBE & U. ZÖPHEL   185

COORDINATION AND INTERACTION   193

Frequency- and phase coordination of rhythmic functions in man.
G. HILDEBRANDT   194

Effect of breathing rate and amplitude on heart rate power spectra
A. PATZAK, C. JOHL & J. EBNER   216

Vegetative rhythms and perception of time
J. STEBEL   223

Coupling during bimanual alternating hand movements in normal subjects and hemi-parkinsonian patients
H. HEFTER, B. BÖSE, M. PIOTRKIEWCZ & H.-J. FREUND   229

CLINICAL ASPECTS 239

Circadian rhythms in cardiovascular diseases
N.L. ASLANIAN 240

Evaluation of the spontaneous variations of cardiovascular variables
K.UEZONO, M.BOTHMANN, G.HILDEBRANDT, R. MOOG &
T.KAWASAKI 247

Circadian blood pressure rhythm in the third trimester of pregnancy
C. MAGGIONI, B. BRODBECK, L. DETTI, G. MELLO &
G. BENZI 254

Predictable profile of blood pressure variability in the course of a healthy human pregnancy
D.E. AYALA & R.C. HERMIDA 263

A circadian noninvasive monitoring of hemodynamic parameters in patients with coronary artery disease
J. LIETAVA, A. DUKAT, I. BALAZOVJECH & Z. MIKES 271

Chronobiological aspects of valvular diseases of the heart in pregnant women
R.M. ZASLAVSKAYA, B.G. VARSHISKY & M.M. TEIBLOOM 277

Loss of the circadian rhythm of plasma concentrations of atrial natriuretic peptide in congestive heart failure
F. PORTALUPPI, L. MONTANARI, L. VERGNANI,
A. D'AMBROSI, M. FERLINI & E. DEGLI UBERTI 288

Circadian variation of heart rate and heart rate variability in short term survivors and non-survivors after acute myocardial infarction
P.VAN LEEUWEN, C.HECKMANN, P.ENGELKE, G.KESTING &
H.C.KÜMMELL 294

Rhythmic abilities in patients with functional cardiac arrhythmias
M. KAYSER & R. RICHTER 299

Changes in pulse-respiration ratio in patients with immunogenic hyperthyroidism (Graves' Disease)
C. HECKMANN, P. ENGELKE, G. KESTING, P. VAN LEEUWEN &
K.H. RUDORFF 305

Respiratory sinus arrhythmia after external cold water application
H.-J. RUDOLPH 310

On semilunar changes in cardiovascular mortality
J. SITAR 315

Do changes of microvascular blood flow of nasal mucosa play a role in occurence of the laterality rhythm of nasal breathing?
M. TAFIL-KLAWE & G. HILDEBRANDT 320

Respiratory patterns in patients with sleep apnea evaluated by
automatic techniques
T. PENZEL, U. MARX, J.H. PETER, H. SCHNEIDER &
P. VON WICHERT — 325

Circadian reactions to nCPAP-treatment
M. VOGEL, G. HILDEBRANDT, R. MOOG & J.H. PETER — 333

Circadian rhythm of erythrocyte sedimentation rate, C-reactive protein
and soluble interleukin-2 receptor in patients with active rheumatoid
arthritis
M. HEROLD & R. GÜNTHER — 339

Circadian variations of cellular immune parameters in healthy
subjects
CHR. GUTENBRUNNER & E. HEINDRICHS — 345

Circadian and circaseptan variations in tuberculin skin reaction
L. PÖLLMANN — 351

Moon cycle and acute diarrheal infections in Bratislava 1988-1990
M. MIKULECKY Jr. & P. ONDREJKA — 356

Circannual pattern of salmonella incidence predicted by
linear-nonlinear rhythmometry
R.C. HERMIDA & D.E. AYALA — 361

Multiple component analysis of plasma insulin in children with
standard and short stature
L. GARCIA, R.C. HERMIDA, A. MOJÓN, D.E. AYALA,
J.R. FERNÁNDEZ, C. LODEIRO & T. IGLESIAS — 370

Sex-dependent circadian and pulsatile secretion patterns of ACTH
and cortisol in normal man.
F. ROELFSEMA & M. FRÖLICH — 378

The clinical use of Na/K oscillations in urine
D. HAJEK — 386

Anti-tumor activity in pineal glands and in urine display similar
circannual rhythmicity
H. BARTSCH & C. BARTSCH — 390

Circannual pattern in uterine cervix cancer screening campaigns
R.C. HERMIDA, J.J. LÓPEZ-FRANCO, D.E. AYALA &
R.J. ARRÓYAVE — 396

Endocrine rhythms in patients with breast and prostate cancer
C. BARTSCH & H. BARTSCH — 405

Mechanisms underlying the pathogenesis of oscillatory activity in pala-
tal myoclonus
H. HEFTER, E. LOGIGIAN, O. WITTE, K. REINERS &
H.-J. FREUND — 411

Circadian patterns in the occurrence of panic attacks
E. BECKER, J. MARGRAF & S. SCHNEIDER 417

CHRONOPHARMACOLOGY AND -TOXICOLOGY 421

Chronopharmacokinetics of amitriptyline in rats after i.v. administration
A. RUTKOWSKA, W. PIEKOSZEWSKI & J. BRANDYS 422

Influence of galanthamine (Nivalin) on synaptic transmission in the rat neocortex in vitro
G. HESS 428

The effect of Nivalin on the facilitation of neocortical responsiveness in comparison with locomotor activity rhythm
M.H. LEWANDOWSKI 432

Daily rhythms of insecticide susceptibility in normal and resistant strains of *Musca Domestica* l. (Diptera)
D. DAHLHELM 439

The influence of some neuroleptic and neuromimetic drugs on the circadian wheel-running activity rhythm in mice under constant darkness condition (DD)
G. BARBACKA-SUROWIAK 445

Rubidium and potassium: Effects on circadian rhythms in hamsters
T.T. BAUER, H. KLEMFUSS, D.F. KRIPKE & B. PFLUG 450

Chronobiological aspects of the prenataltoxic action of chemical substances
R. SCHMIDT 457

Can dopaminergic agents induce coupling of oscillating mechanisms underlying psychomotor performance?
T. RAMMSAYER 463

On the spectrum of the reactive periods studied in patients treated with a cyclic design of pyrogenous drugs
M. WECKENMANN, J. STEGMAIER & E. RAUCH 469

Effectiveness of phytoadaptogens in different seasons
O.N. DAVYDOVA 473

OCCUPATIONAL ASPECTS 477

Circadian rhtymicity in self-chosen work-rate
G. ATKINSON, A. COLDWELLS, T. REILLY & J. WATERHOUSE 478

Sleep-wake pattern type and objective physiological characteristics during night wakefulness
V.A. CHEREPANOVA & A. PUTILOV 485

A questionnaire for self-assessment of individual profile and
adaptability of sleep-wake cycle
A.A. PUTILOV 493

Circadian rhythms of different speed tasks. Influence of sex and
morningness
A. ADAN & M. SÁNCHEZ-TURET 499

Circadian rhythm of teachers' physical activity
M. KWILECKA & K. KWILECKI 506

A study of performance rhythms: A first approximation of the
rhythmicity of different tasks in children.
D. SÁIZ, A. DIAZ & M. SÁIZ 508

Chronobiological differences in performance between trained and
untrained children
H. OSCHÜTZ 513

MONITORING, DATA ANALYSIS 519

A study of the diurnal relationship between oral and axillary
temperatures
A. ADAN & M. SÁNCHEZ-TURET 520

Axillary temperature and chronotype: A pilot diurnal study
H. ALMIRALL 527

Analysis of individual differences in rhythms
J.C. ALDRICH 533

All-purpose experimental data processing package for chronobiology
J. DOMOSLAWSKI 541

Clinorhythmometry: A procedure of periodic-linear regression analysis
for detecting of age-related trends in biological rhythms
P. CUGINI & L. DI PALMA 547

Modelfree exploratory analysis of chronobiological time-series by the
pc-program LOADFUAP
J. PEIL, H. HELWIN & S. SCHMERLING 562

WORKSHOP:
COUPLED OSCILLATORS 569

Coupled oscillators in neurospora differentiation and morphogenesis
L. RENSING 570

Modelling of cell cycle heterogeneity and synchronization with cellular automata
MARKUS,M. & A.SALVADOR 589

Nonlinear oscillations and solitonic excitations in molecular systems
W. EBELING & M. JENSSEN 603

SATELLITE-SYMPOSION:
CHRONOBIOLOGICAL ASPECTS OF PHYSICAL MEDICINE
AND CURE TREATMENT 617

Introduction
G.HILDEBRANDT 619

The time structure of physiological responses to cure treatment
G. HILDEBRANDT 624

Circadian variations in physiological responses to thermal stimuli
P. ENGEL 646

Influence of balneotherapy on skin vasomotion
W. SCHNIZER 660

Circadian variations of physical training
CHR. GUTENBRUNNER 665

Chronobiological aspects of rehabilitation of urological diseases
H.M. SCHULTHEIS & CHR. GUTENBRUNNER 681

AUTHOR INDEX 698

INDEX 703

Style sheets generated with equipment from the COMPUTERLADEN GmbH Marburg

BASIC AND CELLULAR ASPECTS

CYCLIC VARIATIONS OF AMPLITUDE IN CULTURES OF NEONATAL RAT HEART CELLS CONTINUOUSLY MEASURED BY A LASER SCATTERING TECHNIQUE

W. LINKE, W. BOLDT, J. SCHUH & M. WUSSLING

Julius Bernstein Institute of Physiology, Martin Luther University Halle-Wittenberg, Leninallee 6, D-O-4020 Halle/Saale, FRG

INTRODUCTION

Cultures of enzymatically isolated heart cells obtained from neonatal rats are known to reveal various beating rhythms. After 3 days of cultivation a uniform monolayer of myocytes is formed which is called to beat very regularly similar to a clockwork (1). Already in 1965 THARP & FOLK (5) found out rhythmic fluctuations of frequency of interconnected neonatal rat heart cells with periods of 12 and 16-17 hours. The authors measured pulsation rates every 2 hours, which is known to be a very long time distance for such a sensitive model like cultures of neonatal heart cells.

Because of the difficult recording procedure of cell contractions with amplitudes in the range of a few micrometers contractility of isolated myocytes is difficult to investigate directly. It was the aim of this study to get some more information on changes of frequency and to measure amplitudes of cell contractions over a period of several days continuously. This way possible changes of the latter parameter already at the level of a monolayer could be found out (4). A laser scattering technique previously described by WUSSLING & SCHENK (7) was used for data sampling and computation.

MATERIAL AND METHODS

Production and cultivation of neonatal rat heart cells (1-2 days old Wistar rats, ICO) had been carried out in the Institute of Heart and Circular Research Berlin-Buch and are described elsewhere (6). A small group of heart cells (about 60 myocytes) grown on to the bottom of a cultivation flask (FALCON, 50 ml, 25 qcm useful area) was selected for the illumination with a laser beam (He Ne laser, 3 mW, beam diameter 300 µm) and the scattered light was measured by a CCD line using interference phenomena (speckling and scattered light intensity fluctuations). Two PCs served for data sampling and evaluation. The temperature was 37 ± 0.2 °C normally, sometimes 27 ± 0.2 °C. Measurements of parameters of cell contraction mostly started at the 4th day of cultivation at different times of the day and lasted between 24 and 72 hours.

Experiments were subdivided into three groups:
1. Registrations without perfusion of the cultivation chamber and without stimulation of the cell culture. pH of the medium decreased from 7.8 to 7.3 during 3 days of sterile cultivation.
2. Experiments with continuous flow of the medium to eliminate decrease of pH.
3. Experiments with periodic external field stimulation of the whole culture to get constant pulsation rates.

Frequency was plotted every 5 min by calculating the average derived from all beats within 10 seconds. Amplitude was measured every minute using the following calculation procedure (7):

The "amplitude" is considered to be a plane surrounded by the extreme values of successive contractions at each point of the CCD line. The computer utilizes 128 scans in intervals of 20 msec (all together 2.56 sec) to carry out the integral calculus every minute.

Original data series of all experiments were smoothed by a special computer program to eliminate high frequent oscillations because investigated parameters had been found to reveal rhythmic changes with period lengths of at least several hours. Redifferentiation and dedifferentiation processes in the cells during cultivation are the reasons for an increase or a decrease of measured parameters during the time of investigation. These trends were eliminated to get stationary values for trouble-free statistical analyses.

Fast Fourier Transforms served as a basis for the calculation of power spectra. After standardization graphs of different experiments were compared and the major peaks of all spectra were used for documentation in pooled periodograms.

RESULTS

Frequency has been found to be relatively constant during the period of registration, mostly insignificant and irregular changes could be observed (Fig. 1, on the left). Only in 4 out of 14 experiments power spectra showed major peaks with period lengths of 11-12 or 16-17 hours due to rhythmic changes of pulsation rate (Fig. 1, on the right). Lowering of the temperature to 27°C led to irregular beating and bursts were observed. In experiments with perfusion a nearly constant beating behavior could be seen, too.

To the contrary continuous measurements of amplitude showed periodic changes of contractility with period lengths of several hours (Fig. 2, on the left).

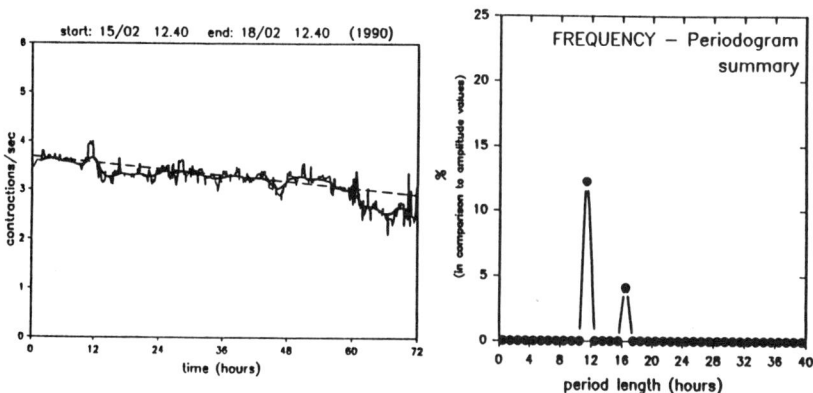

Fig.1: On the left: Typical development of pulsation rate during 72 hours of registration: frequency is relatively constant. Dotted line: simple trend, continuous lines: original values and smoothed graph.
On the right: Pooled periodogram of 4 major peaks (in %, calculated for comparison with amplitude values) obtained from 14 power spectra.

Results of experiments with perfusion and stimulation documented that a decrease of pH from 7.8 to 7.3 as well as periodic external field stimulation of the whole cell culture don't influence these rhythms. Distinct periodic changes of amplitude could also be found in experiments at a temperature of 27 °C.

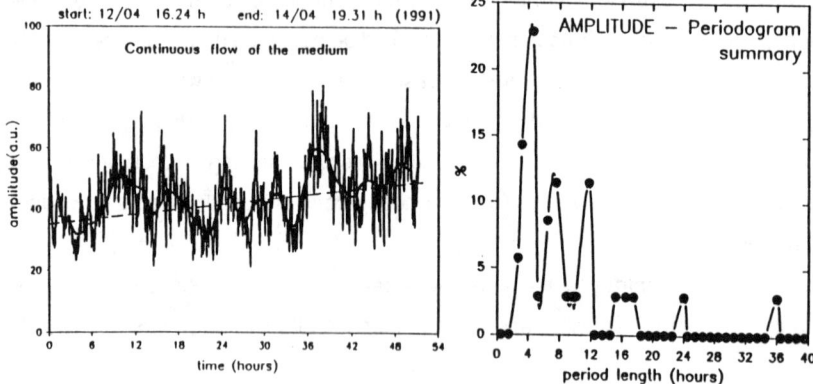

Fig.2: On the left: rhythmic periodic changes of amplitude during prolonged investigation (perfusion with medium). Dotted line: simple trend, continuous lines: original values and smoothed graph.
On the right: Pooled periodogram of 35 major peaks (= 100 %) obtained from 20 power spectra.

The pooled periodogram of all major peaks of power spectra (Fig. 2, on the right) shows results of 20 measurements: 13 major peaks in the range of 3-5, 7 peaks in the range of 6-8, and 4 peaks in the range of 11-12 hours document periodic changes of amplitude with ultradian rhythmicities found in all experiments.

Graphs of original values were investigated to find out correlations of periodic amplitude fluctuations to the solar day-night cycle but coincidences couldn't be observed.

DISCUSSION

Rhythmic changes of pulsation rate in neonatal rat heart cells seem to be very seldom and further investigations are necessary to manifest such rhythmicities.

Cell contractility among other things depends on cAMP content in the muscle cell. This "second messenger" is known to be responsible for phosphorylation of Ca channels and with it for inotropic effects in myocytes (2). LEMMER (3) found a circadian rhythm in basal cAMP content and in the activity of adenylate cyclase (AC) and phosphodiesterase (PDE) in rat heart ventricles, which was accompanied by distinct ultradian rhythms with period lengths of 6-8 and about 12 hours. It is concluded that periodic fluctuations of contractility of isolated neonatal rat heart cells could be due to variable activities of AC and PDE and to an insteady cAMP content within the myocytes endogenously fixed already at the level of a monolayer.

REFERENCES

1. HAVERKAMPF, K., R. JOHNA & H. ANTONI (1989): Rhythm phenomena in cardiac cell culture of neonatal rat. Eur. J. Physiol. 413, Suppl.1, R56
2. KAMEYAMA, M., F. HOFFMANN & W. TRAUTWEIN (1985): On the mechanism of ß-adrenergic regulation of the Ca channel in the guinea-pig heart. Pflügers Arch. 405, 285-293
3. LEMMER, B. (1986): The chronopharmacodynamics and chronopharmacokinetics of beta-adrenoceptors. In: P. SCHOEHNERICH, H.-J. HOLTMEIER & H.S. KRONEBERG (eds.): Cardiovascular Receptors, Stuttgart: Georg Thieme, pp. 125-135
4. LINKE, W., W. BOLDT & M. WUSSLING (1991): A laser scattering technique for investigation of oscillatory phenomena in cultures of neonatal rat heart cells. Eur. J. Physiol. 419, Suppl.1, R109
5. THARP, G.D. & G.F. FOLK Jr. (1965): Rhythmic changes in rate of the mammalian heart and heart cells during prolonged isolation. Comp. Biochem. Physiol. 14, 255-273
6. WOLLENBERGER, A. & G. WALLUKAT (1983): Adrenergic and antiadrenergic activity of cycloheximide in cultured rat heart cells. Biomed. Biochim. Acta 42, 917-927
7. WUSSLING, M. & W. SCHENK (1986): Laser diffraction and speckling studies in skeletal and heart muscle. Biomed. Biochim. Acta 45, 23-27

CHRONOBIOLOGICAL ASPECTS OF COMPUTERIZED CULTURES OF CHLORELLA VULGARIS

H. BÖHM, H. ARNDT & H.-G. HIEKEL

Inst. f. Bioanalytik der Technischen Hochschule Köthen, Lohmannstraβe 23, D-O-4370 Köthen, FRG

INTRODUCTION

There is a lot of scientific results to state that organisms in general are to interpreted as multioscillator systems in which rhythmic processes are an expression of vitality and a fundamental condition for adaptability and dynamic stability. Numerous experiments with plants confirm this fact. To characterize such oscillations we use an unicellular eucaryotic object, the chlorococcalic alga *Chlorella vulgaris*.

MATERIAL AND METHODS

A high temperature strain BÖHM/BORNS 1972/1 of the species *Chlorella vulgaris* BEIJERINCK 1890 was cultivated in an unorganic liquid medium at 37°C using an automated and full computerized illumined fermentor called BIOANALYZER. The fermentor is able to work in batch, turbidostate and chemostate mode by means of integrated optical analyzers at two or more different wavelengths. With the help of this analyzers and a free programmable threshold the computer is steering a precisely working dilutor which realizes very small and uniform dilution steps during continuous cultivation mode in the second phase of our experiment. In the first phase we preferred a batch culture mode with a strain specific light/dark regime and periodic dilution at the beginning of each light cycle. The result of such light/dark cycles controlled by electronic counting and sizing the cells and continued for at least one month was synchrony in cell division and growth. The visible result of this long time procedure: all mother cells divide into new daughter cells within one hour during the dark period. After sporulation the fresh aplanospores will be activated by the now beginning light period setting the physiological clock and initiating different rhythmic processes. Overlapping of successive nuclear cycles is to bee seen coupled with partial divisions of the protoplast till a critical volume of the ripe mother cell is reached and dividing into n daughter cells begins again.

RESULTS AND DISCUSSION

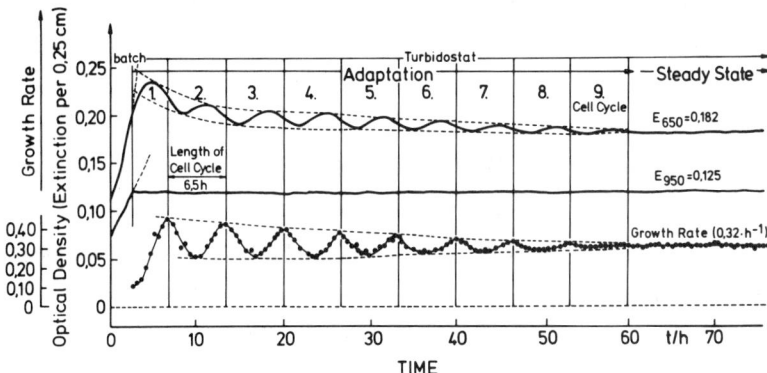

Fig. 1:

The first experiment - see Fig. 1 - starts with a synchronous cell suspension consisting only of fresh aplanospores. At continuous light the culture grows up to a programmed value of optical density. When the threshold is reached the bioanalyzer works as a turbidostat at the extinction near 950 nm representative of the cell growth. The graph derived from this wavelength runs horizontally. But the extinction near 650 nm as a result of light absorption by chlorophyll shows distinct oscillations with decreasing amplitudes but constant length of the periods. Each of this periods is as long as the generation time of *Chlorella vulgaris* under the chosen conditions. After more than sixty hours the oscillations die down and finally they will be replaced by a stable steady state. The picture of the graph founded upon 650 nm is very similar to that which is based upon computer analysis of the dilution steps. The dilution rate shows oscillations too and the length of the periods is identical with that of 650 nm. Like there the oscillations pass over to a definitive steady state at the end of this longtime experiment. But maxima and minima of the oscillations are shifted with a quarter of the length of the periods and the amplitudes are more pronounced.

Let's have a view to the results of a second experiment (see Fig. 2). The initial conditions have changed. We are starting with a homocontinuous culture in turbidostate mode. The control magnitude of the continuously illumined fermentor is the extinction measured at 950 nm representative of the cell growth. All the monitored parameters show to us a steady state of high quality. Electronic counting and sizing illustrates that the empiric distribution of the cells is to 100% equivalent to an exponential distribution which proofs the above mentioned quality of our cell suspension. Later on we applied a small amount of a chemical called CEKAQUAT known as an inhibitor for algal growth and destabilizer of cell suspensions. Both effects are to be seen at once. Growth of the cells stops immediately, and the stability of the cell suspension decreases quickly by conglomeration. Therefore dilution is interrupted for almost three hours. But the biological disaster lasts not too long. The suspension regenerates, and dilution continues but for the present with a reduced number of dilution steps indicating thus a partial growth inhibition, which will be reduced totally within the following fifty hours. After this adaptation period the steady state is restored.

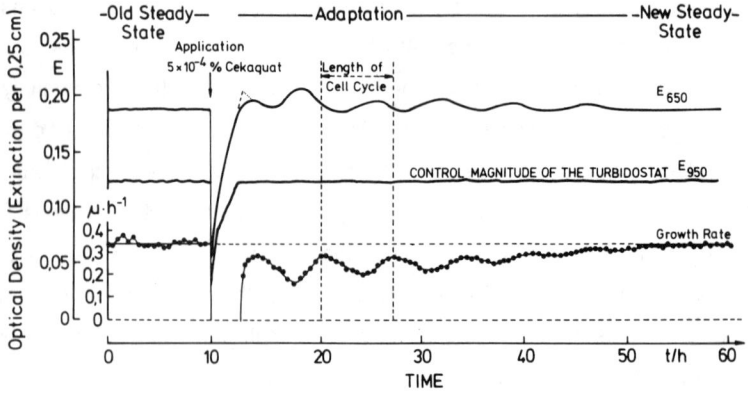

Fig. 2:

Let us now go back to the details of this adaptation process. Again it is to be seen that the measured values of the extinction per 650 nm and the dilution steps dependent on the extinction per 950 nm are oscillating in the same manner like in our first experiment although the initial conditions don't coincide.

CONCLUSIONS

With relation to chronobiological aspects of the two experiments and the results of further investigations we want to present four conclusions:

1. Computerized illumined fermentors of the type BIOANALYZER are able to realize long time experiments which is a necessary precondition to standardize cell suspensions of photoautotrophic microorganisms, to analyze adaptation processes, and to diagnose their rhythmic consequences for growth, development, and metabolism.

2. Really desynchronous cell suspensions don't show rhythmic oscillations under steady state conditions. The rhythmicity of each single cell is existent, of course. But it remains discrete as the main consequence of the typical distribution of the different developmental stages and their individual oscillations within the cell suspension.

3. If desynchronity of a cell suspension will be disturbed in any way - e.g. by a strain-specific light-dark regime or another time setting factor - partial or total synchrony in cell division and growth takes place measurable in partial or total correspondence of the empirical cell distribution with the RAYLEIGH-distribution. An imperative consequence of such a fundamental change is the synchrony of individual rhythms which are measurable at different parameters not used as a control parameter.

4. The transition from desynchronity to synchrony and vice versa are long-time processes with all the phenomenons of stepwise adaptation and connected with the possibility to determine the strain-specific and condition-specific generation time by analyzing the length of the oscillation periods.

CIRCADIAN RHYTHMS IN ECDYSTEROID TITERS IN INSECT HEMOLYMPH

B. CYMBOROWSKI

Department of Invertebrate Physiology, Warsaw University,
Zwirki i Wigury 93, 02-089 Warszawa, Poland

The molting and eclosion behaviors of many so far investigated insects are subjected to circadian timing, which has been analyzed at the neuroendocrine level in several cases involving adult eclosion, larval moltings and also pupations. It has been shown that the release of two neurosecretory hormones, eclosion hormone and prothoracicotropic hormone (PTTH) are involved in triggering these events and is under strict temporal control by circadian oscillator or oscillators. Several experimental techniques, especially transplantation, have localized the oscillators or so called biological clocks timing eclosion hormone and PTTH release to the insect brain (3, 8). On the other hand the final timing of these and many different events might be controlled by a circadian clock(-s) located within the prothoracic glands (2, 4, 5). So far it is obvious that ecdysteroids produced by prothoracic glands (PG) synchronize the cellular and molecular events which occur during the insect molt. But the questions arise what are the ecdysteroids doing during the periods of intermolt and what is the underlying mechanism of circadian rhythmicity of ecdysteroid titers in insects?

In previous papers (1,7) it was shown that in the case of the wax moth (*Galleria mellonella*) larvae the synchronous initiation of molting by transferring from low (18°C) to normal (30°C) temperature leads to remarkable temporal precision in their development. The evidence for circadian control of ecdysteroids titer in the last instar of Galleria larvae was also presented (1).

In further experiments the underlying mechanism of circadian rhythmicity of ecdysteroid titers in this insect was investigated. It was found that brain removal or head-ligature of larvae in which developmental processes were synchronized by transfer from 18 to 30°C, did not abolish circadian rhythms of ecdysteroid titers in the hemolymph (2). The observed rhythms persisted 5 days after treatment (Fig. 1), which suggests that the circadian clock driving these rhythms is not located within the brain.

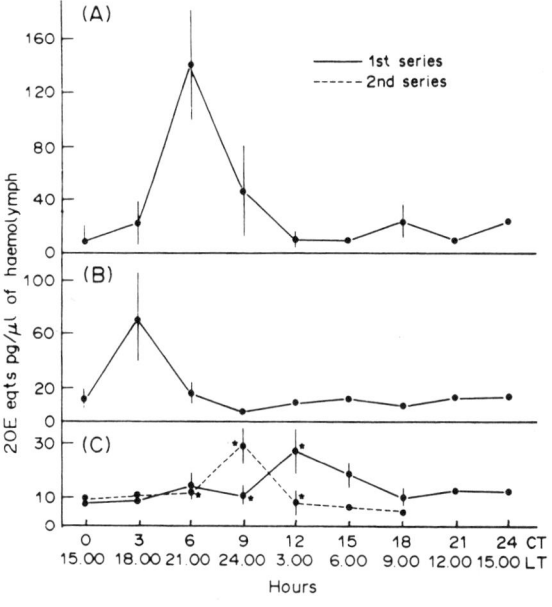

Fig. 1: Daily changes in ecdysteroid titers in the hemolymph of the last instar of *Galleria mellonella* larvae on day 5 after treatment and transfer from 18 to 30°C. A - control larvae; B - brainless larvae; C-larvae ligated behind the head. Samples of hemolymph were assayed by EIA after PORCHERON et al. (6). Each point represents mean (±SD) of 4-7 separate determinations. Data are presented as pg of 20-hydroxy ecdysteroid (20E) equivalents (eqts) per µul of hemolymph. LT - local time; CT - circadian time (hours from treatment and transfer from 18 to 30°C). Asterisks indicate statistical significance at p<0.05. Modified from CYMBOROWSKI et al (2).

In in vitro experiments the prothoracic glands were taken from larvae every 3 hours during 24-h period 4 days after the larvae were transferred from 18 to 30°C. The purpose of this experiments was to find out whether the peak of ecdysteroid titers found in the hemolymph of control larvae (see Fig. 1a) coincided with the maximum of prothoracic gland synthetic activity. As can be seen from Fig. 2 there is indeed such a correlation. The in vitro maximum is the same time that the peak of ecdysteroid titer occurred in the hemolymph of the control larvae transferred from 18 to 30°C four days earlier (Fig. 2b). In the other control group of larvae which were kept at 30°C throughout whole development (wandering larvae) such a peak in ecdysteroid titer could not be found (Fig. 2c).

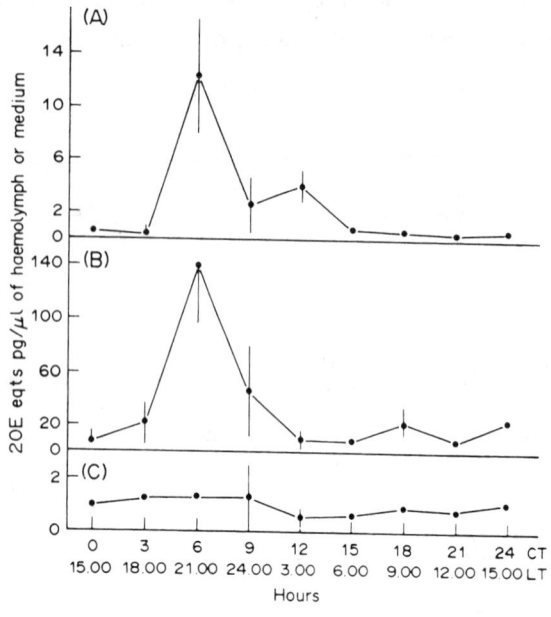

Fig. 2:
Kinetic of synthesis of ecdysteroids by *Galleria* prothoracic glands in vitro at successive hours during a 24-hour period on day 5 after transfer of the larvae from 18 to 30°C (A). Ecdysteroid titers of the hemolymph of the control larvae kept for 3 weeks at 18°C and then transferred to 30°C (B). Samples were taken every 3 hours on day 5 after transfer. Ecdysteroid titers in the hemolymph of wandering larvae reared at 30°C (C).
Aliquots of culture or samples of hemolymph were assayed by EIA. Each point represents mean (±SD) of 6-8 separate determinations. In (A) data are presented as pg of 20-hydroxyecdysone (20E) equivalents (eqts) synthetized and released in culture medium per µl of medium. After CYMBOROSKI et al (2).

The presented results suggest that ecdysteroid synthesis by prothoracic gland shows significant daily increases and decreases, being most active during the time when the peak of ecdysteroids is observed in the hemolymph of control temperature-synchronized larvae. It seems therefore that the clock driving these rhythms is located within the prothoracic glands of *Galleria* larvae. Whether these glands respond promptly to phase shifts of light and/or temperature pulses has, as yet, to be shown. The fact that in the hemolymph of the larvae reared only at 30°C there are no daily rhythms of ecdysteroid titers whereas such rhythms are observed in larvae transferred from 18 to 30°C suggest that the temperature step might play the role of a Zeitgeber (2).

REFERENCES

1. CYMBOROWSKI, B., A. SMIETANKO & J.-P. DELBECQUE (1989): Circadian modulation of ecdysteroid titer in Galleria mellonella larvae. Comp. Biochem. Physiol. 94A, 431-438
2. CYMBOROWSKI, B., M. MUSZYNSKA-PYTEL, P. PORCHERON & P. CASSIER (1991): Haemolymph ecdysteroid titers controlled by a circadian clock mechanism in larvae of the wax moth, Galleria mellonella. J. Insect Physiol. 37, 35-40
3. FUJISHITA, M. & H. ISHIZAKI (1981): Circadian clock and prothoracicotropic hormone secretion in relation to the larval-larval ecdysis rhythm of the saturniid Samia cynthia ricini. J. Insect Physiol. 27, 121-128
4. MIZOGUCHI, A. & H. ISCHIZAKI (1982): Prothoracic glands of the saturniid moth Samia cynthia ricini possess a circadian clock controlling gut purge timing. Proc. natn. Acad. Sci. U.S.A 79, 2726-2730
5. MIZOGUCHI, A. & H. ISCHIZAKI (1984): Further evidence for the presence of a circadian clock in the prothoracic glands of the saturniid moth Samia cynthia ricini: Decapitated larvae can respond to light-dark changes. Devl Growth Differ. 26, 607-611
6. PORCHERON, P., M. MORINIERE, J. GRASSI & P. PRADELLES (1989): Development of an enzyme immunoassay for ecdysteroids using acetylcholinesterase as label. Insect Biochem. 19, 117-122
7. SMIETANKO, A., J.R. WISNIEWSKI & B. CYMBOROWSKI (1989): Effect of low rearing temperature on development of Galleria mellonella larvae and their sensitivity to juvenilizing treatment. Comp. Biochem. Physiol. 92A, 163-169
8. TAKEDA, M. (1984): An "hour-glass" feature in photoperiodic time measurement in Diatraea grandiosella (Pyralidae). J. Insect Physiol. 30, 326-329

DAILY CHANGES OF ß-N-ACETYL-D-GLUCOSAMINIDASE (EC 3.2.1.30.) FROM RAT SUBMANDIBULAR GLAND ASSAYED BY ROCKET IMMUNOELECTROPHORESIS

A. LITYŃSKA

Department of Animal Physiology, Institute of Zoology, Jagiellonian University,
M. Karasia 6, 30-060 Kraków, Poland

INTRODUCTION

ß-N-acetyl-D-glucosaminidase (NAG) is a lysosomal hydrolase that cleaves terminal aminoacetylhexosamines from oligosaccharides, glycoproteins, and glycosaminoglycans. The enzyme is widely distributed in most mammalian tissues and exhibits daily changes of activity in liver and submandibular glands (5). Whether these changes reflect the fluctuation of amount or the properties of the enzyme molecules it remains an open question. Therefore, it appeared interesting to complete results obtained for NAG activity assay with immunochemical quantification. This method allows to obtain rocked shaped precipitates whose height is proportional to the amount of enzyme loaded from crude extracts (4). Thus, the purpose of the present investigation was to study the daily changes of NAG amount by the method of immunoelectrophoresis.

MATERIAL AND METHODS

The animals used were male rats of the Wistar strain kept in windowless room under LD 12/12 (08:00-20:00 light, 20:00-08:00 darkness) conditions, food and water given ad libitum. Group of seven animals each were decapitated at four hour intervals during 24 hour period. The submandibular glands from each animal were weighed, homogenized in buffer (2ml/g tissue), and treated with Triton X-100 (0.2% as a final concentration). Than, the material was centrifuged at 40.000 rpm during 2 hours and supernatant was used for further investigations. The activity of NAG was assayed according to SELLINGER et al. (8) and expressed in μmol/ml/1min, and the amount of protein by the method of BRADFORD (1). Rocket immunoelectrophoresis was performed according to LAURELL (4) in 1% agarose gels in 0.06 M barbital buffer pH 8.8 at 3V/cm for 19 hours. Gel temperature (14^0C) was kept constant with tap water cooling during electrophoresis. In each case 17 μl of antigen (post 40.000 supernatant) and 1.14 $μl/cm^3$ rabbit antiserum was used. At the end of the run, gel plates were washed and pressed before staining for protein with Coomassie Brilliant Blue R-250. Rocket height was settled on graph paper from the top of the application well to the tip of the precipitin rocket, cut out and weighted.

These results are expressed in mg. For the results corresponding to each time point mean, standard deviation (SD), and standard error (SE) were calculated. Student's t-test was used to detect significant differences.

RESULTS

Consistent with the previously reported results the NAG activity varied with the diurnal shape. The highest activity was obtained at 20:00 h what was significantly different ($p<0.05$) from lowest activity at 8:00 h. The same extracts were used for NAG electroimmunoassays. Fig. 1a and 1c show that changes in enzyme activity do not reflect changes in enzyme amount. Antigenically detectable NAG amount showed a bimodal pattern characterized by two maxima at 4:00 h and 12:00 h and minimum at 20:00 h (Fig.1c) ($p<0.02$). As Fig.1 b and 1c show, the amount of antigenic NAG increased and decreased concomitantly with total protein amount.

DISCUSSION

The mechanisms involved in regulation of enzyme circadian rhythmicity has so far been considered in terms of rhythms in protein synthesis or rhythms in mechanisms regulating enzyme activity. The immunological method, independent from enzymatic activity, provided a direct quantification for NAG at different time points. The results of present study showed, that the time-course of antigenically detectable NAG amount was strictly correlated with changes in total protein amount in rat submandibular gland. Bimodal pattern of protein oscillations was in agreement with

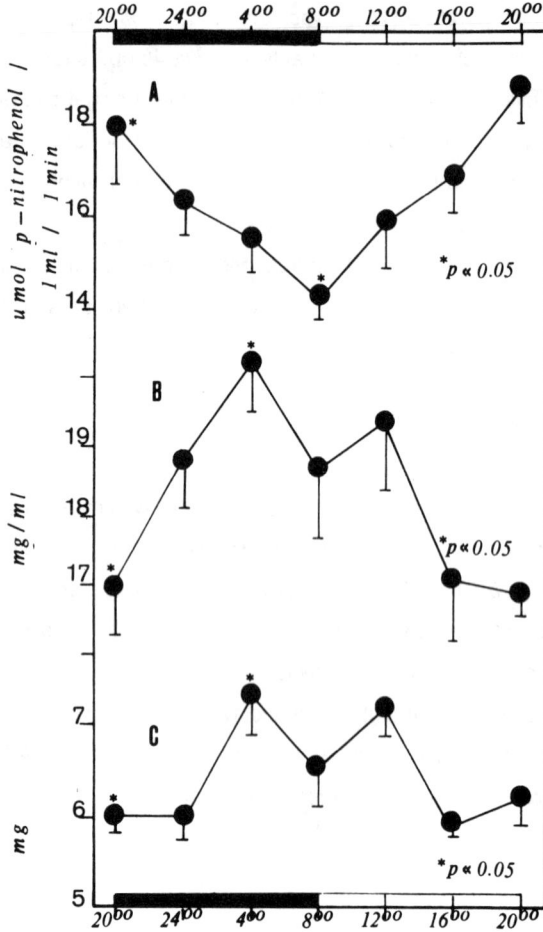

Fig.1:
Temporal pattern of NAG activity (A), protein amount (B), and NAG amount (C). Each point represents the mean value of seven animals -SE. Black bars: dark period. For additional informations see Methods.

the results obtained by ^{14}C-leucine incorporation into rat submandibular gland slices (5). From comparison of enzymatic activity levels and rocket height observed in electroimmunoassay it appears that time-course for NAG activity was not correlated with changes in enzyme antigen amount. These results indicate the existence of temporal changes as well in activity as in amount of NAG in rat submandibular gland. The most remarkable feature of the results is the phase difference between these two oscillations. Most data concerning enzyme circadian rhythms refer to periodic variations in the activity measured in the tissue extracts obtained at different time points. However, changes in enzyme activity might result not only from equivalent changes in number of enzyme molecules but also from other causes.

Studies by GILBERT & TSILIMIGRAS (2) suggest that the oscillations in the amount of protein extractable from culture cells was not primarily responsible for periodic variations in the activities of enzymes in the cell extracts, although it apparently contributed to the complex waveforms of the latter rhythms. From the phase difference it can be concluded that rhythmicity of NAG is controlled not only by the changes of amount of enzyme molecules but also by their properties. The latter can be explained considering the enzyme as a target for the action of effectors or inhibitors which cyclically affect it's catalytic activity. A less conventional explanation is that enzyme molecules by itself could behave as conformational oscillators alternating between phases of high and low sensitivity to ligands (11). In case of NAG, hexosamines may play the role of regulators, but on the other hand, as was shown by ROME & HILL (7), hexosamines produced by lysosomal degradation are rapidly reincorporated into newly synthesized glycoproteins. Increasing attention has been given in recent years to the study of spontaneous changes in conformation of enzymes in solutions (9,10,12). Systematic studies of malate dehydrogenase, under constant conditions using diverse methodologies (enzyme kinetics, proton exchangeability, NMR studies) gave evidence of circadian rhythmicity in continuous shifting between conformational enzyme sub-stages (9,10). These authors also suggest that long period oscillations in enzyme characteristics could be of physiological significance in relation to circadian rhythmicity.

Therefore, further investigations of purified NAG under constant conditions seem very instructive.

REFERENCES

1. BRADFORD, M.M. (1976): A rapid and sensitive method of the quantitation of microgram quantities of protein utilizing the principle of protein-dye binding. Anal. Biochem. 72, 248-254
2. GILBERT, D.A. & C.W.A. TSILIMIGRAS (1981): Cellular oscillations: Relative independance of enzyme activity rhythms and periodic variations in the amount of extractable protein. S. Afr. J. Sci. 77, 66-73
3. LAURELL, G.B. (1981): Electroimmunoassay. In: J.J. LANGONE & H. van VUNAKIS (eds.): Methods in Enzymology 73, Academic Press, New York-Toronto-Sydney, pp. 339-369
4. LAURELL, G.B. (1966): Quantitative estimation of proteins by electrophoresis in agarose gel containing antibodies. Anal. Biochem. 15, 45-52
5. LITYŃSKA, A. & B. WÓJCZYK (1988): Relationship between daily changes of hyaluronidase, ß-N-acetyl-D-glucosaminidase, ß-glucuronidase activities and N-acetylhexosamines distribution in rat liver. In: J. TWARDOWSKI (ed.): Spectroscopic and structural studies of biomaterials. Sigma Press, Wilmslow, U. K., pp. 266-271
6. LITYŃSKA, A., & B. WÓJCZYK (1986): Daily changes of protein synthesis and secretion rates in rat submandibular gland slices. Folia Biol. (Kraków) 34, 199-206
7. ROME, L.H. & D.F. HILL (1986): Lysosomal degradation of glycoproteins and glycosaminoglycans. Biochem. J. 235, 707-713
8. SELLINGER, O., H. BEAUFAY, P. JACQUES & C. DE DUVE (1969): Tissue fractionation study. Intracellular distribution and properties of ß-N-acetyl-glucosaminidase and ß-galactosidase in rat liver. Biochem. J. 74, 405-456
9. QUEIROZ-CLARET, C., Y. GIRARD, B. GIRARD & O. QUEIROZ (1985): Spontaneous long-period oscillations in the catalytic capacity of enzymes in solution. J. interdiscipl. Cycle Res. 16, 1-9
10. QUEIROZ-CLARET, C., R. LENK, O. QUEIROZ & H. GREPPIN (1988): NMR studies of in vitro slow oscillations in enzyme properties and dissipative structures. Plant Physiol. Biochem. 26, 333-338
11. QUEIROZ-CLARET, C. & O. QUEIROZ (1990): Multiple levels in the control of rhythms in enzyme synthesis and activity by circadian clocks: recent trends. Chronobiol. Int. 7, 25-33
13. QUEIROZ-CLARET, C., C. VALON & O. QUEIROZ (1988): Are spontaneous conformational interconversions a molecular basis for long period oscillations in enzyme activity? Chronobiol. Int. 5, 301-309

REGULATION OF PINEAL GLAND FUNCTION IN RATS - THE ROLE OF NONAPEPTIDES

R. RIEMANN & S. REUSS

Department of Anatomy, Johannes Gutenberg-University, D-W-6500 Mainz, FRG

INTRODUCTION

Pineal melatonin synthesis is regulated primarily by the sympathetic system. The differential release of norepinephrine from postganglionic fibers of the sympathetic superior cervical ganglia is thought to mediate a circadian rhythm originally generated in the hypothalamic suprachiasmatic nuclei. In addition, there is increasing evidence that other substances, e.g. neuropeptides, influence pineal function (5,12). Of further interest are the nonapeptides arginine-vasopressin (AVP) and oxytocin (OT) which have been described to occur in intrapineal nerve fibers (1). These substances are synthesized in mammalian hypothalamic nuclei and act upon release from terminal boutons as peptidergic transmitters or as hormones via cerebrospinal fluid or blood circulation.

In particular, electrical stimulation of the hypothalamic paraventricular nuclei (PVN) which increases AVP and OT blood levels diminishes nocturnal pineal melatonin levels (3,4). Interestingly, the intra-arterial application of AVP mimicked these effects (8) thus providing further evidence for a possible role of this peptide for pineal metabolism. Furthermore, it was reported recently that AVP and OT modulate melatonin secretion from rat pineal glands in vitro (10).

One of the best investigated parameters of pineal metabolism is the activity of the melatonin forming enzyme, serotonin N-acetyltransferase (NAT), which is augmented dramatically at night. Similarly, the numbers of pineal "synaptic" ribbons (SR) which are thought to be involved in intercellular communication are well characterized as morphological correlates of a circadian rhythm with increased nocturnal levels.

These pineal parameters, i.e. NAT activity and SR numbers, were chosen as examples to study the possible effects of AVP or OT on the gland's function under in vivo and in vitro conditions.

EXPERIMENTAL APPROACHES AND RESULTS

Several groups of adult male animals of the rat strains Sprague-Dawley (SD, albino), Long-Evans (LE, hooded) or Brattleboro (BB; a genetically AVP-deficient rat strain), housed under constant LD 12:12 conditions, were subjected to the following studies.

In vivo experiments: The methods used in this series have been described previously (5,8). In brief, experimental rats received, into the right common carotid artery, injections of AVP, of 1-deamino[D-Arg8]vasopressin (dDAVP; an AVP analog without vasopressor activity), or of saline only (sham-application) at 00.00-00.30 h. Two hours later all animals were decapitated, pineal glands were quickly removed and further processed for NAT analysis or fixed for electron microscopic analysis, respectively.

AVP or dDAVP reduced NAT activity in BB rats to approximately the half of the control level ($p<0.01$). Application of dDAVP reduced NAT activity in SD rats to 26 % of the control level ($p<0.05$), which is comparable to AVP effects in SD rats (8). Saline injections were without effects in either experiment.

AVP was without effect on SR numbers or intracellular location in BB and SD rats.

In vitro experiments: The methods used for this experimental approach have been described in detail recently (11). In brief, pineals were removed during daytime and placed in culture dishes containing BGJb medium and were incubated for 6 hrs with AVP (10^{-10} M, 10^{-8} M, 10^{-6} M, 10^{-4} M), its major proteolytic fragment VP_{4-9} (10^{-10} M, 10^{-8} M, 10^{-6} M), OT (10^{-5} M), norepinephrine (NE; 10^{-5} M), in combinations of NE with either AVP, VP_{4-9} or OT, or in pure medium (controls). Pineals were then processed for biochemical or electron microscopic analysis, respectively.

Incubation of pineal glands from SD rats with either AVP (10^{-4} M) or VP_{4-9} (10^{-4} M) did not affect NAT activity, while NE (10^{-5}) augmented NAT activity. The

NE-induced elevation of NAT activity was potentiated by AVP at all concentrations used except 10^{-4} M, and by VP_{4-9} at concentrations of 10^{-10} M and 10^{-8} M.

Incubation of SD rat pineals in AVP or VP_{4-9} was without effect on SR numbers.

Incubation of pineals from SD, LE and BB rat with OT alone or in combination with NE did not affect NAT activity when compared to controls, while incubation of the organs in NE alone resulted in a significant augmentation of NAT activity in all three strains. SR numbers were not affected by OT in neither rat strain; however, NE increased SR numbers in BB rats significantly.

DISCUSSION

Recent studies revealed that pineal melatonin synthesis is inhibited by the enhancement of circulating vasopressinergic compounds in intact rats (3,4,8). The present results reveal that the same is true in AVP-deficient BB rats and that this inhibition is not due to the vasoconstrictive action of AVP. Interestingly, the melatonin synthesis-inhibiting effect of PVN-stimulation as observed in SD rats (4) is absent in AVP-deficient BB rats (6) further supporting the view that AVP is the responsible agent under these experimental conditions. However, as BB rats endow functional AVP receptors (9), the enhancement of circulating AVP by i.a. injection of the peptide results in the inhibition of pineal NAT activity in BB (present results) and in SD rats (8).

Furthermore, a secondary vasopressinergic effect which in vivo appears to be masked is seen in vitro. Under these conditions, AVP or its major proteolytic fragment VP_{4-9} do not alter pineal NAT activity. In the presence of NE, however, either peptide potentiates the activity of the rate-limiting enzyme for melatonin synthesis.

It is, however, open how AVP in vitro exerts its influence on the pineal gland. The organ does not endow vasopressinergic receptors, apart from VP_{4-9} binding sites (2). Since AVP and VP_{4-9} produced similar effects in the in vitro experiments presently, there is reason to believe that both peptides act via VP_{4-9}-binding sites which are highly concentrated in the pineal gland (2).

The possible role of AVP in the modulation of pineal function is further supported by the present observation that NE stimulates SR numbers in AVP-deficient BB rats but not in SD or LE strains.

Our results, on the other hand, do not support the previous finding that OT is capable of influencing melatonin synthesis (10). The pineal content of this indolamine was, in the present study, not altered by OT under in vitro conditions. The possibility exists, however, that methodical differences (e.g. incubation vs. perifusion techniques, measurement of content vs. secretion) are responsible for this discrepancy.

With regard to pineal SR numbers, it should be noted that these were neither in vivo nor in vitro influenced by the application of vasopressinergic compounds. These findings support the view that AVP does not play an important role in the regulation of "synaptic ribbons" since SR exhibit circadian changes in number comparable to those found in intact animals (7). There is, in addition, no present evidence for an influence of OT on SR number regulation.

Since peptidergic mechanisms of circadian system regulation are of high interest, further studies are clearly necessary to better understand the possible roles of AVP and OT with respect to pineal function.

REFERENCES

1. BUIJS, R.M. & P. PÉVET (1980): Vasopressin- and oxytocin-fibres in the pineal gland and subcommissural organ of the rat. Cell Tissue Res. 205, 11-17
2. DE KLOET, E.R., T.A.M. VOORHUIS, J.P.H. BURBACH & D. DE WIED (1985): Autoradiographic localisation of binding sites for the arginine-vasopressin (VP) metabolite, VP_{4-9}, in rat brain. Neurosci. Lett. 56, 7-11
3. OLCESE, J., S. REUSS & S. STEINLECHNER (1987): Electrical stimulation of the hypothalamic paraventricular nuclei mimics the effects of the light on pineal melatonin synthesis. Life Sci. 40, 455-459
4. REUSS, S., J. OLCESE & L. VOLLRATH (1985): Electrical stimulation of the hypothalamic paraventricular nuclei inhibits pineal melatonin synthesis in male rats. Neuroendocrinology 41, 192-196
5. REUSS, S. & H. SCHRÖDER (1987): Neuropeptide Y effects on pineal melatonin synthesis in the rat. Neurosci. Lett. 74, 158-162
6. REUSS, S., J. STEHLE, H. SCHÖDER & L. VOLLRATH (1990): The role of the hypothalamic paraventricular nuclei for the regulation of pineal melatonin synthesis: New aspects derived from the vasopressin-deficient Brattleboro rat. Neurosci. Lett. 109, 196-200
7. RIEMANN, R., S. REUSS, J. STEHLE et al. (1990): Circadian variation of "synaptic" bodies in the pineal glands of Brattleboro rats. Cell Tissue Res. 262, 519-522
8. SCHRÖDER, H., S. REUSS, J. STEHLE & L. VOLLRATH (1988): Intra-arterially administered vasopressin inhibits nocturnal pineal melatonin synthesis in the rat. Comp. Biochem. Physiol. 89 A, 651-653
9. SHEWEY, L.M. & D.M. DORSA (1986): Enhanced binding of [^{3}H]arginine^8-vasopressin in the Brattleboro rat. Peptides 7, 701-704
10. SIMONNEAUX, V., A. OUICHOU, J.P.H. BURBACH & P. PÉVET (1990): Vasopressin and oxytocin modulation of melatonin secretion from rat pineal glands. Peptides 11, 1075-1079
11. STEHLE, J., S. REUSS, R. RIEMANN et al. (1991): The role of arginine-vasopressin for pineal melatonin synthesis in the rat: involvement of vasopressinergic receptors. Neurosci. Lett. 123, 131-134
12. YUWILER, A. (1983): Vasoactive intestinal peptide stimulation of pineal serotonin-N-acetyltransferase activity: General characteristics. J. Neurochem. 41, 146-153

IS THERE A CIRCADIAN RHYTHM OF CYTOCHROME P-450 IIE1 ACTIVITY?

D. PANKOW, P. WOLNA & P. HOFFMANN

Martin Luther University, Institute of Industrial Toxicology,
Franzosenweg 1a, D-0-4020 Halle/Saale, FRG

INTRODUCTION

Different forms of cytochrome P-450 are known to catalyze the oxidation of a large number of xenobiotics. For example, alcohols and ketones, nitrosamines, halogenated alkanes, alkenes and ethers as well as aromatic compounds are substrates metabolized via cytochrome P-450 IIE1 (CYP2E1). The aniline hydroxylase activity in the microsomes is one of the indicators of the CYP2E1 activity. Isolated microsomes containing this enzyme show a temporal variation in their activity with high values between 6.00 h and 12.00 h and low values at 17.00 h and at 01.00 h (1). The dichloromethane (DCM) evoked carboxyhemoglobinemia seems to be an in vivo measure of the CYP2E1 activity: After pretreating rats with inducers of CYP2E1 such as isoniazid (isonicotinic acid hydrazide, INH), ethanol, acetone, trichloroethylene, benzene, toluene, 1,3-dimethylbenzene (DMB, m-xylene), pyrazole or imidazole an intensified metabolism of DCM to carbon monoxide (CO) could be observed by measuring the carboxyhemoglobin blood level (COHb). Simultaneous administration of both inducer and DCM was accompanied by an inhibition of the oxidative DCM metabolism. The aim of this study was to determine whether there are diurnal variations in DCM metabolism to CO in rats pretreated with vehicle ("naive rats"), and in rats pretreated with the inducers INH and DMB.

MATERIAL AND METHODS

Male Wistar rats (age: 71 ± 7 days, body mass: 266 ± 23 g; mean ± SD) were housed five or six per cage at 23 ± 2°C with a 12-h light-dark cycle and had free access to standard laboratory diet and tap water. They were treated with
a) INH, 0.36 mmol/kg i.p. in saline (naive rats: 1ml saline/kg i.p.),
b) DMB, 16.3 mmol/kg p.o. in 20% (v/v) Oleum pedum tauri (OPT) (naive rats: OPT, 10 ml/kg p.o.)
at 00.00 h, 04.00 h, 08.00 h, 12.00 h, 16.00 h, 20.00 h, and 24.00 h. Exactly 24 hours after pretreatment the animals received DCM, 6.2 mmol/kg p.o. as a 10% (v/v) solution in OPT. Blood was taken from the retroorbital plexus 6 hours after DCM administration. The COHb level was determined by the palladium-II-chloride method (3). A cosine function

$$y = M + A \cos [\pi/12 (t - \Phi)]$$

was fitted to the data using a least squares method, with
M: mesor, A: amplitude, Φ: acrophase, y: COHb (%), t: time (hours).
Confidence limits were calculated for the parameters M, A and Φ.

Substances:
INH: SERVA, Heidelberg; DMB: Bergakademie Freiberg; DCM: Riedel de Haën, Seelze

RESULTS

The DCM-evoked carboxyhemoglobinemia in naive rats shows a slight but significant circadian rhythm. Pretreatment with INH (Fig. 1) or DMB 24 hours prior to DCM administration results in significantly higher amplitudes and mesors of the COHb level, if compared to rats pretreated with vehicle only. The acrophases in all cases have to be considered as identical (Tab. 1 and 2).

```
——— NaCl    ——— INH      ★ NaCl    ○ INH
     theoret. curves         means (n=6)
```

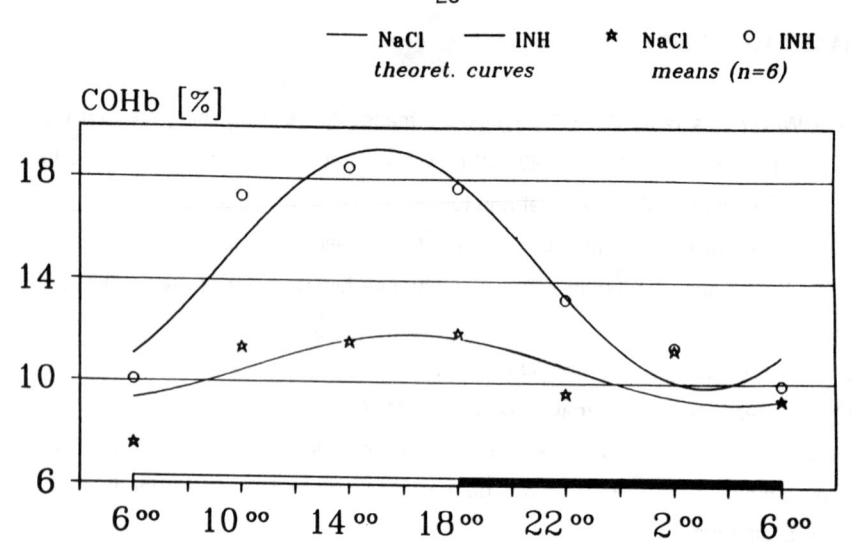

COHb [%]

Fig.1: COHb level after DCM administration in naive and isonazid-pretreated rats.

DISCUSSION

After exposure to DCM there is a circadian rhythm of the COHb formation in both naive rats and, more pronounced, rats pretreated with INH or DMB, which are known for being inducers of CYP2E1. These results correspond with the high aniline hydroxylase activity in the microsomes of the rat liver between 05.00 h and 13.00 h as determined by BÉLANGER & LABRECQUE (1), since COHb-values reach their maxima 6 hours after administration of DCM. The findings support the hypothesis, that the CYP2E1 activity and its induction show a circadian rhythm. In addition the enhanced exhalation rate of CO in nocturnal active rats (2) may play a role.

Supported by the BMFT grant No. 01 HK 0112

REFERENCES

1. BÉLANGER, P.M. & G. LABRECQUE (1989): Temporal aspects of drug metabolism. In: B. LEMMER (ed.): Chronopharmacology. Cellular and biochemical interactions. Marcel Dekker, Inc., New York and Basel, pp. 15-34
2. PANKOW,D., J. HAERTING, W. PONSOLD & J. ADAM (1983): Seasonal variations of carbon monoxide-induced polycythemia in rats. Biomed. Biochim. Acta 42, 255-263
3. SCHMIDT, P. (1988): Biologische Kontrollmethoden in der Arbeitsmedizin. Verlag Volk u. Gesundheit, Berlin, pp. 345-346

MOTILAR ACTIVITY OF INTESTINE OF WHITE RATS AND ITS 24-HOUR RHYTHM

B. HARBACH AND I. MLETZKO

College of Education, Department of Zoology, Kroellwitzer Str.44,
D-O 4050 Halle FRG

One of the endogenous short time rhythms in animals is the motilar activity of intestine. Aim of our investigation:
1. The various spontaneous short time rhythms in the smooth muscle of the different parts of intestine and
2. the variation of contractions in the course of 24 hours.

MATERIALS AND METHODS

The data have been obtained from male wistar rats 80-120 days old. They have grown under standard conditions (light regimen LD 12:12, temperature 20 ±1°C, humidity 65-70 %, food and water ad libitum). Approximately 20 hours before experiments animals have been withdrawn from food. Preparates from gaster, duodenum, jejunum, ileum, caecum, and colon descendens have been examined in intervals of 4 hours per day with the aid of electromechanical transformer system. Frequencies and amplitudes of contractions have been recorded continuously. For statistics have been used U-Test according to Mann and Whitney (14) and Fourier analyses (11).

RESULTS

Each part of gastrointestinal tract has a specific pattern of contraction (Fig. 1). We have found 3 parts with high frequencies, this applies to duodenum (36.8 min^{-1}), jejunum (33.5 min^{-1}), ileum (28.9 min^{-1}). The other parts have lower frequencies such as gaster (5.2 min^{-1}), caecum (3.1 min^{-1}) and colon descendens (6.1 min^{-1}). The amplitudes are very different too. They vary from 2.2 mm/100 (ileum) to 42.4 mm/100

Gaster

Duodenum

Jejunum

Ileum

Caecum

Colon descendens

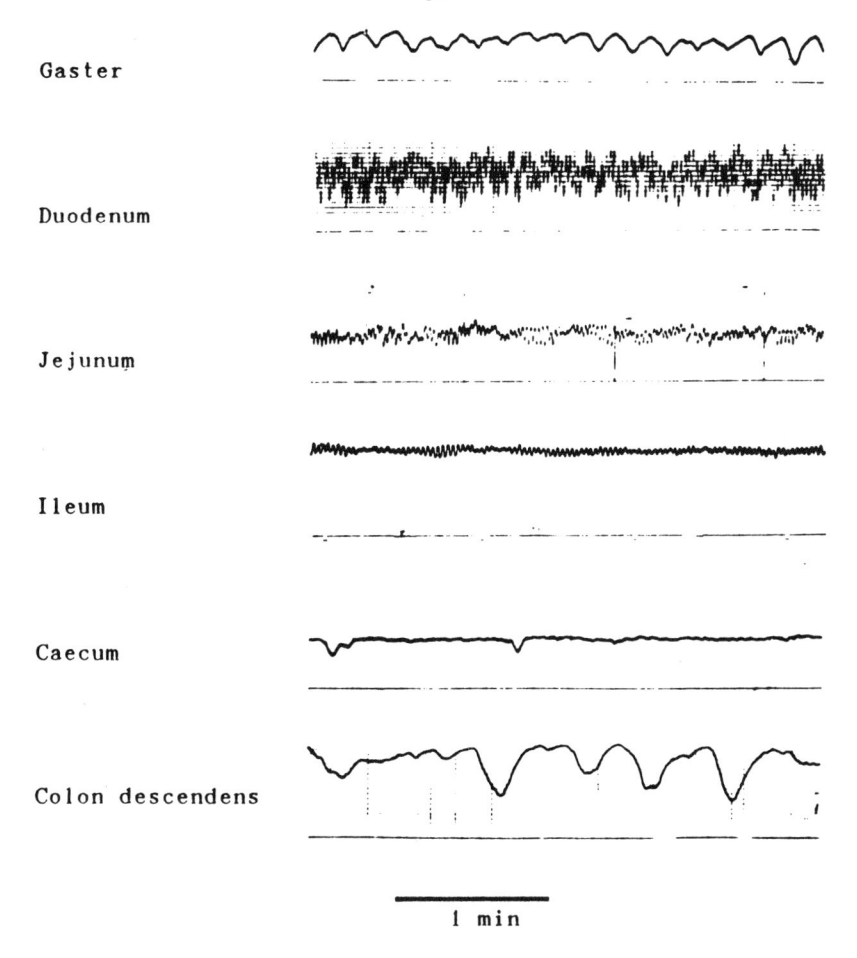

1 min

Fig.1: Recording curves of motilar activity in intestine.

(colon desc.). The different parts of intestine do not show only specific activity in the course of 24-hour-day. The amplitudes of caecum are exceptions. For the amplitudes of this part we could not prove significant differences between the examined day time points. Contractions of caecum show a clear connection to the tonus of preparates and this tonus diminished in the course of each experiment. As an example for the rhythmically changing activity we now describe ileum and colon descendens. Ileum shows vast activity in the night. The maximum of frequencies was found with 30.6 min^{-1} at 22.00 h. The differences are significant to the data from other day times. The minimal frequencies have been registered with 26.9 (min^{-1}) at 14.00 h. The amplitudes show its maximum also in the night at 2.00 h, they reach 6.2 mm/100. At light-off has been found the minimum with 2.25 mm/100. The 24-hour-oscillation dominates in both rhythms in accordance to fourier analysis. The maximum of frequencies has been calculated at 23.30 and of amplitudes at 0.30 h. Especially the amplitudes of colon descendens show a clear bimodal daily rhythm (Fig. 2). The highest contraction

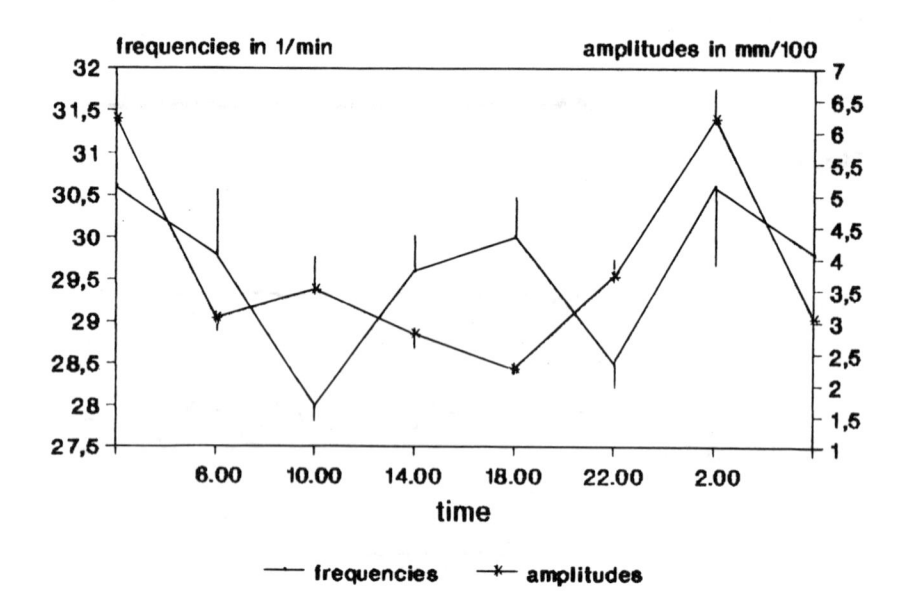

Fig.2: Lieum contraction in daily course

waves appear in the night at 22.00. With average 42.4 mm/100 they have altogether the highest amplitudes from all parts of intestine. These high middle-values have its origin in specific "gigantic waves". Amplitudes from 200 mm/100 and durations up to 30 seconds are characteristical of the gigantic waves. These waves are numerous in the night at 22.00 and rare at 6.00 h. That's why the minimal amplitudes have been registered at this time with 13.1 mm/100, too. The frequencies pass through its maximum (7.6 min^{-1}) at 22.00. According to Fourier analysis the 24-hour oscillation dominates in both parts of intestine.

DISCUSSION

Many authors have determined the various kinds of movements in intestine (4.8). In our investigation we give preference to the definition of YOKOYAMA (1966). He classified the contractions into phasic and tonic ones.

A problem of our experiments is that the longitudinal and the circular muscle layers have not been seperated from each other. There are reasons to believe that both muscle layers work together with temporal and spatial variations.

Each part of intestine has its own specific pattern of contractions. This supports the concept of several pacemakers with different "Zeitgeber" frequencies (3). The pacemakers work for seconds and minute rhythms, the contractions influence and superpose each other. Mechanisms how the oscillations are connected have been already described by v. HOLST (7) by means of fins of fishes as "Magneteffekt" and "Superposition". We think that the cause of the very different amplitudes may be presumed in a different number of contracting cells. Variations of motilar activity of isolated preparates in the course of 24-h day have been proved.

The light-dark changes influence and synchronize the periodic food intake (12). Both factors influence also various rhythms in the gastrointestinal tract. In intestine have been proved several 24-hour rhythms, such as secretin of digestive enzymes (13), the epithelial mitosis (2) or the activity of liver enzymes (9). The daily rhythmically varying contractions of in-vitro intestine preparates are in relation to periodic food intake, digestion and defecation. Highest quantity of food intake happens in the first hours after light-off. After a certain time in gaster (minimal 30 minutes) chymus passes to duodenum, jejunum, and ileum. All parts of intestine become especially active when have been filled with chymus. Then there active transport processes take place. So it is in ileum, it is very active in the night. Already DAVENPORT (4) has found out that after food intake the ileum shows maximal contractions. Daily rhythms in colon descendens are in accordance with filling conditions. High amplitudes in the night guarantee high efficiency of transport. These contractions must take place before defecation. It has been proved in many mammals that the defecation happens in the first hours of activity (1, 5). The increasing activity of our in-vitro preparates to 10.00 h coincide in time with increased locomotor activity by rats at this time (10). Whether these both rhythms are really coupled is unknown. Our results of 24-hour rhythms of motilar activity have been obtained by isolated preparates of intestine, suppose that these rhythms are endogenous.

REFERENCES

1. ALTMANN, D. (1969): Harnen und Koten bei Säugetieren. Ziemsen Verlag
2. ALOV, A. (1963): Daily rhythms of mitosis and relationships between cell works and cell division. Fed. Proc. 22, 357-362
3. BURNSTOCK, G., M.E. HOLMANN & C.L. PROSSER (1963): Electrophysiology of smooth muscle. Physiol. Rev. 43, 482-527
4. DAVENPORT, H.W. (1971): Physiologie der Verdauung. Schattauer, Stuttgart New York
5. GATTERMANN, R. (1984): Chronobiologische Oscillationen des Goldhamsters (Mesocricetus auratus L.). 1. Zirkadiane Rhythmen. Zool. Jb. Physiol. 85, 471-489
6. HARBACH,B. (1989): Die Motilaraktivität des Magen-Darm-Traktes, ihr 24-Stunden-Rhythmus und die tageszeitabhängige Beeinflussung durch Dimethoat am Beispiel der weißen Ratte (Rattus norvegicus BERKENHOUT). Diss. A, PH Halle
7. HOLST, E.v. (1937): Vom Wesen der Ordnung im Zentralnervensystem. Naturwiss. 25, 3-32
8. HUKAHARA, T. (1933): Die Bewegungen des ausgeschnittenen Dünndarmes. Pflüg. Arch. 232, 51-56
9. KAST, J., N.T. YABE, et al. (1988): Circadian rhythm of liver parameters during three consecutive cycles in phenobarbital-treated rats. Chronobiol. Intern. 5, N.4, 363-385
10. MLETZKO, I. (1982): Circadiane Zeitreihen der Motorialaktivität von Landwirbeltieren. Diss. B. Potsdam
11. ORLICK, M. & H.G. MLETZKO (1975): Auswertung biologischer Zeitreihen mittels Fourier- und Autokorrelationsanalyse. Biol. Rdsch. 13, 265-276
12. PHILIPPENS, K.M. (1977): Lichtstandardisierung von Laboratoriumsratten im chronobiologischen Experiment. Nova Acta Leopoldina 46, 271-276
13. RAKHIMOV, K.R. & G.I. SMIRNOVA (1978): Circadian rhythmicity of enzyme activities in small intestine of growing, adult, and aged rats. Fisiol. Shurnal SSSR 7, 970-977
14. WEBER, E. (1961): Grundriß der biologischen Statistik. Fischer Jena

RAT LIVER STRUCTURE:
CIRCADIAN RHYTHM OF SOME MORPHOMETRIC PARAMETERS

C. DOLCI, F. CARANDENTE, L. VIZZOTTO, A. MONTARULI & A. MIANI

Istituto di Anatomia Umana Normale, Cattedra di Cronobiologia, Centro di
Cronobiologia, Università degli Studi di Milano, Facolta di Medicina
Via Mangiagalli 31-7, 20133 Milano-8, ITALY

INTRODUCTION

Many chronobiological studies have been conducted on the hepatic function with the demonstration that the physiological secretion is characterized by periodic variations (5, 6, 7, 12). On this basis concomitant structural variations of this organ have been hypothesized: to verify them we constructed a morphometric model of the rat liver. There are several methods to perform morphometry: the stereological methods allow the statistically validated exploration of the three-dimensional space when only two-dimensional sections of an organ are available, as in our case. We utilized this morphometric model of rat liver to evaluate its circadian rhythmicity. In a previous study (1) we demonstrated circadian rhythm of the mean section area of hepatocytes and the volume fraction of their nuclei; we repeated the experiment to evaluate the periodic pattern of new parameters with a better methodological standardization.

MATERIALS AND METHODS

The study has been conducted on 24 female Wistar rats, four months old, singly housed a month before the experiment with lighting regimen 12:12, constant environmental temperature (22°C ±1) and relative humidity (50%), food and water ad libitum. Animals were randomly divided in 6 balanced groups and killed after ether anesthesia at 6 different time points in 24 hours, four hours apart. In this way we obtained 6 time points almost equally distributed in the 24 hour period. The hepatic tissue was fixed in glutaraldehyde and osmium and embedded in epoxide. This is considered the best way to minimize error due to retraction and axial deformation subsequent to dissection (9, 11).

Measurements were obtained from 1 µm liver sections that were examined at light microscope (1170X) for the evaluation, by stereological methods (10), of the following morphometric parameters:

1) Volume fraction (V_v) of cytoplasm and nuclei of hepatocytes and volume fraction of sinusoids.
   Volumetries were performed by different point counting, according to Delesse formula: $V_v = P_p$ (P_p is the fraction of test points included in the profiles).
2) Nucleus/cytoplasm ratio of hepatocytes.
3) Surface fraction (S_v) of sinusoid, parenchymal and vascular faces of the hepatocytes.
   Surface fractions were calculated by counting the number I_i of the test line intersections with profile boundaries and applying the formula $S_v = 2I_i$.
4) Mean hepatic cell section area. It is the arithmetic mean of the section areas evaluated by a digitalizer.
5) Hepatocyte nuclei size distribution with their number for unit volume and their mean diameter, according to the Schwartz-Saltykov method, modified by DE HOFF supposing spherical nuclei (8).

Chronobiological analysis of the data was performed by single cosinor method (4), that defines:

PR: Percent Rhythm (amount of variability explained by periodic variations).

Mesor: Midline Estimating Statistic of Rhythm and relevant standard error.

Amplitude: it is one half of the extent of rhythmic change. It is expressed as percent of mesor. In brackets the relevant 95% confidence limits.

Acrophase: it is the lag from the reference time, in this case the beginning of light interval, of the crest time in the function approximating the rhythm. It is expressed in HALO (Hours After Light On).

RESULTS

The statistically significant (p <0.05) circadian rhythms of the volume fraction of hepatocyte nuclei, the nucleus/cytoplasm ratio and the mean hepatic cell section area have been demonstrated.

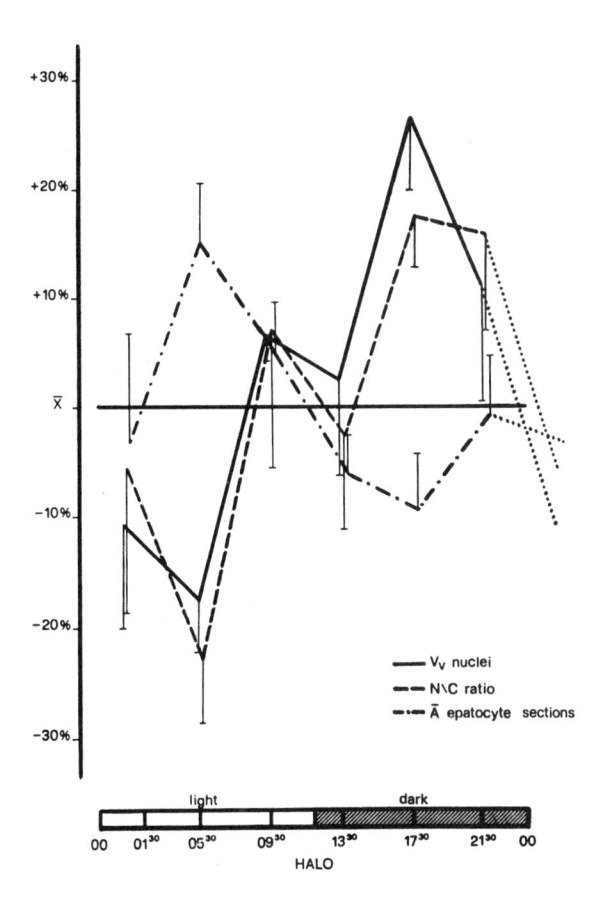

Fig.1: Volume fraction of hepatocyte nuclei (V_v nuclei), nucleus/cytoplasm ratio (N/C ratio) and mean hepatocyte section area evaluated in female Wistar rats. Values are expressed as percent difference of the original data from their mean value and every point is the arithmetic mean of the values obtained from 6 animals with the relevant standard errors. In abscissa the time is reported in HALO (Hours After Light On).

Fig. 1 shows the percent difference of the original data from their mean value at all 6 time points for the three morphometric parameters: the volume fraction of hepatocyte nuclei and nucleus/cytoplasm ratio show the lowest values during the light period and the highest ones during the dark. On the contrary the mean hepatic cell section area has high values during the light period and low ones during the dark.

The cosinor analysis for the demonstration of the periodicity of the phenomena shows the coincidence on time of the volume fraction of hepatocyte nuclei and the nucleus/cytoplasm ratio and the antiphase of the mean hepatic cell section area. The acrophase values are: ϕ = 17.14 (13.40/20.23) HALO for the volume fraction of hepatocyte nuclei, ϕ = 18.37 (13.55/22.41) HALO for the nucleus/cytoplasm ratio and ϕ = 06.02 (01.46/09.42) HALO for the mean hepatic cell section area.

DISCUSSION

We confirmed the existence of circadian periodicity for the mean section area of hepatocytes and the volume fraction of their nuclei, and we demonstrated rhythmicity in the 24 hours also for the nucleus/cytoplasm ratio, not previously put in evidence.

The volume fraction of hepatocyte nuclei and nucleus/cytoplasm ratio show circadian rhythm with acrophase at the half of the dark period (activity phase of the rats). This means that during the dark span nuclei are in average bigger than during the light span. This fact suggests that the highest nuclear transcriptional activity is located during the activity span.

On the contrary the acrophase of the mean hepatic cell section area is located at the half of the light period (rest phase of the rats). This last finding is in a good agreement with our previous study (1) in which the highest storage of glycogen in the rat hepatic cell is just found during the first 6 hours of the light period.

Fig. 2 shows the circadian rhythm of the volume fraction of glycogen respectively in the periportal zone (A), intermediate zone (B) and pericentrolobular zone (C) of the hepatic lobule. It is statistically significant in all three zones with a two-hour delay from the periphery toward the center of the lobule. The acrophase shift of the three zones indicates the subsequent storage from the portal to the centrolobular zone. The highest values of hepatic cell section area and glycogen

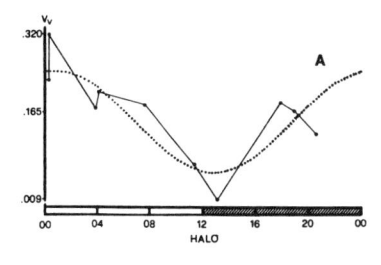

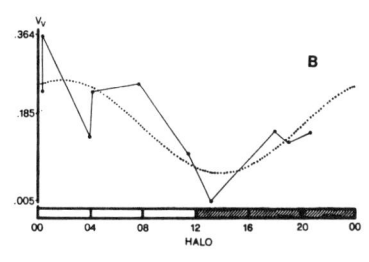

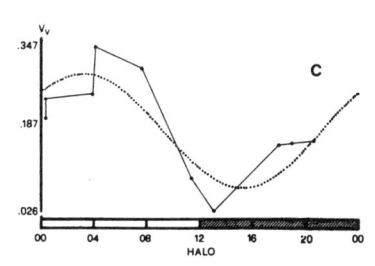

Fig.2:
From our previous study (1), circadian rhythm of the volume fraction of glycogen in periportal zone (A), intermediate zone (B) and pericentrolobular zone (C) of the hepatic lobule of rat. All three circadian patterns are statistically significant, with larger storage during the first hours of the light period and lower storage during the first hours of the dark. A two-hour delay of the acrophase from the periphery toward the center of the lobule indicates the subsequent storage from the portal to the centrolobular zone.

content are almost coincident on time (during light period): the close relationship between structure and function of the organ is put in evidence by the concordance of the circadian timing of the structural and functional variables. Similar findings were reported by HAUS et al. (2) that demonstrated the time relationship between phospholipid, RNA, DNA, glycogen, and mitoses in mouse liver. Also these authors demonstrated the acrophase of glycogen content during the first hours of light.

REFERENCES

1. CASTANO, P., F. CARANDENTE, V.F. FERRARIO et al. (1984): Cronomorfologia del fegato di ratto: aspetti strutturali ed ultrastrutturali. Ric. Sci. Ed. Perm., Suppl. 38, 99-100
2. HAUS, E., F. HALBERG, J.F.W. KUEHL & D.J. LAKATUA (1974): Chronopharmacology in animals. In: J. ASCHOFF, F. CERESA & F. HALBERG (eds.): Chronobiological aspects of endocrinology. Chronobiologia 1, Suppl. 1, 122-156
3. JERUSALEM, C., O. MULLER & H. v. MAYERSBACH (1967): Circadian ultrastructural changes in liver cells. In: H. v. MAYERSBACH (ed.): The cellular aspects of biorhythms. Springer-Verlag, Berlin, pp. 147-154
4. NELSON, W., L.Y. TONG, J.K. LEE & F. HALBERG (1979): Methods for cosinor-rhythmometry. Chronobiologia 6, 305-323
5. OLSHEFSKI, R., D.R. ROSEWALL, W. HRUSHESKY et al. (1983): Chronobiologic aspects of rat liver cholesterol 7α-hydroxylase activity. Chronobiologia 10, 154-155
6. SOLER, G., J.M. BAUTISTA, J.A. MADRID & G.M. SALIDO (1988): Circadian rhythms in enzymatic activity of rat liver arginase and glucose- 6-phosphatase dehydrogenase. Chronobiologia 15, 205-212
7. SPELSBERG, T. & F. HALBERG (1981): Toward a molecular chrono-oncology: circadian rhythms in RNA polymerase activities of rat liver. In: H.E. KAISER (ed.): Neoplasm - Comparative Pathology of Growth in Animals, Plants and Man. Williams & Williams Co., Baltimore, pp. 209-214
8. UNDERWOOD, E.E. (1968): Particle-size distribution. In: R.T. DE HOFF & F.N. RHINES (eds.): Quantitative Microscopy, McGraw-Hill Book Co., New York, pp. 149-200
9. VIZZOTTO, L., V.F. FERRARIO, M.P. MOLINARI TOSATTI et al. (1984): Effetto delle modalità di fissazione ed inclusione sulle caratteristiche morfoquantitative del fegato di ratto. Ric. Sci. Ed. Perm., Suppl. 38, 467-468
10. WEIBEL, E.R. (1980): Stereological Methods. Vol. 2: Theoretical foundations. Academic Press, London
11. WILLIAMS, M.A. (1977): Quantitative Methods in Biology. North-Holland, Amsterdam
12. WOJCZYK, B., A. LITYNSKA & A. KALISZ (1986): Temporal changes of acid phosphatase, arylsulphatase, β-galactosidase and β-N-acetyl-D-glucosaminidase activities in subcellular fractions of rat liver. Chronobiology International 3, 29-37

TIME INTERVALS AND AMPLITUDES OF ULTRADIAN OSCILLATIONS OF CARBON DIOXIDE EMISSION ($\dot{V}CO2$) IN SEVERAL ENDOTHERMIC SPECIES (MICE, RATS, GUINEA-PIGS, QUAIL, CHICKS, MONKEYS AND PREMATURE INFANTS)

V.GOURLET*, M.STUPFEL**, A.PERRAMON***, P.MERAT***, G.PUTET°* & L.COURT°**

French National Institute of Health and Medical Research (INSERM), Research Center, 44,Chemin de Ronde, Le Vésinet, F78110; ***INRA, Jouy en Josas; °*Hôpital Edouard Herriot, Lyons; °CRSSA, Clamart; *INSERM, U169, Villejuif, F94807, France

Continuous recordings of carbon dioxide emission ($\dot{V}CO_2$) in mice, rats, guinea-pigs, quail, and chicks (1), in monkeys (2) and in premature infants (3) show ultradian oscillations of different periods and amplitudes. The general pattern of these recordings seems, at first sight, characteristic of each species. Temporal analysis of the $\dot{V}CO_2$ variations by several statistical processes (Fourier spectral and variance analyses, Enright's chi-square periodogram, run-test) shows that for each species and each individual these variations exhibit episodes of diverse periods and also aperiodic episodes. Further computation was performed in mice, rats, guinea-pigs, quail, and chicks (1) in sampling every 20 minutes the CO_2 concentration values on the continuous CO_2 concentration recordings. The time intervals measured between the $\dot{V}CO_2$ oscillation peaks were not very far from one hour (1.0-1.5h) and variance analysis showed interspecies significant differences; the amplitudes of these $\dot{V}CO_2$ oscillation peaks were also statistically different according to the species and their values ranged from 0.10 in chicks to 0.30 l/h/kg in mice. It was decided to apply these same processes of measurements to $\dot{V}CO_2$ recordings performed in monkeys (2) and premature infants (3) in order to compare together the values of the time intervals and amplitudes of the $\dot{V}CO_2$ oscillations in these 7 species of endotherms and to relate these data to their body weights.

SUBJECTS AND METHODS

The subjects and methods have been previously exposed in detail for the mice, rats, guinea-pigs, quail, and chicks (1). However it will be recalled that the measurements were made under similar environmental conditions: temperature 19-21°C, humidity 40-60% for all the animals, and 34°C and 53% for the premature infants; the ventilation was regulated so that the mean level of the carbon dioxide was inferior to 0.1% in the respiratory chambers for animals, and to 0,5% in the respiratory hood for the premature infants.

The lighting regimen was either LD 12:12 (alternation of 12 hours of 06-18 light (L) and 12 hours 18-06 dark (D), LL continuous light, DD continuous dark; L or LL light intensity was 100 lux at the level of the animals.

The recordings were performed in OF1 male mice (20 in LD 12:12, 18 in LL, and 16 in DD), OFA male rats (37 in LD 12:12, 16 in LL, and 13 in DD), Hartley male guinea-pigs (24 in LD 12:12, 14 in LL, and 12 in DD), Japanese male quail (30 in LD 12:12, 25 in LL, and 16 in DD), male Leghorn and Rhode Island Red Chicks (7 in LD 12:12, 2 in LL, 2 in DD), two male cynomolgus monkeys (in LD 12:12) and 5 male and 2 female premature infants (in LL).

For each animal, measurements were performed from 4 to 40 (mean 9) consecutive days and for each premature infant only during 24 consecutive hours.

Each oscillation were determined on the graphs performed with the CO_2 concentration sampled every 20 minutes on the continuous recordings (1 point every 12 seconds) in eliminating one of two adjacent CO_2 peaks having amplitude differences lower than 20%; in other words the peaks (i.e. the maximum of one oscillation) was at least a 20% increase from a local minimum to the consecutive local maximum. Intervals between oscillations (peaks) and their amplitudes were then computed.

RESULTS

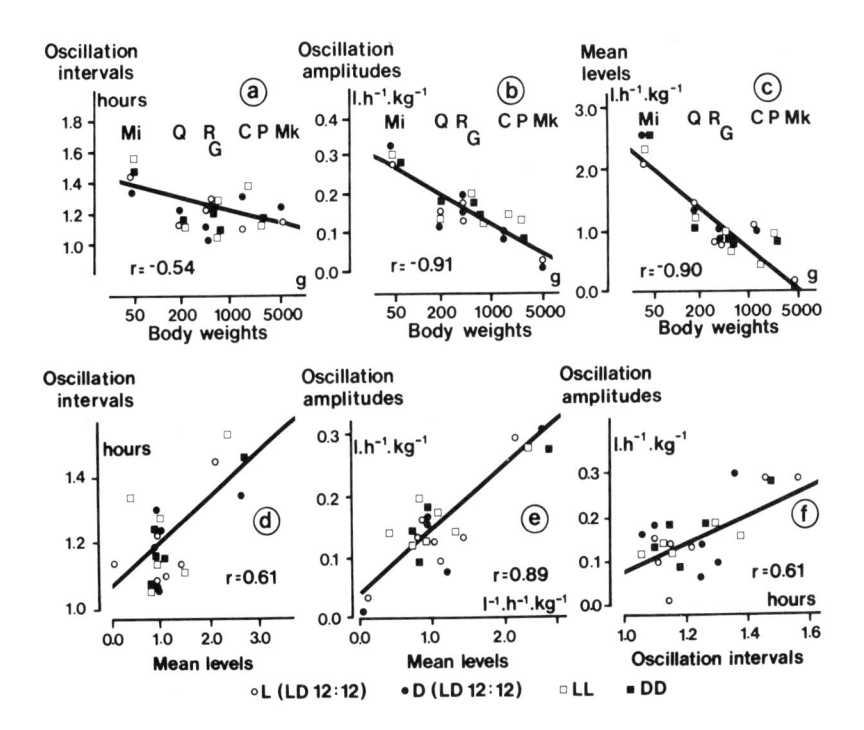

Fig.1: Relationships for mice (Mi), quail (Q), rats (R), guinea-pigs (G), chicks (C), premature infants (P), and monkeys (Mk) between: a) time intervals of ultradian $\dot{V}CO_2$ oscillations - body weights; b) amplitudes (liters per hour and per kilogram bw) of ultradian $\dot{V}CO_2$ oscillations - body weights; c) mean levels of $\dot{V}CO_2$ - body weights; d) time intervals of ultradian oscillations - mean levels of $\dot{V}CO_2$; e) amplitudes of ultradian $\dot{V}CO_2$ oscillations - mean levels of $\dot{V}CO_2$; f) amplitudes of ultradian $\dot{V}CO_2$ oscillations - time intervals of ultradian $\dot{V}CO_2$ oscillations. Each point or square corresponds to data measurements in a mean of individuals either in light (L) - dark (D) alternation of 12 hours (LD 12:12), in continuous light (LL), or in continuous dark (DD); L=100 lux.

Fig. 1 shows the relationships for mice (Mi), quail (Q), rats (R), guinea-pigs (G), chicks (C), premature infants (P), and monkeys (Mk), either in LD 12:12 (white points during L and black points during D), in LL (white squares) and in DD (black squares):

between oscillation time intervals and body weights (a); between oscillation amplitudes and body weights (b); between mean $\dot{V}CO_2$ levels and body weights (c); between oscillation time intervals and mean $\dot{V}CO_2$ levels (d) ; between oscillation amplitudes and mean $\dot{V}CO_2$ levels (e); between oscillation amplitudes and oscillation time intervals (f). It may be seen (a,b,c) that there are but small differences for oscillation time intervals, oscillation amplitudes and mean $\dot{V}CO_2$ levels values in mice, quail, rats, and guinea-pigs in which these measurements were performed in LD 12:12, LL and DD.

Tab.1:
Correlation coefficients in LD 12:12, LL, DD in 7 endothermic species (mice, rats, guinea-pigs, monkeys, premature infants, quail, and chicks) between the ultradian time intervals and amplitudes of $\dot{V}CO_2$ oscillations, $\dot{V}CO_2$ levels and log body weights; CO_2 concentrations are sampled every 20 minutes on CO_2 continuous recordings; number of means of measurements are between parentheses.

Correlation coefficients r		$\dot{V}CO2$ oscillations		$\dot{V}CO2$ levels	Body weights (log)
		Intervals	Amplitudes		
$\dot{V}CO2$ oscillations	Intervals		0.61** (23)	0.61** (23)	-0.54** (23)
	Amplitudes	*** P < 0.001		0.89*** (23)	-0.91*** (23)
$\dot{V}CO2$ levels		** P < 0.01			-0.90 *** (23)

Tab. 1 shows the correlation coefficients between the 4 considered parameters; the numbers of correlated pairs are between parentheses. The ultradian time intervals are significantly correlated to $\dot{V}CO_2$ amplitudes (r =0.61; p<0.01), to $\dot{V}CO_2$ levels (r =0.61; p<0.01) and to log body weights (r =-0.54; p<0.01). The amplitudes of ultradian $\dot{V}CO_2$ oscillations are significantly and strongly correlated to $\dot{V}CO_2$ levels (r =0.89; p<0.001) and to log body weights (r =-0.91; p<0.001). The $\dot{V}CO_2$ levels are strongly correlated to log body weights (r =-0.90; p<0.001).

Multiple correlations a) relating intervals to amplitude - $\dot{V}CO_2$ levels - log body weights is : r =0.66 (p<0.01), b) relating amplitudes to intervals - $\dot{V}CO_2$ levels - log body weights is $r = 0.93$ ($p<0.001$). The use of Spiegel's test (4) to compare these two last correlation coefficients shows that $r = 0.93$ ($n=23$) is significantly ($p<0.01$) greater than $r = 0.66$ ($n=23$). Correlations eliminating log body weights and a) relating time intervals to amplitudes - $\dot{V}CO_2$ levels is $r = 0.64$ ($p<0.01$), b) relating amplitudes to time intervals - $\dot{V}CO_2$ levels is $r = -0.89$ ($p<0.001$). The use of Spiegel's test shows that $r = -0.89$ ($n=23$) is significantly ($p<0.001$) greater than $r = 0.64$ ($n=23$), which confirms that the time intervals of the ultradian oscillations are less correlated than the oscillations amplitudes to $\dot{V}CO_2$ level and log body weights. Furthermore, the 4 considered parameters (intervals, amplitudes, $\dot{V}CO_2$ level, and body weights) explain 93.54% of the oscillation amplitude variance and only 41,09% of the time interval variance.

DISCUSSION

The time intervals of the ultradian oscillations are therefore less correlated than their amplitudes to body weights and $\dot{V}CO_2$ levels. This may be partly the consequence that the range of the time intervals of oscillations (1.0-1.5 hour) is smaller than the range of their amplitudes (0.10-0.30 l/h/kg). However the ultradian rhythmicity of respiratory exchanges, which corresponds to thermodynamic exchanges and are highly depending on body weights, looks to be an amplitude modulation rather than a frequency modulation, as it has been postulated for other types of ultradian rhythms (5). Moreover, it must not be forgotten that the every 20 minute sampling of $\dot{V}CO_2$ values used in these investigations introduces aliasing, and therefore that this comparison of the $\dot{V}CO_2$ time intervals of ultradian oscillations between these seven endothermic species may be only applied to time intervals greater than 40 minutes.

REFERENCES

1. STUPFEL, M., V.GOURLET, A.PERRAMON, P. MERAT & L.COURT(1990): Ultradian and circadian compartmentalization of respiratory and metabolic exchanges in small laboratory vertebrates. Chronobiologia 17, 275-304.
2. STUPFEL, M., J.C.ESTRIES & V.GOURLET (1986): Ultradian respiratory rhythms of mean and high periods (40 min< t < 24 hr) in *Cynomolgus* monkeys In: COURT L., TROCHERIE S.& DOUCET J. (eds.), Traitement du signal en électrophysiologie expérimentale et clinique du système nerveux central, CEA, Fontenay-aux-roses, II, pp.595-601.
3. PUTET, G., M. STUPFEL, V.GOURLET, B.SALLE & L.COURT (1990): Respiratory and metabolic ultradian (40 min < period < 6 h) variations in normal premature infants periodically fed through a gastric tube. Chronobiologia 17, 1-13.
4. SPIEGEL, M.R. (1991): Theory and problems of statistics. Schaum, New York, p. 264.
5. HILDEBRANDT G. (1988): Temporal order of ultradian rhythms in man. Adv. in Biosciences, 1988, 73, 107-122.

SEASONAL DIFFERENCES IN QUALITIVE AND QUANTITIVE INDICES OF MYOCARDIAL ULTRASTRUCTURE IN RABBITS

E.S. MATYEV, V.A. FROLOV, S.M. CHIBISOV & V. MOGYLEVSKY

Patrice Lumumba Peoples' Friendship University, Moscow, U.S.S.R.
Miklukcho-Maklaya str. 8, Dep. Pathophysiology, Russia

It is well known that the activity of the cardiovascular system has rhythmic fluctuations (1, 2, 3, 5). Most investigations are devoted to the circadian heart rhythms (4).

In recent years experimental works on seasonal dynamics of morpho-functional heart status have appeared (6, 7, 8). We carried out an experimental investigation of myocardial ultrastructure in rabbits in altitude hypoxia from the chronobiological point of view.

Morphological study of organs during adaptation to altitude hypoxia in different seasons of the year has revealed that the influence of hypoxia is characterized by destructive changes in mitochondria, nuclei, and cytoplasm. These changes disappeared because of the intracellular regenerative processes. If the frequency of hypoxic factor action exceeds the rate of intracellular reparative processes, it leads to an incompleteness of regeneration which undergoes marked seasonal differences.

In this study we present the electrone-microscopic data on seasonal dynamics of the mitochondrial status of rabbit's myocardium in different conditions of adaptation to altitude hypoxia.

MATERIAL AND METHODS

Experiments were carried out on 280 male rabbits, weighing 2.5 -3.0 kg in summer, autumn, winter, and spring. All the animals were divided to the following groups containing equal numbers of rabbits:

200 m a.s.l. (above sea level) -plain (Moscow)
700 m a.s.l.-low mountain (Bishkek)
1700 m a.s.l.-middle-mountain, aborigen rabbits (Issyk-Kullake)
1700 m a.s.l.-middle-mountain, non adapted rabbits
1700 m a.s.l.-middle-mountain, 2 month adaptation
3200 m a.s.l.-high-mountain, non adaptated rabbits
3200 m a.s.l.-high-mountain, 2 month adaptation

The electron-microscopic investigation was carried out by use of Tesla BS 540 (Chekoslovakia) instrumentation.

We have made seasonal analysis of such indices as the quantity of mitochondria in electronogram (N m EG), area of 1 mitochondrium (S m), the quantity of cristae in 1 mitochondrium (N c m), coefficient of energetic efficacy of mitochondria (KEEM), summarized number of cristae in electronogram (N c EG), and summarized area of mitochondria in electronogram (S m EG).

RESULTS

SPRING: The analysis of spring data is presented in Tab.1.

1. The greatest number of mitochondria in electronogram is observed in plain and in 3200 m a.s.l. without adaptation. The smallest number of mitochondria is in aborigens in 1700 m a.s.l. In this case the difference with the two previous groups is significant.

2. The smallest mitochondria area is observed in 200 m a.s.l. and in adapted animals in 1700 m a.s.l. In aborigens in 1700 m a.s.l. there is the biggest mitochondria area, which is significantly bigger than this index in plain (200 m a.s.l.) and low mountain.

3. The number of cristae per mitochondrium is smallest in plain and greatest in low mountain.

4. KEEM is the highest in aborigens in 1700 m a.s.l. and the lowest in plain animals. (The difference is significant)

5. The number of cristae in electronogram is greatest in rabbits adapted to 3200 m a.s.l. and in aborigens in 1700 m a.s.l. The lowest level of this index is in adapted rabbits in 1700 m a.s.l. and in animals on plain.

6. KEEM of electronogram is maximum in low mountain (700 m a.s.l.), in animals adapted to 3200 m a.s.l., and in aborigens in 1700 m. The lowest level of KEEM of electronogram is observed in plain and in non-adapted on 3200 m.

Tab.1:
The results of quantitive analysis of electronograms of rabbits left ventricle in different seasons of the year.

GROUPS	N m EG	S m	N c m	KEEM m	S m EG	N c EG	KEEM EG
200m aborigens	8.08 ±0.9	0.45 ±0.06	1.83 ±0.42	0.84 ±0.21	3.3 ±0.3	14.33 ±3.15	6.5 ±1.5
700m aborigens	6.38 ±0.71	0.51 ±0.03	5.25 ±0.46	2.74 ±0.32	3.15 ±0.36	31.75 ±3.93	16.23 ±2.24
1700m aborigens	4.94 ±0.55	0.74 ±0,1	4.17 ±0.4	3.22 ±0.57	3.09 ±0.27	20.22 ±2.56	13.03 ±1.96
1700m non adapted	5.47 ±0.49	0.58 ±0.03	3.65 ±0.49	2.2 ±0.35	3.2 ±0.32	18 ±1.66	10.73 ±1.28
1700m adapted	6.59 ±0.52	0.47 ±0.03	1.88 ±0.17	0.89 ±0.09	2.97 ±0.2	11.94 ±1.27	5.47 ±0.58
3200m non adapted	6.8 ±0.92	0.55 ±0.07	2.07 ±0.38	1.13 ±0.28	3.24 ±0.36	13.47 ±2.6	6.99 ±1.76
3200m adapted	8 ±0.87	0.58 ±0.03	3.73 ±0.38	2.2 ±0.29	4.49 ±0.41	28 ±2.27	15.89 ±1,3

We conclude that the highest level of KEEM per mitochondrium and electronogram is observed in aborigens on 1700 m a.s.l. It is determined by high values of mitochondrium area (correlation coefficient between the area per mitochondrium and KEEM is equal to +0.83; $p<0.01$)

In plain animals KEEM is considerably lower and there is no marked dependence on mitochondrium area.

Tab. 2:
The results of quantitive analysis of electronograms of rabbits left ventricle in summer season.

GROUPS	N m EG	S m	N c m	KEEM m	S m EG	N c EG	KEEM EG
200m aborigens	11 ±1.24	0.33 ±0.03	8.28 ±0.64	3.02 ±0.58	3.47 ±0.42	84.39 ±8.7	28.08 ±3.76
700m aborigens	8.06 ±0.66	0.42 ±0.02	3.06 ±0.48	1.34 ±0.23	3.37 ±0.26	21 ±1.77	9.25 ±1.02
1700m aborigens	8.29 ±0.53	0.65 ±0.05	4.7 ±0.45	3.2 ±0.49	5.25 ±0.31	38.9 ±4	25.19 ±3.12
1700m non adapted	7.5 ±0.75	0.59 ±0.05	5.14 ±0.42	2.99 ±0.34	4.1 ±0.3	37.43 ±3.87	20.6 ±1.97
1700m adapted	3.78 ±0.46	1.19 ±0.19	7.44 ±1.25	8.99 ±2.01	4.24 ±0.53	24.22 ±3.28	21.5 ±2.75
3200m adapted	4.9 ±0.46	0.43 ±0.03	6.2 ±0.41	2.65 ±0.24	2.05 ±0.2	29.3 ±2.87	12.48 ±1.21

SUMMER: In summer season the middle-mountain aborigens have a myocardial ultrastructure status similar to spring's one. The KEEM per mitochondrium and electronogram are the highest among all experimental group. As in spring, the KEEM per mitochondrium is determined by the mitochondrium area (the correlation coefficient is equal to +0.84; $p<0.01$), but the number of cristae is also of importance.

The analysis of summer data is presented in Tab. 2.

Tab. 3:
The results of quantitive analysis of electronograms of rabbits left ventricle in autumn season.

GROUPS	N m EG	S m	N c m	KEEM m	S m EG	N c EG	KEEM EG
200m aborigens	8.36 ±0.56	0.67 ±0.03	4 ±0.57	1.74 ±0.24	3.76 ±0.31	33.29 ±5.82	14.82 ±2.46
700m aborigens	7.2 ±1.24	0.66 ±0.04	4.4 ±0.83	2.85 ±0.67	4.56 ±0.79	28.9 ±7.07	14.78 ±3.68
1700m aborigens	8.6 ±0.48	0.42 ±0.05	3.6 ±0.35	1.54 ±0.35	3.59 ±0.25	30.6 ±3.55	12.09 ±1.8
1700m non adapted	7.54 ±1.07	0.72 ±0.15	4.15 ±0.53	3.48 ±0.62	3.89 ±0.44	27.15 ±4.35	15.03 ±2.21
1700m adapted	7.07 ±0.97	0.64 ±0.11	4.04 ±0.55	2.81 ±0.74	3.53 ±0.37	25.5 ±3.67	13.05 ±1.19
3200m non adapted	8.66 ±1.04	0.35 ±0.02	5 ±0.45	1.73 ±0.2	2.95 ±0.38	38.07 ±3.54	13.44 ±1.55
3200m adapted	6.5 ±0.93	0.51 ±0.07	4.3 ±0.7	2.44 ±0.7	3.37 ±0.78	26.7 ±4.15	15.01 ±3.76

AUTUMN: The number of mitochondria in 1 electronogram does not differ significantly in all groups. The mitochondria area is the smallest in non-adapted animals on 3200 m a.s.l. and the biggest on plain and in middle-mountain. The number of cristae in 1 mitochondrium is the highest in non-adapted animals on 3200 m, between other groups there is no significant difference. The biggest KEEM is observed in rabbits without adaptation on 1700 m. The mitochondria area is the largest in low-mountain (700 m) and the smallest in nonadapted animals on 3200 m. There is no significant difference in KEEM values. The results of analysis of autumn data are presented in Tab. 3.

Hence, as far as to the autumn, it can be stated that individual differences in ultrastructure are leveled and it is impossible to say what group of animals has the highest energetic potential of cardiomyocite.

WINTER: Analysis of the data as shown in Tab. 4 allows to make the following conclusions: The number of mitochondria is the same in all groups. The area of a single mitochondrium is the largest in nonadapted rabbits on 3200 m a.s.l. and smallest in rabbits on low-mountain (700 m). The number of cristae in a single mitochondrium is the highest in both groups of animals on 3200 m and the lowest in aborigens on 1700 m. The number of cristae in 1 electronogram is the biggest in animals on low-mountain (700 m) and the smallest in aborigens on middle-mountain (1700 m). KEEM of 1 electronogramm is the highest in both groups of rabbits on 3200 m and the smallest in adapted rabbits on 1700 m and in animals on low-mountain (700 m). So, in winter the highest energetic potential is registrated both in adapted and nonadápted animals on 3200 m altitude.

Tab. 4:

The data of quantitive analysis of the left heart ventricle electronograms of different intact rabbits groups in winter season.

GROUPS	N m EG	S m	N c m	KEEM m	S m EG	N c EG	KEEM EG
200m aborigens	8.43 ±0.74	0.55 ±0.05	3.43 ±0.26	1.91 ±0.25	4.38 ±0.32	28.57 ±3.09	7.53 ±1.25
700m aborigens	6.42 ±0.84	0.46 ±0.05	2.92 ±0.47	1.3 ±0.21	2.75 ±0.28	20.33 ±3.33	8.05 ±1.18
1700m aborigens	6.5 ±0.66	0.69 ±0.07	1.5 ±0.2	0.97 ±0.11	4.32 ±0.44	11 ±2.36	6.63 ±1.15
1700m non adapted	5.63 ±0.54	0.60 ±0.05	4.86 ±0.49	3.15 ±0.57	3.42 ±0.42	27.63 ±3.39	8.64 ±1.25
1700m adapted	5 ±0.49	0.63 ±0.07	2.73 ±0.54	1.62 ±0.31	2.99 ±0.29	13.47 ±2.09	7.16 ±1.02
3200m non adapted	5 ±0.49	0.76 ±0.07	3.27 ±0.35	2.47 ±0.37	3.78 ±0.53	16.27 ±1.99	11.98 ±1.63
3200m adapted	5.73 ±0.69	0.61 ±0.07	3.33 ±0.35	2.15 ±0.38	3.12 ±0.28	17.27 ±2.25	10.65 ±1.65

DISCUSSION

Data shown above indicate that the ultrastructural state of myocardium depends on altitude a.s.l. as well as on the degree of hypoxia adaptation and on the season of the year. Whereas the influence of the hypoxia degree on the mitochondrial state and the level of adaptation to hypoxia can be logically explained by the adaptive reaction of mitochondria to the deficieny of oxygen, the seasonal determination of myocardial ultrastructure remains an area of hypothesis.

Apparently, seasonal differences can be determined by the level of functioning of organism's endocrinal apparatus, which changes from season to season. Possibly, nutritive and thermal factors also are of some value. Surely, the role of helio-geomagnetic factors in different seasons can't be excluded. These questions can be answered only by further investigations based on the data shown in this article.

REFERENCES

1. HALBERG, F., L.E. SHEVING, E. LUCAS et al. (1984): Chronobiology of human blood pressure in the light of static (room restricted) automatic monitoring. Chronobiologia 11, 217-247
2. REINBERG, A. (1976): Advances in human chronopharmacology. Chronobiologia 3, 151-156
3. SMOLENSKY, M.H. (1981): Chronotherapeutics. Chronopharmacology and chronotherapeutics. Tallahassee, pp. 15-30
4. ZASLAVSKAYA, R.M. (1979): Circadian rhythms in patients with cardiovascular disease. Medicina. 165 p. (Russian)
5. KOMAROV, F.I. (1989): Chronobiology and chronomedicine. Moscow, 399 p. (Russian)
6. FROLOV, V.A. (1982): Interrelationship between circadian and circannual rhythms: some indices of heart contractility function and vessel tone. DAN SSSR. 262, No. 3, 753-756. (Russian)
7. MATYEV, E.S. & V.A. FROLOV (1991): Seasonal differences of mitochondrial apparatus by the method electron microscopy. Zdravookchranenie Kirgizii. 1, 26-28. (Russian)
8. CHIBISOV, S.M. & E.S. MATYEV (1991): Variability of amplitude of biological rhythm. Zdravookchranenie Kirgizii 2, 37-41. (Russian)

ARE THERE SEASONAL DIFFERENCES IN RATES OF REGENERATION IN RED-SPOTTED NEWTS, NOTOPHTHALMUS VIRIDESCENS ?

M.F. BENNETT

Department of Biology, Colby College, Waterville, Maine, 04901, U.S.A.

Red-spotted newts, *Notophthalmus viridescens*, are common inhabitants of ponds, streams, and pools in many areas of eastern United States. These amphibians are valuable research animals not only for studies of regeneration (6), but also for investigations of endocrinology, physiological stress, and circadian phenomena (1,2,3,4).

In 1972, SCHAUBLE (5) published the results of observations that suggested seasonal variations, or a possible circannual cycle, in regeneration of *N. viridescens*. She studied the regeneration of forelimbs in newts that lived at 20 +/- 1.0°C and in "constant semidarkness" (5), and concluded that their rates of regeneration were greater during early summer than during other seasons of the year.

Does the rate of tail regeneration in *N. viridescens* also vary seasonally? The study presented in this report was designed to address that question.

MATERIAL AND METHODS

The investigation was begun in September, 1987, and terminated in May, 1990. Eight series of observations, each at least 45 days in length, were made during that time.

The animals for the study, adult male and female *N. viridescens*, were shipped periodically to the Colby College laboratory from western Massachusetts, U.S.A. Upon arrival in the laboratory, the newts were placed in large containers of well-aerated well water at 20 +/- 0.5°C, and were exposed to natural photoperiods. At least three weeks before surgery, they were moved to individual plastic beakers containing 600 ml of well water, and transferred to environmental chambers. There they lived in constant darkness (DD) at 20 +/- 0.1°C.

On day 1 of each series, 60 newts were removed from the chambers, and anesthetized very briefly in 3% aequous ether solution. The tail of each animal was amputated 10 mm posterior to the posterior lips of the cloaca. These newts were returned to the environmental chambers, and left undisturbed except at irregular intervals when they were fed small pieces of lean beef or when their water was changed. Under dim light, the stumps of the regenerating tails were measured with calipers to the nearest 0.01 mm every eight to ten days of a series. The increase in the length of tail in mm per day during 45 days of the respective series was calculated for each surviving newt, and these results were then averaged for each series.

RESULTS AND DISCUSSION

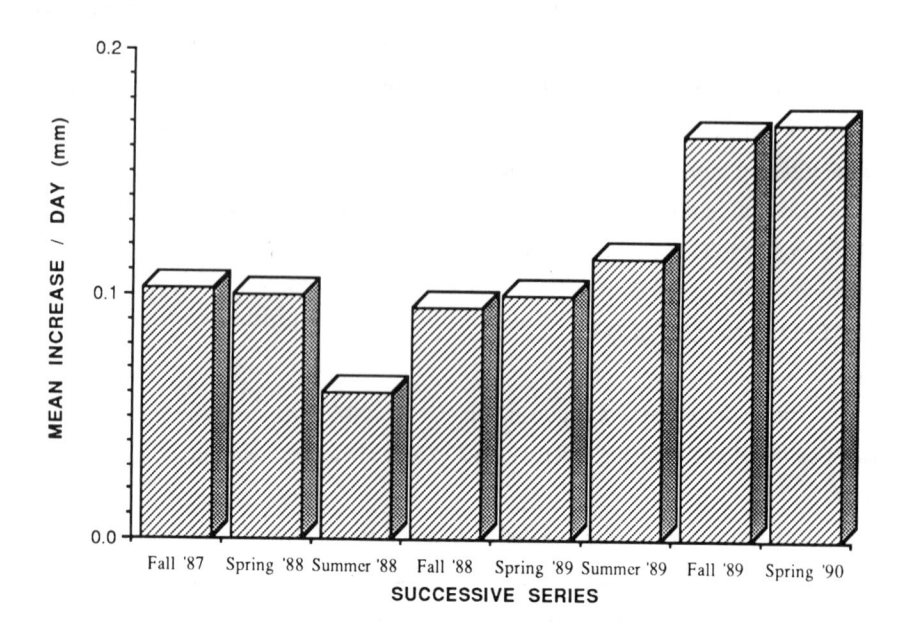

Fig.1: Mean increase per day in length of regenerating tails of *N. viridescens* during eight successive series.

The results of all series, 1987-1990, have been presented in Fig. 1. The mean increases per day, based on measurements of 25 to 35 animals for each series, are plotted on the vertical axis, while the successive series, indicated as Fall, Spring, or Summer, and separated from one another by intervals of three weeks, are indicated on the horizontal axis. The differences among levels of regeneration for the eight series are not significant. There was no evidence of consistent seasonal variations in tail regeneration in *N. viridescens* maintained in DD at 20° +/- 0.1°C.

Comparisons of these results with those of SCHAUBLE (5) that showed obvious and statistically significant seasonal variations in rates of forearm regeneration in the same species lead to interesting questions. Are temporal mechanisms fundamental to forearm regeneration different from those of tail regeneration? Or, are the seasonal differences expressed in constant semi-darkness, as in SCHAUBLE's investigation, not expressed in constant darkness as in the present study? Work in progress may aid in answering these questions.

CONCLUSION

The rates of tail regeneration in *N. viridescens* that were maintained in DD at 20 +/- 0.1°C did not show significant seasonal variations.

ACKNOWLEDGEMENTS

The technical assistance of Adair M. Bowlby, Adam Belanger, and Grace von Tobel, and the financial support of the Natural Sciences Division Grant Committee, Colby College, are gratefully acknowledged.

REFERENCES

1. BENNETT, M.F. (1986): Stress and changes in the blood of newts, *Notophthalmus viridescens*, during early regeneration. J. comp. Physiol. A. 159, 823-826
2. BENNETT, M.F. & K.R. DAIGLE (1983): Temperature, stress and the distribution of leukocytes in red-spotted newts, *Notophthalmus viridescens*. J. comp. Physiol. 153, 81-83
3. BENNETT, M.F., C. REED & R.L. RODEN (1974): Circadian changes in responses of newts, *Notophthalmus viridescens*, to ACTH. J. comp. Physiol. 89, 287-291
4. REED, C. & M.F. BENNET (1973): Circadian effects of hydrocortisone on the blood of the newt, *Notophthalmus viridescens*. J. comp. Physiol. 86, 59-63
5. SCHAUBLE, M.K. (1972): Seasonal variation of newt forelimb regeneration under controlled environmental conditions. J. Exp. Zool. 181, 281-286
6. SICARD, R.E. (ed.) (1985): Regulation of Vertebrate Limb Regeneration. Oxford University Press, New York-Oxford

SEASONAL CHANGES OF THERMOREGULATION IN SPECIES OF DIFFERENT ECOLOGICAL SPECIALIZATION

M.P.MOSHKIN & E.I.ZHIGULINA

Biological Institute, Siberian Division of the USSR Academy of Sciences, Frunze str.II, Biol. Inst., 630091 Novosibirsk, Russia

INTRODUCTION

The studies dealing with seasonal rhythms of thermoregulation in different rodent species, concentrated on quantitative changes of some heat balance parameters (1,2,5). At the same time species-specifity of seasonal rhythms can be determined by an unequal contribution of different parts of the thermoregulatory system in reactions of the organism to external environmental stimuli, that causes their qualitative diversity, the importance of the diversity significantly depending on ecological conditions. Complex study of thermoregulation is of great interest in comparative examining species of different ecological specialization. However, studies dealing with this problem and performed using the same methods of investigation are practically not presented.

The aim of our investigations was to examine seasonal changes of basic characteristics of heat exchange and their reaction to noradrenaline (NA) in the Siberian zokor (*Myospalax myospalax*), rat (*Rattus norvegicus*) and muskrat (*Ondatra zibethica*).

The zokor is a typical representative of burrow animals. It is a permanent resident of holes, where temperature fluctuations are less marked than on the surface. The rat is a terrestrial animal that lives under the conditions of moderate temperature fluctuations.

The muskrat - a small semiaquatic mammal - is exposed to extremely sharp temperature loading, caused by daily and seasonal fluctuations of ambient temperature as well as by transitions from air to water.

MATERIAL AND METHODS

The experiments were carried out in summer and late autumn of 1989-90. The zokors and muskrats were trapped in Kolyvan and Zdvinsk regions, Novosibirsk area, 2-7 days before the experiments. The rats were obtained from the Institute of Cytology and Genetics of the Siberian Division of the USSR Academy of Sciences. The animals were kept individually in cages, water and food ad libitum.

Thermoregulation parameters were studied at the temperature within the thermoneutral zone (27°C) and also after intraperitoneal NA-injection (0.6 µg/kg). Oxygen consumption was measured by means of paramagnetic gasanalyzer MN 5122-194. Heat losses of the body and of the tail were determined by the method of direct calorimetry, the body temperature was determined by means of a thermo-couple. All characteristics of energy exchange were expressed in mW/cm^2 using the coefficient - O_2 ml = 5.82 mW for oxygen consumption. The body surface was calculated from the formula: S=9.13p , where p = body mass in grams, 9.13 p$^{2/3}$ = empirical coefficient for calculating the body surface of the rats in cm^2 (3). The body thermal conductance was determined as HL/Tr - Ta, where HL = heat losses from the body, Tr = rectal temperature, Ta = temperature in the calorimetric chamber. As no reliable differences between males and females were observed, the data on both sexes were combined. Statistical significance of differences was assessed by Student's t-test.

RESULTS

The rats and muskrats demonstrated clear seasonal changes of heat balance (Tab. 1). In autumn these species showed a lower body thermal conductance than in summer. Furthermore, a reduction of the tail heat losses in the muskrat was observed. In the rats the decrease of thermal conductance took place simultaneously with the diminition of energy exchange. In the muskrat such combination was not marked. As a result, the body temperature in the muskrat significantly raised in autumn, and in the rats it remained at the level of summer temperature.

Tabl 1: Seasonal changes of basic thermoregulatory parameters in zokor, rat and muskat at temperature 27°C.

Parameter	Season	Species		
		zokor	rat	muskrat
Oxygen consumption, mW/cm^2	Summer	4.47±0.41(10)	3.90±0.13(38) ‡	5.17±0.34(29)
	Autumn	4.69±0.47(9)	3.46±0.09(37)	5.30±0.34(19)
Body conductance, mW/cm^2·°C	Summer	0.47±0.05(10)	0.40±0.01(38) ‡‡	0.44±0.02(27) |
	Autumn	0.47±0.05(9)	0.33±0.01(37)	0.38±0.02(21)
Tail heat losses, mW/cm^2	Summer	-	2.44±0.53(38)	7.22±0.76(15) ‡
	Autumn	-	1.45±0.30(35)	1.53±1.76(14)
Rectal temperature, °C	Summer	36.7±0.38(9) ‡	37.8±0.12(38)	37.0±0.34(28) ‡
	Autumn	38.4±0.41(8)	37.8±0.13(37)	38.2±0.29(20)

Means+SEM (number of animals);
\| P-value for difference <0.05; ‡ P<0.01; ‡‡ P<0.001

The zokor showed no seasonal changes in oxygen consumption or heat losses from the body surface. We found an increase of rectal temperature in autumn only.

Noradrenaline as the main neuromediator of the thermoregulatory system influences both heat production and heat losses. In the species investigated essential differences of functional organization and seasonal changes of thermoregulatory NA-reaction were observed (Fig.1). The NA-injection in the rats caused simultaneous augmentation of oxygen consumption and heat losses from the body and tail surfaces. The increase of rectal temperature was as low as 0.7°C in summer and 1.4°C in autumn.

In contrast to the rat, the muskrat did not display parallel responses of different parts of the thermoregulatory system. The oxygen consumption increased immediately

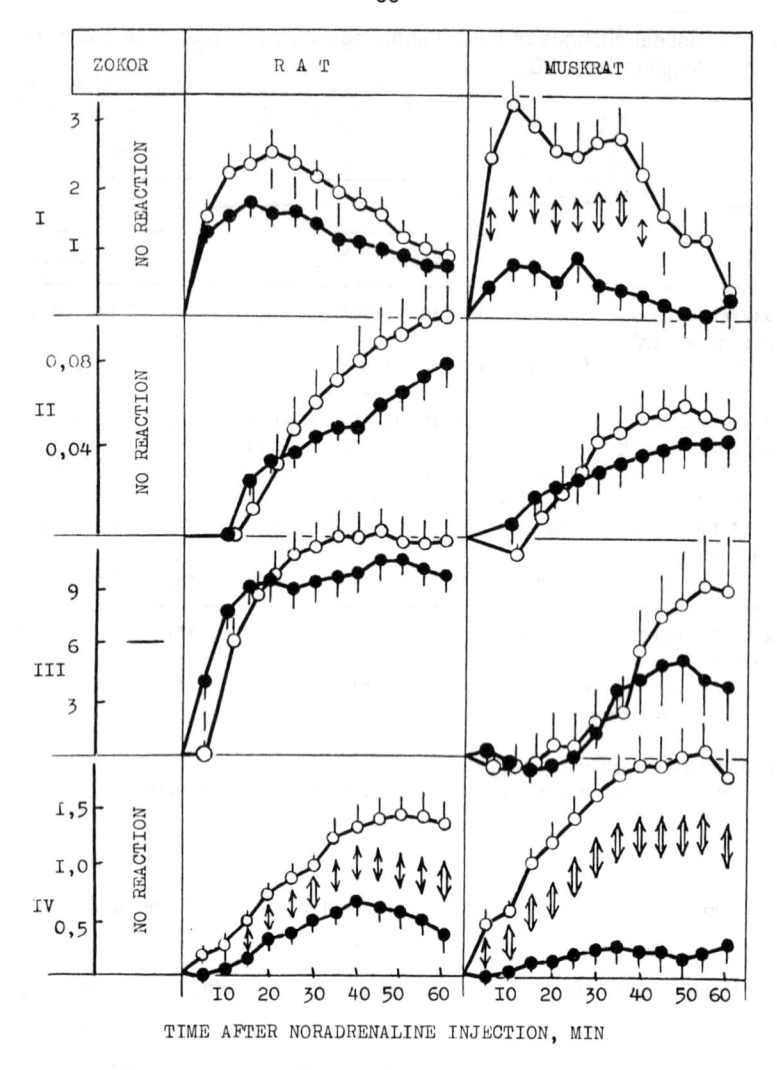

Fig.1: Seasonal differences of thermoregulatory NA-reaction in zokor, rat and muskat. Summer: black circle, winter: white circle. I: deviation of oxygen consumption (mW/cm^2); II: deviation of the body thermal conductance (mE/cm^{2}C); III: deviation of the tail heat losses (mW/cm^2); IV: deviation of rectal temperature (°C), all from the level before NA- injection. (p- value for differences <0.05;<0,001)

after the NA-injection. At the same time heat losses from the tail surface remained constant and raised sharply only at the 40th minute when oxygen consumption began to drop. Body thermal conductance in the muskrats increased after the NA-injection but more slowly than in the rats. The NA-injection resulted in a more significant body temperature increase. The NA-reaction was more expressed in autumn than in summer.

In the zokor the NA-injection did not stimulate changes in oxygen consumption or in other characteristics of thermoregulation.

DISCUSSION

Thus, there is a distinct dependence between seasonal rhythms of the thermoregulatory reaction and specific environmental conditions. Seasonal changes of the basic parameters of heat exchange are more expressed in terrestrial and semiaquatic species in comparison with burrow ones. The thermoregulatory NA-reaction also has seasonal differences. The expressiveness of the changes decreases in the row from semiaquatic to burrow species: muskrat > rat > zokor.

The nature of peripheral responses to hormones and mediators of the sympathoadrenal system depends to a large extent on the correlation of different types of adrenoreceptors in the target organs. Thus, the increase of energy exchange and subcutaneous vasodilation followed by augmentation of heat losses, is realized through β-adrenoreceptors. The activation of α-adrenoreceptors is accompanied by the vasoconstriction effect (4). Apparently, the adaptation to specific environmental conditions leads to the differentiated changes in the activity of α- and β-adrenoreceptors in different parts of the thermoregulatory system in the rodents investigated. Zokors, existing under more stable temperature conditions than semiaquatic or terrestrial species showed reduced thermoregulatory NA-reaction. In rats the sympathoadrenal system stimulated both mechanisms of heat production and heat losses through β-adrenoreceptors. Muskrat, closely connected with water, faces the problem of heat conservation more dramatically. Therefore, the activation of its energy exchange couples with diminishing of heat losses. Obviously, the effect of NA on the mechanisms of heat production is connected with the activation of β-adrenoreceptors, and the activation of α-adrenoreceptors with NA influences subcutaneous circulation.

Our studies have shown that seasonal changes of heat balance and thermoregulatory NA-reaction concern only the quantitative aspect of the reaction, no qualitative changes being observed.

Thus, seasonal changes of heat balance and NA-reaction correspond to specific environmental conditions. Comparative study of species of different ecological specialization will be helpful for analyzing intraspecific variability in animals or man, especially for understanding its adaptive significance.

REFERENCES

1. BASHENINA, N.V. (1977): Adaptational peculiarities of heat exchange in small rodents. Izdatelstvo MGU, Moscow, 294 pp. (in Russian)
2. KLAUS, S., G. HELDMAIER & D. RICQUIER (1988): Seasonal acclimation of bank voles and wood mice: nonshivering thermogenesis and thermogenic properties of brown adipose tissue mitochondria. J. Comp. Physiol. B 158, 157-164
3. MAKHANKO, V.I. & V.N. NIKITIN (1977): Evolution of rates of individual development. Nauka, Moscow, pp. 249-266 (in Russian)
4. PASTUKHOV, Y.F.(1982): Ecological physiology of animals. Nauka, Leningrad, part 3, pp. 92-103 (in Russian)
5. SLONIM, A.D.(1971): Ecological physiology of animals. Visshaya Shkola, Moscow, 448 pp. (in Russian)

LONG-TERM VARIATIONS IN SOLAR AND GEOMAGNETIC ACTIVITIES

J. STRESTIK & I. CHARVATOVA

Geophysical Institute, Czechosl. Academy of Sciences,
Prague, Czechoslovakia

It is well-known that the Earth moves around the Sun. This idea, however, requires a correction. All planets move around the barycentre, the common mass center of all bodies in the solar system. The Sun itself moves around this barycentre in a complicated way. All configurations of the planets are reflected in this motion. The main effect is only due to the four largest planets (Jupiter, Saturn, Uranus, and Neptune), the effect of the others may be neglected.

Imagine that only Jupiter and Saturn exist. In this case the trajectory of the Sun would make a loop, the most distant point being in the moment of mutual conjunction of these planets. This moment reoccurs once every 20 years and the loop is turned through 120° in ecliptical longitude. The motion thus describes a trefoil.

If Uranus and Neptune are taken into account, one can distinguish two cases: In the first these planets are in opposition, their influences are canceled and the Sun moves along a trefoil path, in the Jupiter-Saturn (JS) order (ordered motion). In the second case, Uranus and Neptune disturb the regular system of loops and the motion seems to be chaotic. The periods when the Sun's motion is ordered and is chaotic alternate regularly with a period of approximately 180 years. The most recent periods with ordered motion were centered in 1760 and 1940, the period with chaotic motion was centered in 1820 (2,3).

Solar activity is described by Wolf sunspot numbers. They have been available since 1749 and their annual means have been estimated back to 1500, however, with a decreasing accuracy when going into the past. Their graphical representation is given in Fig. 1a. A pronounced 11-year wave is clearly seen, moreover, one can observe that the height of the maxima varies systematically during recent centuries. In the spectrum a maximum in periods of about 180 years can be seen, together with subsidiary maxima at periods of 90 and 60 years. The main 180-year wave reaches its maxima at periods when the Sun moves around the barycentre in ordered JS-motion, e.g., at the end of the 18th century and in the middle of the 20th century. The 90-year and 60-year waves cause other long-term fluctuations, e.g. a small subsidiary maximum in the middle of the 19th century (4,5,6).

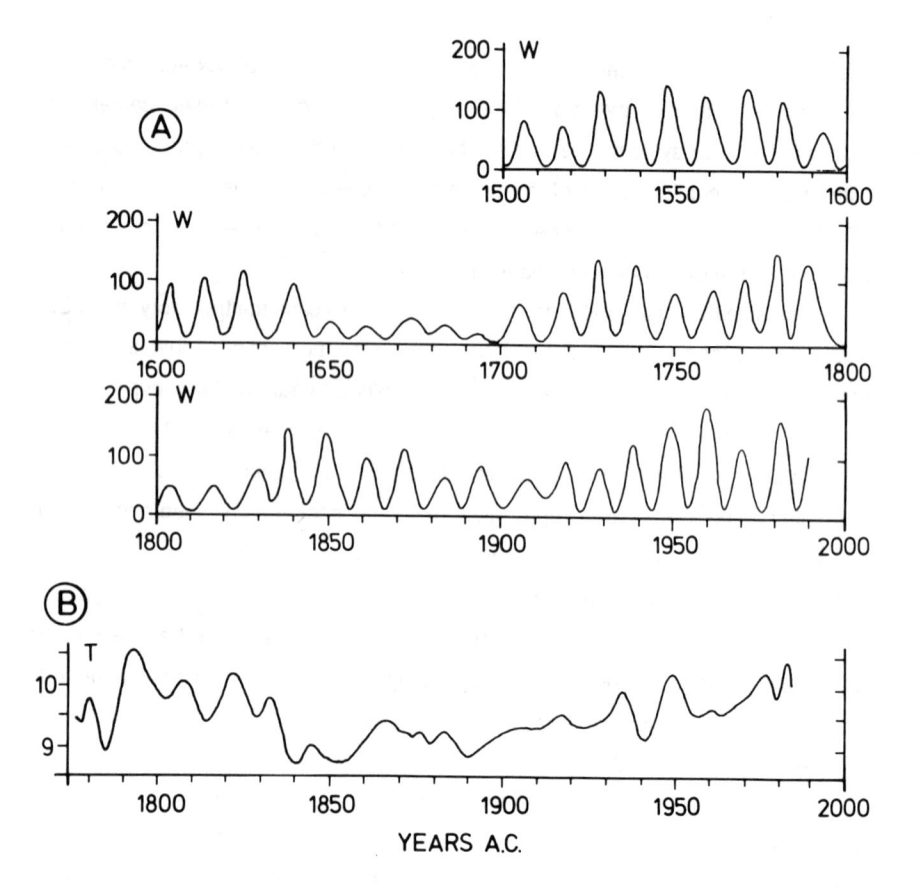

Fig.1: a) Wolf sunspot numbers, 1500 - 1987, b) smoothed mean annual temperatures from Klementinum, 1775 - 1987

The same regularity is observed in all processes which are connected with solar activity. A catalogue of aurorae observed in latitudes < 55° N between 1000 and 1900 A.D. is available. The occurrence of aurorae displays periodicities corresponding to those in the Sun's motion (1,9).

A similar long-term periodicity can also be found in processes which are not evidently connected with solar activity. Very interesting in this respect are the long-term variations in climate.

Mean annual values of the temperature for the Northern Hemisphere and the global temperatures are available since 1850 (8). From the beginning of our century global warming of about 0.5°C has been observed. This warming is often related to the green-house effect, i. e. the warming of the atmosphere due to the increasing content of CO_2 and other industrial pollution. This could create the idea of a nearly catastrophic increase in global temperature.

Longer series of annual temperature means, however, do not confirm this idea. Measurements at some stations are available from as far back as the 18th century. Fig. 1b shows the smoothed series from Klementinum in Prague, starting in 1775. This series, and series of other stations too, indicate that the temperatures at the end of the 18th century were at the present level. In the first half of the 19th century a decline of about 1 to 2 degrees occurred.

These long-term changes are in good correlation with the long-term changes of solar activity. It seems that, in periods with long-term maxima of solar activity (and ordered motion of the Sun), global temperatures are higher than in periods with long-term minima of solar activity (chaotic motion) (7).

This long-term variation is more evident when, due to some appropriate mathematical operations, variations with shorter periods are removed. Smoothing by 21-year running means has proved to be a suitable method for this purpose. The patterns of smoothed Wolf sunspot numbers W, geomagnetic activity indices aa, and global temperatures T are presented in Fig. 2 for the period 1850 - 1980. The forms of all curves are very similar, the correlation coefficient between W and T being 0.85, between aa and T being 0.90.

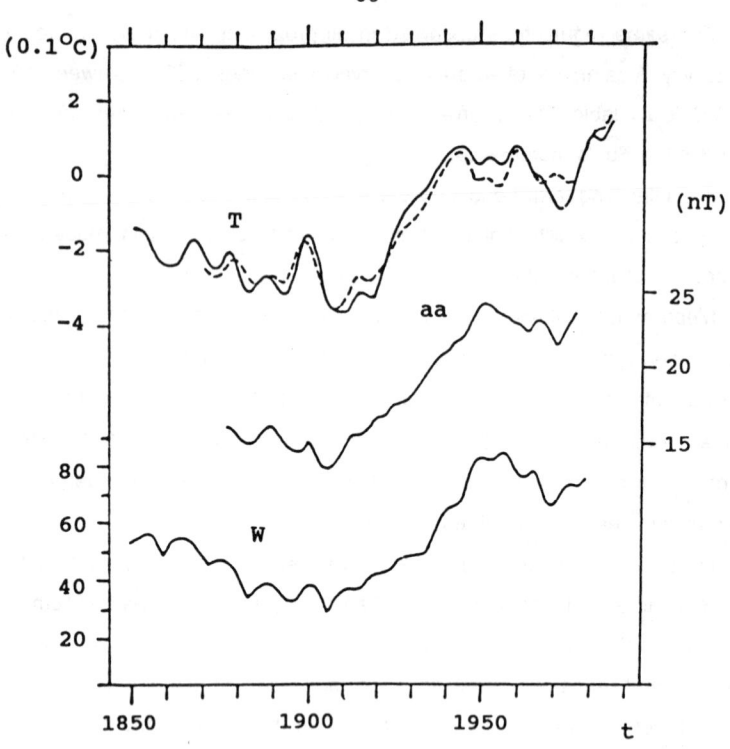

Fig.2: Sunspot numbers (W), geomagnetic activity indices (aa), global temperatures (T - full line) and mean North Hemisphere temperatures (T - dashed line) smoothed by using 21-year running means, 1850 - 1985.

In conclusion a few words on the possible development. The period of ordered motion of the Sun ends at the end of the 20th century and the Sun will begin to move chaotically. This will probably mean a period with lower solar activity and, in consequence, a decrease in global temperatures. This would compensate the systematic increase of temperature due to the green-house effect, and with the combination of both processes the decrease in temperatures would not be so deep as in the 19th century.

After 2040 the Sun will again move in an ordered way, it means, high cycles could occur (2). The global temperature will increase which, together with the green-house effect, might cause many difficulties, if the CO_2 contamination of the atmosphere and other pollution fail to decrease rapidly.

REFERENCES

1. CHARVATOVA-JAKUBCOVA, I., J. STRESTIK & L. KRIVSKY (1988): Studia Geophys et Geod., 32, 70
2. CHARVATOVA, I. (1988): Advances in Space Research 8, No 7, 147
3. CHARVATOVA, I. (1990a): Bull. Astr. Inst. Czechosl. 41, 56
4. CHARVATOVA, I. (1990b): Bull. Astr. Inst. Czechosl. 41, 200
5. CHARVATOVA, I. & J. STRESTIK (1991a): Bull. Astr. Inst. Czechosl. 42, 90
6. CHARVATOVA, I. & J. STRESTIK (1991b): J. Atmosph. Terr. Phys. 53, (in print)
7. CHARVATOVA, I. & J. STRESTIK (1992): Studia Geophys. et Geod. 36, (in print)
8. FOLLAND, C.K., T.R. KARL & K.Y. VINNILOV (1990): Climate Change, IPCC Scientific Assessment Cambridge Univ. Press, Ch. 7, 199-238
9. KRIVSKY, L. (1984): Solar Physics 93, 189

DEVELOPMENT AND AGING

DEVELOPMENT OF CARDIOVASCULAR AND TEMPERATURE RHYTHMS IN NEONATES

U. SITKA[*], D. WEINERT[**], F. HALBERG[***], J. SCHUH[**] & W. RUMLER[*]

Department of Pediatrics[*] and Institute of Zoology[**] Martin Luther University Halle (FRG), Chronobiology Laboratories[***], University of Minnesota, Minneapolis (USA)

INTRODUCTION

Chronobiological investigations are of increasing importance in neonatology also, not only for a better understanding of postnatal adaptive processes, but also for the definition of the normal range of clinically important parameters. These parameters are depending on gestational as well as on postnatal age.

In earlier investigations we analyzed the development of circadian rhythms of rectal temperature in 10 fullterm and 10 preterm infants (9) and of cardiovascular parameters in 26 fullterm and 11 preterm infants (10,12). To verify the obtained differences in ontogenetic changes between temperature and cardiovascular rhythms we expand the previous studies by simultaneous registration of both biological parameters in 14 neonates.

METHODS

With a unit consisting of the blood pressure monitor "Sphygmomanometer BP 107 neonatal" and a biomonitor "BMT 401" we measured simultaneously the systolic, the diastolic, and the mean arterial blood pressure, the heart rate, the core and skin temperature in 14 fullterm babies on the second day of life over a period of 24 to 30 hours in 10 minute intervals. The babies were cared in incubators under neutral temperature and natural light-dark conditions. The behavioral states were registered visually. Neonates were fed every 4 hours. Eight of the babies were investigated at an age of 4 weeks (in the bed) also. Each profile was analyzed individually by the

Cosinor-method (1). Results were compared with the earlier temperature and cardiovascular data.

RESULTS

On the 2nd day of life a statistically significant circadian rhythm could be detected in 9 of the 14 neonates for systolic blood pressure, in 10 for the heart rate, and in all 14 for the rectal temperature. The mean percentages of the circadian rhythm from rhythmical signal are 21.4, 26.2 and 43.9 for systolic blood pressure, heart rate and temperature respectively. The acrophases differ in a wide range especially in the cases of blood pressure and heart rate. Therefore the summarized daily patterns reveal no marked day to night differences with exception of core temperature. In this case higher values were found between 8 a.m. and 8 p.m.

The comparison of the data of eight neonates investigated at the 2nd day as well as on age of 4 weeks showed an increase of the mesor in blood pressure from 76.3 to 78.1 mm Hg and in heart rate from 127.5 to 147.6 beats/min. The circadian amplitudes raised from 2.5 to 3.8 mm Hg for systolic blood pressure and from 0.15 to 0.23°C for temperature. The variance of acrophases decreased, especially in temperature. No day- night differences could be detected, however, in the summarized daily patterns of blood pressure and heart rate at an age of 4 weeks also.

The obtained results confirm with those from previous investigations where core temperature and cardiovascular parameters were analyzed not simultaneously. Therefore we summarized all results in Tab. 1 and in Fig. 1 and added the data from 10 premature babies (gestational age: 31-36 weeks). So the described differences in ontogenetic development of circadian rhythmicity between various parameters will be more obvious. Besides them it demonstrates the influence of gestational age.

The table shows a high degree of statistically significant circadian rhythms in premature babies also. Only in heart rate and rectal temperature at the 2nd day the percentage is lower as compared to fullterms. The mesor of the cardiovascular parameters increases up to the 4th week in both groups. A distinct increase of the circadian amplitude was evident in the core temperature only. In fullterm babies the

Results from 18 fullterms

Results from 10 preterms

Systolic Blood Pressure

Heart Rate

Rectal Temperature

Circadian Acrophases in Degrees

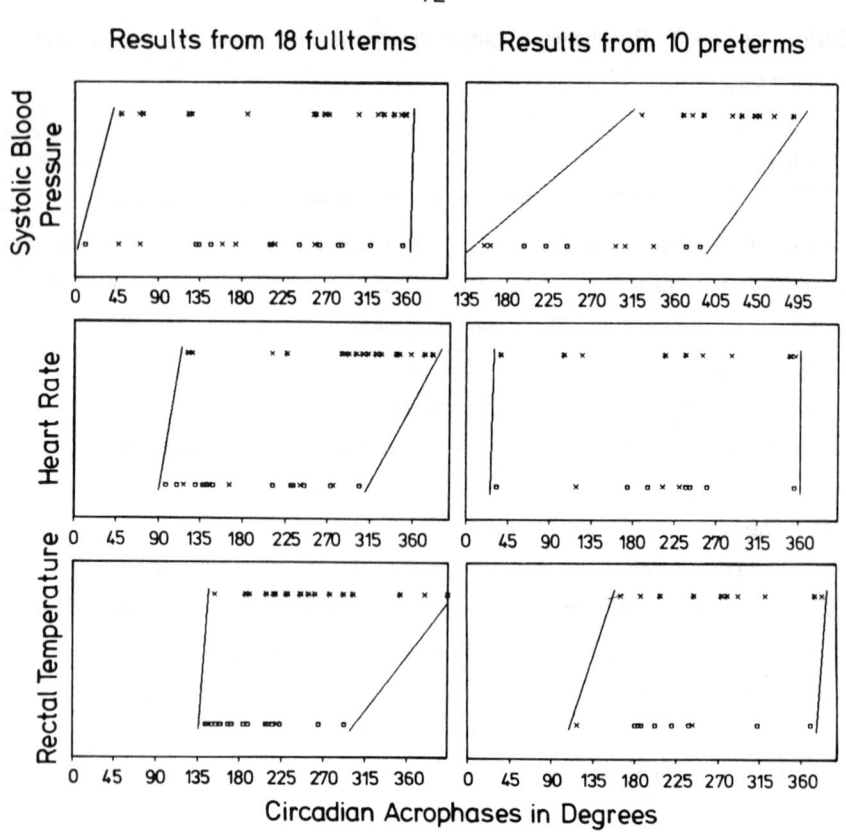

✻ 2nd day of life   □ 4th week of life   x Rhythm non signif.

Fig.1:

Tab. 1: Development of circadian rhythmicity in fullterm and preterm babies from 2nd day to 4th week of life

Parameter		n	2nd day of life			4th week of life		
			CR	Mesor +/- S.E.M.	Amplitude +/- S.E.M.	CR	Mesor +/- S.E.M.	Amplitude +/- S.E.M.
SBP	FT	18	9	73 +/- 2	2.9 +/- 0.5	11	78 +/- 1	3.4 +/- 0.5
	PT	10	6	69 +/- 2	4.2 +/- 0.8	5	75 +/- 3	3.0 +/- 0.5
HR	FT	18	15	128 +/- 2	6.1 +/- 0.5	13	148 +/- 2	6.2 +/- 0.7
	PT	10	5	137 +/- 2	4.8 +/- 0.7	7	150 +/- 2	5.8 +/- 1.3
RT	FT	18	16	37.2 +/- 0.1	0.16 +/- 0.02	17	37.2 +/- 0.2	0.20 +/- 0.02
	PT	10	6	37.0 +/- 0.1	0.11 +/- 0.07	8	37.1 +/- 0.2	0.16 +/- 0.08

SBP - systolic blood pressure, HR - heart rate, RT - rectal temperature, FT - fullterm babies, PT - preterm babies, n - number of investigated babies, CR - number of statistically significant circadian rhythms variance of acrophases, with exception of systolic blood pressure and most pronounced in temperature, decreases from 2nd day to 4th week associated by an positive phase shift. Besides these differences between parameters and between full- and preterm babies in interindividual synchronization no interparametric correlation could be seen.

DISCUSSION

Chronobiological investigations in neonates are necessary not only for a better understanding of physiological states, for instance the sleep-wakefulness and the feeding rhythms. Chronobiological parameters are very important for the determination of pathological conditions also. HALBERG et al. (3) and MAINARDI et al. (5) reported, that neonates with hypertension in their family history differ in their cardiovascular parameters determined by rhythmometry from those without hypertension in family history. A circadian rhythm in infants with a clinically important gastroesophageal reflux was described by DREIZZEN et al. (2). The maximum of reflux frequency and of reflux length can be seen in connection with the evidence of the sudden infant death syndrome. Moreover, the constant environment, e.g. the lack of light-dark changes and the continuous feeding at the intensive care units may have a negative influence on the development of circadian rhythmicity especially on the processes of synchronization with external "Zeitgebers". Besides this practical significance chronobiological investigations are of high theoretical importance.

We found in most cases statistical significant circadian rhythms at the second day of life. The fact that the acrophases of blood pressure, heart rate, and temperature in neonates differ from those in adults could mean that endogenous oscillators are working in neonates already and the obtained rhythms are not only of mothers origin. The mean differences between acrophases of the mother and her neonate were 11 hours in the case of blood pressure and 8 hours for heart rate (10).

PATRICK et al. (7) found a significant positive correlation between the mothers and her fetus daily mean heart rate. MIRMIRAM et al. (6), taking into account the circadian rhythm of deoxyglucose uptake in fetal suprachiasmatic nuclei of the anterior hypothalamus (8) also, postulate, that a biological clock is functioing in preterm infants already. The fetal biological clock is coordinated in utero by the mother, however.

TENREIRO et al. (11) discuss masking effects of randomly external events as a possible cause for the poor presence of circadian heart rate rhythms in neonates. This confirms with our observations. However, the main difference between parameters as well as between fullterm and preterm babies concerns not the presence of statistical significant circadian rhythms, but the degree of internal and external synchronization. It develops later than circadian rhythmicity but earlier in fullterm as compared to preterm babies and in core temperature as compared to cardiovascular parameters.

The lack of a correlation between rhythms of heart rate, blood pressure, and core temperature, evident by different acrophases and by differences in the distribution of ultradian components, is typical for neonates. It can be obtained between several functions even of the same organ and may be characterized as a "physiological dyschrony" (4). The postnatal decrease of the variance of acrophases is the reason for finding circadian rhythms on a population basis at an age of 4 weeks.

REFERENCES

1. BINGHAM, Ch., B. ARBOGAST, G.C. GUILLAUME et al. (1982): Inferential statistical methods for estimating and comparing cosinor parameters. Chronobiologia 9, 397-439
2. DREIZZEN, E., P. ESCOURROU, M. ODIEVRE, et al. (1990): Esophageal reflux in symptomatic and asymptomatic infants: postprandial and circadian variations. J. Ped. Gastroent. Nutr. 10, 316-322
3. HALBERG, F., G. CORNELISSEN, E. HALBERG et al. (1988): Chronobiology of human blood pressure. Medtronic Continuing Medical Education Seminars, 4th ed., pp. 242
4. HELLBRÜGGE, Th.: Physiologische Zeitgestalten in der kindlichen Entwicklung. Nova Acta Leopoldina 46, Nr. 225, 365-387
5. MAINARDI, G., B. TARQUINI, F. HALBERG et al. (1988): Zum Risiko des Neugeborenen später im Leben einen Hochdruck zu entwickeln. Sozialpädiatrie 10, 900-905
6. MIRMIRAN, M., J.H. KOK, M.J.K. DE KLEIE et al. (1990): Circadian rhythms in preterm infants: a preliminary study. Early Hum. Dev. 23, 139-146
7. PATRICK, J., K. CAMPBELL, L. CARMICHAEL & C. PROBERT (1982): Influence of maternal heart rate and cross fetal body movements on the daily pattern of fetal heart rate near term. Am. J. Obstet. Gynecol. 144, 533-539
8. REPPERT, S.M. & W.J. SCHWARTZ (1986): The maternal suprachiasmatic nuclei are necessary for maternal coordination of the developing circadian system. J. Neurosci. 6, 2724-2729
9. SITKA, U., F. NAGEL, W. RUMLER & D. WEINERT (1984): Entwicklung der Zirkadianperiodik der Körpertemperatur bei Neugeborenen. Dt. Ges.-wesen 39, 1334-1339
10. SITKA, U., D. WEINERT, F. RÖPKE et al. (1990): Circadiane und ultradiane Rhythmen cardiovaculärer Parameter im jungen Säuglingsalter. Untersuchungen über die Abhängigkeit vom Gestationsalter. Der Kinderarzt 21, 1393-1399
11. TENREIRO, S., M. CHISWICK, H. DOWSE et al. (1991): Rhythms of heart rate in premature babies in intensive care. J. Interdisc. Cycle Res. 22, 196
12. WEINERT, D., U. SITKA, F. HALBERG et al. (1990): Circadiane und ultradiane Rhythmen cardiovaculärer Parameter im jungen Säuglingsalter. Abhängigkeit vom postnatalen Alter. Der Kinderarzt 21, 1553-1556

CHRONOBIOMETRY OF RHYTHMOGENESIS IN NEWBORN INFANTS

R.SAMMECK[1] & U.STEPHANI[2]

[1] Department of Neuroanatomy, [2] Department of Neuropediatrics, University of Göttingen, Zentrum Anatomie, Abtlg. Neuroanatomie, Kreuzbergring 36, D-W-3400 Göttingen, FRG

Improved monitoring of newborn infants at risk has yielded considerable reduction of perinatal morbidity and mortality. Nevertheless, three major goals of surveillance have not been achieved: 1) The "touchless" monitoring, i.e. surveillance of respiration and heart rate without electrodes, 2) the determination of basal metabolic rates, and 3) identification of the level of maturation of diurnal vigilance. Here a device including first results is presented, that seems to fulfil these needs (1,2).

For monitoring the baby is placed in a crib, which is put on the platform of an electronic balance. The weight data of this balance are processed by a personal computer and plotted by a standard printer. This physico-electronic system, so called CHRONOBIOMETER (CBM) can serve as a complete passive and safe device for monitoring vital rhythmic functions of the heart, respiration, and simultaneously parameters of thermoregulation as well as other kinetic activities of sleeping and waking cycles. Every 100 millisecond the CBM measures the vertical component of the ongoing kinetic activity at a precision of 10 milligram.

Thus, movements initiated in the body are recorded by CBM-monitoring whether they are pulsatile or rhythmic in origin. The baby is in no way disturbed neither by electrodes sticking to the skin, tubes in the nose nor transoxodes.

For a baby at rest the ongoing CBM-recording demonstrates recurrent upward deflections of varying amplitudes, occurring nearly every second. These upward deflections coincide with the upward inspiratory chest movements as can also be shown by simultaneous polygraphic pneumography (Fig.1,left).

The continuous CMB-record of inspiratory and expiratory movements without any artifacts permits quantitative analysis of kinetic activity. The begin of respiratory arrest or apnoea can be determined precisely (Fig.1,right). An arrest of 5 seconds is

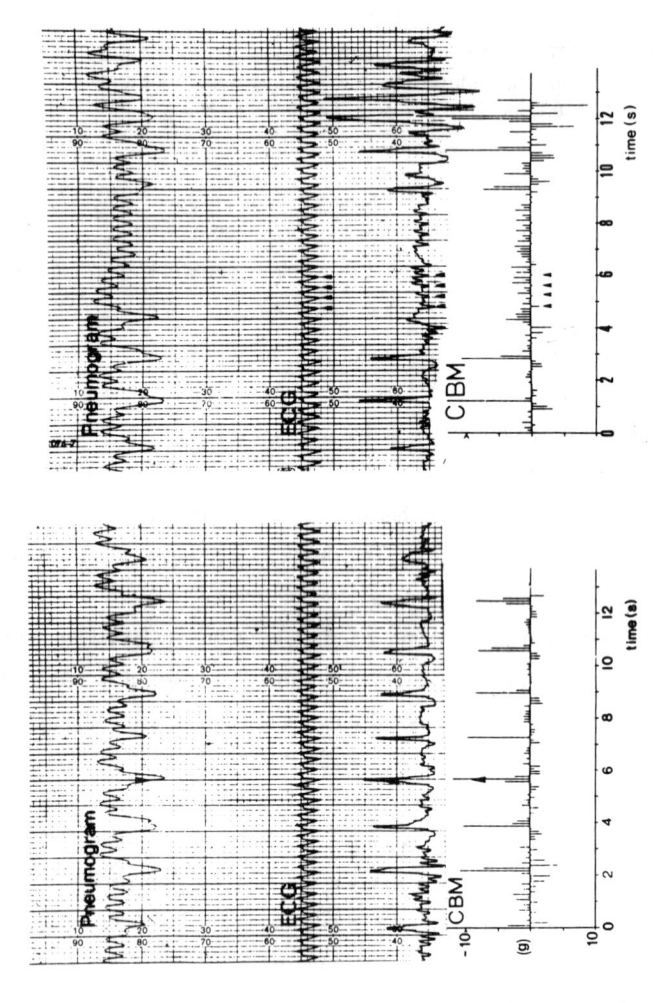

Fig.1: RIGHT: Polygraphic recording of neonate impedance pneumogram, electro-cardiogram, and "GÖTTINGER CHRONOBIOMETER" (CBM) including CBM-original-record (bottom). Recurrent upward deflections of varying amplitudes (ordinate in gram) coincide with inspiratory chest movements (arrowheads).
LEFT: Polygraphic recording of a respiratory pause in a newborn baby at rest. CBM-recording of cardioballistic activities of low-amplitude pulsations (arrowheads) and simultaneous ECG.

shown in this graph. Ongoing cardiac activity is also registered. During apnoeic pauses cardioballistic activity can be discerned by low-amplitude pulsations.These rhythmic pulsations have been confirmed to be of cardiac origin by simultaneous electrocardiographic recordings.

Respiratory movements can clearly be distinguished from heartbeats by the magnitudes of the amplitudes. The magnitude is greater by a factor of five in the case of respiration. In addition to short-term monitoring of cardiac and respiratory activities, the CBM can also provide useful information on the actual developmental status of the baby as shown by various biorhythmic functions on a longer time scale.

Data analysis includes the summing-up of individual measurements plus calculation of mean value. Computerized differences of mean values serve as a measure of net-weight loss, which we call Delta g for gram. Data analysis includes further calculation of kinetic total activities.

Respiratory bodyweightloss is the result of oxygen uptake of inhaled air and release of carbondioxide and water. Delta g can be described as an indirect measure of energy exchange as well as bodyheatloss. Hence change in Delta g may serve as a parameter of the effectiveness of respiratory performance.

In a healthy sleeping neonatal infant computerized measurements of Delta g approximate 1 g per kilogram of bodyweight per hour. Increasing values of Delta g are registered during the arousal reaction, when the baby awakes and begins to increase motor activities.

Thus CBM-monitoring can serve as a protocol of the ability of the newborn to synchronize various biorhythmic autonomous functions (Fig.2). This composite plotting represents CBM-recordings of the 3rd, 4th, and 8th postnatal night of a baby, born by caesarean section.

Focussing attention to CBM-recording during 4th postnatal night, which has been interrupted twice due to feeding, a biorhythmic pattern of kinetic activity followed by a spell of sleeping behavior is seen. This cycle repeats approximately every hour. This ultradian rhythm of kinetic activity is not yet present in the slope of Delta g. However, four nights later, CBM recording of autonomous thermokinetic functions

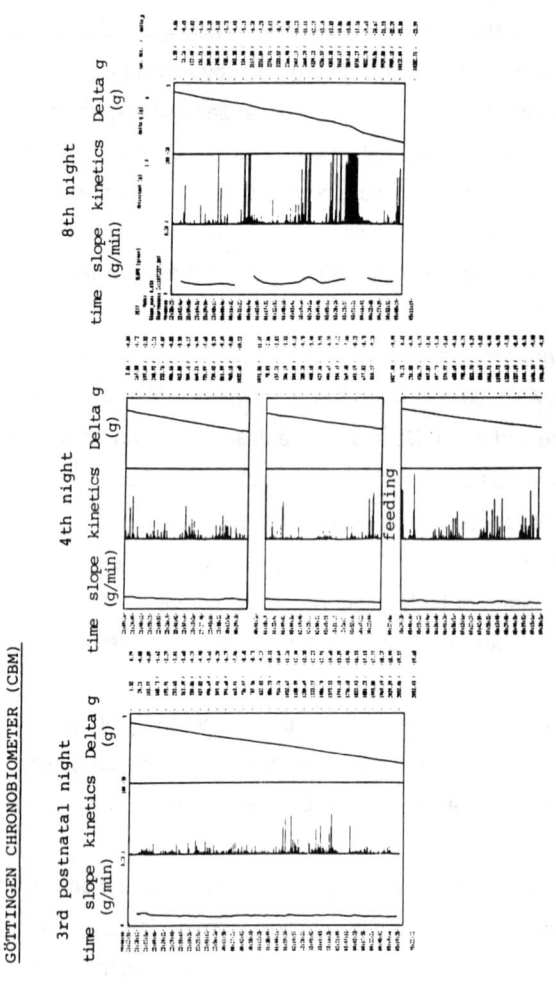

Fig.2: CBM-longterm recording of thermokinetic rhythmogenesis in a term-born by caesarean section (inhalation anesthesia) during 3rd, 4th, and 8th postnatal night. Local time is printed out to the left of each nightplotting; total kinetic activities and net-weightloss Delta g to the right, each summed up in 15 min intervals. CBM-recording of the 4th night (middle) has been interrupted at 00:41:31 and 04:37:23 h by feeding. Vertical plotting of Delta g (right), slope of Delta g (left), and kinetic activities (middle). Series analysis reveals ultradian rhythmogenesis of kinetic activities with period lengths of approximately 1 h. Ultradian rhythmogenesis of autonomous thermoregulatory functions appear later during 8th postnatal night of CBM-recording, synchronized with ultradian kinetic activities of increased amplitudes.

during the 8th postnatal night - to the right - clearly demonstrate ultradian cycles. In addition both thermokinetic rhythms appear to be synchronized. This onset of synchronization of thermoregulation and motor activity appears to be delayed in term-caesarian born - when compared to mature spontaneous born babies. In healthy neonates synchronized rhythmogenesis of thermokinetic functions can be recorded at earlier stages of postnatal development by CBM-monitoring.

REFERENCES

1. SAMMECK, R. & W. GIBB (1987): "GÖTTINGEN CHRONOBIOMETER": A non-invasive monitoring system in biomedical research and risk assessment. In: Schering Symposium Proceedings, Berlin, pp. 131-134
2. SAMMECK, R., W. GIBB, P. HEIDEMANN & U. STEPHANI (1987): Das "GÖTTINGEN CHRONOBIOMETER" zur Früherkennung von Apnoen im Neugeborenen- und Säuglingsalter. In: G. JORCH (ed.): Der plötzliche Kindstod. Minister für Arbeit, Gesundheit und Soziales des Landes NRW, pp. 123-127

DYNAMIC VARIATION OF CIRCADIAN AMPLITUDES AND ACROPHASES DEPENDING ON AGE

J. SCHUH, D. WEINERT, G. D. GUBIN[*] & I. P. KOMAROV[*]

Inst. of Zoology, Martin-Luther-University Halle, FRG;
[*]Medical Institute, Dept. of Biology, Tjumen, CIS

INTRODUCTION

In the course of ontogeny all rhythmic charateristics including the relations of circadian and ultradian components in the frequency spectra undergo great dynamic variations. This may be caused by the distinct morpho-functional maturation and aging of the organismic system the different requirements to the environment as well as the changing valence of Zeitgebers.

Insomuch each ontogenetic stage is characterized by a distinct temporal organization and it should be expected that the distinct stages of temporal order may contribute to a different functional fitness and adaptability of the organism according to his ontogenetic stage (4).

In former investigations we found in experiments on laboratory mice and rats that the time needed for resynchronization (T_R) and reconstitution after Zeitgeber-shifts or stress is quite different depending on age. T_R is markedly prolonged in the groups of old aged animals as compared to the adults and juvenile ones (7).

The lower potency for resynchronization and reconstitution in the old aged animals could be caused by a weakening of the interparametric coupling and external synchronization. To evaluate the possible underlying conditions we analysed on parameters of the carbohydrate metabolism the course of circadian acrophases and amplitudes in rats and respiratory parameters of adult and senescent man.

MATERIAL AND METHODS

Laboratory female rats at the age of 34, 86, 200, 430, 645, 820 d were caged in groups of eight animals (L:D = 12:12; L = 250 lx; L = 6.00-18.00 h; temp.=19-21°C; food and water ad libitum). Blood was sampled every six hours and blood glycogen, blood glucose, insuline, amylase, aldolase, pyruvate, lactate, and liver glycogen were measured. Circadian amplitudes and acrophases were calculated by Cosinor-method. The variance between the individuals and the measured time-related samples were calculated by analysis of variance. Respiratory parameters of man were sampled from 8-14 individuals in each group of 25-35 and 75-90 years of age. (Time of measurement: 7.00; 11.00; 15.00; 19.00; 23.00 h).

RESULTS AND DISCUSSION

Fig. 1 demonstrates the changing relations of circadian acrophases in groups of juvenile (34 d), adult (430 d) and senescent rats (820 d), calculated by the Cosinor-method. The circadian amplitudes of the adults are set as 100 %. The amplitudes increase from the juvenile state up to the adult and decrease up to the senescent group.

Similar results were obtained in man demonstrated on respiratory parameters (Fig. 2). A significant decrease of circadian amplitudes from the adult to the senescent stage occure. In both cases (Fig. 1,2) the acrophases of the single parameters change in a different way.

The changes of circadian amplitudes for five parameters of carbohydrate metabolism and rectal temperature in six classes of age are demonstrated in Fig. 3. Taken the circadian amplitudes as 100 %, the amplitudes from the juveniles rise up to average of 44 % and decrease in the old aged in a similar way (Fig. 3, left side). From a methodological point of view we calculated the variances based on differences between the measured values depending on the time of day as well as the interindividual variances because decreasing amplitudes - calculated by Cosinor - could be a consequence of non-ideal sinus-shaped rhythms or differences between the individuals.

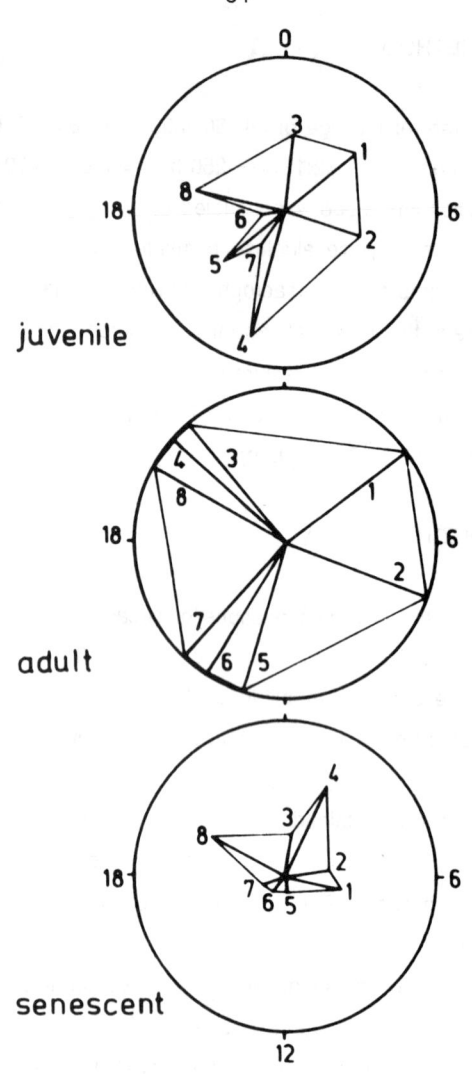

Fig1: Circadian acrophases and amplitudes.
1 = liver glycogen; 2 = blood glycogen; 3 = blood glucose; 4 = insulin; 5 = amylase; 6 = aldolase; 7 = pyruvate; 8 = lactate.

The variances characterizing the changes in the course of day are demonstrated at the right side of Fig. 3 (above). The increase up to the adults and decrease in the following stages confirm the described changes of amplitudes. The dependence of the measured parameters on the circadian time is most prominent in the adults.

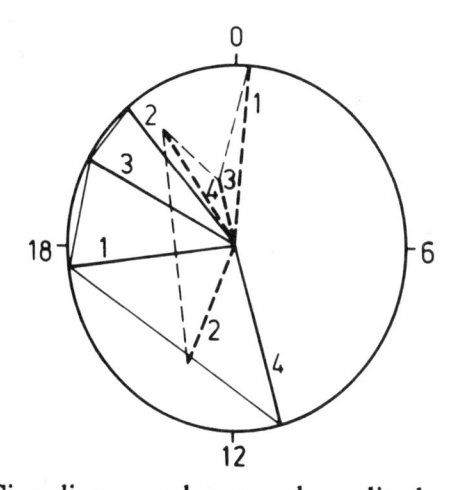

Fig.2: Circadian acrophases and amplitudes.
adult ——— ; senescent - - -;
1 = respir. frequency; 2 = respir. vol.;
3 = respir. vol./min.; 4 = vital capacity.

The lower part of Fig. 3 shows the percentage of interindividual variances (right side) from the total variance of the time series. They demonstrate nearly a reciprocal course as compared to the amplitudes. From the first weeks of age it will be observed an improvement of interindividualy coordination up to the adults and a change to the worth in the old aged. This could be caused by a deterioration of the external and internal coupling mechanism as well as by differences of the biological age.

For aging animals and man phase-shifts, decreasing amplitudes and an increase of ultradian rhythms are well described (1,2,3,5,6,8,9,10). High amplitudes and relative stable acrophases in the adult stage mark the organism with respect to its temporal structure in the state of best functional fitness and adaptability. Decreasing circadian amplitudes in aging individuals may involve - together with changes in other rhythmic characteristics - a weakening or a loss of the inter

Changes of Circadian Amplitudes and Variances in Rats Depending on Age

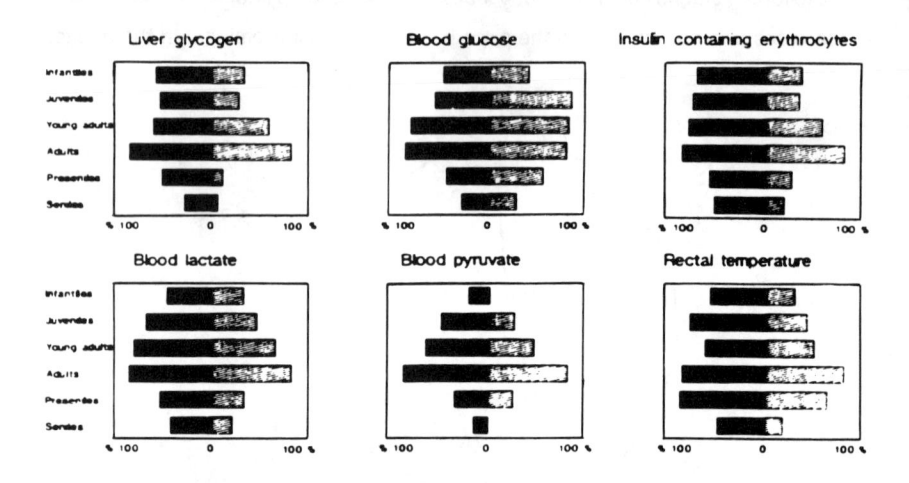

Liver glycogen Blood glucose Insulin containing erythrocytes

Blood lactate Blood pyruvate Rectal temperature

Changes of Circadian Amplitudes and Interindividual Variances in Rats Depending on Age

Liver glycogen Blood glucose Insulin containing erythrocytes

Blood lactate Blood pyruvate Rectal temperature

left side circadian amplitudes
right side variances

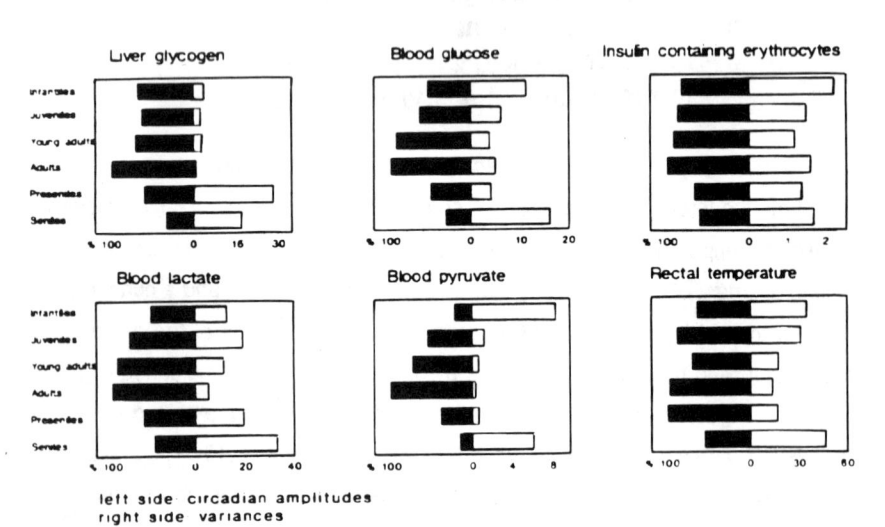

Fig.3:

parametric coordination and external synchronization. This findings will confirm the different potency for resynchronization and reconstitution the rhythmic order after Zeitgeber-shifts and streses (7). Moreover, the directed changes of rhythmic variables in the course of ontogeny may characterize the stages of age in a distinct qualitative way and contribute to a better definition of the biological age.

REFERENCES

1. ASCHOFF, J., U. GERECKE & P. WEVER (1967): Desynchronization of human circadian rhythms. J. appl. Physiol. 17,450-457.
2. CAHN, A.A., G. FOLK & P.E. HUSTON (1968): Age comparison of human day-night physiological differences. Aerospace Med. 39; 608-610.
3. DAVIS, F.C. (1981): Ontogeny of circadian rhythms. In: J.ASCHOFF, Handbook of behavioral neurobiology, vol. 4, Plenum Press New York, 257-274.
4. GUBIN, G.D., A. DUROV, O.A. VORONOV & P.I. KOMAROV (1987): Issledovanije izmenenija circadiannych bioritmov v ontogeneze zivotnych i celoveka. Z. evol. bioch. fiziol. 23; 629-633.
5. GUBIN, G.D. & D. WEINERT (1991): Bioritmy i vozrast. Uspechy fiziol. nauk 22; 77-96.
6. RENSING, L. (1989): Chronobiologie des Alterns: Veränderungen der zeitlich-periodischen Ordnung. Geront. 22; 73-78.
7. SCHUH, J., D. WEINERT & G.D. GUBIN (1991): The time of resynchronization as an indicator of the adaptive potency. Chronobiology and Chronomedicine. Peter Lang, Frankfurt/M., Bern, New York, Paris; 131-135.
8. STUPFEL, M., V. GOURLET & L. COURT (1986): Effect of aging on circadian and ultradian respiratory rhythms of rats synchronized by an LD 12:12 lighting (L = 100 lx). Gerontology 32, 81-90.
9. WEINERT, D. & J. SCHUH (1984): Untersuchungen zur Entwicklung der Zeitstruktur ausgewählter Parameter im Verlaufe der postnatalen Ontogenese. I. u. II.; Zool. Jb. Anat. 111, 133-153.
10. WEINERT, D., F.E. ULRICH & J. SCHUH (1986) Ontogenetische Änderungen des Tagesmuster der Plasmainsulinkonzentration und ihre Beziehung zur Nahrungsaufnahme. Biomed. Biochim. Acta 46, 387-395.

THE EFFECTS OF THYROID STATE ON RHYTHM COHERENCE: INTERACTIONS WITH AGING.

D.L. McEACHRON & N.T. ADLER

Department of Psychology, University of Pennsylvania,
Philadelphia, PA 19104, USA

INTRODUCTION

Circadian dyschronism has been suggested as a factor in aging and longevity (1). Elderly humans display an increased variability in rhythm parameters (4) and older rats also show a decrease in activity rhythm coherence, displaying less stable entrainment and a greater proportion of daytime activity compared with younger adult animals.

In a recent report, HERLIHY et al. (2) proposed that decreases in thyroid activity may increase longevity in rats. This study presents data which suggest that these two physiological conditions which have been associated with aging may also be associated with each other.

METHODS

The data presented here formed part of a larger study investigating the interaction of lighting and thyroid state on activity rhythms in rats. Twenty-four male rats, 12 thyroparathyroid-ectomized (TPX) and 12 sham-operated (Sham) were placed in individual running wheel activity cages and housed in groups of six in light-tight and sound-proved cabinets. Activity was monitored by microswitches which automatically recorded wheel turns and transferred this data to an IBM PC-XT computer. Lighting was provided either by 15 Watt incandescent bulbs (~15 lux) during light-dark (LD) cycles or 7.5 Watt red safelight bulbs (variable) when exposed to constant conditions (RR). Food and water (2% calcium lactate added) were provided ad lib.

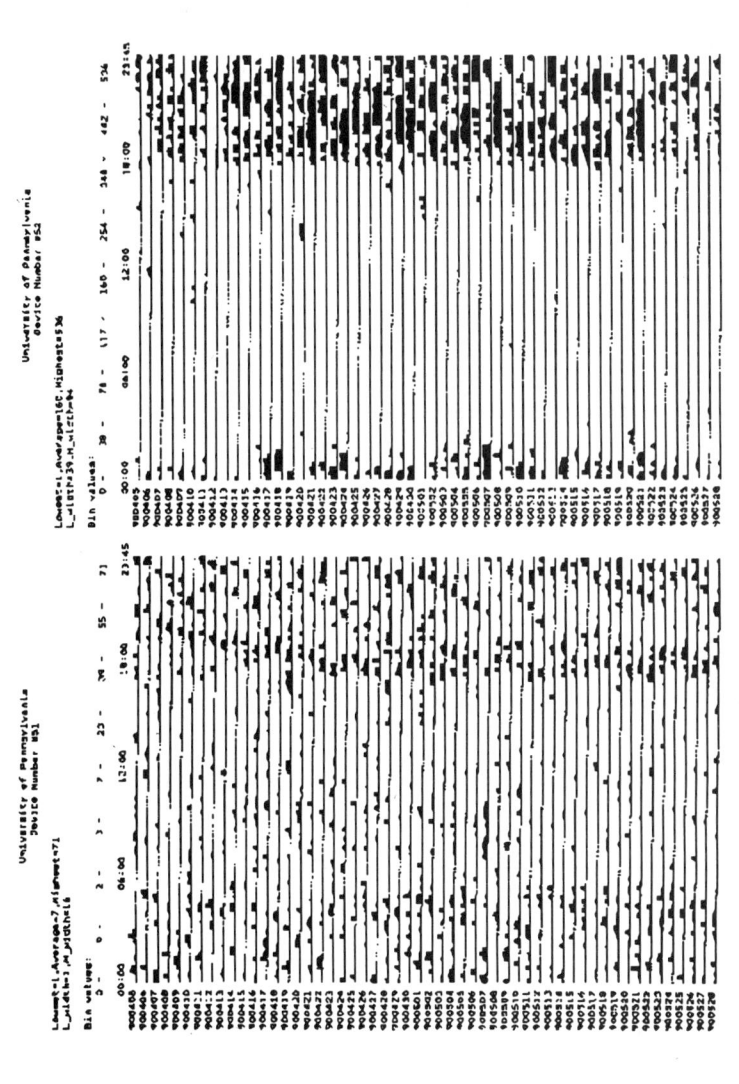

Fig. 1: Comparison of Sham and TPX rats exposed to a LD 12:12 cycle. The computer-generated actogram on the left is from a Sham rat while the actogram on the right shows a TPX rat. Note the cleaner and more coherent pattern of the TPX animal.

All rats were exposed to the following protocol: 2 weeks of LD 16/8; 4 weeks of LD 8/16; 6 weeks of RR at 2.6 lux; 4 weeks of RR at 0.3 lux; and 14 weeks of RR at 2.6 lux. At this time, 1/2 of the rats were sacrificed. The remaining animals were then exposed to a 12:12 LD cycle for 15 weeks. The results presented are from an analysis of the initial eight weeks of reexposure to the LD cycle.

Data were analyzed by a Chi-square periodogram procedure using the automated analysis package TAU developed by SCHULL. The two parameters which were measured were period and proportion of activity associated with the light and dark phases of the LD cycle.

RESULTS

One TPX animal died during this phase of the experiment leaving 6 Sham and 5 TPX animals for the analysis.

Three of six Shams did not fully entrain to the 12:12 LD cycle, showing periods others than 24 hours. Additionally, three of six Sham also showed numerous peaks in the periodogram of longer duration than 24 hours. In fact, only 1 of the 6 Shams showed a significant period of 24 hours without secondary peaks of longer duration.

On the other hand, only 1 of 5 TPX animals was found to have a significant period other than 24 hours (23.90 hours) as well as secondary peaks in the periodogram.

The proportion of total running-wheel activity displayed by Shams during the light phase of the LD cycle was 23% ±4.5% (SEM) while the same proportion in the TPX animals was 12% ±2.7% (SEM) (Fig. 1). This difference was significant ($p<0.05$, Student's t-test, one-tailed).

CONCLUSIONS

We have previously reported that TPX animals appear to display clearer and more coherent rhythms under a variety of conditions (3). This is the first time, however, we have specifically examined this factor in older animals. Despite of the very small number of animals examined here, it does appear that TPX animals are not only more rhythmically coherent in experiments done a few months after surgery, but that these rats retain this stability as they age.

In previous experiments, it was noted that TPX animals appear younger than control animals at time of sacrifice (5). TPX animals are smaller, more active, and their fur appears less yellow, an effect which was also observed in this experiment. It remains to be proven as to whether or not these observations, which can be associated with aging processes, actually translate into an increase in longevity for TPX rats.

Even if this relationship between thyroid state and longevity can be demonstrated, the causal factor or factors remain to be identified. HERLIHY et al. (2) would link the change in thyroid state with reduced food intake as the causal factor. Another possibility is that the increase in rhythm stability itself is the reason. These hypotheses are not mutually exclusive and investigations into these relationships should prove quite fruitful in the future.

ACKNOWLEDGEMENTS

This work was supported in part by a grant from the National Institutes of Health (HD04522) to Dr. Adler and by support from the University of Pennsylvania.

REFERENCES

1. EHERT, C.F., K.R. GROH & J.C. MEINART (1978): Circadian dyschronism and chronotypic ecophilia as factors in aging and longevity. In: H.V. SAMIS Jr. & S. CAPOBIANCO (eds.): Aging and Biological Rhythms, Advances in Experimental Medicine and Biology, Vol. 108. Plenum Press, New York, pp. 185-213
2. HERLIHY, J.T., C. STACY & H.A. BERTRAND (1990): Long-term food restriction depresses serum thyroid hormone concentrations in the rat. Mech. Ageing Develop. 53, 9-16
3. McEACHRON, D.L., J. LEVINE & N.T. ADLER (1990): Evidence that the pacemaker controlling activity rhythms is shortened in male thyroparathyroidectomized (TPX) rats: Similarities to the effects of estradiol in females. In: D.K. HAYES (ed.): Chronobiology : Its Role in Clinical Medicine, General Biology, and Agriculture, Part B , Alan R. Liss, Inc., New York, pp. 13-24
4. SCHEVING, L.E., R. CHESTER, F. HALBERG & J.E. PAULY (1974): Circadian variations in residents of a "senior citizens home". In: L. SCHEVING, F. HALBERG & J.E. PAULY (eds.): Chronobiology. Igaku Shoin Ltd., Tokyo, Japan, pp. 353-357
5. SCHULL, J., J. WALKER, K. FITZGERALD et al. (1989): Effects of sex, thyro-parathyroidectomy, and light regime on levels and circadian rhythms of wheelrunning in rats. Physiol. Behav. 46, 341-346

ONTOGENY AND 24-h RHYTHMS

H.-G. MLETZKO & I. MLETZKO

University of Rostock and College of Education Halle-Köthen, Goldberger Str. 12, D-O-2600 Guestrow, FRG

Ontogeny - the lifespan from the two-cell stage of zygote until the natural death - is subdivided into distinguishable periods. They are characterized by considerable changes of biorhythmical parameters corresponding with the genetic time table of life. Therefore it is not correct to speak about *the* daily rhythm of a species. Instead of this simplified view it is necessary to determine the distinguishing marks of rhythms along ontogeny.

In the following there is given a suggestion for the sufficient separation of life-phases from each other.

1. Period from formation of gametes to zygote - genomeszenz.
2. Period from two-cell stage of zygote to final leaving of egg-skins or from two-cell stage of a soma-cell to separation into autochthone organisms (clones). - embryoneszenz.
3. Period from birth (slipping) to formation of ripe gametes in the body - filialeszenz.
4. Period from ripe gametes in the body up to the ending of ripening - parenteszenz.
5. Period from ending of ripening gametes to death - seneszenz.

Each part of ontogeny has its own endogene marks of spatial - temporal structure. Chronobiological invariances have not been found along the ontogeny.

In particular this problem plays an enormous role in biological long-time experiments with relatively short-living animal species, such as the laboratory rat too (2, 3). There are phases in life with rapidly going changes of rhythms and these phases are connected with "physiologischer Dyschronie" (HELLBRÜGGE;1) of several rhythms and are easily to disturb. The consequence of this are disorders of biorhythms and its interactions. Such phases of transitions are embryogenomphase, filialembryophase, parentfilialphase and senesparentphase. In contrast to these the biorhythmical characteristics remain relatively stable in the parenteszenz - the middle phase of ontogeny.

Our investigations concern the 24-h rhythms of body weight in white rats. For this purpose male Wistar rats living in LD 12:12 have been weighed every four hours at an age of 15, 25, 50, 100 and 200 days after birth. Each group contained 24 individuals.

A 24-h oscillation of body weight is typical in all age-groups. In the course of filialeszenz and of parenteszenz the phases of 24-h rhythm change considerably. First the maximum of body weight is in the light span, while the mothers are very active in the dark span, and the young rats show the minimum of body weight. The pattern of oscillation has the form of a bigeminus. But already in 25 days old rats the maximum is going into the dark span, between 23.00 and 6.00 h the pattern is becoming more similar to a plateau. At the age of 50 days the oscillation becomes sinusoidale with its maximum at 6.00 h. In the further course of ontogeny a subsidiary maximum appears around 22.00 h. At 200 days old rats it becomes the main maximum and a subsidiary maximum appears at 10.00 h. Later this maximum of body weight is growing in relation to the main maximum.

The successive 24-h oscillations of body weight are overlaid by the rapid rise of absolute body weight. This phenomenon is visible in Fig. 1 as the changing ascent angle of curves on three consecutive days (mean). The two youngest animal groups show an $\tan \alpha$ of $\pi/4$, their body weight by 400 percent in relation to the oldest in our investigation.

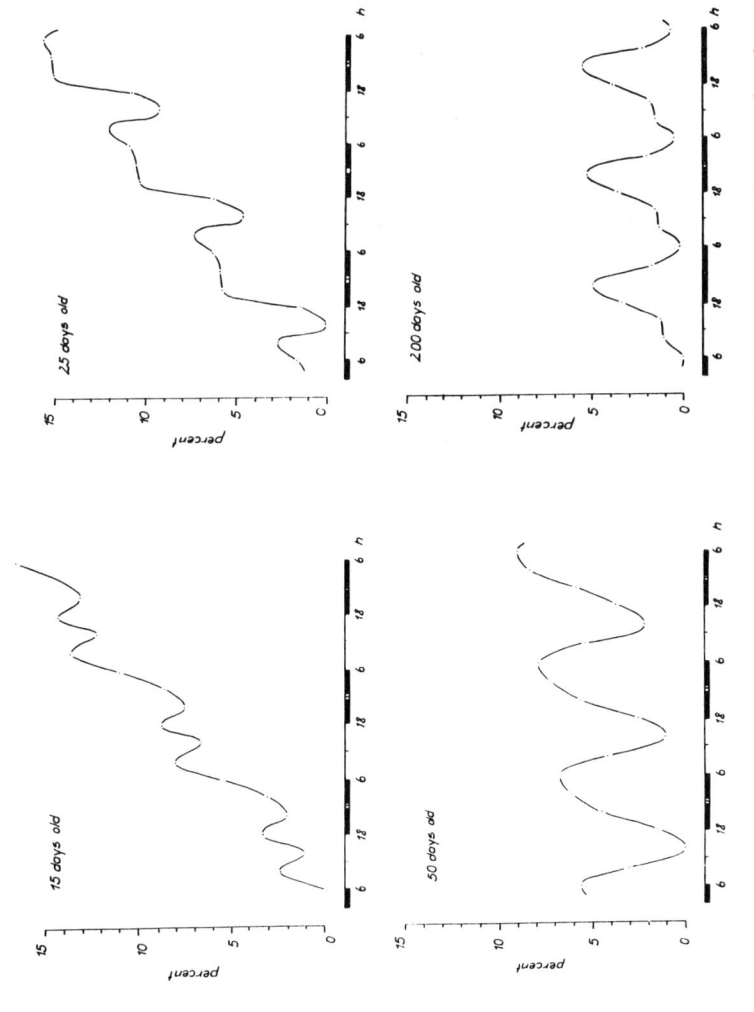

Fig.1: 24-h rhythm of body weight and middle ascent in the course of three days in male white rats of various age.

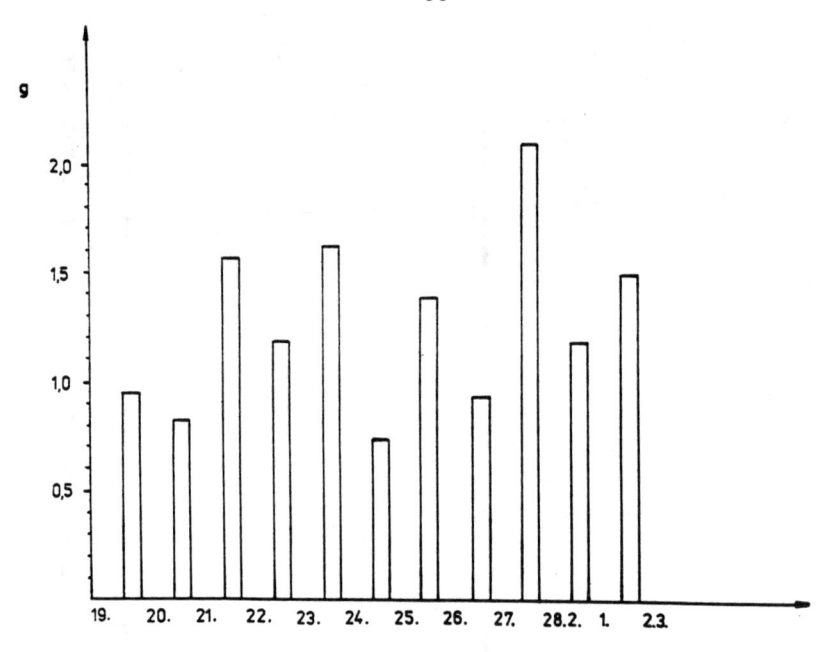

Fig. 2: Daily rate of body weight rise of rats, measurements beginning four days after birth (16th of February).

Fig. 1 and 2 show that growth is not a monotonous function of time, as it followed from $y(t) = Ae^{-b} - ct$.
SCHARF (5) has described the process with tan h:
$y = A_0 + A_1 \tan h\, c\,(t - \tau)$.
But this equation does not reflect the periodically happening pushes of growth rise. Changing speed of growth process requires a corrective term (trend):

$y_1 = f_1(t) = Ae^{\alpha t} = Z_0 e_{\alpha t}$ and
$y_2 = f_2(t) = A_0 + A_1 \cos \omega t + A_2 \sin \omega t$ then
$y = y_1 + y_2 = Z_0 e_{\alpha t} + A_0 + A_1 \cos \omega t + A_2 \sin \omega t$

But this trigonometrical function in form of a sigmoid does not yet describe the overlapping oscillations with various frequencies. The diurnal pushes of growth (2) as well as the two-daily oscillations of growth rise (4) can only be described in small limits.

Starting and end terms of measured parameters are genetically determined within the borders of biological variance just as the way of changes of biorhythmical properties of biological functions along ontogeny.

REFERENCES

1. HELLBRÜGGE, T.F. (1977): Physiologische Zeitgestalten in der kindlichen Entwicklung. Nova Acta Leopoldina NF 225, 46, 365-388
2. MLETZKO, H.-G. (1977): Chronobiologische Oscillationen ausgewählter Parameter der weißen Ratte (Biorhythmik). Diss. B, Univ. Halle
3. MLETZKO, I. (1982): Circadiane Zeitreihen der Motorialaktivität von Landwirbeltieren (Biorhythmik). Diss. B, Potsdam
4. MLETZKO, I., H. GRILL, H.-G. MLETZKO & I. KÖPPER (1985): Zur Postembryogenese der Laborratte unter biorhythmischem Aspekt. Wiss. Z. PH Halle, XXIII, 2, 55-57
5. SCHARF, J.H. (1979): Der Schritt von der heuristischen Wachstumskurve zum Biophysikalischen Modell. Gegenbaurs morphol. Jahrb. 125, 586-610

SPECTRAL ANALYSIS OF HEART RATE VARIABILITY IN PREMATURE NEWBORNS

N. HONZÍKOVÁ, B. FIŠER & E. KONVIČKOVÁ

Department of Physiology, Faculty of Medicine, Masaryk University, Komenskeho nam, 2, 662 43 Brno, Czechoslovakia

INTRODUCTION

Recently, considerable attention has been directed towards spectral analysis of heart-rate variability in infants. The studies have been focussed upon two principal groups: healthy infants and those at risk of sudden infant death syndrome (SIDS). Even though we can prove similar oscillations in healthy newborns as in adults (in the frequency range 0.11-0.8 Hz, i.e. 7- 50 cpm - range of respiratory sinus arrhythmia; 0.1 Hz peak, i.e. 6 cpm - the 10-s rhythm which is associated with baroreflex; in the low frequency range under 0.09 Hz - fluctuations having various origins), the quantitative difference is substantial (3). The specific influence on heart-rate variability in newborns is associated with their periodic breathing and frequent apneas (5) and with transition in various sleep stages (1,2).

After an ontogenetic study, we focussed upon the heart-rate variability in premature newborns. Only a few investigations of the heart rate and its variability have been done in newborns suffering from respiratory distress syndrome (RDS) (4). This syndrome affects preterm infants and it is often the cause of their death. It is due to a deficiency of the surfactant which begins to be produced between 28 and 32 weeks of gestation. The surfactant minimizes the surface forces in alveoli and it prevents their collapse. The heart-rate variability reacts on changes in depth and frequency of respiration, but it is also an image of the maturity of the autonomous nervous system which controls activity of the heart.

We decided to carry out a study in the hope that the analysis of heart rate variability might contribute to the prediction of the risk of preterm newborns.

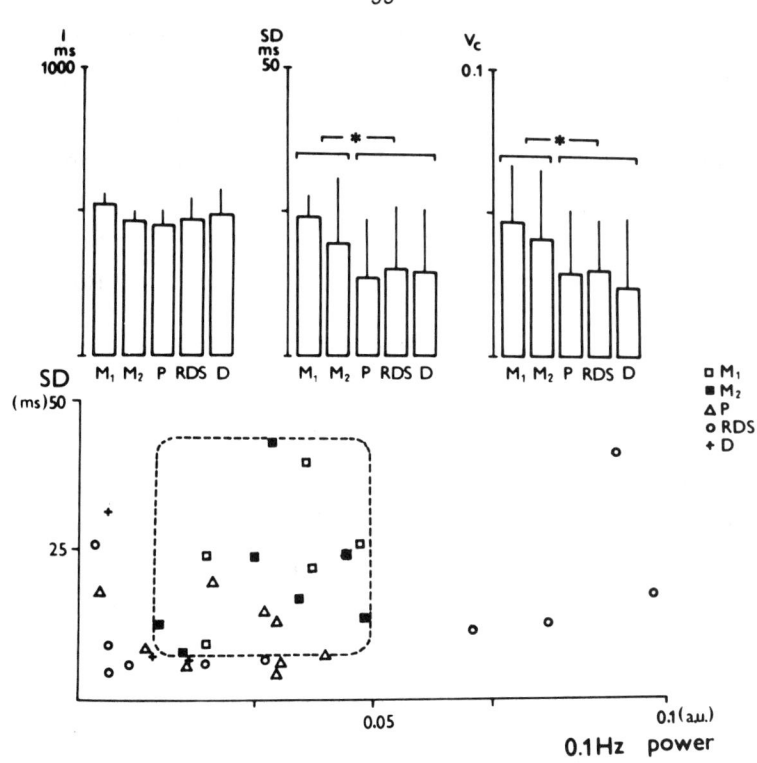

Fig.1: Above: The mean RR intervals (I), standard deviations (SD), and variation coefficients (VC) in the following groups: M1 - mature newborns weighing over 3.300 g, M2 - mature newborns from 3.300 g to 2.500 g, P - premature newborns under 2.500 g, RDS - newborns with respiratory distress syndrome, D - newborns with RDS who later died. The mean value, and SD of each group are indicated, as well as significant differences between the groups of mature and premature newborns (by asterisks).
Below: Scatter diagram of the relation between the amplitude of 0.1 Hz peak (abscissa) and the standard deviation of RR intervals (ordinate) in each newborn (arbitrary units).

METHODS

The ECGs of 35 newborns were recorded during the first 24 hours after delivery, immediately after feeding the children. Three-minute records were used for the calculation of the mean cardiac interval, the standard deviation of cardiac intervals and the power spectra of cardiac intervals variability. The newborns were divided into five groups. The newborns who did not suffer from RDS were divided according to the birth-weight, as it is usual in obstetrics: the first group - newborns weighing over 3.300 g, the second group - newborns from 3.300 g to 2.500 g, and the third group - premature newborns under 2.500 g. The fourth group contained newborns later suffering from RDS, the last small group included 3 newborns with RDS who later died.

RESULTS

The average values and standard deviations of the mean RR intervals, standard deviations, and variation coefficients of RR intervals in each group of newborns are shown in Fig. 1. While no significant difference in the mean RR intervals between these groups was observed, a significant difference in the variability of RR intervals was found.

The differences between the spectra of all groups of the newborns are shown in Fig. 2. While the peak at 0.1 Hz occurred in both groups of mature newborns, in premature ones the curve fell smoothly in this frequency range. The total shift of frequency of fluctuations to the lower frequencies was most expressive in newborns who died later. But there were large individual differences so that power spectra could not sufficiently differentiate between clinical groups of newborns. We tried to increase the sensitivity of this analysis by using a combination of calculated parameters of variability.

A correlation between the standard deviation and the peak at 0.1 Hz was not found in any group. This fact suggests a possible increase of discrimination when both - SD and 0.1. Hz peak - are used together. In the plot of SD against the 0.1 Hz peak (Fig. 1), after drawing a frame in which all mature children are contained, we can see that only three premature newborns are in the same field and none of the newborns with RDS and newborns who died later. In these newborns we found mostly low variability of heart rate together with small or lacking 10-s rhythm, or one of these parameters. Exceptionally both were extremely high.

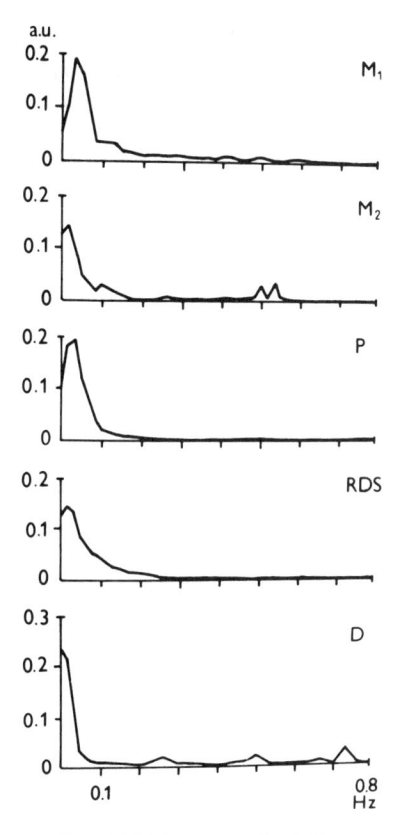

Fig.2: Averaged power spectra of RR intervals. Definitions as arbitrary units.

DISCUSSION

The major finding that emerges from this analysis is the lack of the physiological heart rate variability in premature newborns, especially in newborns who are at risk of RDS. Mostly, it is an overall reduction of variability, but sometimes it is also an increase in low-frequency oscillations. According to the approach of synergetics it may be interpreted as the decreased order and increased chaos in the regulation of circulation.

From the point of view of clinical use, the results are similar to those obtained by spectral analysis in SIDS (6). There are significant differences between newborns at risk of RDS and other newborns, but the specificity and sensitivity of this test is low. JENKINS et al. (4) needed to calculate thirteen measurements of short term and long term heart-rate variability for differentiation between clinical groups of infants. This study indicates that more recordings of children with subsequent fatal history are needed to establish a definite conclusion.

REFERENCES

1. DE HAAN, R., J. PATRICK, G.F. CHESS & N.T. JACO (1977): Definition of sleep state in the newborn infant by heart rate analysis. Am. J. Obstet. Gynecol. 127, 753-758
2. HARPER, R.M., B. LEAKE, J.E. HODGMAN & T. HOPPENBROUWERS (1982): Developmental patterns of heart rate and heart rate variability during sleep and waking in normal infants and infants at risk for the sudden death syndrome. Sleep 5, 28-38
3. HONZÍKOVÁ, N., E. KONVIČKOVÁ, B. FIŠER & Z. FIALOVÁ (1990): Ontogenese der Herzfrequenzvariabilität beim Menschen. Wiss. Zeitschrift der Humboldt-Univ. zu Berlin, R. Med. 39 (2), 93-94
4. JENKINS, J.G., R.H. MITCHEL & B.G. MCCLURE (1983): Heart rate variability in newborn infants. Automedica 4, 263-270
5. NUGENT, S.T. & J.P. FINLEY (1983): Spectral analysis of periodic and normal breathing in infants. IEEE Trans. Biomed. Eng. 30, 672-675
6. SCHECHTMAN, V.L., R.M. HARPER, K.A. KLUGE et al. (1988): Cardiac and respiratory patterns in normal infants and victims of the sudden infant death syndrome. Sleep 11, 413-424

EVALUATION OF A NEONATAL CARDIOVASCULAR RISK SCORE FROM BLOOD PRESSURE AND HEART RATE CHARACTERISTICS

J.R. FERNÁNDEZ[1], R.C. HERMIDA[1], D.E. AYALA[1], A. MOJÓN[1], A. REY[2], J. RODRÍGUEZ-CERVILLA[2], J.R. FERNÁNDEZ-LORENZO[2] & J.M. FRAGA[2]

[1] Bioengineering & Chronobiology Laboratories, E.T.S.I. Telecomunicación, University of Vigo, Apartado 62, Vigo, Spain.
[2] Hospital General de Galicia & Medical School, University of Santiago, Santiago de Compostela, Spain.

INTRODUCTION

Chronobiologic methods (6) recognize variability as the source of information to be extracted by special hardware and software for use in screening and prevention as well as in diagnosis and cure. Along these lines, rhythm characteristics (as quantitative endpoints of a component of a biologic time series, formulated algorithmically as a recurrent phenomenon and demonstrated as being periodic by the use of inferential statistics) could complement traditional statistical parameters (e.g. means, ranges, and standard deviations) for the detection at birth of the risk of developing a high blood pressure (BP) later in life (4,7,8).

At this point, most of the attempts done to distinguish, at an early age, those persons with an elevated risk for development of high BP from those without such risk have not succeeded. Previous studies have been based mostly on casual BP measurements, not specific in terms of time or rhythm stage (2). The approach presented herein (using hardware for automatic systematic around-the-clock sampling and software for quantification of rhythm dynamics and regression and/or discriminant analysis of these indices (7)) is the better alternative. By this approach, one can determine a neonatal index of cardiovascular risk that can be used to evaluate the results of preventive or therapeutic measures instituted on the pregnant woman and may provide indications for intervention on the newborn, if the index is responsive to measures during pregnancy, by contrast to the unalterable family history (10). This index could be based on characteristics of BP and heart rate (HR) variability during the first two days after birth.

Neonatal monitoring of BP and HR should be cost-effective irrespective of whether it can improve on or replace inferences drawn from the family history alone. Only a neonatal index for the early recognition of risk can assess the success or failure of preventive interventions on the pregnant women (such as dietary supplements of calcium and/or magnesium, weight control, etc., to which the family history obviously cannot respond) and the possibly harmful effects (on the baby) of the administration of drugs such as beta-adrenergic agonists or corticoids.

Credit for the demonstration with rhythmometric methodology (irrespective of the family history of cardiovascular disease risk) of a circadian rhythm in the BP of the human newborn is due to KELLEROVÁ (9). She did not, however, seek or report any differences as a function of the familial background of cardiovascular disease and/or a high BP. HALBERG et al. (4) demonstrated a circadian rhythm in BP and HR on an individualized basis for some but not all neonates monitored automatically for 48 hours at 30-minute intervals during the first week of life. Moreover, the circadian amplitude, and a linear trend were more prominent in a group of neonates with a positive family history of high BP and cardiovascular disease as compared to a group with a negative such family history (3,4,5).

We have examined the possibility that differences in rhythm characteristics may be indeed apparent at birth, so that those characteristics can then be used 1) to predict for neonates, from BP readings, the likelihood of high BP in later life, and 2) to test preventive measures that, if validated, could be implemented before actual damage occurs.

SUBJECTS AND METHODS

Starting shortly after birth, the systolic and diastolic BP and HR of 127 newborns were automatically monitored at about half-hour intervals for 48 hours with a Nippon Colin (Komaki City, Japan) device at the Hospital General de Galicia, Santiago de Compostela, Spain. On the basis of questionnaires given to the parents inquiring notably about the medical and familial history, the newborns were assigned a cardiovascular risk score (CRS). This risk scale was developed on the basis of two principles: 1) presence of overt cardiovascular disease was assigned a risk score twice as large as the risk associated with the presence of an elevated BP, and four times as large as the risk associated with the presence of obesity. And 2) the risk is balanced across 2 generations, those of the newborn's parents and of his grandparents; it is additive within each generation and across generations, with a maximal risk of 2 per generation. The risk scale thus spans from 0 to 4, with minimum incremental step of .125. Such a CRS is used because even when environmental or other factors may also play a role in the predisposition of the neonate to elevated BP, in the first two days after birth the primary factor for the predisposition to elevated BP later in life comes from heredity (1,7,8).

BP and HR values outside ±3 standard deviations from the individual series mean were removed, and the remaining data were fitted by linear least-squares with a 24-hour cosine curve (6). Circadian characteristics (the rhythm-adjusted mean or mesor, amplitude, and linear parameters computed from the amplitude-acrophase pair) and descriptive indices of location and dispersion (mean, 90% range, 50% range, standard deviation (SD) and circadian range, defined as the difference between the maximal and minimal hourly mean values after stacking all data to cover a single idealized 24-hour span) for the three circulatory variables were first individually correlated with CRS, and then used to compute a linear prediction function for CRS by the use of multiple regression analysis (7,8,12).

Three different stepwise procedures were carried out to build a regression model for the neonatal CRS. Methods of stepwise regression use a convenient computational algorithm to limit possible models to a relatively small number, providing a systematic technique for examining only a few subsets of each size (12). A path through the possible models is chosen, looking first at a subset of one size, and then looking only at models obtained from preceding ones by adding or deleting variables. The basic algorithms for stepwise regression used here were forward selection, backward elimination, and stepwise.

RESULTS

Results from correlation analysis indicate statistically significant linear relations between CRS and indices of variability or dispersion, notably circadian range, SD, circadian amplitude and 90% range of systolic and diastolic BP and HR, and lack of significant correlation between CRS and average values or indices of location (means, mesors and even 50% ranges). The higher correlations with CRS are obtained with a model-independent index of dispersion, the circadian range.

The obtained characteristics were then used as predictors to compute a linear prediction function for CRS. All three stepwise procedures of multiple regression concluded with the same result. The best model for the prediction of CRS includes the circadian amplitudes of systolic and diastolic BP and the circadian range of systolic BP and HR. The regression coefficients for the model including those four predictors are indicated in Table 1. No other variable can be further included in the model since their t statistics are less than the predetermined value (that yielding a p value less than .05 for the number of degrees of freedom considered) to stop the stepwise procedure. The expected average prediction error for the computation of CRS in a new neonate using the model in Table 1 is 0.232.

Tabl.1: Linear model for the prediction of a neonatal cardiovascular risk score obtained by multiple regression analysis on data from 127 newborns.

Variable*	Estimate	Std. Error	t value	P value
CA of SBP	.12637	1.6653E-2	7.59	<.001
CA of DBP	-.13737	2.5974E-2	-5.29	<.001
CR of HR	.00470	9.3907E-4	5.01	<.001
CR of SBP	.00723	2.3179E-3	3.12	<.001

Degrees of freedom = 123; Residual mean square = 5.3869E-2; Root mean square= .2321; Coefficient of determination (R2) = .8734; Adjusted R2 =.8693. * CA: Circadian amplitude, obtained by the linear least-squares fit of a 24-hour cosine curve to data from each newborn. CR: Circadian range, defined as the difference between the highest and the lowest mean values of hourly classes, after stacking all data from that individual to cover a single idealized 24-hour span. SBP: Systolic blood pressure (in mm Hg). DBP: Diastolic blood pressure (in mm Hg). HR: Heart rate (in beats/min.).

Fig. 1 represents the relation between the CRS computed from the questionnaires (horizontal axis) and the predicted CRS from the linear model indicated in

Table 1 (vertical axis) for the 127 neonates contributing data in the computation of the prediction function. The multiple correlation coefficient between the predicted and the computed CRS is 0.663 (P<0.001).

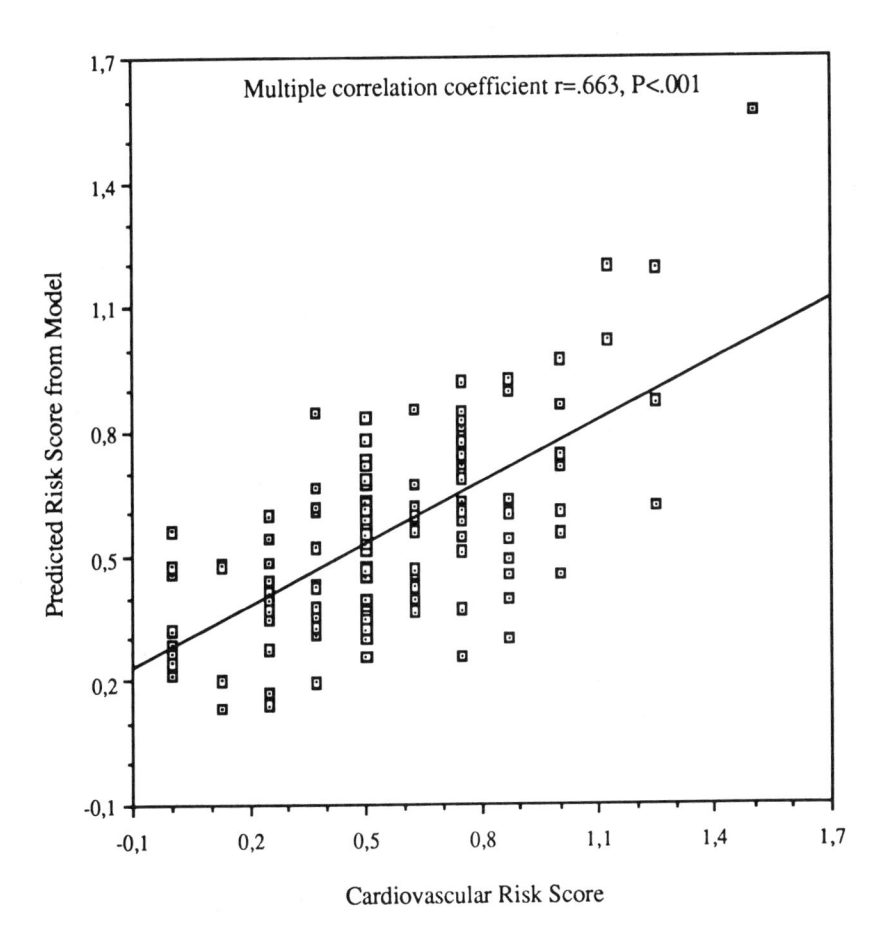

Fig.1: Predicted risk scores from linear model in Table 1 in relation to cardiovascular risk score obtained from family history questionnaires for 127 Spanish neonates.

The linear model from Table 1 was also used to predict CRS for a new group of 23 neonates, monitored according to the same scheme and conditions as the original reference group of 127 newborns. The correlation between the CRS computed from the model and the obtained from the family history of each of those 23 newborns is 0.565 (p=0.005). The average absolute prediction error in computing CRS for those 23 neonates was 0.180, even smaller than the expected value of 0.232 noted above for the model in Table 1.

DISCUSSION

Relationships existing between different sets of variables and their relative power with respect to cardiovascular indices at birth were here investigated by methods of least-squares rhythmometry and multiple regression analysis (7,8). A CRS was computed for each neonate and the obtained discrete variable used for building a prediction model based on only four predictors (Table 1). In comparison with similar risk scores used by others (1), obesity was now included in the evaluation of the CRS since it is associated with coronary heart disease, primarily because of its influence on BP, blood cholesterol and precipitating diabetes (10). Results indicate a high improvement by the use of the modified risk score. Even when the selected model was the same considering or not obesity as a contributing factor and the estimated regression coefficients were not statistically different between both analyses, the coefficient of determination was highly improved (from 0.6282 to 0.8734) when the score for obesity was added in the evaluation of the neonatal CRS.

The tabulated result already represents a first testable battery, even when it may have to be modified as a function of the coexisting risk of diseases other than high BP. Future work in the identification of linear models for prediction of CRS should include, among other endpoints, ultradian characteristics of BP and HR (including parameters of harmonic components with periods of 12, 8 and/or 6 hours, statistically significant for some of the neonates here studied), in order to reduce the prediction error in the evaluation of neonatal CRS.

Such remaining tasks notwithstanding, the results here described represent a step forward in the identification of a neonatal index of cardiovascular risk. A CRS computed from the model in Table 1 could now be used for the evaluation of the effects on the newborn of preventive intervention on the pregnant women (e.g., calcium, aspirin, and/or magnesium supplementation, weight control., etc.), if the index is proved to be in fact responsive to such intervention, by contrast to the unalterable CRS obtained according to family history. That index could also need to be correlated with morphologic or functional measures from echocardiography that demonstrate target organ involvement at birth and/or during tracking. In fact, in 14-year-old children, the circadian amplitude of diastolic BP correlates with the interventricular septum thickness, notably in children with a positive family history of high BP (11).

ACKNOWLEDGEMENTS

This research was supported in part by Dirección General de Investigación Científica y Técnica, DGICYT, Ministerio de Educación y Ciencia (PA86-0229 & PB88-0546); Consellería de Educación e Ordenación Universitaria, Xunta de Galicia (XUGA-709CO388); and Nippon Colin Co., Komaki City, Japan.

REFERENCES

1. CORNELISSEN, G., R. KOPHER, P. BRAT et al. (1989): Chronobiologic ambulatory cardiovascular monitoring during pregnancy in Group Health of Minnesota, Proc. 2nd Annual IEEE Symposium on Computer-Based Medical Systems. Minneapolis, MN, June 26-27, pp. 226-237
2. HALBERG, F. & H. FINK (1985): Juvenile human blood pressure: need for a chronobiologic approach. In: G. GIOVANNELLI, M. NEW & S. GORINI (eds.): Hypertension in Children and Adolescents. Raven Press, New York, pp. 45-73
3. HALBERG, F., R.C. HERMIDA, G. CORNELISSEN et al. (1985): Toward a preventive chronocardiology, J. Interdiscipl. Cycle Res. $\underline{16}$, 260
4. HALBERG, F., G. CORNÉLISSEN, C. BINGHAM et al. (1986): Neonatal monitoring to assess risk for hypertension, Postgrad. Med. $\underline{79}$, 44-46
5. HALBERG,F., R.C. HERMIDA, G. CORNÉLISSEN et al. (1987): Further steps toward a neonatal chronocardiology. In: G. HILDEBRANDT, R. MOOG & F. RASCHKE (eds.): Chronobiology & Chronomedicine. Basic Research & Applications, Peter Lang, Frankfurt am Main, pp. 288-292
6. HERMIDA, R.C. (1987): Chronobiologic data analysis systems with emphasis in chronotherapeutic marker rhythmometry and chronoepidemiologic risk assessment. In: L.E. SCHEVING, F. HALBERG & C.F. EHRET (eds.): Chronobiotechnology and Chronobiological Engineering. Martinus Niijhoff, NATO ASI Series E, No. 120, Dordrecht, The Netherlands, pp. 88-119
7. HERMIDA, R.C., J.M. FRAGA, D.E. AYALA et al. (1989): Evaluation of a neonatal prediction function for cardiovascular risk assessment, Proc. 1989 IEEE Int. Conf. Systems, Man, & Cybernetics. Cambridge, MA, Nov. 14-17, pp. 1124-1129
8. HERMIDA, R.C., J.M. FRAGA, A. REY et al. (1990): Linear model for prediction of a neonatal cardiovascular risk score. In: A. REINBERG, G. LABRECQUE & M. SMOLENSKY (eds.): Annual Review of Chronopharmacology, Vol. VII, Pergamon Press, Oxford, pp. 145-148
9. KELLEROVÁ, E. (1981): Physiological responses of blood pressure & heart rate in neonates and infants. Adv. Physiol. Sci. $\underline{9}$, 367-375
10. PAYNE, W.A. & D.B. HAHN (1986): Understanding Your Health. Times Mirror/Mosby College Publishing, St. Louis
11. SCARPELLI, P.T., G. CORNÉLISSEN, F. HALBERG et al. (1987): Echocardiographic correlations of chronobiologic parameters. In: Proc. 3rd Eur. Mtg. Hypertension; Milan, June 14-18, p. 505
12. WEISBERG, S. (1985): Applied Linear Regression. 2nd edit., John Wiley & Sons, New York

ENTRAINMENT, PHOTOPERIOD, LIGHT EFFECTS

A HYPOTHESIS ON THE EVOLUTIONARY ORIGINS OF PHOTOPERIODISM BASED ON CIRCADIAN RHYTHMICITY OF MELATONIN IN PHYLOGENETICALLY DISTANT ORGANISMS

R. HARDELAND, B. PÖGGELER, I. BALZER & G. BEHRMANN

I. Zoologisches Institut, Universität Göttingen, Berliner Str. 28, D-W 3400 Göttingen, FRG

Photoperiodism as a means for temporal orientation within the year represents a widely spread phenomenon found in phylogenetically distant taxa, such as angiosperms, insects, sauropsids and mammals (4,5,13,15,20). Recently, we have detected the existence of a photoperiodic mechanism in a dinoflagellate, *Gonyaulax polyedra* (2).

In mammals, the determination of daylength involves the circadian rhythm of melatonin, a substance acting as a mediator of darkness (1,13). In sauropsids (10,17) and in insects (19) melatonin also exhibits circadian rhythmicity; it seems to transduce the signal 'darkness' in these organisms as well, but its particular role in photoperiodism still remains to be elucidated. Moreover, melatonin has been shown to influence reproduction of planarians (11); in this case again the relationship to photoperiodism in a strict sense has to be clarified. Although only a few data are available on invertebrates, they may be taken as an indication for a broader phylogenetic basis of both melatonin periodicity and photoperiodism. This view has been supported by our findings that melatonin occurs also in *Gonyaulax*, in concentrations as high as in vertebrate pineals (3), and exhibits circadian oscillations in this unicell, showing a prominent nocturnal maximum (12).

LABILITY OF MELATONIN UNDER ILLUMINATION

It is well-known that indole compounds and, in particular, melatonin can be peroxidized by superoxide anions (18). The product formed by cleavage of the pyrrole ring is a substituted kynuramine, as shown in Fig. 1a. This type of degradation can occur both by the action of indoleamine dioxygenase (7,8) and non-enzymatically. Especially this latter possibility should not be underrated, since the reaction with the free radical is greatly enhanced through the catalysis by complexed iron (18), e.g. free or protein-bound iron porphyrins.

We have studied the degradation of melatonin under the influence of cellular material from *Gonyaulax* by fluorometrically discriminating between melatonin and products. Our results clearly demonstrate that cell homogenates from the dinoflagellate contain compounds capable of catalyzing the cleavage of the pyrrole ring. Fig. 1b shows the fluorescence spectra of the product and of melatonin. With regard to both excitation and emission spectra this product exhibits the typical characteristics of a kynuramine, and it is clearly different from quinolines, which can be formed in the course of further oxidation processes.

The formation of the substituted kynuramine does not require the presence of an enzymatic activity, since it is also observed in preparations in which proteins have been precipitated by ethanol and boric acid, but it depends on the presence of light-induced superoxide anions. Therefore, this reaction occurs only upon exposure of the samples to light (Fig. 1c). Already moderate illumination intensities of visible light at 240 lx, even in the absence of mentionable amounts of UV, lead to a considerable degradation of melatonin, whereas in darkness the compound remains fairly stable. Without the catalysis by components of the cell extract, the decay of melatonin is negligible, even upon exposure to light. These results lead to conclusions of potential importance for our understanding of the cyclicity in melatonin concentrations: (a) Melatonin acts as a radical scavenger. (b) The reaction of melatonin with superoxide anions is highly potentiated by cellular components and, hence, much higher within than outside the cell. (c) The stability of melatonin depends on irradiation and, therefore, on the light-dark cycle.

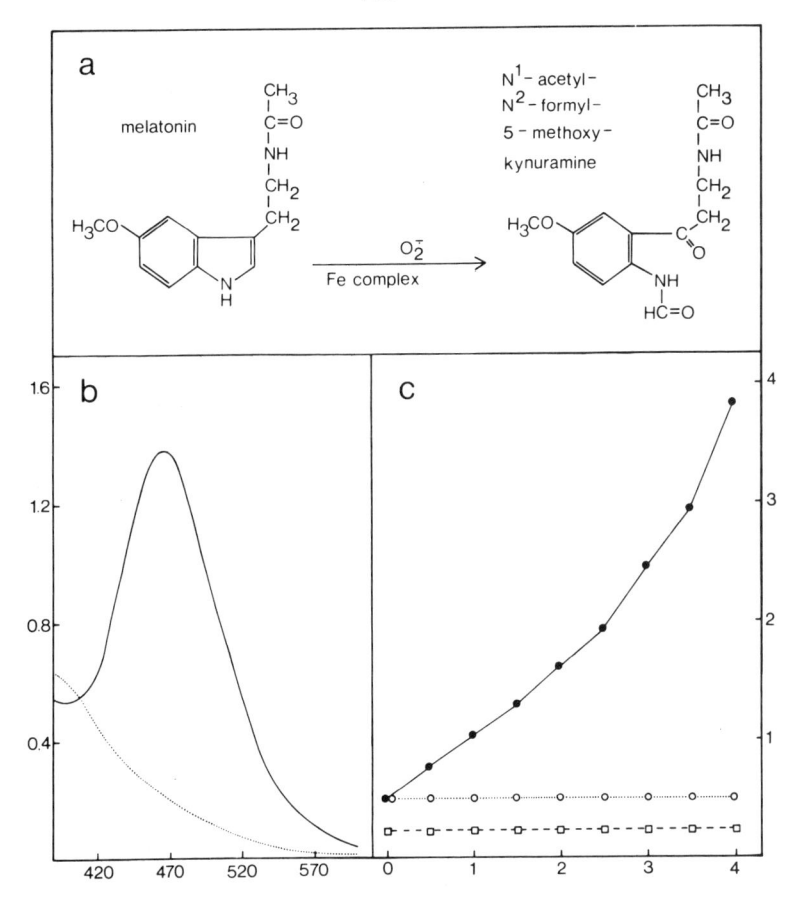

Fig.1: Non-enzymatic degradation of melatonin by superoxide anions. a. Formation of substituted kynuramine, as catalysed by complexed iron. b. Fluorescence spectra of melatonin (dotted line) and substituted kynuramine (solid line) formed under the influence of light and *Gonyaulax* extract (ordinate: fluorescence units; nm of emitted light). Cell homogenates in 0.2 M tris/HCl buffer, pH 8.5 containing 0.003 M dithioerythritol were centrifuged at 2.800 X g for 10 min, mixed with melatonin, and protein was precipitated with 6-fold volume of 1% boric acid in ethanol (final concentration of melatonin: 0.98 mM); after sedimentation of protein, the supernatant was exposed to light (240 lx). Spectra were measured in an AMINCO-BOWMAN spectrofluorometer with ellipsoidal condensing system, using the ratio mode (excitation 348 nm). c. Dependence of non-enzymatic melatonin degradation on light exposure and presence of *Gonyaulax* extract (solid line: extract + light; dotted line: extract without light; dashed line: light without extract; ordinate: fluorescence units; abscissa: hours). Procedure as in b (excitation 350 nm; emission 470 nm).

A POSSIBLE ANCIENT ROLE OF SUPEROXIDE ANIONS IN DIURNAL CYCLICITY OF EARLY EUKARYOTES

Presumably one of the largest catastrophes which ever occurred in the biosphere during earth history originated from the 'invention' of photosystem II with the consequence of O_2 production. Although O_2 itself is not really harmful to organisms, the light-induced production of aggressive oxygen radicals such as superoxide anions and hydroxyl radicals represented the real and severe problem for the organisms, which may have been extinguished in large quantities, and which were only able to survive after having developed managing strategies for eliminating the free radicals. One of these is based on the activities of superoxide dismutase and catalase/peroxidase, but the presence of radical scavengers might have been an additional requirement. As demonstrated above, melatonin can, in fact, be used as a superoxide anion scavenger (cf. also (18)). Indoleamines can be utilized for purposes of protection from radicals also by recent organisms. In analogy to melatonin, its precursor serotonin can already react with superoxide anions. In yeasts, it is highly effective in preventing UV-induced damage of DNA, it increases the susceptibility to short-wave irradiation, and defect mutants of the serotonin pathway exhibit greatly enhanced mutation rates (6,9,16). The utilization of melatonin as a radical scavenger may be of advantage as compared to that of serotonin, since the former reacts more easily with superoxide anions, especially at a physiological pH (18).

In the presence of oxygen, the cyclicity of light and darkness causes a cyclicity of enormous amplitude in the concentration of free radicals, especially with regard to the light-induced superoxide anions. It is our aim to direct the attention to the possibility that the periodic occurrence of these highly aggressive and life-threatening substances may have exerted an extremely strong selection pressure favoring the origination of the circadian oscillator. One could, however, pose the question of why aerobic prokaryotes possessing superoxide dismutase and catalase or peroxidase have not evolved this type of rhythmicity. The reason may be sought in smallness and short generation time of prokaryotic cells, in which a low-frequency oscillator might represent a kind of ballast or would perhaps perturb - or be perturbed by - a high-frequency cell division cycle. Nevertheless, the idea of a cycle of free radicals as a major driving force in the evolution of circadian rhythmicity may particularly be attractive, as it implies an evolutive process related to the development of aerobic life during early eukaryotic history and, therefore, offers a common explanation for the existence of circadian oscillations in animals, fungi, and plants. Moreover, it assumes negative selection by an extremely strong, potentially cytocidal factor which generated the initial evolutionary pressure rather than a moderate positive selection advantage based on anticipation, the strength of which may be comparably weak in heterotrophic unicellulars so that its significance is somehow hard to conceive.

ON THE POSSIBLE PHYLOGENETIC ORIGINS OF MELATONIN RHYTHMS AND PHOTOPERIODISM

Due to their potency as radical scavengers, indoleamines and, in particular, melatonin may have been utilized as protective agents already during early evolution. Also in later stages of life history and even in recent organisms any transiently light-exposed cell producing melatonin should be expected to experience a decrease in the concentration of this indoleamine upon irradiation and hereby caused formation of superoxide anions. So if one assumes as a primary phylogenetic step an <u>aperiodic production</u> of melatonin for purposes of eliminating free radicals, a light-dark cycle would inevitably lead to a <u>periodic pattern</u> in the <u>concentration</u> of melatonin, with a minimum during photophase due to the reaction with superoxide anions catalyzed by cellular components (Fig. 2). Under the assumption of a constant rate of melatonin production the resulting temporal pattern would be more or less rectangular.

As a result of such an exogenously determined periodicity, high melatonin concentrations would be always associated with darkness. Thus, it is highly suggestive to suppose that this substance may have been finally utilized by cells as a metabolic indicator of darkness. A compound exhibiting this particular time course should be suitable as a mediator of scotophase. In principle, this holds regardless of whether melatonin rhythms may have been evolved mono- or polyphyletically. The instability of melatonin in light-exposed cells, in *Gonyaulax* as well as in a retinal cell or in a pinealocyte of primitive vertebrates, would in any case lead to a temporal pattern favoring the role as a metabolic representative of scotophase.

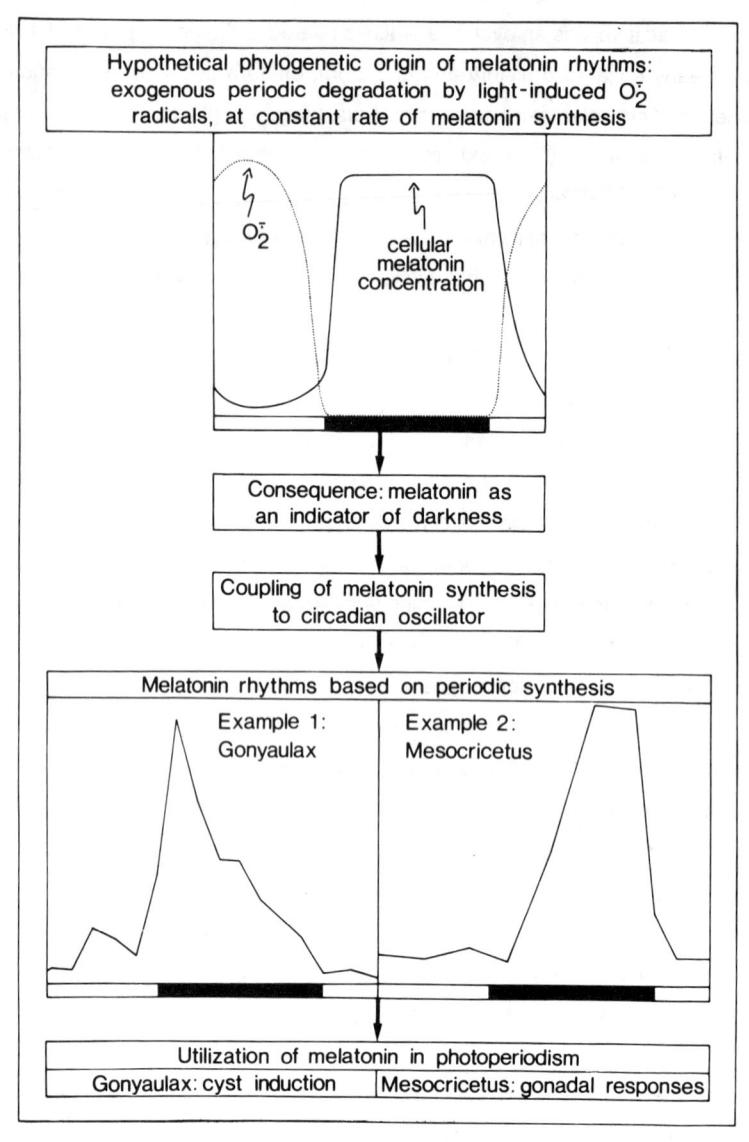

Fig.2: A phylogenetic model on the possible origins of melatonin rhythms and melatonin-related mechanisms of photoperiodism. Melatonin rhythm in *Gonyaulax* after PÖGGELER et al. (12) (HPLC data), in *Mesocricetus* after REITER (14).

The next evolutionary step may have been that of a coupling of melatonin formation to the circadian oscillator (Fig. 2). As a result, the melatonin rhythm could be based on periodic synthesis rather than degradation. Consequently, the rhythm of the indoleamine has become subject to control by the cell; this may be regarded as a development towards perfection of a mechanism which has to be reliable also under varying environmental conditions, e.g. in different illumination intensities due to changes of weather, which would otherwise result in changes of radical formation and, therefore, varying melatonin concentrations within the photoperiod.

As a consequence of an autonomous melatonin rhythm due to periodic synthesis, the temporal pattern could become different from the rectangle type. As in the case of Gonyaulax, the circadian maximum can be fixed to the beginning, or, as in Mesocricetus, to the end of scotophase (Fig. 2). This deviation from the rectangular pattern, which may be regarded as a kind of narrowing of the signal, might be advantageous for photoperiodic purposes, especially in terms of an internal coincidence mechanism.

The rhythmicity of melatonin, as observed in phylogenetically different organisms such as dinoflagellates, insects, and vertebrates, provides a basis for the evolution of photoperiodism, which may have developed even polyphyletically. It is remarkable that a unicell, Gonyaulax, does not only exhibit a high-amplitude circadian rhythm of this indoleamine (Fig. 2; (12)) and that melatonin is also able to mimic short-day effects leading to cyst formation (2). The common biochemical basis of photoperiodism seems, therefore, to be much broader than ever believed.

REFERENCES

1. ARENDT, J. (1986): Role of the pineal gland in seasonal reproduction function in mammals. Oxford Rev. Reprod. Biol. 8, 266-320
2. BALZER, I. & R. HARDELAND (1991): Induction of cyst formation by short photoperiods and 5-methoxylated indoleamines in a dinoflagellate, Gonyaulax polyedra. Eur. J. Cell Biol. 54, Suppl. 32, p. 57
3. BALZER, I., B. PÖGGELER & R. HARDELAND (1990): 5-Methoxylated indoleamines in the dinoflagellate Gonyaulax polyedra: presence, stimulation of bioluminescence, and circadian variations in responsiveness. Eur. J. Cell Biol. 51, Suppl. 30, p. 52

4. BÜNNING, E. (1973): The Physiological Clock. Springer, N.Y. - Heidelberg. - Berlin
5. FARNER, D.S. & R.A. LEWIS (1971): Photoperiodism and reproductive cycles in birds. In: A.C. GIESE (ed.): Photophysiology, Vol. VI, pp. 325-370
6. FRAIKIN, G. Y., E.V. IVANOVA & M.G. STRAKHOVSKAYA (1986): Inhibition of thymine dimer formation in DNA by serotonin. Photobiochem. Photobiophys. 12, 289-294
7. HAYAISHI, O. & R. YOSHIDA (1979): Rhythms and physiological significance of indoleamine 2,3-dioxygenase. In: M. SUDA, O. HAYAISHI & H. NAKAGAWA (eds.): Biological Rhythms and their Central Mechanism, Elsevier/North-Holland, Amsterdam - N.Y. - Oxford, pp. 133-141
8. HIRATA, F., O. HAYAISHI, T. TORUYAMA & S. SENOH (1974): In vitro and in vivo formation of two new metabolites of melatonin. J. Biol. Chem. 249, 1311-1313
9. IVANOVA, E.V., M.E. POSPULOV, M.G. STRAKHOVSKAYA & G.Y. FRAIKIN (1982): The competitive action of UV of different wave lengths on the survival of Saccharomyces cerevisiae. Mikrobiologiya 51, 761-764
10. MENAKER, M. (1982): The search for principles of physiological organization in vertebrate circadian systems. In: J. ASCHOFF, S. DAAN & G.A. GROOS (eds.): Vertebrate Circadian Systems. Springer, Berl. - Heidelbg. - N.Y., pp. 1-12
11. MORITA, M. & J.B. BEST (1984): Effects of photoperiods and melatonin on planarian asexual reproduction. J. Exp. Zool. 231, 273-282
12. PÖGGELER, B., I. BALZER, R. HARDELAND & A. LERCHL (1991): Pineal hormone melatonin oscillates also in a unicellular, the dinoflagellate Gonyaulax polyedra. Naturwissenschaften, in press
13. REITER, R.J. (1980): The pineal and its hormones in the control of reproduction in mammals. Endocr. Rev. 1, 109-131
14. REITER, R.J. (1982): Chronobiological aspects of the mammalian pineal gland. In: H. v. MAYERSBACH, L.E. SCHEVING & J.E. PAULY (eds.): Biological Rhythms in Structure and Function. Alan R. Liss, N.Y.
15. SAUNDERS, D.S. (1976): Insect Clocks. Pergamon, Oxford
16. SOBOLEV, A.S. (1973): Mechanics of protective effect of serotonin. Doctoral Thesis, Moscow
17. TAKAHASHI, J.S. (1982): Circadian rhythms of the isolated chicken pineal in vitro. In: J. ASCHOFF, S. DAAN & G.A. GROOS (eds.): Vertebrate Circadian Systems. Springer, Berl. - Heidelbg. - N.Y., pp. 158-163
18. UEMURA, T. & K. KADOTA (1984): Serotonin- and melatonin-dependent light emission induced by xanthine oxidase. In: H.G. SCHLOSSBURGER, W. KOCHEN, B. LINZEN & H. STEINHART (eds.): Progress in Tryptophan and Serotonin Research. Walter de Gruyter, Berlin, pp. 673-676
19. WETTERBERG, L., D.K. HAYES & F.HALBERG (1987): Circadian rhythms of melatonin in the brain of the face fly, Musca autumnalis De Geer. Chronobiologia 14, 377-381
20. WITHROW, R.B. (ed.) (1959): Photoperiodism and Related Phenomena in Plants and Animals. AAAS, Wash.

NEUROPEPTIDES IN THE HYPOTHALAMIC SUPRACHIASMATIC NUCLEI: SEASONAL CHANGES IN A MAMMALIAN CIRCADIAN OSCILLATOR

S. REUSS

Department of Anatomy, Johannes Gutenberg-University, Saarstr. 19-21,
D-W-6500 Mainz, F.R.G

INTRODUCTION

The suprachiasmatic nuclei (SCN) work as endogenous pacemakers responsible for the generation and entrainment of circadian rhythms in mammals. They are connected to the visual system via the retino-hypothalamic (4) and geniculo-hypothalamic (5) tracts and play an eminent role in the generation and entrainment of rhythms (8). A considerable number of morphological investigations on these structures, which are located in the basal hypothalamus bilateral to the third ventricle, have shown that different, at least partly topographically segregated subpopulations of neurons are found within its borders which produce a variety of neuroactive substances (1,6,7,10).

A predominant portion among these substances are neuropeptides. It is reasonable to believe that they are, as transmitters or modulators, involved in the mechanisms that generate and maintain rhythms. In addition to circadian, infradian rhythms such as seasonal changes are of interest to the chronobiologist.

As an animal model to study the rhythmic system, the Djungarian hamster (*Phodopus sungorus*) is widely accepted. Endocrine parameters such as reproduction in this species are regulated by photoperiod. The animals exhibit striking seasonal rhythms of body mass, gonadal activity and fur color (2). Despite the importance of the hypothalamus for the physiology of *Phodopus sungorus*, our knowledge of morphological parameters of hypothalamic nuclei in these animals is still limited. In particular, only little information about the distribution of neuropeptides and the mechanisms of their regulation is available.

To characterize the distribution of neuropeptides in the hypothalamus of Djungarian hamsters, studies are carried out in our laboratory in male and female animals held under either long or short photoperiods. The aim was to provide an analysis of immunoreactivity for arginine-vasopressin (AVP), bombesin (BBS), cholecystokinin (CCK), corticotropin-releasing factor (CRF), luteinizing hormone-releasing hormone (LHRH), neuropeptide Y (NPY), neurotensin (NT), oxytocin (OT), peptide histidine-isoleucin (PHI), somatostatin (SS), and vasoactive intestinal polypeptide (VIP). In the present paper, the attention is focused on the suprachiasmatic nuclei.

MATERIAL AND METHODS

Adult female and male Djungarian hamsters (*Phodopus sungorus*), held under either long photoperiod (light/dark (LD) 16:8, lights on at 04:00 h) or short photoperiod (LD 8:16, lights on at 08:00 h), were killed by perfusion with a fixative at the middle of the light period (11:30-12:30 h). Brain regions containing the SCN were sectioned serially in the coronal plane at 40 µm. Adjacent sets of sections were incubated in primary antisera and processed for routine immunohistochemistry. Care was taken that sections from hamsters held under different lighting conditions were treated simultaneously under identical conditions. For further details, the interested reader is referred to previous publications (6,7).

RESULTS

The SCN of *Phodopus* as seen in Nissl-stained sagittal and frontal sections is given in Fig.1.

The immunocytochemical analysis of cell and fiber distribution showed distinct patterns of immunoreactivity for the antisera employed. It was evident that some of the substances are seen in perikarya, while others occur only in fibers in the SCN. In particular, cell bodies were stained with antisera to AVP, BBS, CCK, NT, PHI, SS, and VIP. SS-LI was relatively weak; in contrast, AVP- and VIP-LI was strong. In these cases, also the numbers of immunoreactive fibers were high. Furthermore, NT as well as the satiety peptides BBS and CCK were seen in fairly high amounts, especially in the medial aspect of the nucleus. NPY-LI in the SCN region was detected in fibers and terminals only; the staining, however, was very strong. In contrast, LHRH-LI in the SCN was relatively scarce and was seen predominantly in fibers. CRF and OT were not found at all within the borders of the nucleus.

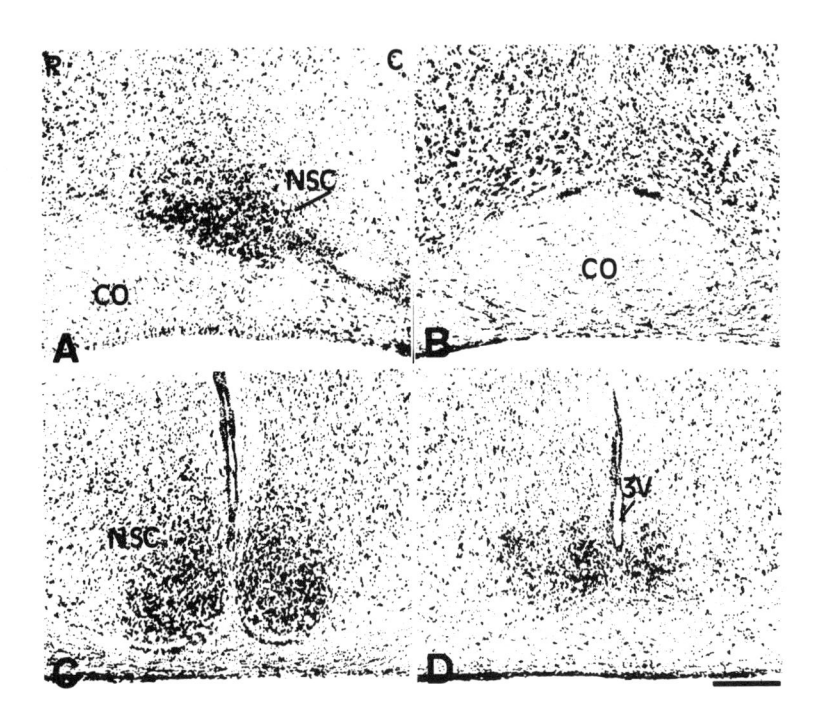

Fig.1: Nissl-stained sections of the *Phodopus* hypothalamus in the plane of the SCN. (A) sagittal section, (B)-(D) frontal sections of the rostral (B), medial (C) and caudal (D) aspects of the nucleus. Abbreviations: C caudal, CO optic chiasm; NSC suprachiasmatic nucleus; R rostral, 3V third ventricle. Calibration bar, 200 μm (A-D).

With regard to photoperiodic regulation, it was observed that some of the peptides investigated (BBS, SS) were seen in relatively constant amounts. In contrast, staining for other substances (CCK, NT, and, to a minor degree, AVP) was decreased in both number and intensity in short photoperiod animals, while PHI, VIP (and LHRH) immunoreactivity was augmented in both quantity and quality under these conditions. The distribution of NT and BBS in the SCN are given in Fig.2 as examples for neuropeptide-LI which are (NT) or are not (BBS) influenced by photoperiodic changes.

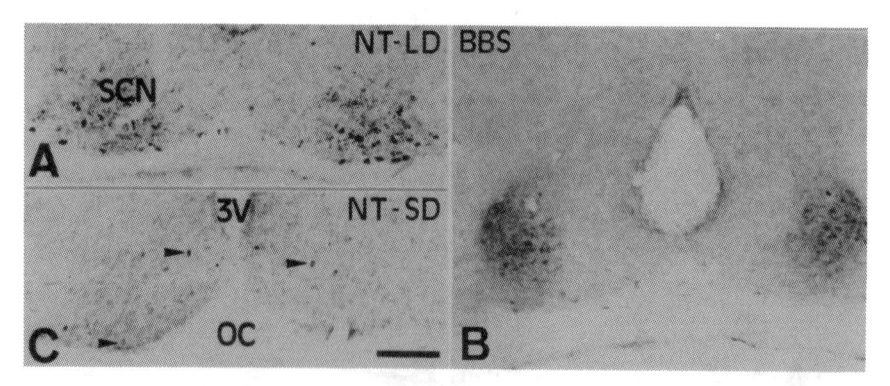

Fig.2: Distribution of neurotensin- (NT) and bombesin- (BBS) LI in medial sections of the SCN of *Phodopus* held under long (A,B) or short (C) photoperiods. Note the numerical decrease of NT immunoreactive neurons (arrowheads) in short-day animals (C). The plane of this section corresponds to the one shown in (A). In contrast to NT, BBS-LI is not significantly altered between LD and SD animals. Abbreviations: LD long day, SD short day, OC optic chiasm, 3V third ventricle. Calibration bar, 100 µm (A-C).

DISCUSSION

The understanding of the neural mechanisms of circadian rhythmicity requires advanced knowledge of the functional organization of the SCN. The present study provides evidence that multiple neuropeptides are synthesized in the *Phodopus* SCN and that some of these substances are regulated by photoperiod, i.e. exhibit seasonal changes.

The importance of BBS, VIP, and PHI for the mechanisms of rhythm regulation is underlined by the fact that cell bodies staining for these substances were restricted to the SCN and were not observed in other hypothalamic nuclei. In addition, with the exception of very few positive cells in the PVN, CCK-LI perikarya in *Phodopus sungorus* were found in the SCN only. Furthermore, the immunoreactive staining for most of the peptides is strongly suggesting that the contents of these neuropeptides are considerably high in the *Phodopus* SCN.

A considerable part of neuropeptidergic perikarya are located in the ventral aspect of the nuclei which are known to receive retinal efferents. Although there is only one report in rats describing retinal afferents to innervate VIP-LI cell bodies in the SCN (3), and a direct retinal projection to immunoreactive cells in the *Phodopus* SCN has not been demonstrated yet, there is evidence for photic regulation of the expression of some of the peptides since, in the present study, amount and intensity of immunoreactivity for CCK, NT, PHI, VIP, (AVP and LHRH) are altered in hamsters held under short-day conditions. These changes in immunoreactivity are thought to reflect long-range phases in activity rather than circadian effects since all animals were killed at the same time of the day (i.e. the middle of the light period under both long- and short-day conditions). It is, however, open whether the observed changes are due to alterations in synthesis or in secretion of these peptides. Interestingly in the ground squirrel SCN, immunoreactivity to AVP, NPY, VIP, and substance P exhibit striking differences between hibernating and non-hibernating animals (9).

In conclusion, the present results reveal occurrence and differential regulation by photoperiod of neuropeptide-LI in the SCN of *Phodopus sungorus*. While some neuropeptides in this hypothalamic nucleus apparently are not under the influence of daylength, the synthesis and/or secretion of others appear to be regulated by light. Photoperiodic manipulation, therefore, may thus be an interesting technique for the investigation of peptide regulation in the hypothalamus.

ACKNOWLEDGEMENT

The author thanks U.Göringer-Struwe, I.v.Graevenitz, P.Hödl and A.Thomas-Semm for expert technical assistance. This study has been financially supported by the Naturwissenschaftlich-Medizinisches Forschungszentrum (Mainz).

REFERENCES

1. CARD, J.P. & R.Y. MOORE (1984): The suprachiasmatic nucleus of the golden hamster: Immunohistochemical analysis of cell and fiber distribution. Neuroscience 13, 415-431
2. HOFFMANN, K. (1979): Photoperiod, pineal, melatonin and reproduction in hamsters. Progr. Brain Res. 52, 397-415
3. IBATA, Y., Y. TAKAHASHI, H. OKAMURA et al. (1989): Vasoactive intestinal peptide (VIP)-like immunoreactive neurons located in the rat suprachiasmatic nucleus receive a direct retinal projection. Neurosci. Lett. 97, 1-5
4. MOORE, R.Y. (1983): Organisation and function of a central nervous system circadian oscillator: the suprachiasmatic hypothalamic nucleus. Fed. Proc. 42, 2783-2788
5. PICKARD, G.E. (1982): The afferent connections of the suprachiasmatic nucleus of the golden hamster with emphasis on the retinohypothalamic projection. J. Comp. Neurol. 211, 65-83
6. REUSS, S. (1991): Photoperiod effects on bombesin- and cholecystokinin-like immunoreactivity in the suprachiasmatic nuclei of the Djungarian hamster (*Phodopus sungorus*). Neurosci. Lett. 128, 13-16
7. REUSS, S., E.C. HURLBUT, J.C. SPEH et al. (1989): Immunohistochemical evidence for the presence of neuropeptides in the hypothalamic suprachiasmatic nucleus of ground squirrels. Anat. Rec. 225, 341-346
8. RUSAK, B. & I. ZUCKER (1979): Neural regulation of circadian rhythms. Physiol. Rev. 59, 449-526
9. SCHINDLER, C.U. & F. NÜRNBERGER (1990): Hibernation-related changes in the immunoreactivity of neuropeptide systems in the suprachiasmatic nucleus of the ground squirrel, *Spermophilus richardsonii*. Cell Tissue Res. 262, 293-300
10. VAN DEN POL, A.N. & K.L. TSUJIMOTO (1985): Neurotransmitters of the hypothalamic suprachiasmatic nucleus: immunocytochemical analysis of 25 neuronal antigens. Neuroscience 15, 1049-1086

ON THE MECHANISM OF CYST INDUCTION IN GONYAULAX POLYEDRA: ROLES OF PHOTOPERIODISM, LOW TEMPERATURE, 5-METHOXYLATED INDOLEAMINES, AND PROTON RELEASE

I. BALZER & R. HARDELAND

I. Zoologisches Institut, Universität Göttingen, Berliner Str. 28, D-3400 Göttingen, FRG

Several species of dinoflagellates are known to form cysts under conditions which are, from an ecological point of view, considered to be unfavorable (DALE 1983, 1,19). We have recently discovered that also *Gonyaulax polyedra* can enter such a dormant stage (3). According to morphological criteria, the resting cells which we observed were of the type of so-called 'thin-walled' cysts (14,19). Transformation of the *Gonyaulax* mastigote cell into a cyst (cf. Fig.1) is a complex process including cessation of movement, loss of flagella, retraction from the theca, intracellular rearrangements of organelles, and formation of a spherical cyst wall (BALZER, BEHRMANN & HARDELAND, unpubl.).

In this paper, we demonstrate that encystment of *Gonyaulax polyedra* can be induced by short-days in combination with lower temperatures. This effect can be mimicked by 5-methoxylated indoleamines, which are present in this organism (5,6,16), and by electroneutral protonophores.

MATERIALS AND METHODS

Gonyaulax polyedra was grown in a modified f/2 medium (12), at 20°C, in LD 12:12. Photoperiodic experiments were carried out at either 20 or 15°C, under various LD schedules, and at an illumination intensity of 400 lx. Cells were transferred to experimental conditions at CT 3.

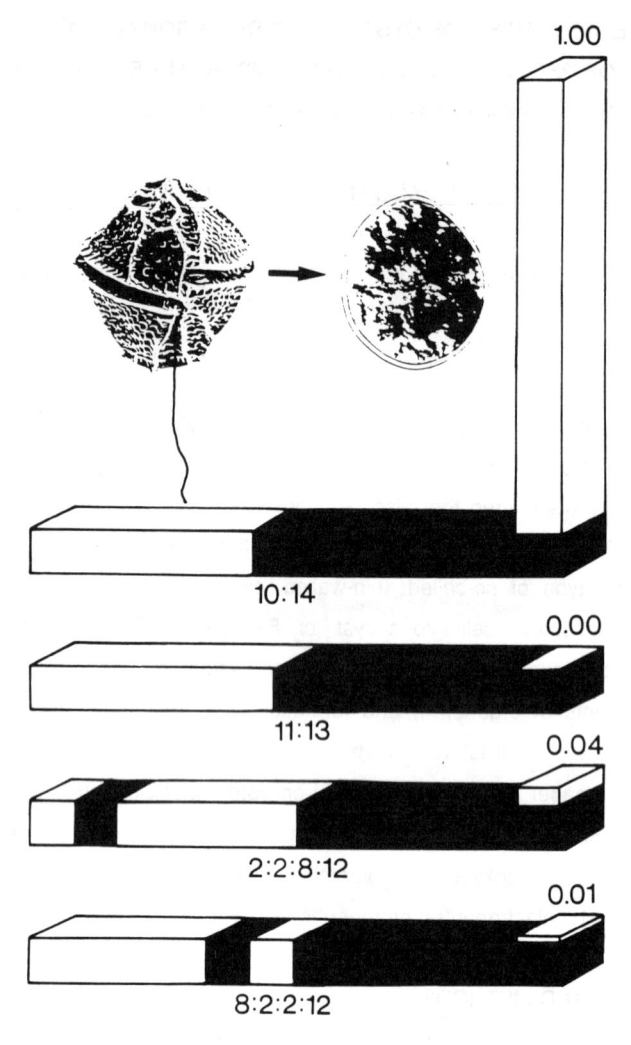

Fig.1: Photoperiodic cyst induction and its suppression by light breaks. Experiments were carried out at a reduced temperature of 15°C. Photographs show mastigote cell and thin-walled cyst. Horizontal bars: lighting schedules; vertical bars: ratio of cysts formed within 4 days by total cell number.

Measurements of bioluminescence were carried out at 20°C, in a scintillation spectrometer, as previously described (10). In these experiments, including concomitant screenings of cyst formation, cells were transferred at CT 12 to DD. Bioluminescence data are presented as 'stimulation factors' (cf. 4).

Stock solutions of agents were prepared directly before use as follows: 5-methoxytryptamine: 0.2 M in DMSO; monensin: 0.5 mg/200 µl ethanol; nigericin: 0.5 mg/200 µl DMSO/ethanol 1:1; 1,2-dihydro-4-hydroxy-6-methoxy-N-methylquinoline: 1 mM in culture medium. Solutions were further diluted with culture medium to give the desired molarities.

1,2-Dihydro-4-hydroxy-6-methoxy-N-methyl-quinoline (DHMMQ) was synthesized by our co-worker H.-P. KUBIS (13) on the basis of a procedure described by BACK (2) for preparation of methoxylated kynuramines. The product was identified by both mass spectroscopy and NMR.

RESULTS AND DISCUSSION

Cysts can be induced by photoperiodic or pharmacological treatments. The efficiency of the respective method depends, however, on temperature. A temperature lower than 16°C seems to represent a precondition for cyst induction by short-days and also by melatonin. At 20°C, even an LD of 6:18 remains without effect. After exposure to 15°C, as shown in Fig.1, an LD of 10:14 causes all cells to encyst within, at latest, 4 days. With regard to the critical day-length, photoperiodic time measurement appears to be relatively precise, since already in LD 11:13 no cyst formation is observed.

Decreasing the length of photoperiod does not lead to encystment simply by light deficiency, but rather represents a short-day effect in the strict sense. This can be demonstrated by night-interruption experiments, in which the overall duration of lighting is 10 h/day (e.g., LDLD 2:2:8:12 or LDLD 8:2:2:12; Fig.1). Although the amount of light was small as in the cyst-inducing LD 10:14, the number of cysts has remained relatively low, also over extended observation periods. Therefore, the temporal distance between beginning of the first and end of the second lighting periods have been decisive for the behavior of the cells, which obviously have interpreted these lighting schedules as long-days.

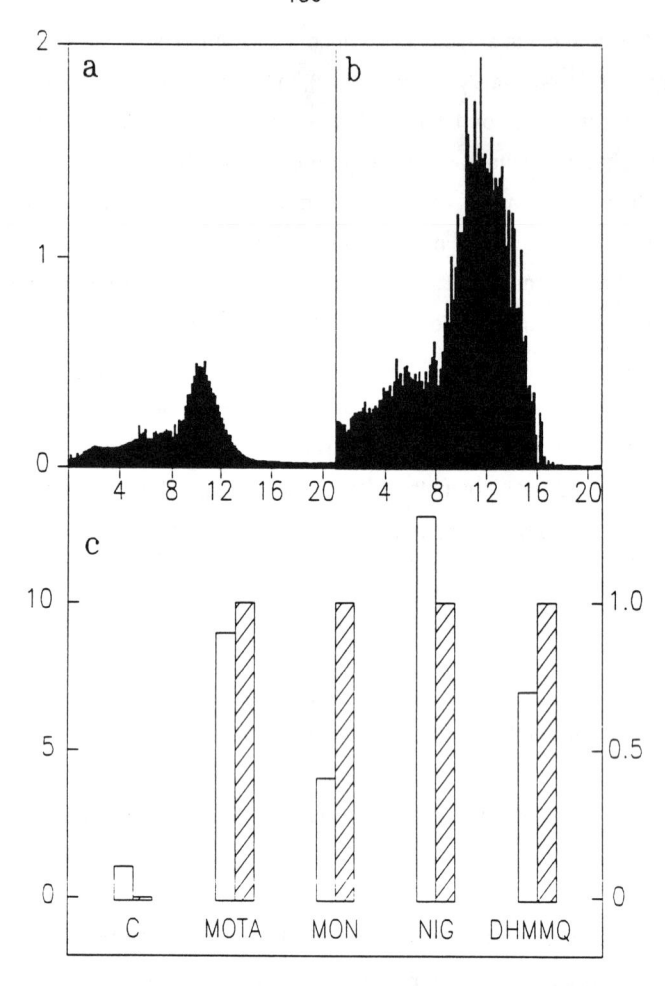

Fig.2: Stimulation of bioluminescence as an early event in cyst induction. Experiments were carried out at the rearing temperature of 20°C. a: Circadian rhythm of bioluminescence in DD (control); b: effect of 2 X 10-7 M monensin on bioluminescence, added at CT 12; c: comparison of effects on bioluminescence and cyst formation; white bars: bioluminescence; hatched bars: cyst ratio (cf. Fig.1); C: control; MOTA: 2 X 10^{-5} M 5-methoxytryptamine; MON: 2 X 10^{-7} M monensin; NIG: 2 X 10^{-7} M nigericin; DHMMQ: 5 X 10^{-5} M 1,2-dihydro-4-hydroxy-6-methoxy-N-methyl-quinoline. Ordinates:(upper: bioluminescence, in millions of cpm; lower left: stimulation factor of bioluminescence; lower right: cyst ratio; abscissae: h in DD.

On the other hand, at 15°C a sub-critical photoperiod of 11 h can be transformed into a subjective short-day by a single addition of 10^{-4} M melatonin given 1 h before the end of the photoperiod. It is worth noting that the concentration applied may be different from that reaching the cells, since melatonin is rapidly degraded after exposure to light, due to the reaction with light-induced superoxide radicals (cf. 11). Therefore, the threshold of the inducing concentration in the cell may be, in fact, much lower.

An analogue of melatonin, 5-methoxytryptamine, seems to be more potent than melatonin in inducing cyst formation (cf. Fig.2). Within a few hours and even in LD 12:12 and at 20°C this substance leads to quantitative encystment. The potency of 5-methoxytryptamine to induce cyst formation may be related to another action previously described, namely the stimulation of bioluminescence (4,5,6). Light emission can be enhanced up to 55-fold at a concentration of 10^{-4} M and minor stimulations are dectable down to 2×10^{-8} M.

We suspected that the increase of bioluminescence might be related to cyst formation. In *Gonyaulax*, bioluminescence is controlled by the transfer of protons from acidic vacuoles to the cytoplasmic side of the vacuole membrane, where the protons cause the release of luciferin from luciferin-binding protein (9). An acidic vacuole requires the permanent expense of ATP to keep a high proton gradient. A resting stage may be unable to maintain this proton gradient, because of the greatly reduced ATP production. It might be economic for the cell to release the protons from the vacuole prior to entrance into dormancy. Therefore, proton release, as detectable by stimulation of bioluminescence, could represent an early step during transformation into a cyst.

The decrease of cytoplasmic pH induced by the indoleamine as an early event in cyst formation might be causal for the later steps of encystment. In fact, we observed that stimulation of bioluminescence was followed by cyst formation also with other agents, such as the quinoline DHMMQ (Fig.2). This view is supported by experiments with two electroneutral protonophors, the Na^+/H^+ antiporter monensin, and the K^+/H^+ antiporter nigericin showing again the typical sequence of enhanced bioluminescence and encystment (Fig.2). With both substances, cyst induction was possible down to concentrations of 10^{-7} M (data not shown). This interpretation would fit well the more general experience that dormant or inactive cellular stages are associated with a lower intracellular pH in many different organisms (15), e.g. in unfertilized eggs of sea urchins (18,21) or frogs (20), in cysts of the shrimp *Artemia salina* (8), in spores of several Bacillus species (17) and of the yeast *Pichia pastoris* (7).

REFERENCES

1. ANDERSON, D.M., A.W. WHITE & D.G. BADEN (eds.) (1985): Toxic Dinoflagellates. Elsevier, N.Y.-Amsterd.-Oxford
2. BACK, W. (1970): Über die Synthese von o-Acylamino-ß-dimethylaminopropiophenonen. Arch. Pharmaz. 304, 27-31
3. BALZER, I. & R. HARDELAND (1991a): Induction of cyst formation by short photoperiods and 5-methoxylated indoleamines in a dinoflagellate, Gonyaulax polyedra. Eur. J. Cell Biol. 54, Suppl. 32, p. 57
4. BALZER, I. & R. HARDELAND (1991b): Stimulation of bioluminescence by 5-methoxylated indoleamines in the dinoflagellate Gonyaulax polyedra. Comp. Biochem. Physiol. 98C, 395-397
5. BALZER, I., B. PÖGGELER & R. HARDELAND (1989): Indoleamines in the circadian organization of Gonyaulax: evidence for presence and effects on bioluminescence. J. Interdiscipl. Cycle Res. 20, 169
6. BALZER, I., B. PÖGGELER & R. HARDELAND (1990): 5-Methoxylated indoleamines in the dinoflagellate Gonyaulax polyedra: presence, stimulation of bioluminescence, and circadian variations in responsiveness. Eur. J. Cell Biol. 51, Suppl. 30, p. 52
7. BARTON, J.K., J.A. den HOLLANDER, T.M. LEE et al. (1980): Measurement of internal pH of yeast spores by 31P nuclear magnetic resonance. Proc. Natn. Acad. Sci. USA 77, 2470-2473
8. BUSA, W.B. (1982): Cellular dormancy and the scope of pH_i-mediated metabolic regulation. In: R. NUCCITELLI & D.W. DEAMER (eds.): Intracellular pH: Its Measurement, Regulation, and Utilization in Cellular Functions. Alan Liss, N.Y., pp. 417-426
9. DUNLAP, J.C., W. TAYLOR & J.W. HASTINGS (1981): The control and expression of bioluminescence in dinoflagellates. In: K.H. NEALSON (ed.): Bioluminescence: Current Perspectives. Burgess, Minneapolis, pp. 108-124.
10. HARDELAND, R. (1980): Effects of catecholamines on bioluminescence in Gonyaulax polyedra (Dinoflagellata). Comp. Biochem. Physiol. 66C, 53-58
11. HARDELAND,R., B.PÖGGELER, I.BALZER & G.BEHRMANN (1992): A hypothesis on the evolutionary origins of photoperiodism based on circadian rhythmicity of melatonin in phylogenetically distant organisms. In: GUTENBRUNNER,Chr., G. HILDEBRANDT & R. MOOG (eds.): Chronobiology & Chronomedicine, eds. . Peter Lang, Frankf./M. - Bern - N.Y. - Paris, 115-123

12. HOFFMANN, B. & R. HARDELAND (1985): Membrane fluidization by propranolol, tetracaine and 1-aminoadamantane in the dinoflagellate, Gonyaulax polyedra. Comp. Biochem. Physiol. 81C, 39-43
13. KUBIS, H.-P. (1991): Chronobiologische Untersuchungen an Neurospora crassa im Zusammenhang mit der Rolle von Tryptophan-Metaboliten. Diploma Thesis, Göttingen
14. MATSUOKA, K., Y. FUKUYO & D.M. ANDERSON (1989): Methods for modern dinoflagellate studies. In: T. OKAICHI, D.M. ANDERSON & T. NEMOTO (eds.): Red Tides. Biology, Environmental Science, and Toxicology. Elsevier, N.Y.-Amsterd.-Lond., 461-479
15. NUCCITELLI, R. & J.M. HEIPLE (1982): Summary of the evidence and discussion concerning the involvement of pH_i in the control of cellular functions. In: R. NUCCITELLI & D.W. DEAMER (eds.): Intracellular pH: Its Measurement, Regulation, and Utilization in Cellular Functions. Alan Liss, N.Y., 567-586
16. PÖGGELER, B., I. BALZER, R. HARDELAND & A. LERCHL (1991): Pineal hormone melatonin oscillates also in a unicellular, the dinoflagellate Gonyaulax polyedra. Naturwissenschaften, in press
17. SETLOW, B. & P. SETLOW (1980): Measurements of the pH within dormant and germinated bacterial spores. Proc. Natn. Acad. Sci. USA 77, 2474-2476
18. SHEN, S.S. & R.A. STEINHARDT (1978): Direct measurement of intracellular pH during metabolic derepression of the sea urchin egg. Nature (Lond.) 272, 253-254
19. TAYLOR, F.J.R. (ed.) (1987): The Biology of Dinoflagellates. Blackwell, Oxford
20. WEBB, D.J. & R. NUCCITELLI (1982): Intracellular pH changes accompanying the activation of development in frog eggs: Comparison of pH microelectrodes and 31P-NMR measurements. In: R. NUCCITELLI & D.W. DEAMER (eds.): Intracellular pH: Its Measurement, Regulation, and Utilization in Cellular Functions. Alan Liss, N.Y., 293-324
21. WINKLER, M.W. (1982): Regulation of protein synthesis in sea urchin eggs by intracellular pH. In: R. NUCCITELLI & D.W. DEAMER (eds.): Intracellular pH: Its Measurement, Regulation, and Utilization in Cellular Functions. Alan Liss, N.Y., 325-340

ARE NEUROSECRETORY CELLS INVOLVED IN THE CIRCADIAN CONTROL OF LOCOMOTOR ACTIVITY OF DROSOPHILA MELANOGASTER?

C. HELFRICH-FÖRSTER

Institut für Botanik, Auf der Morgenstelle 1, D-W 7400 Tübingen, FRG

INTRODUCTION

To understand the organization of multicellular circadian systems it is important to find out where the underlying pacemakers are located and by which pathways the pacemakers control locomotor activity. Among insects best understood in this respect are cockroaches and crickets. In both cases the pacemakers controlling locomotor activity could be traced to the optic lobes and rhythmic signals are mediated electrically through neuronal pathways (1,2).

Nevertheless a humoral involvement can not be excluded: In both insect groups locomotor activity becomes arrhythmic when the neurosecretory cells of the pars intercerebralis are destroyed (3,4), and in a cricket an arrhythmic animal could be made rhythmic by transplanting the brain of a rhythmic donor into its abdomen (5).

In *Drosophila* it is not yet known whether electric coupling through neuronal pathways is involved, but humoral pathways seem to participate in the coupling between pacemakers and overt rhythm: by transplantation of a brain from a rhythmic fly into the abdomen of an arrhythmic per^0 mutant rhythmicity could be induced in some of the recipients (6). Candidates for mediating the rhythmic information from the donor brain to the recipient are neurosecretory cells. This possibility would be in agreement with the finding that, in LD, neurosecretory cells of the pars intercerebralis show a bimodal rhythm of neurosecretion (7) which is paralleled by a bimodality in locomotor activity.

In the present study the neurosecretory system in the brain of the wildtype (Canton S) was compared with the following mutants:

The period mutants per^l, per^s, per^0, and the blind mutants *sine oculis* (*so*), *small optic lobes*; *sine oculis* (*sol;so24*) and *disconnected* (*disco*).

In *so* the optic lobes are reduced to about 20%, the flies show normal locomotor activity rhythms (8). In *sol;so* the optic lobes are reduced to less than 5%, the flies show a split locomotor activity rhythm (9). In *disco* the photoreceptor cells are disconnected from the brain (10). They have either tiny rudiments of the optic lobes (unconnected phenotype) or normal sized but grossly disorganized optic lobes (connected phenotype). In both phenotypes circadian rhythmicity is severely disturbed or completely absent (11,12). In the three strains used (attXyf/wdisco(2), disco1656/FM7a, and disco(1)/FM7) the *disco* mutation is linked to marker mutations which allow to distinguish them from non disco-phenotypes in the same stock. The latter served as internal control.

METHODS

Flies were grown on standard corn meal food under LD 12:12 at $20^{\circ}C$. At circadian time CT0-CT6 mutant and wildtype adults of the same age (10 to 50 days) were dissected for histological studies. They were processed together to allow the comparison of the staining intensity. The brains, after removal of fat body and trachea, were fixed in Bouin's fluid. They were stained as a whole with paraldehyde fuchsin according to DOGRA & TANDAN (13).

RESULTS

Distinct bilaterally distributed groups of neurosecretory cells could be stained and were designated as groups I to V (Fig. 1):

- One large group in the pars intercerebralis containing 10-14 cells (group I),
- a small group (1-2 cells) lateral to the pars intercerebralis (group II; an additional single cell lying at the distal border of the dorsal protocerebrum was included in this group),

- two groups in the posterior part of the brain, one of which located close to the mushroom body (group III, 1-3 cells),
- the other more ventral in the neighborhood of the oesophagus (group IV, 1-3 cells),
- and one group in the lateral anterior part between medulla and protocerebrum (group V).

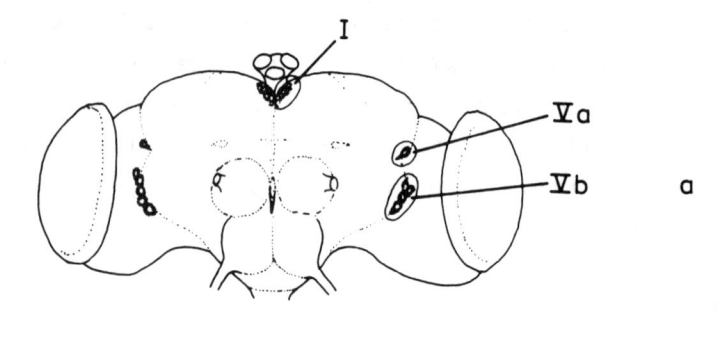

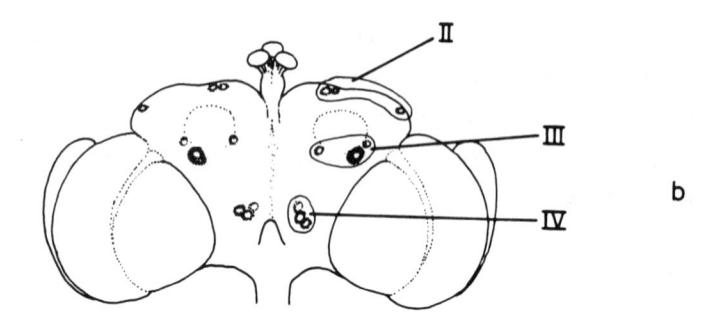

Fig.1: Schematic presentation of neurosecretory cells in the brain of *Drosophila melanogaster* as detected by paraldehyde fuchsin staining.
a) frontal view of the brain (groups I and V)
b) caudal view of the brain (groups II,III,IV)

Of group V, a single neurosecretory cell (Va) lies more dorsally and close to the anterior optic tract and 3-8 cells (Vb) are arranged in a cluster or a row along the boundary between medulla and protocerebrum at the level of the oesophagus.

All neurosecretory cells appear "doughnut-shaped", due to stained cytoplasm and unstained nuclei. The staining intensity increased with the age of the flies. The cells of group V stained most intensively.

The neurosecretory cell groups I to IV appeared normal in all of the mutants studied. For group V no difference in the number of neurosecretory cells were found between wildtype, per^0, per^s, per^l and so. In the mutant sol;so, however, the number of cells in group Vb is reduced (maximally 3 cells found). In 15.6% of the flies no cells of group Vb were detected. Cell Va was always present. The three disco strains did not show any differences in staining and were, therefore, evaluated as one strain. Cells of group Vb were normal in control flies but poorly detectable in those bearing the disconnected phenotype. In 46.9% of the flies no cells were detected. In the remaining 53.1% maximally 3 cells were stained. The cell Va was detectable in 50% of the cases. No differences in staining were found between the connected and unconnected phenotype.

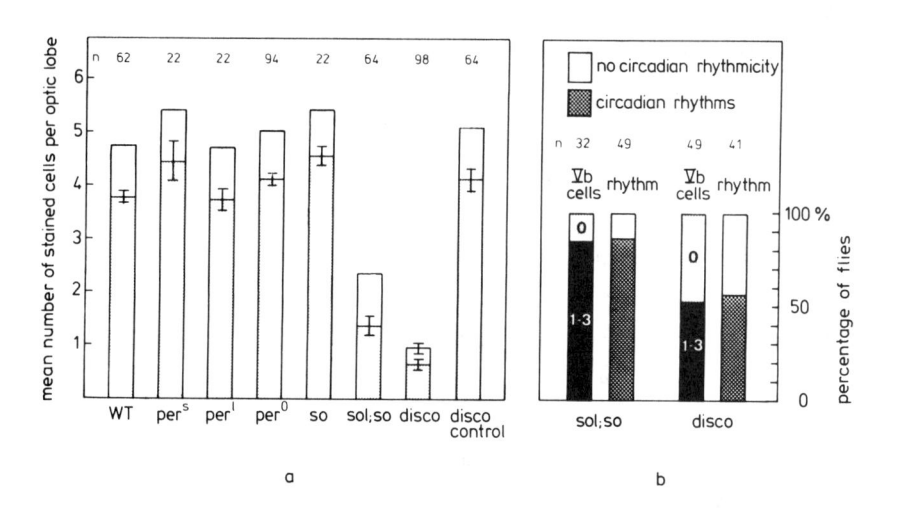

Fig.2: a) Mean number of stained neurosecretory cells in group Va (white area) and Vb (grey area) for all strains studied (±SE); n: number of optic lobes.
b) Percentage of flies with 1-3 (black area) or without (fair dotted area) neurosecretory cells of group Vb for sol;so and disco. The percentage of flies with (dark dotted area) or without (white area) clear circadian components in their locomotor activity are shown for comparison (data derived from (9) and (12)). n: number of flies.

Fig. 2a summarizes the mean number of stained cells in Va and Vb for all strains. Fig. 2b shows the percentage of flies with or without cells of group Vb for *sol;so* and *disco*.

DISCUSSION

In *Drosophila* five distinct groups of neurosecretory cells were stained by paraldehyde fuchsin. No obvious differences in location and number of stained cells were found for group I-IV between wildtype and the mutant strains studied.

Clear differences were, however, found in the number of stained cells in group V between wildtype flies and the mutants *sol;so* and *disco*.

For *sol;so*, in 15.6% of the flies no cells of group Vb were found and in the remaining 84.4% the number of stained cells is reduced. Interestingly recordings of locomotor activity showed that 14.3% of *sol;so* flies where arrhythmic and 85.7% showed a split activity rhythm (9).

For *disco*, in 46.9% of the flies no cells of group Vb could be detected and in the remaining 53.1% their number is reduced. Again these findings are paralleled by results obtained for locomotor activity: about 44% of *disco* flies had no circadian rhythmicity, whereas in the remaining 56% some circadian components are present, as shown by MESA analysis (12).

The striking congruence between number of neurosecretory cells in group Vb and expression of locomotor activity rhythm in both mutants (Fig. 2b) might indicate an involvement of these cells in the circadian system. This is strengthened by the fact that they are located in the same position (between medulla and protocerebrum) in which a group of cells could be labelled by an antibody against the period protein (14). In the mutant *disco* the period protein is expressed in other tissues (e.g. glia cells and photoreceptor cells), but poor labelling by the period antibody was found between medulla and protocerebrum (14).

Thus the question arises whether the neurosecretory cells of group V are identical with the period protein containing cells between medulla and protocerebrum, or whether there is at least some kind of interaction between them. For the expression of rhythmic locomotor activity, both, a cycling period protein and the presence of the neurosecretory cells of group V might be necessary: In *disco* mutants the period protein is expressed but in about 47% of the flies the neurosecretory cells of group Vb are completely absent. In per^0 mutants the neurosecretory cells are present, but the period protein is missing. In both cases arrhythmicity is found.

The relationship between the cycling period protein and the neurosecretory cells is still unknown. In the scorpion the efferent neurosecretory fibers in the visual system serve as the pathways for circadian signals (15) and they could be labelled by the antibody against the period protein (FLEISSNER, personal communication).

Further studies are needed to show whether, in *Drosophila*, period protein containing cells and neurosecretory cells are identical and what role they play in the circadian system.

REFERENCES

1. COLWELL, S.C. & T.L. PAGE (1990): A circadian rhythm in neural activity can be recorded from the central nervous system of the cockroach. J. Comp. Physiol. A166, 643-649
2. TOMIOKA, K. & Y. CHIBA (1986): Circadian rhythm in the neurally isolated lamina-medulla-complex of the cricket, Gryllus bimaculatus. J. Insect Physiol. 32, 747-755
3. NISHIITSUTSUJI-UWO, J. et al. (1967): Central nervous system control of circadian rhythmicity in the cockroach. I. Role of the pars intercerebralis. Biol. Bull. 133, 679-696
4. SOKOLOVE, P.S. & W. LOHER (1975): Role of the eyes, optic lobes, and pars intercerebralis in locomotory and stridulatory circadian rhythms of Teleogryllus commodus. J. Insect Physiol. 21, 785-799
5. CYMBOROWSKI, B. (1981): Transplantation of a circadian pacemaker in the house cricket, Acheata domesticus L. J. interdiscipl. Cycle Res. 12, 133-140
6. HANDLER, A.M. & R.J. KONOPKA (1979): Transplantation of a circadian pacemaker in Drosophila. Nature 279, 236-238
7. RENSING, L. (1966): Zur circadianen Rhythmik des Hormonsystems von Drosophila. Z. Zellforsch. 74, 539-558
8. HELFRICH, C. & W. ENGELMANN (1983): Circadian rhythm of the locomotor activity in Drosophila melanogaster and its mutants "sine oculis" and "small optic lobes". Physiol. Entomol. 8, 257-272
9. HELFRICH, C. (1986): Role of the optic lobes in the regulation of the locomotor activity rhythm of Drosophila melanogaster. Behavioral analysis of neural mutants. J. Neurogenetics 3, 321-343
10. STELLER, H., K.F. FISCHBACH & G.M. RUBIN (1987): Disconnected: a locus required for neuronal pathway formation in the visual system of Drosophila. Cell 50, 1139-1153

11. DUSHAY, M.S. et al. (1989): The disconnected visual system mutations in Drosophila melanogaster drastically disrupt circadian rhythms. J. Biol. Rhythms 4, 1-27
12. DOWSE, H.B. et al. (1989): High resolution analysis of locomotor activity rhythms in disconnected, a visual system mutant of Drosophila melanogaster. Behav. Genet. 19, 529-542
13. DOGRA, G.S. & B.K. TANDAN (1964): Adaptation of certain histological techniques for in situ demonstration of the neuro-endocrine system of insects and other animals. Quart. J. Microsc. Sci. 105, 455-466
14. ZERR, D.M. et al. (1990): Circadian fluctuations of period protein immunoreactivity in the CNS and the visual system of Drosophila. J. Neurosc. 10, 2749-2762
15. FLEISSNER, G. (1983): Efferent neurosecretory fibers as pathways for circadian clock signals in the scorpion. Naturwiss. 70, 366-367

ON THE CIRCADIAN RHYTHMS IN MICROPHTHALMIC MICE

H. HILBIG

Institut für Hirnforschung, Emillienstrasse 14, D-O 7010 Leibzig, FRG

In the present paper we report results obtained on microphthalmic mice (mi/mi).This mutation was first described by HERTWIG (1). It is caused by irradiation with X-rays. Heterozygoses of this strain appear normal with grey skins und iris pigmentation, but they produce litters in which typically between 2 and 5 offspring exhibit lack of pigmentation, anomalies of the skeleton, curled whiskers and bilateral colobomatous microphthalmia. These are mi/mi homozygoses. They can be found in the litter by the lack of irispigmentation after birth. Fig. 1 shows such animals on pd 1. All littermates together were reared under natural conditions and developed a normal ontogenesis of reflex-, open-field and swimming behavior. The acoustical, the olfactory and the tactile senses were normally developed.

On pd 15, the day of eye opening there was the question to answer if the microphthalmic animals are blind. The result of our EEG experiments shows clearly that no potentials are evoked in the visual cortex by a flash. Also there is no response in the visual cortex of mi/mi mice on an acoustical signal (Fig. 2).

In our research on the visual system of these mice we found very differentiated changes in the Corpus geniculatum laterale, pars dorsalis (CGLd), in the superficial layers of superior colliculus and in the visual cortex.

In the suprachiasmatic nucleus we found no degenerating axon terminals of mi/mi mice after enucleation. There were no connections between the retina and the suprachiasmatic nucleus and between the retina and the other visual nuclei.

On pd 21, after weaning, we made groups whether microphthalmic mice only or control mice only. The locomotor activity of these groups were registered by the Animex system.

In both cases regular circadian patterns were found, but only the pattern of the healthy control mice was synchronized by the day-night-periodicity (Fig.2).

Fig.1: Microphthalmic (above) and control mouse on pd 1, note the iris pigment!

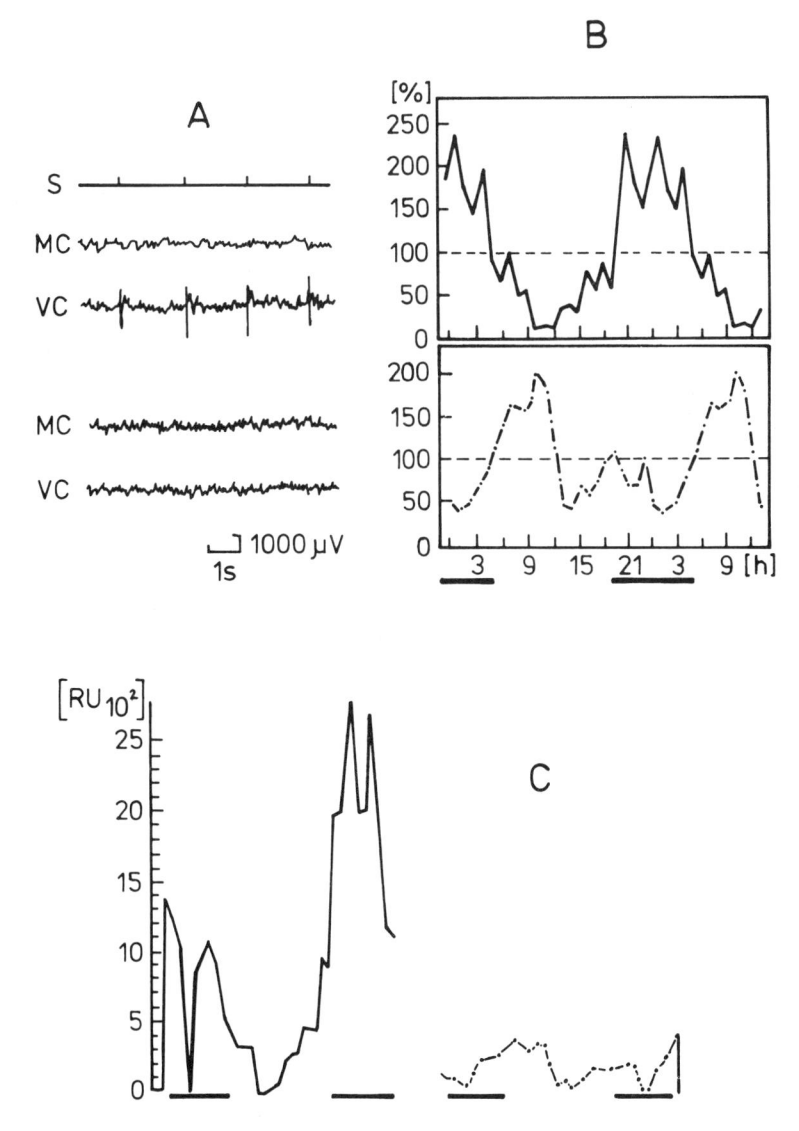

Fig.2: A) EEG recordings in the Visual cortex (VC) and in the Motorcortex (MC) as a reference, S = Signals of light; below: microphthalmic, above: control mouse.
B) Pattern of the circadian rhythm in % of the 100%-level in control (above) and microphthalmic mice (below).
2C) Original registration units of locomotor activity in control (left) and microphthalmic mice (right).

The blind mi/mi mice had a free-running pattern with a spontaneous period of about 25 hours.

The level of activity was 4 times lower than that of the control mice. The results indicate that the suprachiasmatic nucleus shows a normal function in mi/mi-mice. We expect a fully normal circadian pattern of locomotor activity in a mixed group, that means in a group of mi/mi and control mice. The perception of light is necessary for its influence as a Zeitgeber but it could be influenced by social contacts too.

An explanation of the reduced amplitude of locomotor activity can be the lower level of Na,K-ATPase and AChE in the brain of microphthalmic animals. We found a decrease between pd 20 and pd 25 in mi/mi-mice and an increase in controls.

REFERENCES

1. HERTWIG, P. (1942): Neue Mutationen und Kopplungsgruppen bei der Hausmaus Z. indukt. Abstamm.- Vererb.- Lehre. 80, 220-246

CONSEQUENCES OF MODIFIED ZEITGEBER- CONDITIONS DURING JUVENILE PHASE FOR THE CIRCADIAN SYSTEM OF ADULT MICE

D. WEINERT, H. EIMERT & J. SCHUH

Martin-Luther-University Halle-Wittenberg, Institute of Zoology, Domplatz 4, D-O 4020 Halle/Saale, FRG

The ontogenetic development of the circadian system may be understood as a process containing components of maturation as well as adaptive ones (12). In the past main attention was paid to endogenous components however (6,7). It was shown that circadian rhythms are of hereditary nature and that during ontogenesis a genetic program is realized. But exogenous factors may influence the development of the circadian system also (1, 3-5, 9). We found in previous investigations that the development of the circadian system can be injured by modified zeitgeber conditions during juvenile period (11). To clarify, whether and in what a way such environmental conditions influence the time order of adult organisms, we carried out the following investigations.

MATERIALS AND METHODS

Investigations were carried out on laboratory mice of our own outbred stock HaZ:ICR. Breeding of animals took place under standardized conditions (L:D=12:12, L: 7.00 h - 19.00 h; temperature: 21 ± 2°C; relative humidity: 55-60%). Food and water were available ad libitum.

Beginning after weaning (on the 21st day of life) up to an age of 13 weeks groups of 7 female mice were kept under the following conditions (3 groups each):

I. L:D=12:12, L: 7.00 h - 19.00 h, feeding: ad libitum (control)
II. L:D=12:12, L: 19.00 h - 7.00 h, feeding: ad libitum
III. L:D=12:12, L: 7.00 h - 19.00 h, feeding: 19.00 h - 7.00 h
IV. L:D=12:12, L: 7.00 h - 19.00 h, feeding: 7.00 h - 19.00 h.

Beginning with the 13th week of age up to the 18th week all animals were kept under normal conditions (I) and than exposed to a LD-inversion (once doubling of dark phase) to obtain possible differences in the resynchronization process.

In the course of the whole experiment the locomotor activity was registered by means of infrared light barriers.

To describe and to quantify the resynchronization process the coefficients of correlation between daily patterns of single days and a mean chronogramm from 7 days before LD-inversion were calculated (10).

RESULTS AND DISCUSSION

The daily patterns of motor activity are shown in Fig. 1. In 13 weeks old mice they differ markedly according to raising conditions. Only animals raised under standard conditions (I) show the typical bimodal activity pattern of laboratory mice described already by ASCHOFF (2). Raising under inverse lighting regimen (II) results in a monophasic pattern, e.g. without a distinct morning peak. Under conditions of restricted feeding the activity rhythms is masked by the rhythm of food intake. So mice are active during the whole dark period and lights on induces an additional increase (III). When food is available only during light time (IV), animals feed mainly at the beginning and the end of light time. This is reflected in the activity pattern as well.

After five weeks under standard conditions the patterns reveal no differences depending on previous environmental conditions. Differences were obtained, however, in resynchronization processes after inversion of lighting regimen (Fig. 2). The times of resynchronization depending on environmental conditions I-IV consist: 6, 5, 7 and 9 days. In mice raised under restricted feeding additionally a delay in begin of resynchronization was obtained.

During the 3rd week after LD-inversion the circadian activity patterns are the same ones as before taking into account the time shift of course. Only in the case of group II raised under inverse lighting regimen the pattern is more similar to that of 13th week of life.

I. LD, ad libitum feeding

II. DL, ad libitum feeding

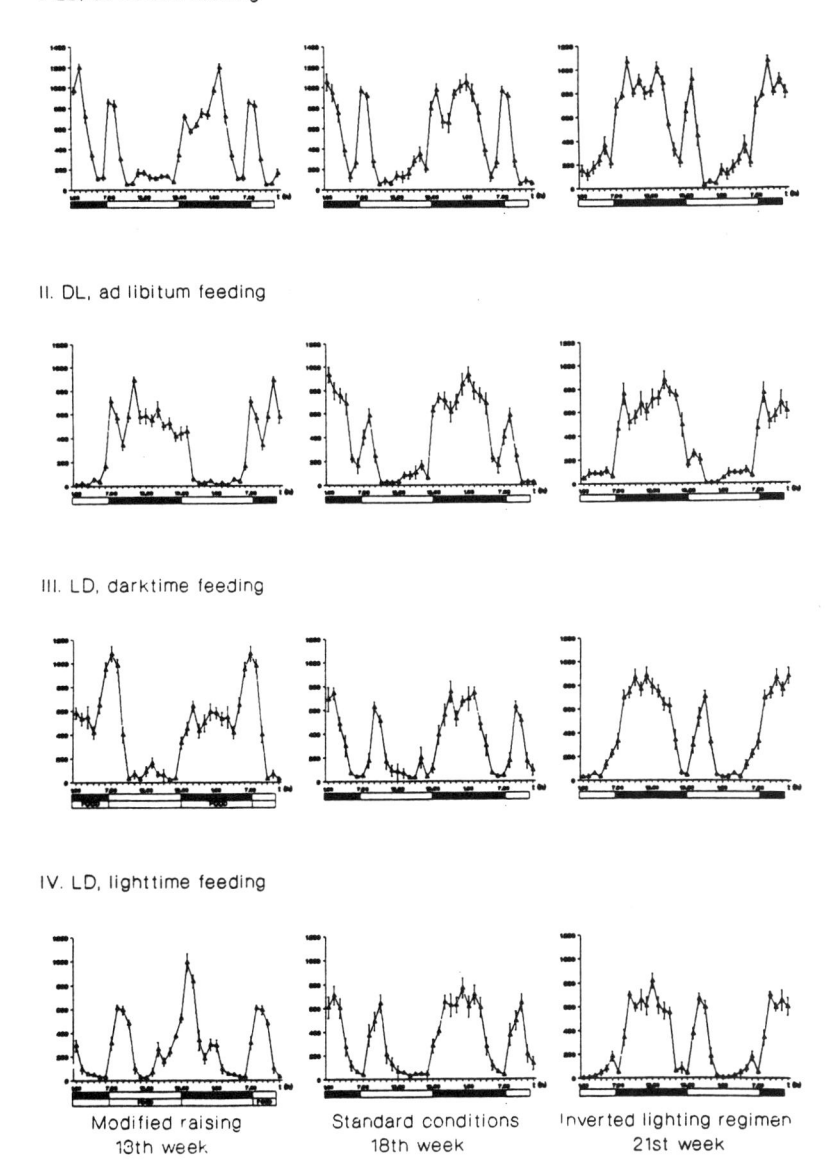

III. LD, darktime feeding

IV. LD, lighttime feeding

| Modified raising 13th week | Standard conditions 18th week | Inverted lighting regimen 21st week |

Fig. 1: Locomotor activity rhythms of laboratory mice during the 10th week under differing raising conditions, the 5th week under standard conditions and the 3rd week after LD-inversion.

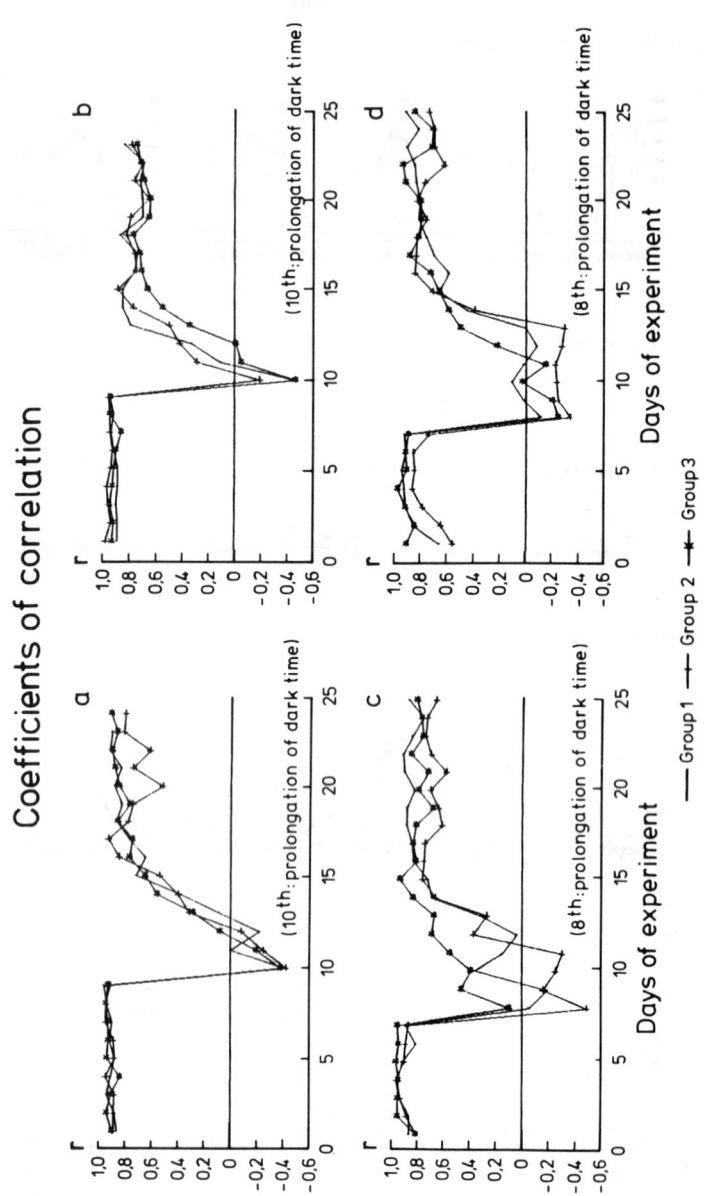

Fig. 2: Resynchronization of locomotor activity rhythms after inversion of lighting regimen depending on environmental conditions during juvenile phase (a-d correspond to I-IV)

Results confirm previous investigations that the time of resynchronization is a very sensitive parameter to obtain differences in the circadian system even if they are not reflected in the activity pattern (8). Reasons for differences in adaptive behavior are obscure yet. From investigations on cockroaches raised in non-24 h light cycles it is obvious that in adults the freerunning period as well as the response to light (phase response curve, range of entrainment) differ significantly from those in control animals (3, 4). Similar results were obtained in crickets raised in 24-h cycles but with differing L/D-ratios (9). In our case we suppose that the mechanisms of endogenous coupling and of exogenous synchronization are influenced. Therefore we want to characterize more detailed, e.g. by investigating several rhythmic functions, the time order of adult mice raised under modified zeitgeber conditions.

REFERENCES

1. ANDERSON, V.N. & G.K. SMITH (1987): Effects of feeding and light cycles on activity rhythms of maternally isolated rat pups. Physiol. Behav. 39, 169-181
2. ASCHOFF, J. (1962): Spontane lokomotorische Aktivität. Handb. Zool. 8, 30. Lief., 1-74
3. BARRETT, R.K. & T.L. PAGE (1989): Effects of light on circadian pacemaker development. I. The freerunning period. J. Comp. Physiol. A 165, 41-49
4. BARRETT, R.K. & T.L. PAGE (1989): Effects of light on circadian pacemaker development. II. Responses to light. J. Comp. Physiol. A 165, 51-59
5. CAMBRAS, T. & A. DIEZ-NOGUERA (1991): Evolution of rat motor activity circadian rhythm under three different light patterns. Physiol. Behav. 49, 63-68
6. DAVIS, F.C. (1981): Ontogeny of circadian rhythms. In: ASCHOFF, J. (ed.): Handbook of behavioral neurobiology, vol. 4: Biological rhythms. New York, Plenum Press
7. HELLBRÜGGE, Th. (1977): Physiologische Zeitgestalten in der kindlichen Entwicklung. Nova Acta Leopoldina 46, Nr. 225, 365-387
8. SCHUH, J., D. WEINERT & G.D. GUBIN (1991): The time of resynchronization as an indicator of the adaptive potency. In: SUROWIAK, J. & M.H. LEWANDOWSKI (eds.): Chronobiology & Chronomedicine. Peter Lang, Frankfurt/Main-Bern-New York-Paris, pp. 131-135
9. TOMIOKA, K. & Y. CHIBA (1989): Photoperiod during post-embryonic development affects some parameters of adult circadian rhythm in the cricket, Gryllus bimaculatus. Zool. Science 6, 565-571
10. WEINERT, D. & J. SCHUH (1987): Zur Stabilität biologischer Rhythmen in Abhängigkeit vom postnatalen Alter. Ergebnisse von Resynchronisationsexperimenten. In: SCHUH, J., R. GATTERMANN & J.A. ROMANOV (eds.): Chronobiologie-Chronomedizin. Wissenschaftliche Beiträge d. Martin-Luther-Universität Halle-Wittenberg 1987/36 (P30), pp. 415-420
11. WEINERT, D., F.-E. ULRICH & J. SCHUH (1989): Postnatale Änderungen der Tagesperiodik unter normalem und inversem Lichtregime. Zool. Jb. Physiol. 93, 257-269
12. WEINERT, D., G.D. GUBIN & J. SCHUH (1990): Biologische Rhythmen und Lebensalter. Wiss. Zeitschr. HUB, R. Math./Nat. Wiss. 39, 404-412

THE VALENCE OF PHOTOPERIOD AND FEEDING AS 'ZEITGEBERS' OF THE KIDNEY RHYTHMS OF PEPTIDE-CLEAVING ENZYMES

D. BALSCHUN, S. JANKE & J. SCHUH[*]

Institute of Neurobiology and Brain Research, O 3090 Magdeburg;
*Department of Biosciences, Martin-Luther-University, D-O 4020 Halle, FRG

INTRODUCTION

In previous studies we could demonstrate that the activities of peptide-cleaving enzymes in the kidneys of mice undergo daily rhythmic changes up to 100% and more (2,3). The investigated enzymes dipeptidyl peptidase IV (DPP IV, EC 3.4.14.5) and microsomal alanyl aminopeptidase (mAAP, EC 3.4.11.2) belong to the dominant proteins of the microvillar membrane of proximal tubules. They have the capacity to cleave a variety of peptides such as substance P and angiotensin III by complementary action. The rhythms of DPP IV and mAAP were found to be mutually synchronized and mostly, the rhythm of the right kidneys resembled that of the left. The data led us to hypothesize that light acts only as a weak zeitgeber of the investigated enzyme rhythms.

The present study aimed to examine 1) whether a 12/12 regime of food access is more potent to entrain the kidney rhythms of peptidases than a 12/12 photoperiod and 2) whether the influence of these putative zeitgebers differs between young and old animals.

MATERIALS AND METHODS

Female mice of the ICR strain were raised in groups of eight under 12 hour light: 12 hour dark cycles (LD 12:12) after weaning. Animal rooms were illuminated by fluorescent lighting from 07.00 to 19.00 h. Food and water were provided ad libitum.

14 days prior to experiment food access was restricted to 07.00 -19.00 h (day feeding; DF) or to 19.00 - 7.00 (night feeding; NF). Half of the DF- and NF-groups

were subjected to an inversion of the photoperiod (DL 12:12). Therefore, four combinations of photoperiod and feeding resulted: 1) L/D & DF, 2) L/D & NF, 3) D/L & DF, 4) D/L & NF.

At the age of 13 and 60 weeks, respectively, one group of animals of each 'zeitgeber' combination was killed by decapitation every two hours. The kidneys were rapidly removed, frozen on dry ice and stored at -20°C until further processing as described elsewhere (3).

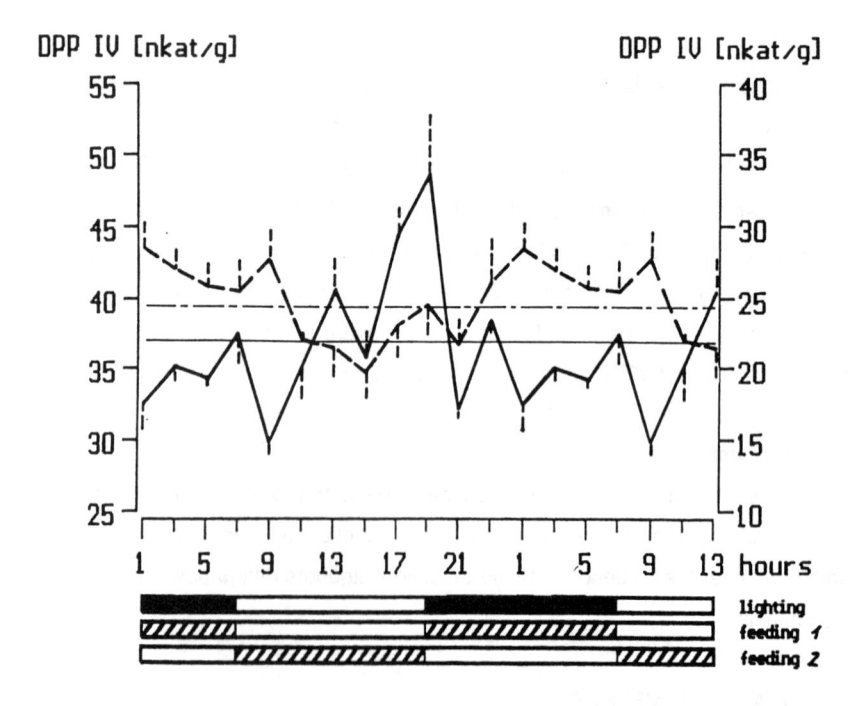

Fig.1: The effect of a reciprocal feeding but equal lighting regimen on the DPP IV activity in the right kidney of 13 week old females.
Left ordinate, solid line: night feeding (feeding 1);
Right ordinate, broken line: day feeding (feeding 2);
Horizontal lines indicate daily mean, vertical bars SEM.
The acrophases of the circadian spectral components were calculated to occur at 13.45 h ± 79 min and 05.27 h ± 68 min, respectively.

DPP IV and mAAP activity were measured spectrophotometrically as endpoint method with artificial chromogenic substrates. Protein was determined using the biuret method. Time series analysis was performed by a spectral cosinor method.

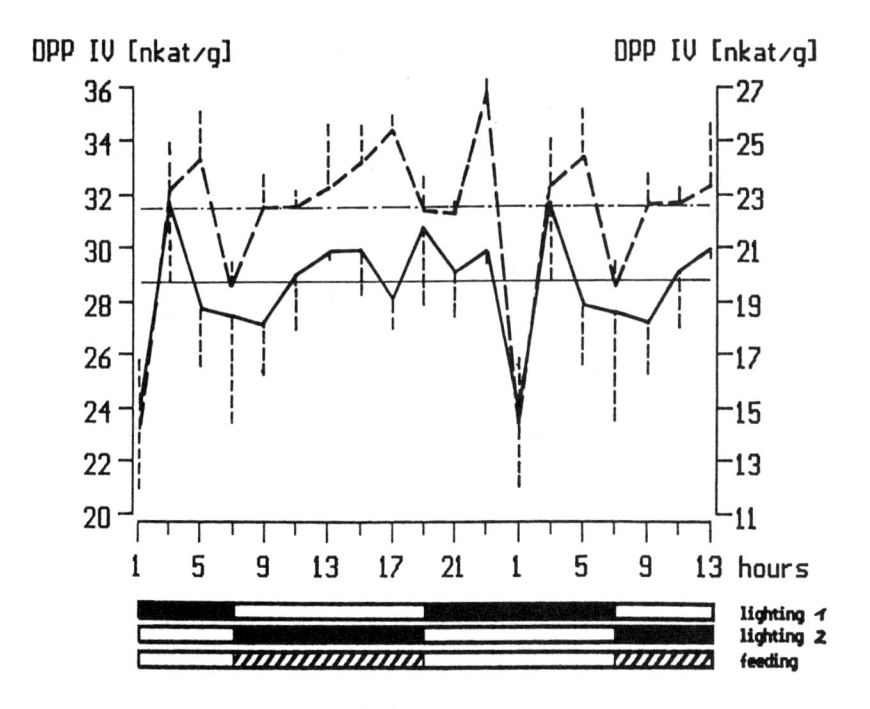

Fig.2: The effect of a reciprocal lighting but equal feeding regimen on the DPP IV activity in the right kidney of 60 week old female mice.
Left ordinate, solid line: LD (lighting 1);
Right ordinate, broken line: DL (lighting 2);
Horizontal lines indicate daily mean, vertical bars. SEM: Spectral cosinor revealed dominant ultradian components in both time series.

RESULTS AND DISCUSSION

Inversion of the feeding regimen caused much stronger effects than inversion of the photoperiod. This applied to the enzyme rhythms in the right and left kidneys of the young as well as of the old animals. As illustrated by the example in Fig.1 a normal photoperiod of 12/12 (LD) and night feeding resulted in high enzyme activities during the astronomic day. The same lighting but inverse feeding regimen (day feeding) led to nearly inverse patterns, however with clearly differing mesor and amplitude. The DPP IV and mAAP rhythms reacted on the inversion of the feeding regimen in a similar way. (Only DPP IV chronograms of the right kidneys are shown as examples.)

The consequences of opposite photoperiods but the same feeding schedule are depicted in Fig.2. Day feeding resulted in chronograms with a relatively high portion of activity during the astronomic night, irrespective of the chosen lighting regimen, i.e. the patterns of enzyme activity were very similar although the lighting conditions were inverse. Differences could be found only in mesor and amplitude.

Rhythmic changes of peptidases are scarcely documented (5,6,9,10). HOFFMANN & HARDELAND (6) reported an unimodal rhythm of gamma-glutamyl-transpeptidase (GGT) activity in the kidneys of rats fed ad libitum. SAITO (9) and FRANCIOLINI et al. (5) could not detect rhythmic changes of leucine aminopeptidase (LAP) and cathepsin D, respectively, in rat kidney. Detailed studies on the temporal regulation of kidney enzymes are not available at all.

The results of the present investigation demonstrate that even a gently restricted feeding of 12 hours food access acts as 'entraining' force on the daily rhythms of peptide-cleaving enzymes in the kidneys. This corresponds well with the findings of STEPHANSON et al. (10) and SUDA & SAITO (12) demonstrating a strong synchronizing effect of a restricted food access on the rhythms of intestinal enzymes. However, whereas the maxima of intestinal enzyme activities occurred around the feeding time, the peptidases in the kidney brush border display an nearly inverse phasing, i.e. night feeding resulted in high enzyme activity during the day and vice versa.

The clear difference in phasing of intestinal and renal enzyme rhythms has to be considered with respect to their location and functions. DPP IV and mAAP belong to the dominant membrane proteins of kidney brush border. No other plasma membrane was found to contain such an abundance of peptidases as that of the proximal tubules (11). The physiological function of the microvillar peptidases within the kidney is not completely understood. Due to their complementary substrate specifity DPP IV and mAAP are able to cleave a variety of polypeptides and proteins. Experimental evidence supports the view (7) that both peptidases participate in the renal reclamation of amino acids and di- and tripeptides. This function as well as the metabolic position of the kidney as a final organ in the utilization of food can account for the clear phase difference between feeding time and peak values of DPP IV and mAAP activities in the kidneys.

To decide whether the stable phase relations of the peptidase rhythms to the feeding time are caused by entrainment (A) or masking (B) is impossible as yet. The numerous studies concerning with the effects of restricted feeding on different behavioral and physiological parameters lead to the conclusion that the portion of mechanism A or B contributing to a putative entrainment or to the appearance of anticipatory activity depends on the parameter, the species and even the strain investigated (1,4). In the rat, restricted feeding entrains the intestinal enzyme activity as well as other physiological parameters such as plasma corticosterone [see (12) for references]. On the contrary, the phase shifts in the freerunning rhythms of electrolyte excretion were shown to be caused by masking (8). ABE et al. (1) could show that a restricted feeding for two hours was able to entrain the freerunning wheel-running activity of CS mice but not of C57BL/6J mice.

The phase dynamic of the kidney rhythms of peptidases under ad lib feeding found in previous investigations (3) and the results reported here, lead us to suppose that the found synchronization by restricted food access reflects entrainment rather than masking. However, this has to be examined in further investigations.

REFERENCES

1. ABE, H., M. KIDA, K. TSUJI & T. MANO (1989): Feeding cycles entrain circadian rhythms of locomotor activity in CS mice but not in C57BL/6J mice. Physiol. Behav. 45, 397-401
2. BALSCHUN, D., S. JANKE & J. SCHUH (1991): Rhythms in the activity of peptide-cleaving enzymes: Current state and outlook. In: J. SUROWIAK & M.H. LEWANDOWSKI (eds.): Chronobiology and Chronomedicine; Basic Research and Applications, Peter Lang, Frankfurt, pp. 121-125
3. BALSCHUN, D., Z. LOJDA & J. SCHUH (1988): Rhythmic changes of dipeptidyl peptidase IV activity in the blood and in the kidney of mice: Biochemical and histochemical investigations. Biol. Zentr.bl. 107, 181-188
4. BOULOS, Z. & M. TERMAN (1980): Food availability and daily biological rhythms. Neurosci. Biobehav. Rev. 4, 119-131
5. FRANCIOLINI, F., A. BECCIOLINI, V. CASATI et al. (1979): Circadian activity of rat kidney enzymes. Experientia 35, 582-583
6. HOFFMAN, J. & R. HARDELAND (1981): Diurnal rhythmicity in rat kidney gamma-glutamyltranspeptidase activity. Arch. Int. Physiol. Biochem. 89, 245-247
7. MIYAMOTO, Y., V. GANAPATHY & F.H. LEIBACH (1988): Role of dipeptidyl-peptidase IV in renal handling of peptides. In: G.M. BERLYNE, N.Y. BROOKLYN & S.G. PISA (eds.): Contributions to Nephrology 68. S. Karger, Basel, pp. 1-5
8. POULIS, J.A., F. ROELFSEMA & D. VAN DER HEIDE (1989): The role of feeding time in the evolution of urinary rhythms in rats. J. interdiscipl. Cycle Res. 20, 81-96
9. SAITO, M. (1972): Daily rhythmic changes in brush border enzymes of the small intestine and kidney in rat. Biochem. Biophys. Acta 286, 212-215
10. STEPHANSON, N.R., F. FERRIGNI, K. PARNICKY et al. (1975): Effect of changes in feeding schedule on the diurnal rhythms and daily activity levels of intestinal brush border enzymes and transport systems. Biochem. Biophys. Acta 406, 131-145
11. STEPHANSON, S.L. & A.J. KENNY (1987): Metabolism of neuropeptides; Hydrolysis of the angiotensins, bradykinin, substance P and oxytocin by pig kidney microvillar membranes. Biochem. J. 241, 237-247
12. SUDA, M. & M. SAITO (1979): Coordinative regulation of feeding behavior and metabolism by a circadian timing system. In: M. SUDA, O. HAYASHI & H. NAKAGAWA (eds.): Biological rhythms and their central mechanisms. Elsevier/North-Holland Biomedical Press, pp. 263-271

DIFFERENTIAL EFFECTS OF EXOGENOUS MELATONIN ON CORRELATES OF OESTRUS IN DOMESTIC RABBITS

R. HUDSON

Institute of Medical Psychology, University of Munich, Goethestr. 31, D-W-8000 Munich 2, FRG

In the European rabbit (*Oryctolagus cuniculus*) reproductive activity is stimulated under both natural and laboratory conditions by long photoperiod (summarized in 6). Although it is now generally accepted that in mammals the pineal gland mediates such photoperiodic effects, there is still considerable uncertainty as to how this is achieved. In particular, it remains unclear to what extent the pineal hormone melatonin is involved, and if so, where and in what way it acts (9,10). As the reproductive system of female rabbits is highly sensitive to changes in daylength (6), it was decided to investigate the effect of administering melatonin on two non-invasive and readily quantifiable measures of oestrus, namely pheromone emission and chin-marking behavior.

Newborn rabbits are completely dependent on a pheromone on the mother's ventrum to locate nipples and suckle. Using the nipple-search response as a bioassay, it has been shown that in non-breeding does pheromone emission parallels seasonal changes in daylength and is stimulated by experimental long day and suppressed by experimental short day within 1 to 2 weeks (5,6). These changes are accompanied by alterations in the frequency of chinning, a behavior in which secretion from a sub-mandibular skin gland is deposited on protruding objects by rubbing the chin over them (8). Both parameters are suppressed by ovariectomy and stimulated by estradiol replacement within a few days (5,7).

This study therefore aimed to test whether the inhibitory effect of short photoperiods on these parameters can be simulated by melatonin administration.

METHODS

Animals: A total of 28 sexually mature female domestic rabbits was used. They were caged separately at 21° ± 2° C and provided with food and water ad libitum.

Test Procedures: Each day each doe was tested in the early afternoon in the following two ways: Pheromone emission was measured by holding a pup of 2-10 days of age to the doe's ventrum for a maximum of 10 sec and taking its ability to locate a nipple as evidence of pheromone emission. If pups are able to locate nipples at all they will do so in this time.

Chinning was assessed by placing the doe in an arena 1 m in diameter and containing three 15-cm high bricks. The number of times the doe rubbed its chin on the bricks during a 10-min period was recorded. To control for possible non-specific effects of melatonin on animals' general alertness, on the last two days of treatment does were given a second test after two of the bricks had been exchanged for one's chin marked by sexually active donor males, and the number of marks directed to each brick recorded (8).

Melatonin Treatment: Experimental subjects were implanted subcutaneously with 5 mg melatonin (Sigma) in a 50 mg beeswax pellet, and control does with beeswax only. Three pellets were implanted at weekly intervals and then all removed after the fourth week.

Experiment I: Does melatonin suppress pheromone emission and chinning under long day? After one month testing in 16L/8D, 8 experimental and 8 control does were implanted and tested for a further four weeks. The pellets were then removed and testing continued for a further month.

Experiment II: Does melatonin suppress the stimulatory effect of returning does to long day? After one month testing in 8L/16D, 6 experimental and 6 control does were implanted, returned to 16L/8D and tested for four weeks. Pellets were then removed and testing continued for a further month.

Statistical Analysis: The Mann-Whitney U test was used to compare mean scores for the experimental and control does for the week before treatment, for the final week of treatment, and for the final week of the experiment three weeks after the pellets had been removed. An alpha value of 0.05 was taken as the level of significance.

RESULTS

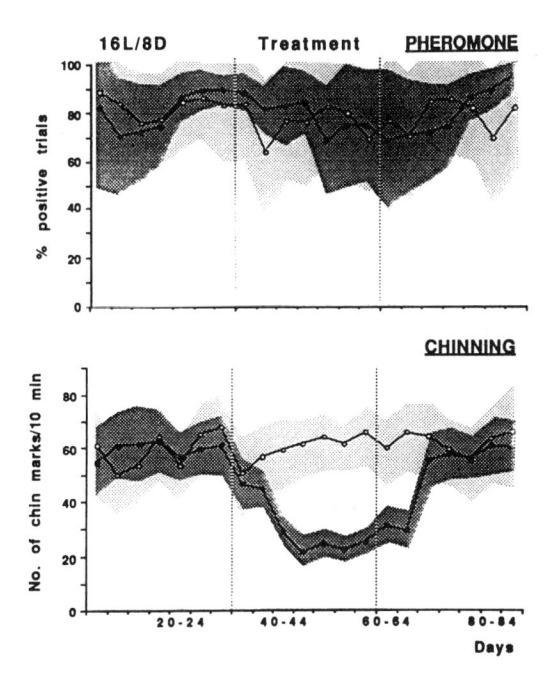

Fig.1:
Effect of melatonin administered to female rabbits under long day. Filled circles and dark shading give the means and SEMs for animals receiving melatonin implants (n = 8), and open circles and pale shading for controls receiving the vehicle alone (n = 8).

Exp. I: Melatonin in Long Photoperiod

As shown in Fig. 1, the melatonin treatment had no detectable effect on pheromone emission and no significant difference between the scores of the experimental and control groups before, during or after treatment was found. In contrast, whereas no significant difference in chinning activity for the two groups was recorded before or after treatment, during melatonin administration scores were significantly lower for the experimental animals.

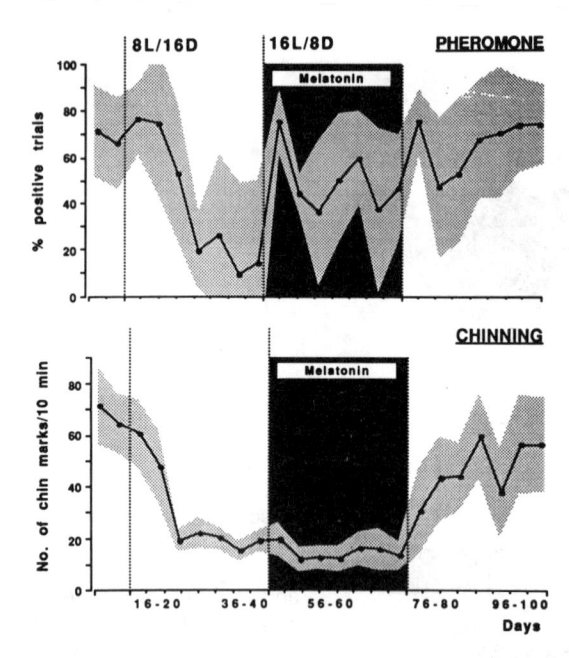

Fig.2:
Effect of melatonin administered to female rabbits returned to long day (n = 6).

Exp. II: Melatonin and Return to Long Photoperiod

As shown in Fig. 2, melatonin failed to suppress pheromone emission in does after the return to 16L/8D, and although for simplicity the data of the control animals are not shown, no significant difference in the scores of the two groups were recorded before, during or after treatment. In contrast, chinning activity failed to recover until the melatonin had been removed, and whereas no difference in mean scores for the two groups was recorded before or after treatment, scores during treatment were significantly lower for the experimental animals.

When tested with bricks marked by donor bucks, both groups responded with an increase in chinning over the regular test for that day although an increase of 40% for the treated animals was significantly greater than 20% for the controls. However, the distribution of marks across the three bricks was very similar, with does from both groups directing on average three times as many marks to the brick of buck A than to that of B or the old brick.

DISCUSSION

Reports on pineal activity in the rabbit are consistent with findings for other mammals (2,3), and with melatonin's putative role as a neuroendocrine transducer of photoperiod. Melatonin levels are highest at night, size and amplitude of the peak depends on the duration of the scotophase (11), and melatonin implanted in wild, long-day bucks results in rapid testicular regression, although it fails to prevent testicular recrudescence (1). However, the present study provides only partial support for melatonin having such a signal function. Whereas in both experiments the inhibitory effect of melatonin on chinning mimicked the effect of short photoperiod, this was not the case for pheromone emission - normally a most sensitive indicator of daylength (6).

Although the doses used here were possibly too low to suppress emission, given that changes in daylength of just 1 hour are normally sufficient to produce clear effects (6), this seems unlikely. Conversely, suppression of chinning due to artefactual effects such as illness also seem unlikely given that does responded to the bricks from donor bucks in the same vigorous way as untreated, short-day females (8). A more likely explanation for the failure to obtain the full photoperiodic effect is the use of implantation rather than daily injection as the mode of administration. BOYD (1), however, induced testicular regression using implants, and FARRELL et al.(4) failed to suppress ovulation in does by melatonin injection. Further, the dissociation in the effects of treatment found here accords with the earlier finding that while pheromone emission is enhanced during pregnancy or following progesterone administration, chinning is suppressed (6).

In conclusion, should these findings be replicated using other procedures then it would seem that melatonin may not be regarded as the sole mediary of the photoperiodic response but, at least in female rabbits, that different pineal products (2,9) may affect the reproductive system differentially or act in synergism.

ACKNOWLEDGEMENT
This work was supported by the DFG (Hu 426/1).

REFERENCES

1. BOYD, I.L. (1985): Effect of photoperiod and melatonin on testis development and regression in wild European rabbits (*Oryctolagus cuniculus*). Biol. Reprod. 33, 21-29
2. BRAINARD, G.C., S.A. MATTHEWS, R.W. STEGER et al. (1984): Day:night variations of melatonin, 5-hydroxyindole acetic acid, serotonin, N-acetyltransferase, tryptophan, norepinephrine and dopamine in the rabbit pineal gland. Life Sci. 35, 1615-1622
3. DE GALLARDO, M.R.G.P. & J.H. TRAMEZZANI (1975): Hydroxyindole-O-methyl-transferase activity in the pineal gland of the rabbit. J. Neural. Transm. 36, 51-57
4. FARRELL, G., D. POWERS & T. OTANI (1968): Inhibition of ovulation in the rabbit: Seasonal variation and the effects of indoles. Endocrinol. 83, 599-603
5. HUDSON, R. & H. DISTEL (1984): Nipple-search pheromone in rabbits: Dependence on season and reproductive state. J. Comp. Physiol. A 155, 13-17
6. HUDSON, R. & H. DISTEL (1990): Sensitivity of female rabbits to changes in photoperiod as measured by pheromone emission. J. Comp. Physiol. A 167, 225-230
7. HUDSON, R., G. GONZALES-MARISCAL & C. BEYER (1990): Chin-marking behavior, sexual receptivity and pheromone emission in steroid treated, ovariectomized rabbits. Horm. Behav. 24, 1-13
8. HUDSON, R. & T. VODERMAYER (in press): Spontaneous and odour-induced chin-marking in domestic female rabbits. Anim. Behav.
9. REITER, R.J. (1980): The pineal and its hormones in the control of reproduction in mammals. Endocrine Rev. 1, 109-131
10. TAMARKIN, L., C.J. BAIRD & O.F.X. ALMEIDA (1985): Melatonin: A coordinating signal for mammalian reproduction? Science 227, 714-720
11. YOUNGLAI, E.V., S.F. PANG & G.M. BROWN (1986): Effects of different photoperiods on circulating levels of melatonin and N-acetylserotonin in the female rabbit. Acta Endocrinol. 112, 145-149

WEATHER-INDUCED INTERDIURNAL VARIATIONS OF L/D-PROPORTION AND OF BRIGHTNESS AFFECTS MAN IN SAME MANNER AS FLIGHTS IN DIRECTION OF LATITUDES

L. KLINKER & H. PIAZENA

Schulstr. 1, D-W-2381 Busdorf/Schleswig, FRG

INTRODUCTION

Former researches (3) indicate a negative influence of interdiurnal variations of daylight by weather changes on human regulation. As a consequence we started measurements about the intensity of such variations. For one year the times at which outdoor brightness exceeds or falls below certain levels between 500 and 5000 lux were recorded by a luxmeter in the morning and in the evening.

Moreover the global-radiation (W/m^2) was continuously recorded by a pyranometer. These data were converted into brightness (lx) values (5). An empirical model was developed for computation of brightness as a function of solar radiation on clear days (6). Now these data were compared with the actual times at which brightness exceeds physiologic relevant values at days with different cloudiness.

RESULTS

Interdiurnal variations of l/d-proportion and of brightness

As a function of thickness of clouds the actual points of time at which brightness exceeds 1000 lx and 5000 lx, respectively, varies considerably from day to day. The interdiurnal differences between cloudy and clear days did not show annual variations (4). The mean and the extreme differences are presented in Tab. 1.

Tab.1: Time differences between the occurence of selected intensities of brightness at cloudy and at clear days

B(lx)	time differences			
	morning		evening	
	mean	max	mean	max
1000	20'	100'	30'	140'
3000	30'	160'	40'	170'
500	40'	180'	50'	180'

Compact clouds as observed on rainy days change the duration of effective day-length up to 4 hours for intensities ≥1000 lx and up to 6 hours for ≥5000 lx. That is in the same order as the day-length variations arising by flights between Europe and North-America. For example an L/D-proportion from 12/12 during equinox may be modified to 6/18.

Similar modifications of L/D-proportion by weather-changes are induced indoors (4).

A further effect of weather-changes are strong variations of brightness. Some possible differences are presented in Tab. 2.

Tab 2:

time (CET)	brightness (10^3 lx)			
	summer		winter	
	max	min	max	min
12.30	110	6	20	3

In summer there are variations between 6000 and 110.000 lx at noon, in winter between 3000 and 20.000 lx.

Meteoropathological events

Most of former research in the problem of meteoropathological disturbances suffers from the neglect of uni-modal and bi-modal annual rhythms in man. Morbidity, mortality as well as physiological parameters underlie annual variations with different monthly fixation of minima and maxima. Since most meteorological parameters have their own seasonal variations, such investigations will be influenced by automatic correlations if they are extended over the whole year as usual. It is necessary to analyze such time-series for each month separately or at least for the different seasons.

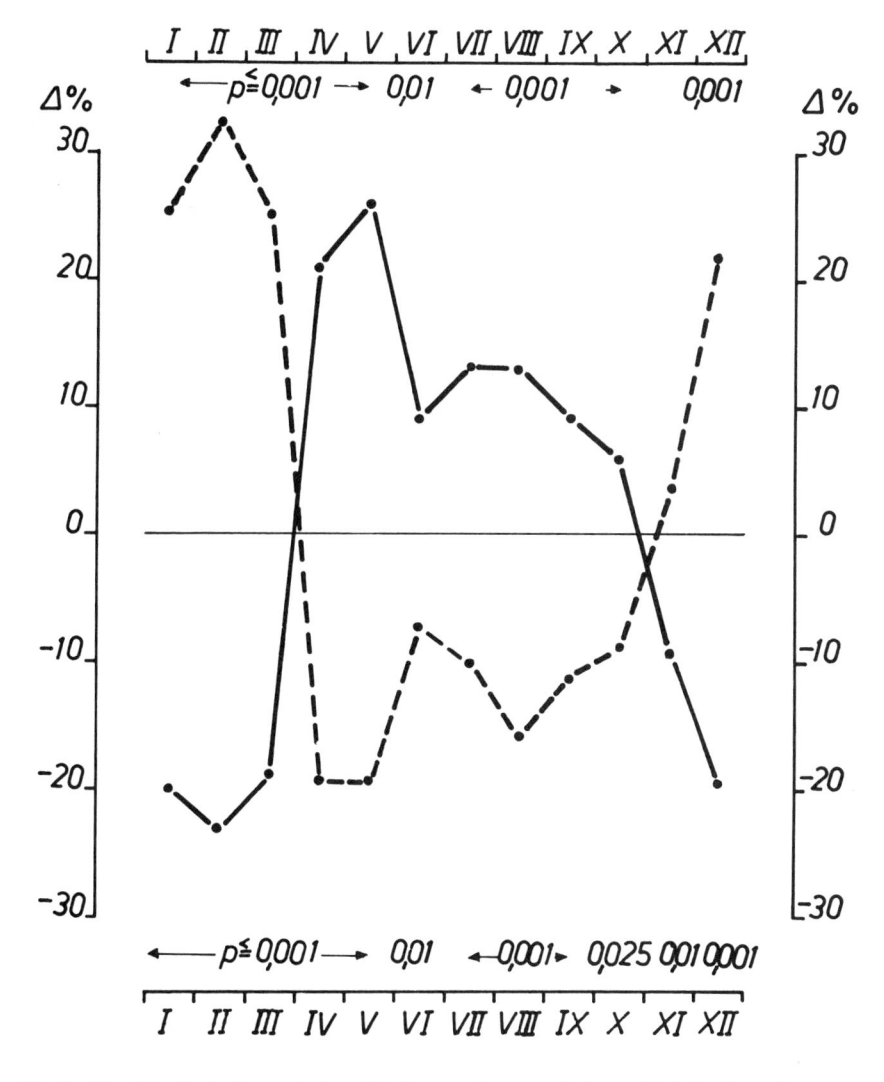

Fig.1: Monthly deviations of dental inflammations at dark and bright days from expected value "e" valid for a random distribution of such events.

A research of 4.619 dental inflammations under the premises mentioned before had following results (Fig.1): During summer there are significant (p>=0.001) increases of inflammations above expected values at dark days and decreases below expected values at bright days. During winter an opposite behavior takes place.

The analysis of further 16 long-time series leads to similar results: Inflammations, colics, convulsions, suicides, vegetative disturbances, acute respiratory disturbances and embolisms are very often light-induced (3). During the bright months in summer, the seldom dark days and during the dark months in winter the seldom bright days lead to significant increases of such events in Central-Europe.

CONCLUSIONS

Following ASCHOFF et al. (1) artificial variations of frequency of the zeitgeber light cause phase-shifts between two or more circadian oscillators in man. Moreover, WEVER et al. (7) and CZEISLER & ALLAN (2) show that essential phase-shifts of the light-sensitive circadian oscillators in man may occur under the influence of changes in brightness.

Under natural conditions - as demonstrated above - there are great interdiurnal changes of L/D-proportion including variations in frequency of the zeitgeber light as well as simultaneous fluctuations of brightness depending on weather conditions. All factors may act together to produce fluctuating phase-shifts between two ore more circadian oscillators possibly leading to internal desynchronisation and negative consequences for the coordination of vital functions in man. Under this aspects the day-to-day weather changes produce variations in light regime similar to flights in direction of latitudes.

REFERENCES

1. ASCHOFF, J., E. PÖPPEL & R. WEVER (1969): Circadiane Periodik des Menschen unter dem Einfluß von Licht-Dunkel-Wechseln unterschiedlicher Periodik. Pflüg. Arch. 306, 58-70
2. CZEISLER, A. & J.S. ALLAN (1987): Rapid phase shifting in humans requires bright light. Chronobiologia XIV, 187
3. KLINKER, L. (1989): Der Einfluß des natürlichen Tageslichtes auf die menschliche Regulation. Z. ges. Hyg. 35, 196-202
4. KLINKER, L. & H. PIAZENA (1991): in preparation
5. NEUMANN, B. (1987): Untersuchungen zur Anwendbarkeit von Luxmetern zur Charakterisierung des solaren Strahlungsangebotes. Abschlußbericht TAM d. MD d. DDR
6. PIAZENA, H. (1989): Untersuchungen der solaren Beleuchtungsverhältnisse in städtischen Wohngebietsstrukturen. Forschungsbericht des MD d. DDR
7. WEVER, R. et al (1983): Bright Light Affects Human Circadian Rhythm. Pflüg. Arch. Eur. J. Physiol. 341, 85-87

SERUM β-GLUCURONIDASE ACTIVITY AFTER LIGHT EXPOSURE

A. FALKENBACH, U. UNKELBACH, J. CRONIBUS & U. SEIFFERT

Physische Therapie, Zentrum der Inneren Medizin, Klinikum der Johann Wolfgang Goethe Universität, Th.Stern Kai 7, D-W 6000 Frankfurt am Main 79, FRG

INTRODUCTION

Several diseases, in which seasonal changes in disease activity are well known, e.g. rheumatoid arthritis (6) and systemic lupus erythematodes, show increased serum levels of lysosomal enzymes (3). The causal role of the lysosomes in joint disease has extensively been discussed by WEISSMANN (8) and EBERHARD et al. (2).

The release of lysosomal enzymes is stimulated by a variety of factors, such as antigen/antibody formation, immune complexes etc. (1,4). In vitro studies revealed that lysosomal enzymes are released after exposure to ultraviolet light, too (9).

To evaluate whether changes in sunlight composition might contribute to changes in the release of lysosomal enzymes in man, serum β-glucuronidase activity was determined before and after exposure to ultraviolet or infrared light. β-glucuronidase is a marker enzyme of the lysosomes. A new enzyme kinetic method was used for laboratory determination of the serum activity.

METHOD

20 young healthy volunteers (7 male, 13 female) were included into the study. Mean age was 28 years. By use of the method first described by WUCHERPFENNIG (10) the individual minimal erythema dose (MED) was determined. The next day all subjects were exposed to light with the exposure time beeing 80 % of the MED. For irradiation all volunteers were divided into two groups. Group I was exposed to ultraviolet light on a common solarium in a supine position with simultaneous ventral and dorsal irradiation. The spectrum used for irradiation and the technical data are

shown in Fig. 1. As controls, group II subjects were exposed to infrared light. The device used for infrared irradiation was kept at such a distance that only a slight cutaneous sensation of warmth was achieved, but no rise in body temperature could occur.

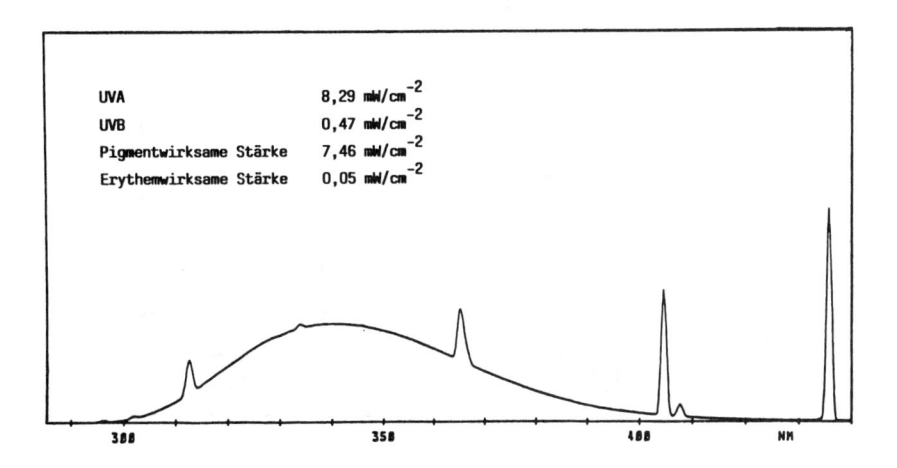

UVA	8,29 mW/cm^{-2}
UVB	0,47 mW/cm^{-2}
Pigmentwirksame Stärke	7,46 mW/cm^{-2}
Erythemwirksame Stärke	0,05 mW/cm^{-2}

Fig 1:

The day before and the day after irradiation venous blood was taken from a peripheral vein between 8 and 9 am. Serum β-glucuronidase activity was determined by use of a new enzyme kinetic assay as described by SEIFFERT et al. (7). The results before and after irradiation were analysed by two tailed Student t-test.

RESULTS

In the UV exposed group the serum β-glucuronidase activity (mean ± standard deviation) increased from x_b = 1.85 (± 0.81) U/L before to x_a = 2.22 (± 0.78) U/L after irradiation (0.5 < p < 0.1 in paired Student t-test). In group II, β-glucuronidase activity changed from x_b = 1.76 (± 0.76) U/L to x_a = 1.88 (± 0.58) U/L.
The data are shown in Fig. 2. At the 95 % level the changes were not significant statistically.

β-Glucuronidase
before --> after UV and IR

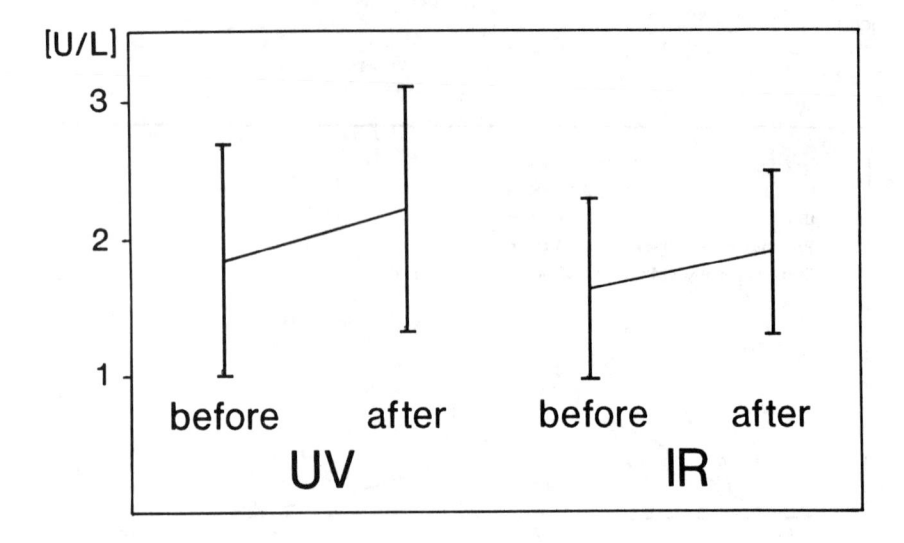

Fig.2

DISCUSSION

After a single exposure to ultraviolet light with 80 % of MED an increase of serum β-glucuronidase activity was seen, which, however, was not of statistical significance. The reason for the increase is rather speculative. Ultraviolet light might have caused changes in cutaneous tissue, which resulted in repair mechanisms and antigen/antibody formation, so that a release of lysosomal enzymes was stimulated. Alternatively, a direct demage to the lysosomal membrane, possibly by peroxydation of lipids (5), might have caused the increase of β-glucuronidase activity.

The simultaneously enhanced concentration of glucocorticoids, which has been reported to occur after UV irradiation, must be taken into account in the interpretation of the results. β-glucuronidase activity might have been more pronounced (and statistically significant) without the simultaneous release of glucocorticoids stabilizing the lysosomal membrane (8). A study on the influence of a serial ultraviolet irradiation has already been launched.

REFERENCES

1. CARDELLA, C.J., P. DAVIES & A.C. ALLISON (1974): Immune complexes induce selective release of lysosomal hydrolases from macrophages. Nature 247, 46-48
2. EBERHARD, A., U. LAAS, O. VOJTISEK & H. GREILING (1972): Zur Bedeutung und Herkunft von lysosomalen Enzymen in der Synovialflüssigkeit bei chronischen Gelenkerkrankungen. Z. Rheumaforschg 31, 105-118
3. FALKENBACH, A., U. UNKELBACH, R. GOTTSCHALK & JP. KALTWASSER (1991): β-Glucuronidase-Konzentration im Serum als potentieller Indikator der Krankheitsaktivität bei rheumatoider Arthritis (RA) und systemischem Lupus erythematodes (sLE). Med. Klin. 86, 465-468
4. FERENCIK, M. & J. STEFANOVIC (1979): Lysosomal enzymes of phagocytes and the mechanism of their release. Folia Microbiol. 24, 503-515
5. MEFFERT, H. (1975): Vermittlung von UV-Wirkung auf die menschliche Haut durch Lipidperoxydation. Zschr. Physiother. 27, 421-424
6. PILGER, A. (1970): Chronische Polyarthritis und Klima. Med. Klin. 65, 1363-1365
7. SEIFFERT, U.B., U. UNKELBACH & WH. SIEDE (1988): Enzyme kinetic assay for β-glucuronidase in serum of men. Enzyme 40, Suppl.1, 62
8. WEISSMANN, G. (1966): Lysosomes and joint disease. Arthritis Rheum. 9, 834-840
9. WEISSMANN, G. & J. DINGLE (1961): Release of lysosomal protease by ultraviolet irradiation and inhibition by hydrocortisone. Exp. Cell Res. 25, 207-210
10. WUCHERPFENNIG, W. (1933): Eine automatische Sektorenanalyse zur genauen Bestimmung der Erythemschwelle des UV. Strahlentherapie 48, 391-396

NEITHER 0-HOUR NOR 12-HOUR TEMPORAL RELATIONS OF 5-HTP AND L-DOPA INHIBIT HIBERNATION IN THE THIRTEEN-LINED GROUND SQUIRREL, SPERMOPHILUS TRIDECEMLINEATUS

J.T. BURNS & K.A. ROTHENBERGER

Department of Biology, Bethany College,
Bethany, West Virginia, 26032, USA

INTRODUCTION

A yearly change has been described in the thirteen-lined ground squirrel, *Spermophilus tridecemlineatus*, in sexual development, in the tendency to hibernate (4,11) and in metabolic activity (1). In this species various manipulations of the photoperiod do not effect the circannual cycle of hibernation and its associated physiological changes (5,9,11). MROSOVSKY (6) has proposed that a sequence of physiological stages may occur in this species and that fluctuations in environmental temperature may allow the progression from one stage to the next (3). Thus, the zeitgeber involved in synchronizing the thirteen-lined ground squirrel to seasonal changes and the role of possible circannual rhythms remain to be elucidated.

The function of temporal synergisms of injected hormones (corticosterone and prolactin) and neurotransmitter precursors (5-HTP and L-DOPA) as regulators of seasonality in vertebrates has been investigated extensively by Albert H. Meier and colleagues. The induced seasonal conditions were thought to be the net result of temporal interactions among all the neural and hormonal expressions of two circadian neural oscillations that may be reset by either hormones or 5-HTP and L-DOPA (12). Striking results have been reported for a wide variety of vertebrates (for review see 3,7,8,12). Of particular interest, a series of timed daily injections of prolactin and corticosterone in a 12-hour relation partially inhibited and in a 20-hour relation completely inhibited hibernation in the autumn in the thirteen-lined ground squirrel (2). Although the tendency to hibernate was not studied, various time relations of daily injections of 5-HTP and L-DOPA were found to reset the circannual cycles of the scotosensitive or scotorefractory Syrian hamster with regard to testis weight or uterus weight as well as thyroxine and luteinizing hormone serum levels (12).

Therefore, we studied the thirteen-lined ground squirrel to find out if a series of injections of 5-HTP or L-DOPA would modify the seasonal pattern of autumnal weight gain and the onset of hibernation.

MATERIALS AND METHODS

For use in the experiment, 25 adult thirteen-lined ground squirrels, *Spermophilus tridecemlineatus*, were caught in September. Beginning on October 9, the ground squirrels were maintained on continuous light (110 lux) and with a room temperature of approximately 20 degrees C. Purina rat chow, apples, and water were provided ad libitum. The ground squirrels were weighed before the beginning of injections (October 19) and after the injection period (November 3).

The ground squirrels were divided into 5 groups (with males and females in each group) which beginning on October 21 received 13 daily injections of saline, 5-HTP, L-DOPA, and 5-HTP and L-DOPA in either a 0-hour or 12-hour relation. The saline-injected group received 0.1 ml of 0.9% at 0600 and 1800 EST, the 5-HTP-injected group received 10 mg of 5-HTP in 0.1 ml of 0.9% saline at 0600, the L-DOPA-injected group received 10 mg of L-DOPA in 0.1 ml of .9% saline at 0600, the 0-hour and 12-hour groups received similar injections of 5-HTP followed by L-DOPA in either a 0-hour (both injections at 0600) or a 12-hour relation (5-HTP at 0600 and L-DOPA at 1800).

After receiving 13 days of injections, on November 3 the ground squirrels were put into a dark, cold (5 - 15 degrees C) environmental chamber and were checked every other day for hibernation. Active ground squirrels were given food. If an alive ground squirrel had a rectal temperature within 5 centigrade degrees of the ambient temperature, it was considered to be in hibernation.

RESULTS

As summarized in Table 1, similar seasonal weight gain occurred in each of the five groups of thirteen-lined ground squirrels and no significant differences were found in weight gain between the groups according to the Student's t-test. Weight gains of approximately 25% were found in the thirteen-lined ground squirrels no matter whether they received saline, 5-HTP, L-DOPA, or 5-HTP and L-DOPA in a 0-hour relation or in a 12-hour relation. Additionally, all ground squirrels in the experiment hibernated regardless of the treatment they received. The groups did not differ significantly according to the Student's t-test in the mean daily percent occurrence of hibernation, which had a mean value of about 85%.

Tab. 1: Seasonal weight gain and incidence of hibernation in drug-treated thirteen-lined ground squirrels.

	% Weight Gain	Daily % Hibernating
Saline	25.4 ± 3.5	88.0 ± 6.6
5-HTP	25.2 ± 6.6	73.3 ± 16.7
L-DOPA	35.5 ± 8.4	80.0 ± 17.2
0-hour relation of 5-HTP and L-DOPA	24.5 ± 6.7	84.8 ± 9.5
12-hour relation of 5-HTP and L-DOPA	22.8 ± 5.3	89.3 ± 7.5

No significant differences were found among the groups in either % weight gain or daily % hibernating according to the Student's t-test.

DISCUSSION

The results provided no evidence for a resetting of circannual rhythms of body weight or of the tendency to hibernate in the thirteen-lined ground squirrel by means of timed daily injections of the neurotransmitter precursors 5-HTP and L-DOPA injected in 0-hour and 12-hour relations in autumn.

These time relations were selected because of their known effects in the scotosensitive Syrian hamster (12). In that study there was a dramatic inability of short photoperiods to cause gonadal regression in both male and female scotosensitive Syrian hamsters following repeated injections of 5-HTP and L-DOPA given in a 12-hour relation, whereas injections given in a 0-hour relation had little effect. There is an inverse relation between gonadal development versus fattening and hibernation in the thirteen-lined ground squirrel (4) and other hibernators (for review see (10)). Thus, the 12-hour relation of injections of 5-HTP and L-DOPA might have been expected to enable autumnal gonadal growth and reduce seasonal fat stores in the thirteen-lined ground squirrel and to cause an inhibition of hibernation.

However, individual species differences have been found for the specific temporal relation of 5-HTP and L-DOPA which induces a particular physiological state (8) and consequently species differences may be expected. Other temporal relations should be tested in the autumn and additional experiments should be conducted during other seasons.

ACKNOWLEDGMENTS

We thank Matt Zilich for substantial help in catching and caring for the animals, Ted Cox of Chilicothe, Ohio, for permission to collect the ground squirrels, Mr. & Mrs. Pete Bailer for housing the ground squirrels during the quarantine period, and Dr. Michael J. Miller for veterinary assistance. Financial support was given by the Bethany College Gans Research Fund and the University of Kentucky Faculty Scholars Program.

REFERENCES

1. ARMITAGE, K.B. & E. SHULENBERGER (1972): Evidence for a circannual metabolic cycle in Citellus tridecemlineatus, a hibernator. Comp. Biochem. Physiol. A, 42, 667-688
2. BURNS, J.T. & A.H. MEIER (1988): The phase relation of corticosterone and prolactin injections as a factor in the inhibition of hibernation in the thirteen-lined ground squirrel, Citellus tridecemlineatus. Ann. Rev. Chronopharmacology 5, 57-60
3. BURNS, J.T. & R.W THIELE (1990): The effect of corticosterone on the pigeon cropsac response to prolactin. Ann. Rev. Chronopharmacology 7, 65-68
4. FOSTER, M.A., R.C. FOSTER & R.K. MEYER (1939): Hibernation and endocrines. Endocrinol. 24, 603-612
5. JOHNSON, G.E. & E.L. GANN (1933). Light in relation to the sexual cycle and to hibernation in the thirteen-lined ground squirrel. Anat. Rec., Suppl. 57, 28
6. JOY, J.E. & N. MROSOVSKY (1985): Synchronization of circannual cycles: a cold spring delays the cycles of thirteen-lined ground squirrels. J. Comp. Physiol. A, 156, 125-134
7. MEIER, A.H. & N.D. HORSEMAN (1978): Circadian mechanisms in reproduction. In: N. ALEXANDER (ed.): Animal models for research on contraception and fertility. Harper & Row, New York, pp.33-45
8. MEIER, A.H. & J.M. WILSON (1985): Resetting annual cycles with neurotransmitter-affecting drugs. In: B.K. FOLLETT, S.S. ISCHII & A. CHANDOLA (eds.): The endocrine system and the environment. Japan Scientific Societies Press, Tokyo, 149-156
9. MORRIS, L. & P. MORRISON (1964): Cyclic responses in dormice, Glis glis and ground sqiurrels; Spermophilus tridecemlineatus, exposed to normal and reversed yearly light schedules. Science in Alaska: Proc. 15th Alaskan Sci. Conference AAAS. pp.40-41
10. WANG, L.C.H. (1982): Hibernation and the endocrines. In: C.P. LYMAN, J.S. WILLIS, A. MALAN & L.C.H. WANG (eds.): Hibernation and torpor in mammals and birds. Academic Press, New York, pp.206-236.
11. WELLS, L.J. (1935): Seasonal sexual rhythm and its experimental modification in the male of the thirteen-lined ground squirrel (Citellus tridecemlineatus). Anat. Rec. 62, 409-447
12. WILSON, J.M. & A.H. MEIER (1989): Resetting the annual cycle with timed daily injections of 5-hydroxytryptophan and L-dihydroxyphenylalanine in Syrian hamsters. Chronobiol. Int. 6, No. 2, 113-121

AN ENDOGENOUS CIRCADIAN RHYTHM IN THE SWIMMING ACTIVITY OF THE BLIND MEXICAN CAVE FISH

S. CORDINER & E. MORGAN

School of Biological Sciences, The University of Birmingham
Edgbaston, Birmingham, UK

INTRODUCTION

Astyanax mexicanus is a common characin found in rivers and subterranean pools in Mexico and parts of Texas. Its taxanomic status has been the subject of some debate. Geologically isolated since the early pleistocene (MYERS, 1966; in (2)) and lacking eyes and body pigment, the cave-dwelling fish was originally regarded as a separate species, *Anoptichthys jordani*; only to be shown subsequently to interbreed experimentally with the pigmented river fish whose name it now shares (Sadoglu, 1982).

Although now bred commercially, the hypogean form is of interest to the chronobiologist in that many of the geophysical variables which characterize the solar day are precluded from its ancestral caverns. Dial changes in light and darkness, the predominant zeitgeber for the entrainment of circadian rhythms in epigean organisms are unlikely to be experienced by cave-dwelling specimens of *A. mexicanus* in their natural state, and the waters in which these are normally found are presumably thermostable over the dial period (4). The selective advantage of endogenous circadian rhythmicity in such a habitat is unclear, and it seems reasonable to expect that the endogenous time-keeping mechanism of the hypogean from many have regressed during the period of its reproductive isolation. As in many cave-dwelling animals, the visual system has indeed regressed (6) and experiments reported by ERCKENS & MARTIN (2) suggest that, although these fish are still photo- responsive to 24h cycles of light and darkness, the endogenous circadian control system may system may have degenerated. Recently however it has become apparent that such dial cycles can entrain an endogenous rhythm of locomotor activity in the eyeless hypogean form of A. mexicanus and the results of some preliminary experiments are reported here.

MATERIALS AND METHODS

Juvenile fish of indistinguishable sex and of unknown genotype, with an average weight of 1g and an average fork length of 3.5 cm were obtained from a local aquarist. The fish were kept individually in circular Perspex tanks, 18 cm diameter x 9.5 cm high, with a raceway 2.5 cm wide, filled to a depth of 7 cm with water at a temperature of $24° \pm 1°C$ and held in a controlled environment room. Illumination was provided by a single 40W daylight fluorescent tube, situated 50 cm above the tanks, which were shielded with opaque blue polythene sheet to diffuse the light and to reduce its intensity. Illumination at the water's surface was 50 lux and LD 12:12 photoperiods were generated using a Smiths electronic timeswitch.

The fish were fed on powdered fish-flake at irregular intervals of between one to seven days, and at a different time each day to avoid possible entrainment to the time of feeding.

Locomotor activity was measured with a paired infra-red photocell system which straddled the raceway, with the sensors positioned approximately one and four centimeters above the bottom of the tank. Each sensor was connected to a data-logger (Birmingham University; Biochemistry Electronics) which recorded the number of beam interruptions and summated the data over a selected recording interval. At the end of each recording session the data were transferred to a BBC microprocessor and to the University mainframe computer for time-series analysis.

RESULTS AND DISCUSSION

Juvenile mexicanus were found to swim continuously in the recording tanks, but overhead illumination in a LD 12:12 regimen imposed a diurnal variation on their vertical distribution in the water column. The fish spent less time near the surface or in the supper part of the tank during the light period than during the dark, when the swimming excursions were more evenly distributed throughout the tank. Comparison of the total swimming activity recorded by both upper and lower sensors during the dark with that recorded during the light period suggests that the fish were also generally less active during the photophase. Fig. 1 shows the activity recorded at the lower sensor over a period of six days in the entraining regimen, and during a further six days after 44 days in constant light.

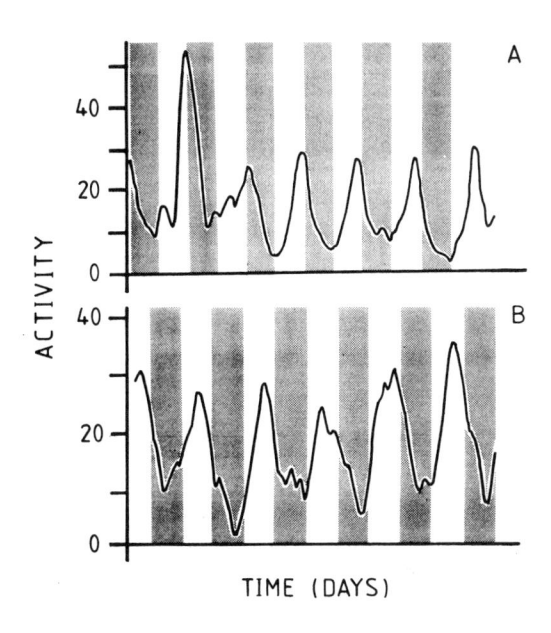

Fig.1:
Retraced computergraphic print out showing the activity pattern of juvenile *Astyanax mexicanus* in a LD. 12:12 lighting regimen (A), and after 44 days in constant light (B). Activity was recorded every 60 mins and the data smoothed using a 5 point moving average. Shaded regions indicate the time of darkness (A) or expected darkness (B).

A similar periodic change in distribution was evident following transfer to constant light (Fig.1B) and MESA (1) and Periodogram analysis (3) of these and other data indicated a free-running period of about 24h. This periodicity was often masked by apparently random, infradian oscillations in the overall activity level, and array analysis (5) suggests that the period of the underlying oscillations may be labile, sometimes exceeding and sometimes being less than 24h. Thus the bouts of benthic swimming activity sometimes drift earlier and sometimes later relative to the subjective photophase. Fortuitously, the activity pattern of the fish illustrated in Fig. 1B had almost reverted to its original phase relationship with the erstwhile entraining regimen.

The irregular feeding routine suggests that light, rather than food availability was the main zeitgeber. Prior to the experiments, the juvenile fish had been fed sporadically but mainly during the light phase, when they came to the surface to take the powdered food. In contrast they preferred the lower part of the tank during the light phase of the imposed photic regimen, and during the subjective light phase following transfer to constant conditions, i.e. a behaviour pattern opposite to that expected if feeding, rather than light was the entraining agent.

These results are at variance with those of ERCKENS & MARTIN (2) who concluded that the hypogean forms of *A. mexicanus* they studied showed, at best, only a weak endogenous circadian rhythm, and there are a number of possible explanations for the difference between the two sets of results. The suggestion that the time-keeping mechanism is lost during the later stages of development in the hypogean form has recently been shown to be untenable, as adult specimens show circadian entrainment to stimuli other than light (N. ZAFAR, unpublished data), and an alternative explanation, that the light threshold for entrainment is more readily exceeded in young fish, is more plausible. Both juvenile and adult specimens respond to changes in light intensity, and the eye rudiments and pineal are potential receptors and pathways for photic entrainment. In juveniles of the hypogean form these are more superficially located than in the adult. Age dependent changes in feeding habits from surface to benthic foraging may also affect their distribution but perhaps more significant is the possible difference in genotype of the fishes used in the two studies. The adult specimens used by ERCKENS & MARTIN (2) were inbred from wild stock collected from the Cueva de El Pachon, in the Sierra de El Abra region of Mexico. In contrast, despite their morphological similarities with the natural hypogean form (cf. SADOGLU, 1982) the juvenile specimens used in the current investigation were of uncertain genotype, bred commercially and the possibility of interbreeding with the epigean ancestral stock at some time during this process cannot be precluded. Possible genetic outcrossing apart, the selective advantage of maintaining an endogenous circadian rhythm of swimming activity in a relatively constant cave environment is unclear, and the rhythmic swimming pattern described above may well result from the entrainment of a physiological vestige, as ERCKENS & MARTIN (1982) suggest.

REFERENCES

1. DOWSE, H.B., J.C. HALL & J. RINGO (1987). Circadian and ultradian rhythms in period mutants of Drosophila melanogaster. Behaviour Genetics 17: 19-35.
2. ERCKENS, W. & MARTIN, W. (1982). Exogenous and endogenous control of swimming activity in Astyanax mexicanus (Characidae; Pisces) by direct light response and by a circadian oscillator. II. Features of time-controlled behaviour of a cave population and their comparison to a epigean ancestral form. Z. Naturforsch, 37c, 1266-1273.
3. HARRIS, G.J. & E. MORGAN (1983). Estimates of significance in periodogram analysis of damped oscillations in a biological time series. Behaviour Analysis Letters 3: 221-230.
4. MITCHELL, R.W., W.H. RUSSELL & W.R. ELLIOT (1977). Mexican eyeless characin fishes, Genus Astyanax: Environment, distribution and evolution. Special Publications, The Museum, Texas Tech. University, No.12, 89pp.
5. PALMER, J.D. & B.E. WILLIAMS (1986). Comparative studies of Tidal Rhythms 1. The characterization of the activity rhythm of the Pliant-Pendulum Crab, Helice crassa. Mar. Behav.Physiol. 12: 197-207.
6. ZILLES, K., B. TILLMANN & R. BENNEMANN (1983). The development of the eye in Astyanax mexicanus (Characidae, Pisces), its blind derivative, Anoptichthys jordani (Characidae, Pisces) and their crossbreds. Cell Tissue Res. 229 : 423-432.

NONLINEAR DYNAMICS CAUSED BY PHASE SHIFTS

G. TSCHUCH & T. SÜSSMUTH

Zoologisches Institut, FB Biologie der Martin-Luther-Universität,
Domplatz 4, D-O 4010 Halle, FRG

In rhythmic systems without memory it is possible to analyze the influence of zeitgebers to an oscillator by computer simulations using the natural phase response curve. At first we'll introduce and define the used parameters as follows (Fig.1). The unperturbed period of the oscillator should be T_0. This is at the same time the reference for all measurements of phases and times. If a stimulus happens at time t_s the period T will be influenced depending on the phase ϕ defined by $t_s - t_0$ in relation to T_0. Phase response plots are often diagrams with normalized $T(\phi)$ versus ϕ or diagrams with the resulting new phase ϕ_{new} versus the old phase ϕ_{old}. We can calculate the new phase as $[\phi_{old} - T(\phi_{old}) / T_0]$ mod 1. The operation "modulus 1" should suggest that you have to add or substract 1 until the phase is in the interval 0 to 1.

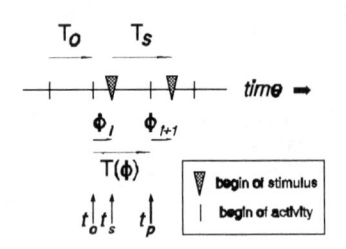

Fig. 1

In Fig. 2 we see an example for a typical phase response from type 1 plotted in two different graphs. In type 1 resetting the stimulus has less influence as in type

0. The one extreme for type 1 response would be "no influence" with a horizontal line in the upper plot and a linear dependence with an increase of 1 in the lower plot. The other extreme is a fixed phase ϕ_{new} independent of ϕ_{old}. In the upper graph we had then to draw a decreasing line.

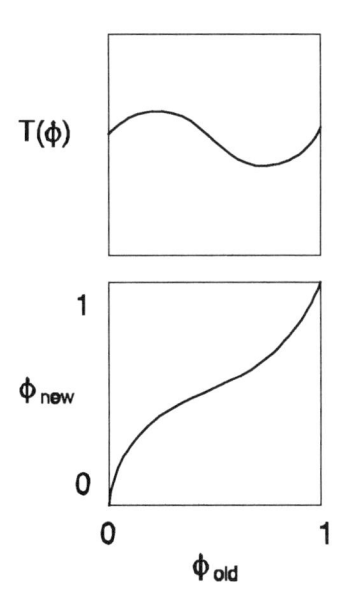

$T(\phi)$

ϕ_{new}

ϕ_{old}

Fig. 2

The present paper deals with periodic stimuli (Fig. 1). In this case it is possible to calculate the next phase ϕ_{i+1} as [(present phase ϕ_i - perturbed period $T(\phi_i)$ - stimulus period T_s) / T_0] mod 1. With this iteration a so called POINCARÉ map is defined.

The computer simulations were e.g. done with a phase response curve of type 1 like in Fig. 1 merely shifted for 0.1 along the x-axis. At first the phases of 200 iterations are plotted for a ratio of stimulus period to unperturbed period between 0.5

and 2. This plot contains no information of the resulting perturbed period, but we could recognize regions with phase locking. Because of nonlinearity of the process there are not only lockings with 1:1, 2:1 and so on, but also lockings with e.g. 3:2, 4:3. In this regions it is possible to find alternations between 2, 3 or more different phases. In other regions deterministical chaos occurs.

In the region of phase locking near $T_s/T_0 = 1$ the resulting perturbed oscillator period increases and decreases linear with increasing or decreasing stimulus period. In other narrow regions of phase locking we can find period doubling, so called bifurcations, and higher splits.

A simple modification of the computer program allows to investigate interactions between two oscillators. In this case the stimulus period is not constant but it is additionally disturbed by the oscillator with the same phase response. Here some phase lockings are wider as in the example with the fixed stimulus period. Besides, especially in the regions around 1:1 and 3:2 intervals with nearly constant pase angles occur.

By calculating the perturbed periods we found an interesting phenomenon around 1:1 locking. The oscillator with the higher frequency is in a wide range not influenced, but the oscillator with the lower frequency shifts to shorter periods.

At present we test the usefulness of such simulations on the chorus of male crickets. Two singing males influence their calling rates reciprocally and so we have a useful model with periods lower 1 second. Fortunately the phase response is of type1.

In summary we can notice that linear approaches are not practicable for the investigation of rhythmically influenced oscillators.

REFERENCE

1. WINFREE, A.T. (1980): The Geometry of Biological Time. Springer, New York-Heidelberg-Berlin

CIRCADIAN AND CIRCANNUAL ACTIVITY RHYTHMS OF MICROTUS BRANDTI (RADDE, 1861)[1]

A. STUBBE & U. ZÖPHEL

Institute of Zoology, Martin-Luther-University Halle,
Domplatz 4, D-O-4020 Halle/Saale, FRG

INTRODUCTION

The vole *Microtus brandti* is a characteristic species of the steppe zones of Central Asia, especially Mongolia and China. It's a very social animal living in big colonies. A high degree of day-time activity was observed by field investigations.

First we have looked in which group size *Microtus brandti* shows it's normal activity pattern. It was found, that it must be a group with more than three animals. The next step was the investigation of the activity rhythms of the voles around the year influenced by a simulated photoperiod. The results of the third stage are now produced and will be published in the next future. There not only the annual photoperiod but also the varying temperature around the year had influence on the activity rhythms of the voles.

MATERIAL AND METHODS

The activity rhythms were recorded by light barriers. For studying the circadian activity rhythms animals were kept under standardized environmental conditions (~21°C, 50-60% humidity, L:D=12:12, light intensity about 300-400 Lux). Investigations were carried out in animal groups consisting of two up to five and in single males.

Circannual activity rhythms were investigated under the same climatic conditions as above, only the light-dark relation was varying in accordance with the natural

---

[1]Results of the Mongolian-German Biological Expeditions since 1962, No. 218

photoperiod. The animal groups consisted of six males or three males and three females, so called "mixed" groups.

RESULTS

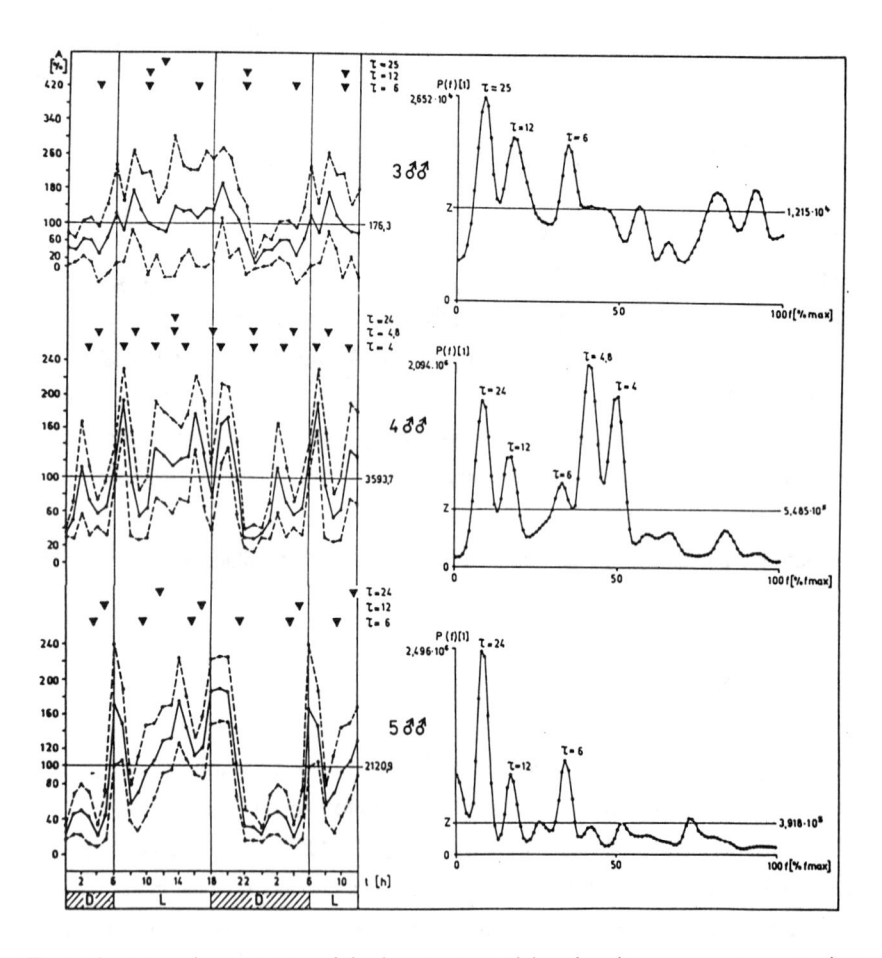

Fig.1: Average chronograms of the locomotor activity of male groups represented as the deviation from the whole average value (=100%) over seven days with dispersion (broken lines) and acrophases ( ) -left- ; and power spectra -right-.

Single males are displaying a high variety of individual activity patterns with a lot of activity peaks and high dispersion in the average chronograms. We can distinguish three main forms of locomotor activity patterns:
-animals, more active at light-time,
-animals, dividing their locomotor activity more or less equaly around the day,
-animals, more active at dark-time.

By investigations of animal groups consisting of two up to five males the pattern of locomotor activity will be clearer, if you take three or more animals (Fig. 1, left). Then it consists of three activity peaks with predominant activity during day-time. If you have two or three males, the period length of circadian rhythm seems to be longer then 24h (25h - 27h). Four males have already a manifest circadian oscillation but first in a group size of five animals it dominates harmonious ultradian components (Fig. 1, right).

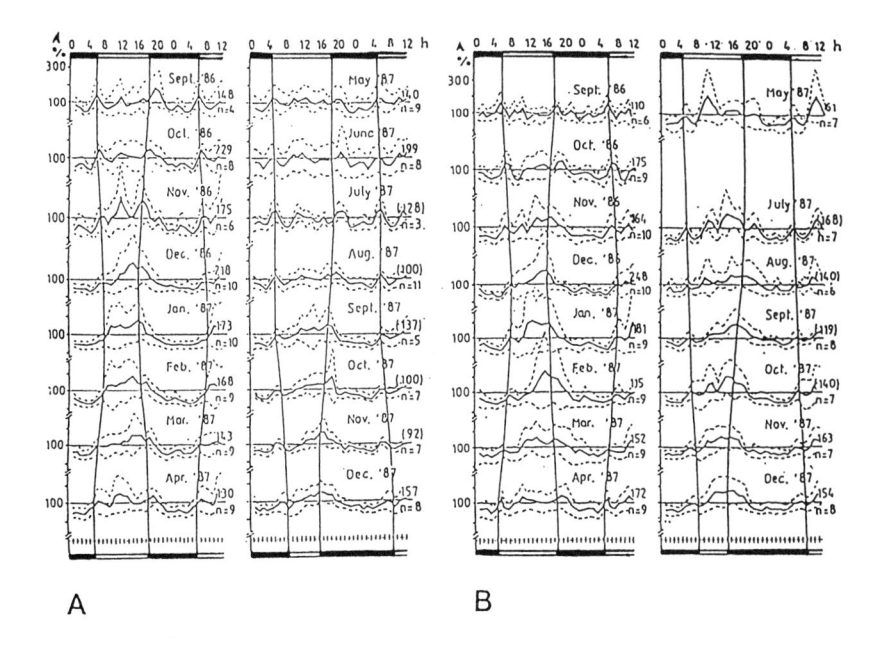

A   B

Fig.2: A - Monthly average values of the locomotor activity of male groups with standard deviation (broken lines) around the year, n = number of investigated animal groups, figure on the middle line = whole average value of the month.
B - The same as 2A, only for mixed groups.

Hence social synchronization can be suggested. For these reasons we have enclosed in our programme of circannual rhythm studies only animal groups consisting of six males or three males and three females.

Fig.2A shows the activity patterns of male and Fig.2B these of mixed groups as monthly average chronograms with standard deviation (broken lines) through the year.

By influence of a simulated annual photoperiod the above found basic activity pattern will be changed in the following way:

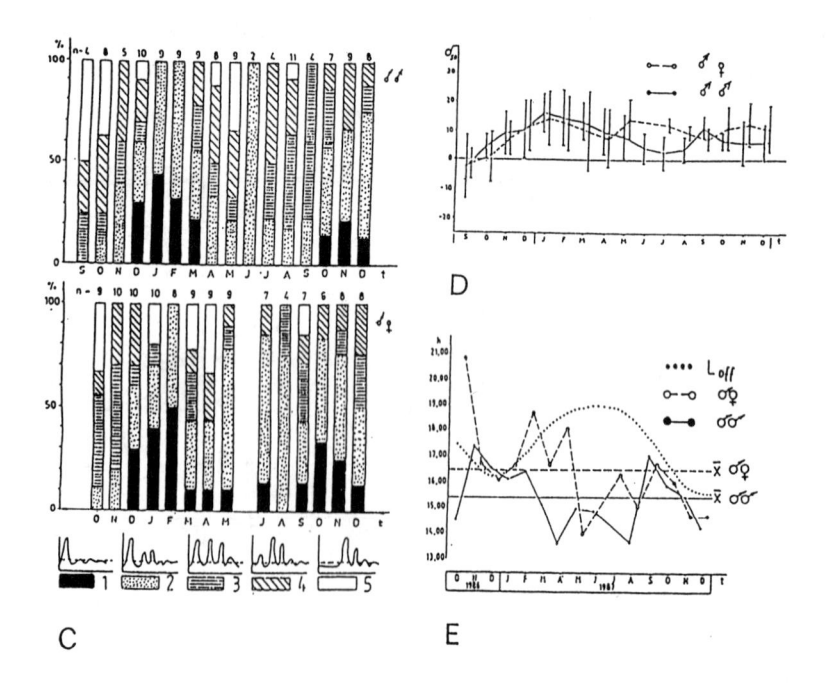

Fig.2: C - Monthly percentage of power spectra types around the year for male (above) and mixed groups (below).
D - Monthly average δ_{50}-values with standard deviation around the year.
E - Monthly average values of acrophases of the circadian oscillation in relation to L_{off} around the year.

1. Under long-day conditions, especially in midsummer, the pattern is typical polymodal and we found:
   - distinct peaks around sunrise and sunset,
   - two up to five peaks between sunrise and sunset,
   - clearly decreasing activity in the first half of the night,
   - a small increase of activity at the beginning of the second nighthalf.

2. Under short-day conditions the pattern will be unimodal, especially by fusion with the approaching peak of sunset.

The dominating daily activity will be kept through the year. A numerical value for this phenomenon can be the δ_{50}-index of HALLE & LEHMANN (6). If $\delta_{50} > 0$, then the activity is higher in light-time and if $\delta_{50} < 0$ higher in dark-time. Fig. 2D shows the average δ_{50}-values with standard deviation for all investigated animal groups. Male groups seem to have a higher degree of daily activity in short-day conditions, whereas in midsummer the activity is divided more equally between day and night. In mixed groups this tendency can be overlayed by changing activity patterns of females in gravidity from April to September, the reproduction time. Hence mixed groups have also a high level of daily activity under long-day conditions.

By Fourier-analysis we can show the following points. Through the year we have observed different types of power spectra (Fig. 2C):

Type 1: circadian component is predominant, ultradian components are stochastical (black columns),

Type 2: circadian component dominant, ultradians subordinated (pointed columns),

Type 3: circadian and ultradian components at the same level (lined columns),

Type 4: ultradian components dominant, circadian component present (diagonal lined columns),

Type 5: ultradian components predominant, circadian absent (white columns).

The Fig. 2C displays the percentage of power-spectra types each month:

- In short-day conditions (December to February) the share of circadian oscillation is reinforced.
- At times with rapidly changing light-dark relations (September/October and March/April/May) we found a higher percentage of ultradian rhythms, which can dominate or displace circadian components.
- In midsummer (long-day conditions) we see again more circadian oscillations, but their percentage is lower than in winter.

The acrophases of the circadian oscillation (first maximum) are placed normally at the second half of light-time (Fig. 2E shows that in relation to L_{off}). Under short-day conditions the acrophase lies nearer to L_{off}, under long- day conditions we can find it more frequently at the begin of afternoon.

DISCUSSION

If *Microtus brandti* shows its normal activity pattern only, when living in groups of more than three animals, social synchronization should be suggested. Fourier analysis reveals a dominant circadian and several subordinated harmonious ultradian components in groups of five animals. Similar results were found by SCHEIBE et al. (11) in their investigations on sheeps.

Animal groups kept under annual varying LD and constant temperature are also more active during light-time, but the circadian oscillation is clearly predominant only under short-day conditions. These are manifesting stronger the circadian component; but light conditions like around the equinoxes increase the ultradian oscillations. Similar notes are to be find in KATINAS (7). So we have observed disturbed circadian order up to missing the circadian oscillation and predominant ultradian components during spring and autumn. Also ERDAKOV (3) reported on disturbed rhythmicity during the transition time from summer to autumn for several vole species. In midsummer the activity levels of *Microtus brandti* in light- and dark-time can be nearly equal. The power spectra show circadian rhythms but also ultradian components.

The results of other authors concerning activity phase changes are often contradictory. So OSTERMANN (9), ERKINARO (4,5), ROWSEMITT et al. (10) by laboratory investigations and BÄUMLER (1) by field studies reported on a real phase change of vole activity, whereas LEHMANN & SOMMERSBERG (8) and

BLUMENBERG (2) couldn't observe this. STEBBINS (12) asserted, that the activity-phase change is influenced by the latitude. In our investigations *Microtus brandti* didn't show a full phase change of activity like ERKINARO (5) found it for some other *Microtus*-species. Only the activity niveau varied depending on the season.

In accordance with ROWSEMITT et al. (10) we're interpreting the seasonal changes of activity patterns and power spectra as consequences of adaptation to the natural living conditions of *Microtus brandti*, especially to the climatic factors.

REFERENCES

1. BÄUMLER, W. (1975): Activity of some small mammals in the field. Acta theriol. 20, 365-377
2. BLUMENBERG, D. (1986): Telemetrische und endoskopische Untersuchungen zur Soziologie, zur Aktivität und zum Massenwechsel der Feldmaus, *Microtus arvalis* (Pall.). Z. f. Angew. Zool. 73, 301-344
3. ERDAKOV, L.N. (1984): Organisacija ritmov aktivnosti grysunov. Novosibirsk
4. ERKINARO, E. (1969): Der Phasenwechsel der lokomotorischen Aktivität bei *Microtus grestis* (L.), *M. arvalis* (Pall.) und *M. oeconomus* (Pall.). Aquilo, Ser. Zool. 8, 1-29
5. ERKINARO, E. (1972): Seasonal change in the phase position of circadian activity rhythms in some voles, and their endogenous component. Aquilo, Ser. Zool. 13, 87-91
6. HALLE, S. & U. LEHMANN (1987): Circadian activity patterns, photoperiodic responses and population cycles in voles. I. Long-term variations in circadian activity patterns. Oecologia 71, 568-572
7. KATINAS, G.S. (1981): Zur Charakterisierung ultradianer Biorhythmen des Funktionszustandes einiger Gewebe bei adaptiven Reaktionen. In: SCHUH, J., K. HECHT & J.A. ROMANOW (eds.): Chronobiologie-Chronomedizin. Abh. Akad. Wiss. DDR 1981, 1979, Nr. 1N, 511-526
8. LEHMANN, U. & C.W. SOMMERSBERG (1980): Activity patterns of the Common Vole, *Microtus arvalis* - Automatic Recording of Behavior in an Enclosure. Oecologia 47, 61-75
9. OSTERMANN, K. (1956): Zur Aktivität der heimischen Muriden und Gliriden. Zool. Jb. (Physiol.) 66, 355-388
10. ROWSEMITT, C.N., L.J. PETTERBORG, L.E. CLAYPOOL, et.al. (1982): Photoperiodic induction of diurnal locomotor activity in *Microtus montanus*, the montane vole. Can. J. Zool. 60, 2798-2803
11. SCHEIBE, K., R. SINZ & G. TEMBROCK (1977): Chronophysiologische Synchronisationsanalyse in der Tierproduktion. Wiss. Beitr. MLU Halle 1977/40 (PG), Chronobiologie 76, 281-291
12. STEBBINS, L.L. (1972): Seasonal and latitudinal variations in the circadian rhythms of Red-backed Vole. Arctic 25: 216-224

COORDINATION AND INTERACTION

COORDINATION OF BIOLOGICAL RHYTHMS. FREQUENCY- AND PHASE COORDINATION OF RHYTHMIC FUNCTIONS IN MAN.

G. HILDEBRANDT

Institut für Arbeitsphysiologie und Rehabilitationsforschung,
Philipps-Universität Marburg, Robert-Koch-Str. 7a, D-3550 Marburg, FRG

Modern chronobiology has pointed out, that the rhythmical functions of the organism establish a complex time structure, in which frequencies and phases are ordered to ensure optimal cooperation of the various functions. However, the question for the underlying mechanisms of this co-operation cannot be answered in general. We must consider, that within the broad spectrum of biological rhythms different preconditions exist for a co-operation between different rhythms.

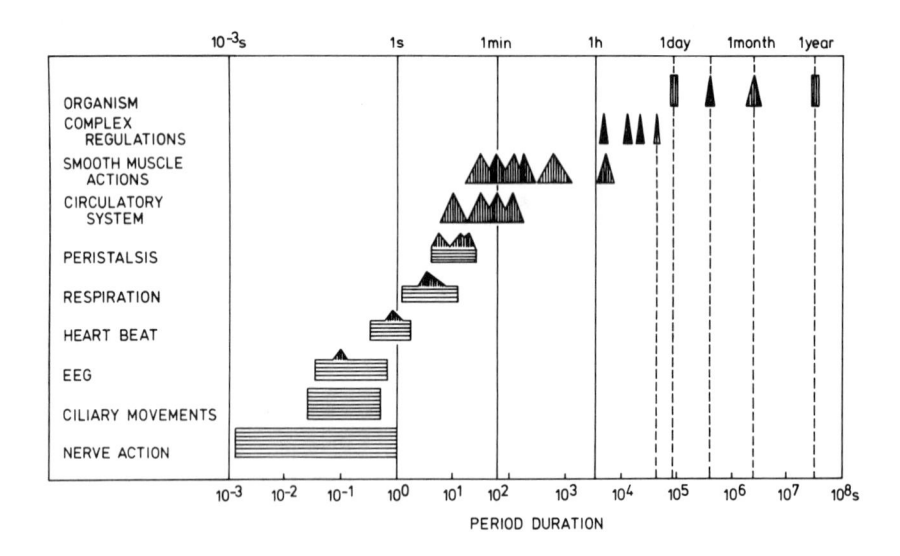

Fig.1: Spectrum of human rhythmical funtions according to their period duration. Horizontally hatched areas indicate the range of frequency change under functional effort. Vertically hatched triangles indicate the statistical variability of frequencies in repose. (From 14).

Looking at the spectrum of period lengths of rhythmical functions in man (Fig.1), we can derive that with increasing period lengths the rhythms become increasingly complex. This begins with cellular high-frequency rhythms, going on to organ and systemic rhythms up to the circadian variations, which already involve the entire organism, and finally to the reproductive functions and rhythms of whole populations. More and more single functions are comprised to a synchronized coaction with increasing period duration. At the same time, rhythms are increasingly controlled by hormonal mechanisms, whereas in the high-frequency section nervous control is dominating.

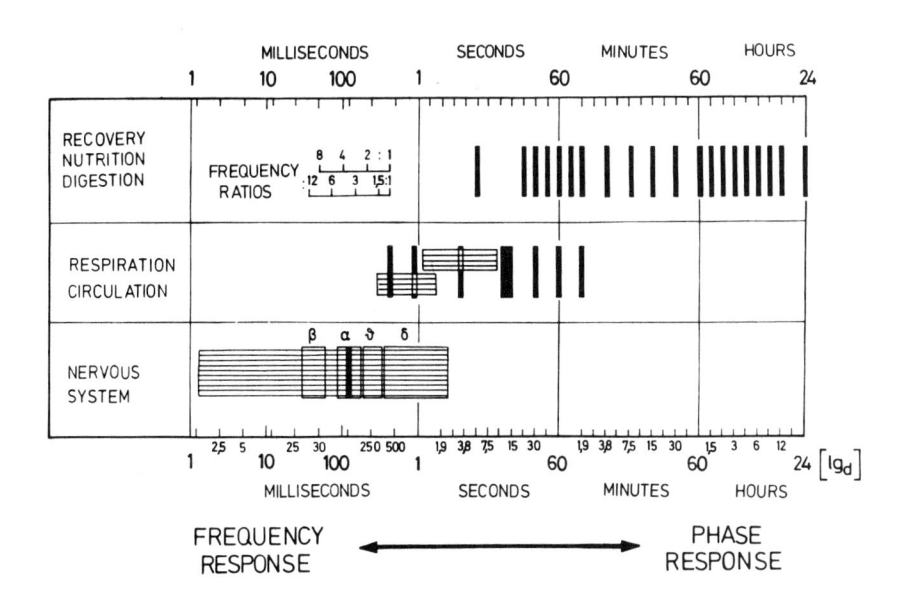

Fig.2: Frequency behavior and modes of response of the endogenous rhythmic functions in different parts of the spectrum. Black vertical bars indicate the preferred frequency bands, horizontal hatching indicates the range of frequency modulation. For further explanation, see the text. (After 15, modified)

From this survey we can derive, that the longer-wave rhythms must be much more suited for synchronizing effects from each others, whereas the faster rhythms tend to act independently from other rhythms in respect to their phase relationships.

This can also be proved, if we look at the functional behavior of frequencies of the various rhythmical functions in the different parts of the spectrum (Fig. 2). According to the functional significance three different parts of the spectrum may be considered. In the longer-wave section we may discover mainly rhythms of the metabolic system, e.g. nutritional, digestive, and recovery functions. By contrast, the short-wave section contains the rhythmic actions of the nervous system, forming an information system. In the intermediate range we find a predominance of rhythms serving the transportation and distribution functions, mainly of the circulatory, respiratory, and intestinal system, as well as locomotor rhythms.

As symbolized by the horizontal hatching, the rhythms of nerval action exhibit the greatest variability of frequency. These rhythms are responsible for all the transmission and communication in the information system by portraying the momentary degree of excitation by means of frequency modulation. It is only during sleep, that the rhythmicity of the central nervous system becomes partially synchronized into certain frequency bands of the EEG. Thus we may speak of a *frequency-responding system*.

By contrast, the rhythms of the longer-wave section prefer distinct frequency bands, which are ordered into integer ratios. In the figure, the black bars mark the preferred frequency bands or frequency norms, respectively. Some of them have proved to dispose of stabilizing mechanisms, as for instance with regard to temperature. The abscissa represents a logarithmic scale, hence the frequency ratios, as indicated in the left upper part, apply to all parts of the spectrum. In order to maintain this temporal order, the rhythmic functions tend to respond to interferences by modifying their phase relations, whereby changes in frequency only occur by "jumping" into other preferred bands of the harmonic system. Therefore, these rhythms of the metabolic system are characterized by phase responses, leading to frequency multiplication or demultiplication, respectively.

In the middle wave-length section the rhythmical functions of the transportation- and distribution system exhibit both: frequency-modulation response to functional loads as well as setting of frequency norms by coordinating the rhythmical actions to a larger harmonic time structure of integer frequency ratios, based on the preference of phase response.

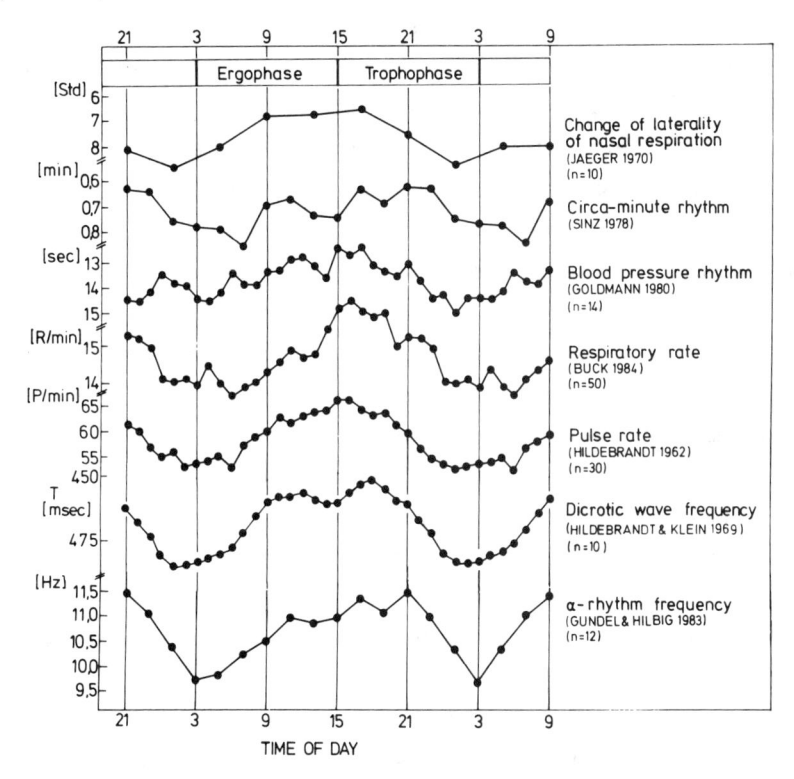

Fig.3: Daily courses of the mean frequencies of various ultradian rhythms. A compilation of data from the literature. (From 16).

Concerning the interactions of the various rhythmical functions, as based on these structural principles, we must expect slower rhythms to be capable to act upon the faster rhythms preferably by modulating their frequencies. This can be exemplified already by the wellknown respiratory modulations of heart rate (respiratory arrhythmia) or - more general - by the circadian variations of the frequency of various rhythmic functions like (Fig. 3): blood-pressure rhythm, circa-minute rhythms, arterial dicrotic wave length or alpha frequency in the EEG as well as the laterality rhythm of nasal respiration. The amplitudes of circadian frequency modulations are similar, amounting to about 10-12 %.

On the other hand, faster rhythms can act upon slower ones by originating phase responses, leading to certain phase coupling phenomena. However, this mostly occurs by reciprocal interactions whereby both of the rhythms leave their original phase position and period length in a compromise. The resulting coaction must depend on the specific response curves of the rhythms concerned.

Under natural conditions, the long-wave rhythms like the circadian rhythm are the most stable ones in respect to frequency and phase position because of the external entrainment by the environmental zeitgebers. However, in isolation experiments under free running conditions, ASCHOFF (1), WEVER (33), and their group were able to show disturbances of the internal synchronization between different components of the circadian system, free running with different period lengths and synchronizing to one compromising frequency according to the strength of the coacting oscillators.

MENZEL (23) for the first time pointed out (1955), that patients suffering from sleep disorders exhibit a prominent 12-hour period of vegetative functions instead of the normal 24-hour rhythm. Furthermore, he observed in patients, that with increasing loss of functioning the daily courses were superimposed by submultiple periods of the circadian rhythm, mainly 12-, 8-, 6-hour periods.

The preference of whole-number relationships to the 24-hour period was confirmed by results obtained by pergressive FOURIER analysis (2). In the mean time many observations of various authors could confirm MENZEL's concept of a harmonic system of frequency multiplication in the circadian system.

However, up to now it has not been considered that the various frequency multiples of the 24-hour period keep a fixed phase position within the 24-hour day and by this prove a strong connection to the synchronized circadian system. Fig. 4 shows a few examples of dominant 12-hour periods within the daily courses of various parameters. Physiological functions like body temperature or respiration rate as well as vigilance parameters or even death rate of cardiac patients exhibit in principle identical phase positions, passing through maximal ergophases around 9.00 and 21.00 hours and maximal trophophases around 3.00 and 15.00 hours.

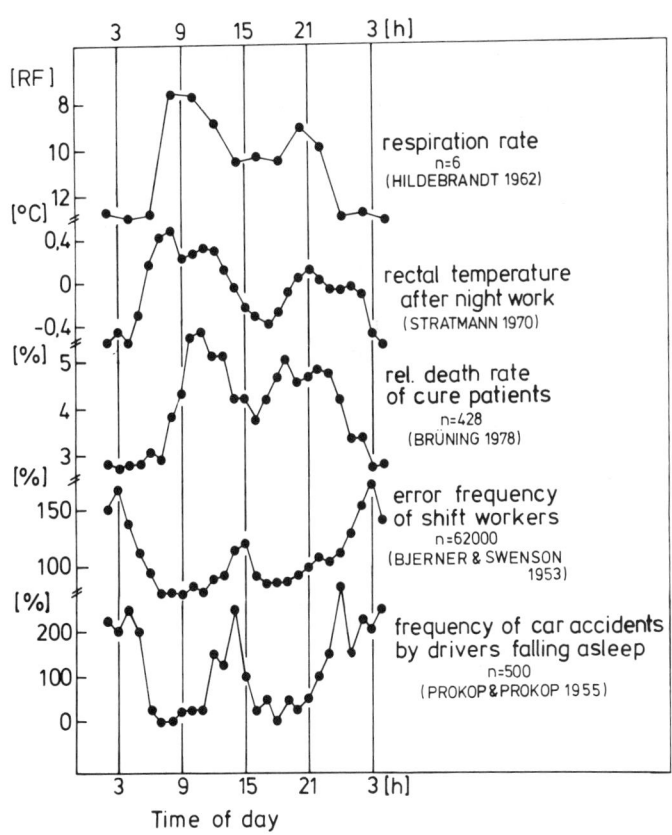

Fig. 4: Mean daily courses of various physiological and pathological parameters, exhibiting a superimposition of 12-hour periods. A compilation of data from the literature.

Respective findings can be demonstrated for 8- and 6-hour ultradian periods. Furthermore, the phase-binding influence of the synchronized circadian system is to be found also in a stable phase position of the 90-min Basic Rest Activity Cycle (REM-Non REM-Cycle). Fig. 5 shows the courses of pain threshold during forenoon of three subjects who were studied each three times at different intervals up to 46 days. As you can see, no significant phase difference can be detected between the different individual curves. This was true also in the majority of 10 more subjects, as controlled 3 times by PÖLLMANN (un-published) of our group.

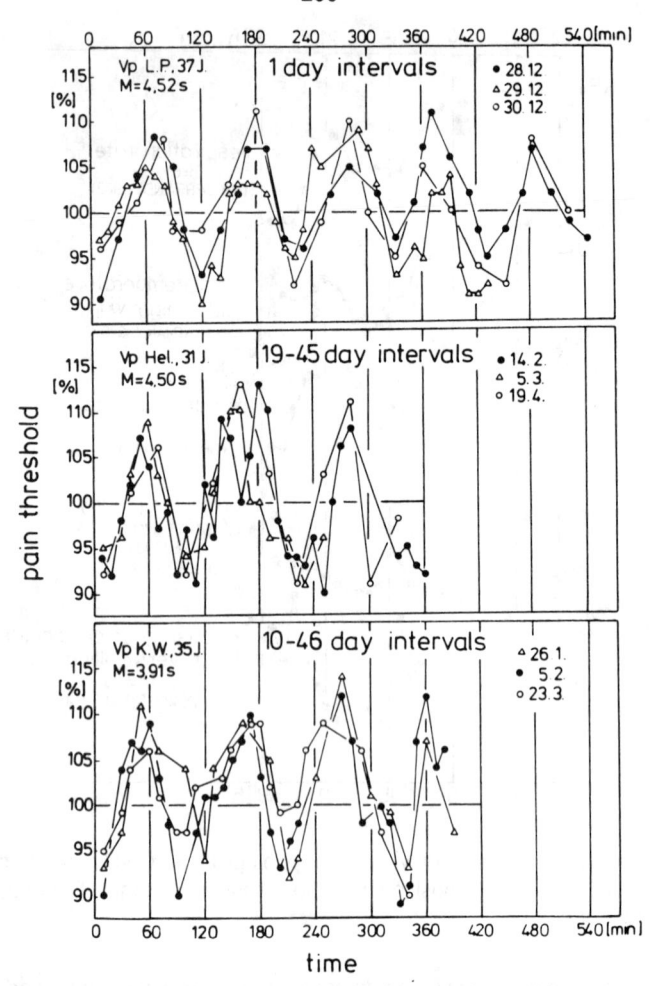

Fig.5: Spontaneous variation of the pain threshold in healthy subjects as measured by the minimum cold application time during forenoon. Each subject was controlled at three different days with different intervals. (From 20).

These findings show, that the phase coupling influence of the synchronized circadian system can be extended up to a very broad range of multiple ultradian frequencies. Here the dominance of phase coupling effects allows changes in frequencies, by this maintaining distinct phase relationships between the rhythmic functions.

Concerning the mechanism of phase coupling in this part of the period spectrum, we cannot decide up to now, whether or not external synchronizing factors are involved, however, we can assume, that maxima of responsiveness of the participating rhythms determine the resulting phase coaction. But especially in man, studies in this field are missing.

Of course, the strict phase binding influence of the synchronized basis of the circadian system diminishes stepwise with higher frequencies of the rhythmical functions. Nevertheless, the preference of whole number frequency relationships between the rhythms can be demonstrated also in the transportation and distribution functions in the middle part of the spectrum, irrespective of different underlying mechanisms.

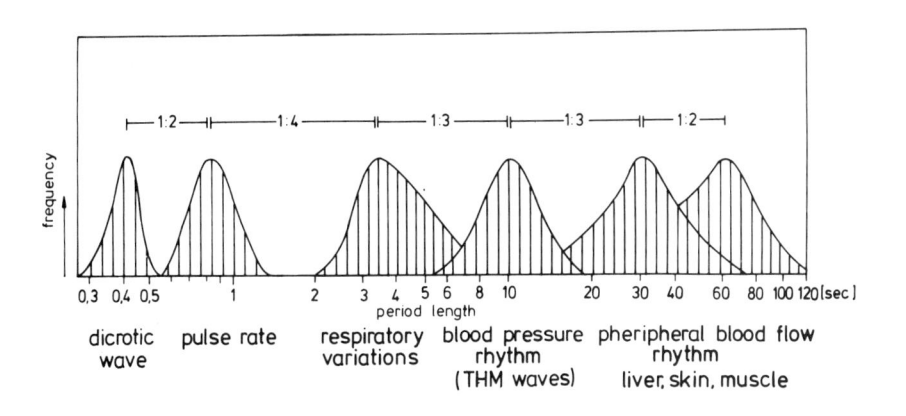

Fig. 6: Frequency distributions of the period lengths of various respiratory and circulatory rhythms in man. (From 14, modified).

Fig. 6 shows a compilation of histograms of period lengths of various respiratory and circulatory rhythms, resulting from numerous studies in healthy resting subjects. There is a great variability of period lengths in the different functions, however, a statistical preference of whole number frequency relationships can be detected.

During night sleep the variability decreases to form very strict whole number frequency relationships, as shown, for instance, by the findings of RASCHKE (29) in our laboratory.

Rhythmic functions of about 1 min period length, so called 1-min rhythms, occur mainly in the smooth muscle system, e.g. in the intestinal tract, blood vessels or uterus motility, respectively (10). They are, on the one hand, synchronized by central nervous control, inducing constant phase relationships of the blood flow rhythms in different organs (Fig. 7). On the other hand, multiple frequencies can be observed, indicating that mechanisms of phase coupling also determine the coaction of the different rhythmic functions.

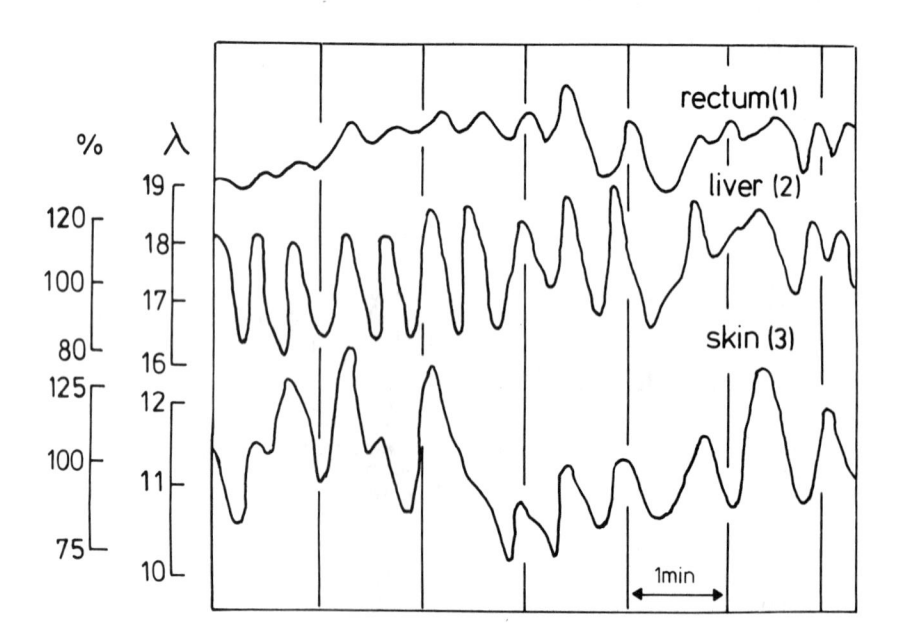

Fig.7: Spontaneous rhythmical course of blood flow of various organs in man, as measured simultaneously by heated thermocouple probes. (From 13).

Already WEITZ & VOLLERS (32) demonstrated by pressure registrations in different sections of the colon, that peristaltic activity is coordinated into integer frequency ratios and certain phase relationships. The rhythm of human stomach peristalsis is three times faster than the 1-min rhythm of the spontaneous activity of the fundus (10), and the rhythm of peristalsis of the stomach exhibits a frequency ratio of 1:4 as compared to the peristalsis of the duodenum (5).

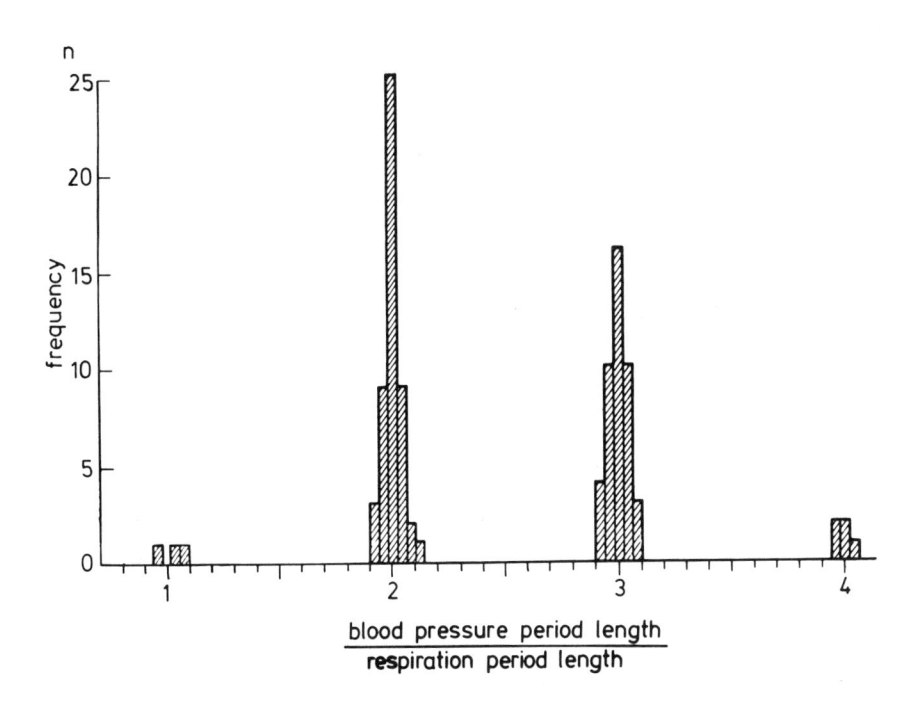

Fig.8: Frequency distribution of the ratio between the period lengths of blood pressure waves and respiration in a dog after blockade of N. phrenicus. (From 22).

These findings indicate, that internal peripheral mechanisms are involved to harmonize rhythmical functions of different own frequencies.

On the other hand, the coordination between blood pressure rhythm and respiration represents an example for the fact, that phase coupling between rhythms of different frequencies, leading to a preference of whole number frequency relationship, can be produced by central nervous mechanisms without any participation of peripheral influences.

During night sleep the preferred ratio between these rhythms is 4:1, but at daytime a preference of 3:1 can be observed too. In animal experiments KOEPCHEN & THURAU (22) were able to prove, that after cutting the N. phrenicus in order to remove the mechanical effect of respiration, the blood pressure rhythm still exhibited strict phase coupling to the central respiration rhythm (Fig. 8). Irradiations between the activities of respiratory and circulatory centers in the reticular formation are responsible for the co-action of certain phases of the both rhythms. Respective evidence we gained in human experiments as performed with GOLENHOFEN (11). Moreover, it could be shown, that internal phase coupling is promoted by trophotropic states of the autonomous system e.g. recumbancy or sleep.

For coupling effects, which are produced without any participation of direct mechanical interactions nor reflectoric mechanisms, v. HOLST (21) coined the term "magnetic effects", originally applied to phase coupling of locomotor rhythms in animals.

Most extensive studies have been performed to elucidate the co-action of respiration and heart beat and their coordination with other rhythimc functions in the middle part of the period spectrum.

Respiration shows phase coordination not only with locomotor rhythms, but also - as already shown - with blood pressure rhythm, moreover with eyelid blinking, swallowing, sucking, and mastication (Fig. 9).

For instance, 80 % of swallows occur during the expiratory phase of respiration. Recent studies in narcosis by NISHINO & HIRAGA (25) showed, that central nervous mechanisms are responsible for a proper co-action. The same holds true for the coordination between sucking and respiration in babies, as already demonstrated by PEIPER (26).

In 1953 we first observed, that in healthy adults the frequency ratio of heart beat and respiration converged during night sleep very closely to 4:1. This finding was confirmed by several authors. Fig. 10 shows recent results from PÖLLMANN (27) from our group. 86 healthy adults were grouped according to the 24-hour means of this ratio and controlled throughout 36 hours. During the night hours the mean curves of the five groups converge to the whole number ratio of 4:1, and the range of variance decreases to a minimum.

```
                    Minute Rhythm of
                  / Tissue Blood Flow          Metabolic Motion – Sensory Motility – Locomotion
                    Repetitive Sneezing
                    (Stomach Peristalsis)
                    Blood Pressure Rhythm
  Respiratory       Eyelid Blinking Rhythm
  Rhythm            Swallowing
                    Sucking
                    Mastication
  Cardiac Rhythm    Tapping
                    Walking
                    Pedalling
                    Trotting
```

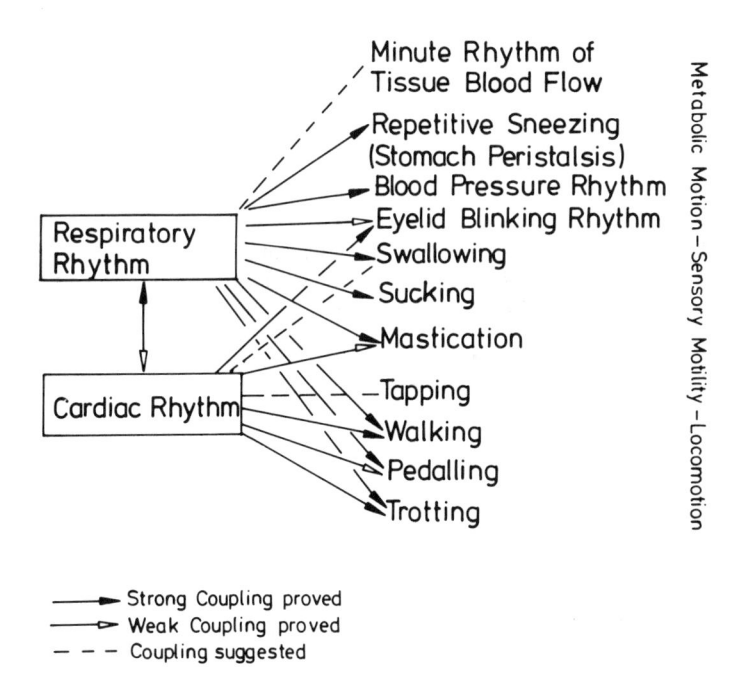

──► Strong Coupling proved
──▻ Weak Coupling proved
─ ─ ─ Coupling suggested

Fig.9: Phase coordination of cardiac and respiratory rhythms with other rhythmic functions. (From 17).

The question arises, if this ratio between the two functions is merely caused by a respective relationship of capacities according to the groundplan of the organism, coming to appearance under undisturbed conditions. In joint studies with DAUMANN (18) we could observe, that also under exhausting physical loads this ratio between

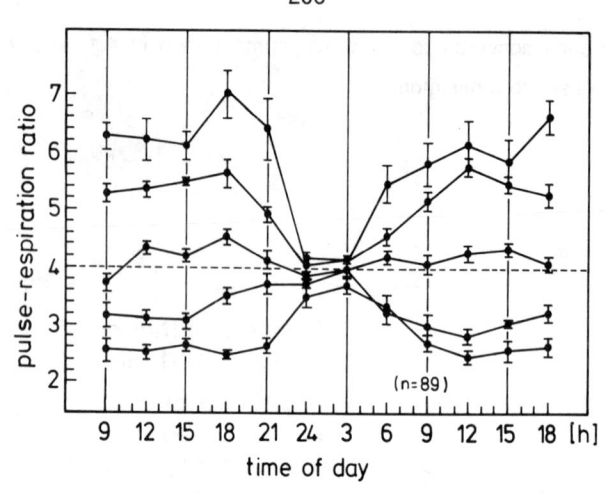

Fig.10: Mean course of pulse-respiration frequency ratio in five subgroups of healthy adults with different 24-hour levels under resting conditions. Brackets indicate standard errors. (Data from PÖLLMANN, un-published).

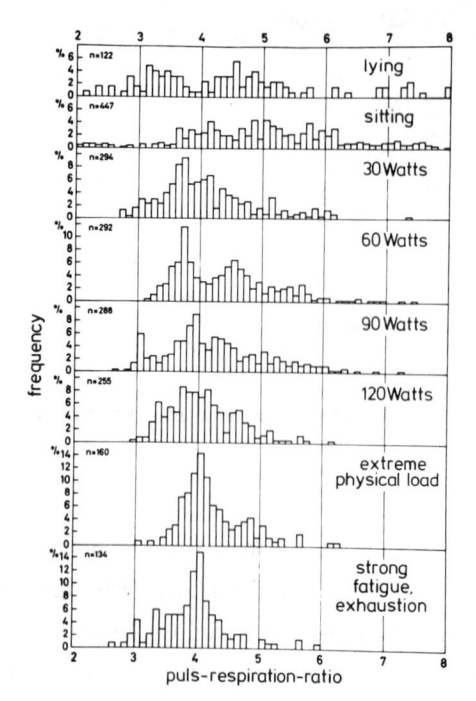

Fig.11: Frequency distributions of the pulse-respiration frequency ratio of healthy adults in various conditions of physical load. (From 18).

(Fig.11). By cutting trough the left vagus in rabbits it could be proved, that this phase coupling was arranged from the afferent impulses from the heart beat, triggering the onset of inspiration (3). However, in animal experiments also some evidence was found that the respiratory rhythm is able to trigger heart action, but only under artificial conditions (12).

In joint studies with STORCH (30) and STUTTE (31), exclusively performed in human beings, it could be shown, that phase coupling between heart beat and respiration mainly concerns the onset of inspiration, not of expiration. In healthy adults the coupling rate (χ^2-values of the histograms) for the onsets of inspiration reached high significance, whereas at the same time the onsets of expiration reached significant coupling rates only, if the durations of expiration exhibited a low variance, causing an automatic accumulation of trigger points within the heart revolution.

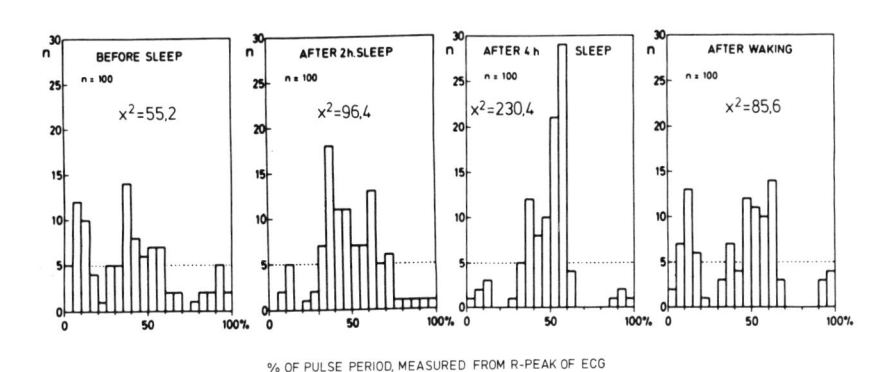

Fig.12: Frequency distributions, each of 100 inspiratory onsets within the heart period, which is measured from R-peak to R-peak of ECG and divided into 20 equal classes of 5 %, before and during night sleep, as well as after waking. Healthy subject. For evaluation of rate of phase coupling the χ^2 values are given, which reach their limit of significance at 30.1 (p=0.05). (Data from 30).

In further studies the coupling rate between heart beat and onset of inspiration was followed up during night sleep. As shown by an example (Fig. 12), the coupling rate increases during the first hours of sleep, indicating that phase coupling must be involved in the coordination of the frequencies as expressed by the convergence to the integer ratio 4:1. The respective histograms of the onsets of inspiration over the heart period show, that mostly three trigger points can be marked off. Hence different afferences from the heart beat and its effects in the arterial system form the coupling patterns (31). However, with increase of the coupling rate the third trigger point becomes dominant, located about 600 msec after the R-peak of the ECG.

A significant positive correlation could be observed between the coupling rate and the frequency ratio of heart beat and respiration, whereby whole number ratios lead to maxima of the coupling rate (30).

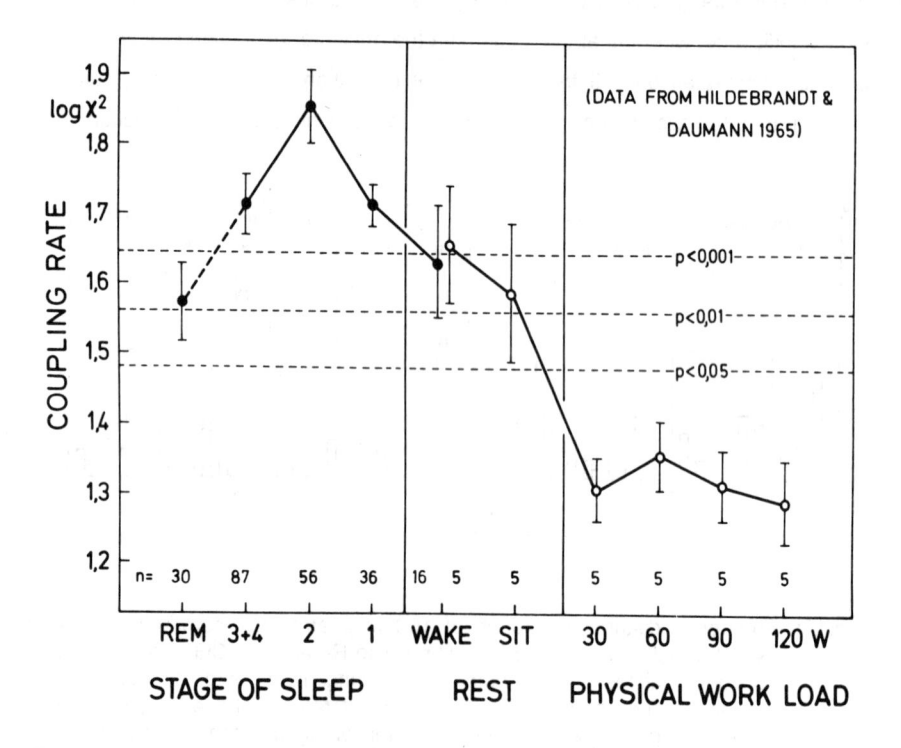

Fig. 13: Mean coupling rates between onset of inspiration and heart beat in healthy subjects during different sleep stages, waking, and sitting, as compared to different amounts of physical work load. (From 29).

RASCHKE & HILDEBRANDT (29) measured the coupling rate between heart beat and respiration at different sleep stages. It could be shown that maximal coupling rates occur during sleep, whereas coupling rates decrease during light sleep and disappear during physical load (Fig.13), as already observed by HILDEBRANDT & DAUMANN (18).

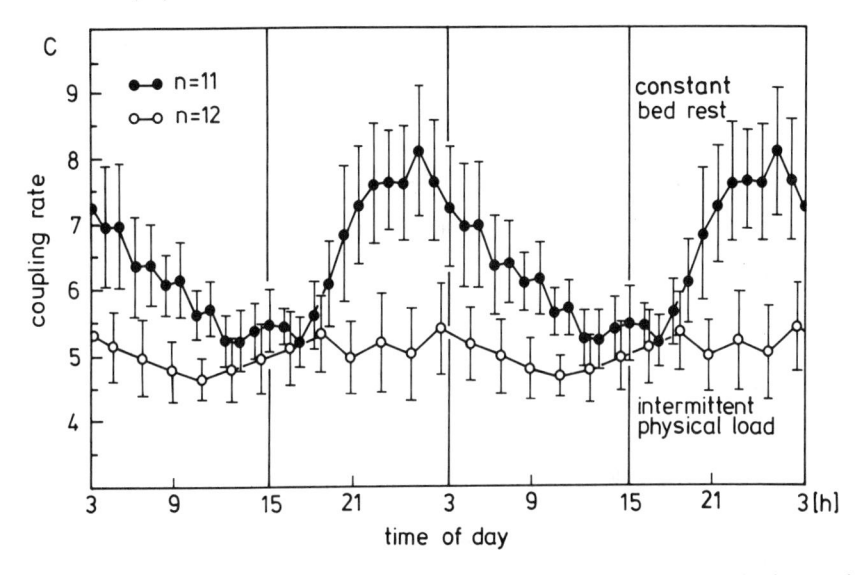

Fig.14: Mean daily course of the coupling rate between onset of inspiration and heart beat in healthy subjects under constant bed rest (filled circles) and under 2-hourly intermittent work loads on a bicycle ergometer (open circles). Brackets indicate standard errors. (From 7).

ENGEL et al. (7) controlled the coupling rate between heart beat and respiration throughout 24 hours under strict resting conditions as well as in subjects who were wakened every 2 or 4 hours respectively for a short dosed physical work load. As shown in Fig. 14, the nightly increase of coupling rate is completely prevented by the intermittent loads in 2-hour intervals. It needs about 4 hours of sleep to reestablish phase coupling.

Furthermore, studies in trained subjects and comparison between supine and standing position confirmed, that various trophotropical conditions are able to increase phase coordination between the both rhythms (c.f. 31). Finally, ENGEL et al. (8) using different narcotic drugs were able to show, that the reticular formation of the brain stem plays an important role in the trigger mechanism of the afferent impulses to the respiratory center.

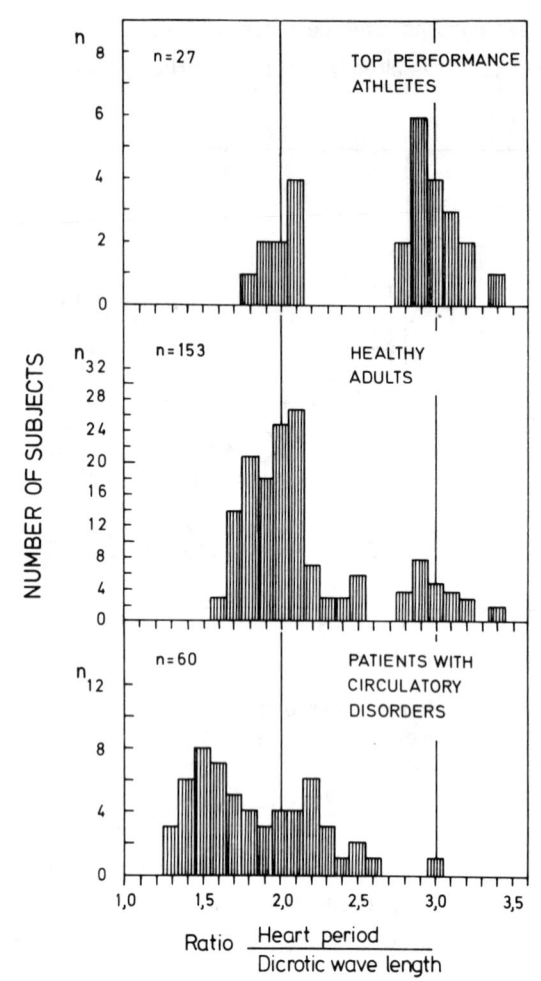

Fig. 15:
Histogram of the mean ratios between heart period and dicrotic wave length under resting conditions in top-performance athletes (top), healthy adults (center), and patients suffering from circulatory disorders (bottom). (From 9).

In contrast to the interactions of slower rhythms, whereby both partners can have equal rights to exert phase coupling impulses, phase coupling between heart beat and respiration is dominated by the afferent impulses of heart beats. However, we found an example for a further step to a merely one-sided mode of phase coupling. This concerns the coordination between heart beat and arterial dicrotic waves.

As demonstrated in Fig. 15, in joint studies with GADERMANN & JUNGMANN (9) we were able to show, that the ratio between the period duration of the heart rhythm and the wave length of the dicrotic wave, as standing in the arterial system, prefers the whole number ratio of 2:1 in healthy subjects. This means, that the ejection phase of the heart rhythm coincides with a certain phase of the arterial pressure oscillation. MILLAHN (24) and ECKERMANN (6) have estimated, that this economical co-action can save about 30 % of the myocardial energy consumption. Trained subjects exhibit much stronger coordination including the ratio of 3:1 because of the bradycardia resulting from physical training. However, in patients suffering from functional circulatory disorders this economical coordination is lost (see lower part of the figure).

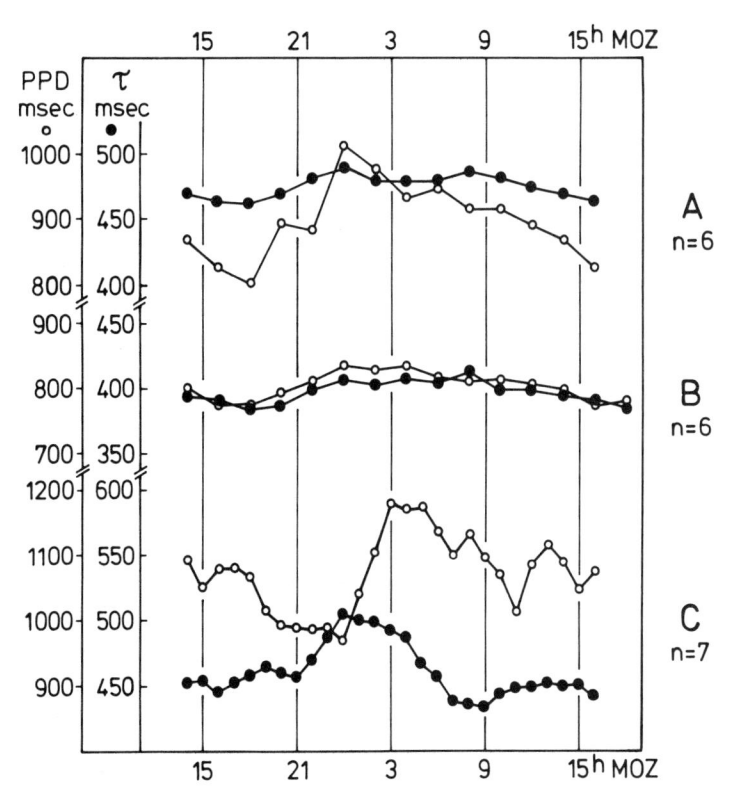

Fig. 16: Mean daily course of pulse period (PPD) and dicrotic wave length (Tau) in three groups of healthy subjects exhibiting different ratios. Ordinates are scaled according to the normal ratio of 2:1. (From 19).

If one studies the circadian behavior of this coordination, it becomes evident (Fig. 16), that the circadian variation of the dicrotic wave length is very stable and identical in three groups of subjects, whereas the daily course of the period length of the heart rhythm differs considerably. In the ordinates both period lengths are plotted according to the normal ratio of 2:1. In the middle group both curves parallel throughout the day, indicating that the normal integer ratio is kept all over the day. In the other groups both curves do not run together, but meet only during the early night hours, and this irrespective of whether the heart period is to long or to short during the day. This indicates, that the resonance wave of the arterial system represents the guiding rhythm of this coordination, and - presumably by means of reflex mechanisms - couples the heart beat into the optimal phase co-action. The practical significance of this chronobiological approach to the circulatory system is not sufficiently evaluated up to now.

short-wave rhythms ←————————→ long-wave rhythms

phase response

frequency response

loss of phase coupling	phase coupling (rel. coordination)	synchronization (absol. coord.)
frequency modulation	preference of integer frequency ratios	frequency multiplication

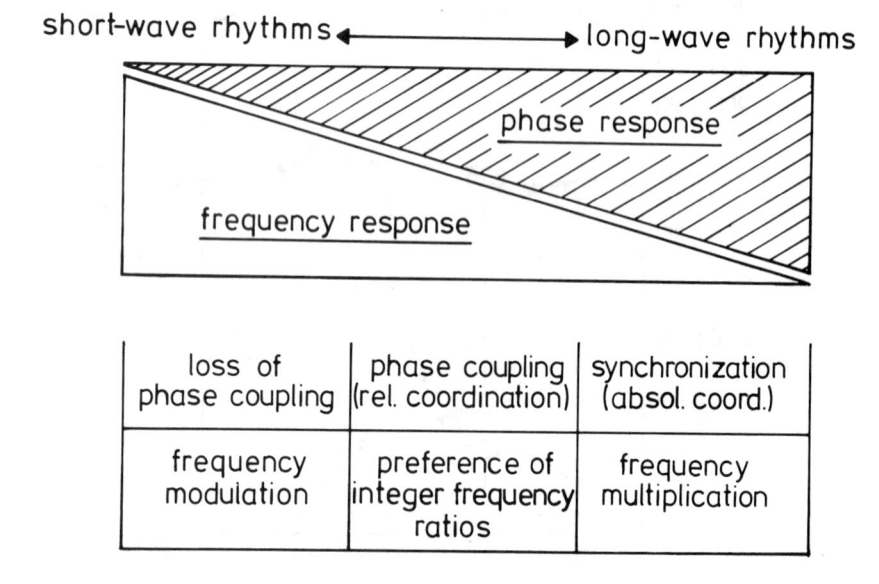

Fig. 17: Schematic figure of the different modes of response and related phenomena in the different parts of the rhythm spectrum. For further explanation, see the text.

In this overview, I will not touch the mechanisms of coordination of the locomotor rhythms as well as of ciliar transport or synchronization phenomena in the nervous system. These problems must be dealt with in special lectures.

Coming back to the spectrum of biological rhythms in man (Fig. 17) the examples discussed demonstrate, that there exists a polar structure concerning the contrast between a preference of phase response of the longer-wave length rhythms and the preference of frequency response mainly in the short-wave section of the spectrum. All steps of transition between the both extremes can be observed. The highest degree of phase response is implemented in the external and internal synchronization of long-wave rhythms including the strong phase coordination of their multiple frequencies.

In the middle part of the spectrum a weaker phase coupling originates the phenomenon of a more or less statistical preference of certain phase coherences as well as dominance of integer frequency ratios.

The rhythms of the short-wave section - mainly of the information system - seem to be characterized by a more or less complete loss of phase coupling. This is the essential precondition for an unrestricted frequency response in this part of the spectrum, leading to frequency modulations.

Up to now, there is very little insight into the response curves of the different rhythms, particularly in man. However, phase coordination or phase control represents an important aspect of autonomous regulation of bodily functions. Therefore, future research should be focussed on this field, which lets expect important results, useful for better understanding of biological orders and for medical practice as well.

REFERENCES

1. ASCHOFF, J. (1978): Features of circadian rhythms relevant for the design of shift schedules. Ergonomics 21, 739-754
2. BLUME, J. (1965): Nachweis von Perioden durch Phasen- und Amplitudendiagramm mit Anwendungen aus der Biologie, Medizin und Psychologie. Westdeutscher Verlag, Köln-Opladen
3. BUCHER, K. & P. BÄTTIG (1960): Zur Bedeutung der Vagi für die pulssynchrone Atmung. Helv. physiol. pharmacol. Acta 18, 219-224
4. BUCHER, K. & K.E. BUCHER (1977): Cardio-respiratory synchronisms: Synchrony with artificial circulation. Res. exp. Med. 171, 101-108
5. DAVENPORT, H.W. (1971): Physiologie der Verdauung. Eine Einführung. Schattauer Verlag, Stuttgart-New York
6. ECKERMANN, P. (1969): Untersuchungen an einem Kreislaufmodell mittels Analogrechner. Inaug.-Diss. Rostock
7. ENGEL, P., G. HILDEBRANDT & E.D. VOIGT 1969): Der Tagesgang der Phasenkoppelung zwischen Herzschlag und Atmung und seine Beeinflussung durch dosierte Arbeitsbelastung. Int. Z. angew. Physiol. 27, 339-355
8. ENGEL, P., A.J. JAEGER & G. HILDEBRANDT (1972): Über die Beeinflussung der Frequenz- und Phasenkoordination zwischen Herzschlag und Atmung durch verschiedene Narkotika. Arzneimittelforsch. (Drug Res.) 22, 1460-1468
9. GADERMANN, E., G. HILDEBRANDT & H. JUNGMANN (1961): Über harmonische Beziehungen zwischen Pulsrhythmus und arterieller Grundschwingung. Z. Kreislaufforsch. 50, 805-814
10. GOLENHOFEN, K. (1987): Endogenous rhythms in mammalian smooth muscle. In: G. HILDEBRANDT, R. MOOG & F. RASCHKE (Eds.): Chronobiology & Chronomedicine. Basic Research and Applications. Peter Lang Verlag, Frankfurt-Bern-New York-Paris. pp. 26-38
11. GOLENHOFEN, K. & G. HILDEBRANDT (1958): Die Beziehungen des Blutdruckrhythmus zu Atmung und peripherer Durchblutung. Pflügers Arch. ges. Physiol. 267, 27-45
12. GOLENHOFEN, K. & H. LIPPROSS (1969): Mechanische Kopplungswirkungen der Atmung auf den Herzschlag. Pflügers Arch. ges. Physiol. 309, 159-166
13. GRAF, K., W. GRAF & S. ROSELL (1959): Zusammenhänge der Durchblutungsrhythmik in Haut-, Muskel- und Intestinalstrombahn des Menschen. Pflügers Arch. ges. Physiol. 270, 43
14. HILDEBRANDT, G. (1967): Die Koordination rhythmischer Funktionen beim Menschen. Verh. Dtsch. Ges. Inn. Med. 73, 922-941
15. HILDEBRANDT, G. (1986): Functional significance of ultradian rhythms and reactive periodicity. J. interdiscipl. Cycle Res. 17, 307-319
16. HILDEBRANDT; G. (1987): The autonomus time structure and its reactive modifications in the human organism. In: L. RENSING, U. van der HEIDEN & M.C. MACKEY (Eds.): Temporal Disorder in Human Oscillatory Systems. Springer Verlag, Berlin-Heidelberg-New York-London-Paris-Tokyo. pp. 160-175
17. HILDEBRANDT, G. (1988): Temporal order of ultradian rhythms in man. In: W.Th.J.M. HEKKENS, G.A. KERKHOF & W.J. RIETVELD (Eds.): Trends in Chronobiology. Pergamnon Press, Oxford-New York etc. pp. 107-122

18. HILDEBRANDT, G. & F.J. DAUMANN (1965): Die Koordination von Puls- und Atemrhythmus bei Arbeit. Int. Z. angew. Physiol. einschl. Arbeitsphysiol. 21, 27-48
19. HILDEBRANDT, G. & H.R. KLEIN (1969): Untersuchungen über den Tagesgang der Koordination von Herzryhthmus und arterieller Grundschwingung. Arch. Kreislaufforsch. 59, 235-259
20. HILDEBRANDT, G. & L. PÖLLMANN (1987): Chronobiologie des Schmerzes. Heilkunst 100, 340-358
21. HOLST, E. von (1939): Die relative Koordination als Phänomen und als Methode zentralnervöser Funktionsanalyse. Ergebn. Physiol. 42, 228-306
22. KOEPCHEN, H.P. & K. THURAU (1958): Untersuchungen über Zusammenhänge zwischen Blutdruckwellen und Ateminnervation. Pflügers Arch. ges. Physiol. 267, 10
23. MENZEL, W. (1955): Therapie unter dem Gesichtspunkt biologischer Rhythmen. In: H. LAMPERT (Hrsg.): Ergebnisse der physikalisch-diätetischen Therapie. 5, 1-38. Steinkopff, Dresden-Leipzig
24. MILLAHN, H.P. (1962): Das Verhältnis von Pulsperiodendauer zur Dauer der arteriellen Grundschwingung bei Jugendlichen. Z. Kreislaufforsch. 51, 1155-1159
25. NISHINO, T. & K. HIRAGA (1991): Coordination of swallowing and respiration in unconscious subjects. J. Appl. Physiol. 70, 988-993
26. PEIPER, A. (1938): Das Zusammenspiel des Saugzentrums mit dem Atemzentrum beim menschlichen Säugling. Pflügers Arch. ges. Physiol. 240, 312
27. PÖLLMANN, L. (1991): Data unpublished.
28. RASCHKE, F. (1982): Analysis of the frequency and phase relationships of circulatory and respiratory rhythms during adaptive processes. In: G. HILDEBRANDT & H. HENSEL (Eds.): Biological Adaptation. Georg Thieme Verlag, Stuttgart-New York. pp. 52-63
29. RASCHKE, F. & G. HILDEBRANDT (1982): Coupling of the cardio-respiratory control system by modulation and triggering. In: Th. KENNER, R. BUSSE & H. HINGHOFER-SZALKAY (Eds.): Cardiovascular system dynamics - models and measurements. Plenum Press, New York-London. pp. 533-542
30. STORCH, J. (1967): Methodische Grundlagen zur Bestimmung der Puls-Atem-Koppelung beim Menschen und ihr Verhalten im Nachtschlaf. Med. Inaug.-Diss. Marburg/Lahn
31. STUTTE, K.H. (1967): Untersuchungen über die Phasenkoppelung zwischen Herzschlag und Atmung beim Menschen. Med. Inaug.-Diss. Marburg/Lahn
32. WEITZ, W. & W. VOLLERS (1926): Über rhythmische Kontraktionen der glatten Muskulatur an verschiedenen Organen (Magen, Darm, Harnblase, Scrotum, Penis, Uterus, Milz und Gefässe). Z. ges. exper. Med. 52, 723-746
33. WEVER, R. (1979): The circadian system of man. Results of experiments under temporal isolation. Springer Verlag, New York-Heidelberg-Berlin

EFFECT OF BREATHING RATE AND AMPLITUDE ON HEART RATE POWER SPECTRA

A. PATZAK, C. JOHL & J. EBNER

Institut für Physiologie, Humboldt-Universität zu Berlin,
Hessische Str. 3-4, D-O 1040 Berlin, FRG

INTRODUCTION

The Fast Fourier Transformation of the heart rate fluctuations is an effective method for the estimation of the rhythm of the heart as a result of the sympathetic and parasympathetic modulation of the heart rhythm generator (1). The mechanism of the effect of respiration on the heart rate, so-called respiratory sinus arrhythmia (RSA), remains unclear. A number of different hypothesis have been provided to explain this phenomenon. Following factors can be significant: (a) changes of the filling volume of the heart; (b) the baroreceptor reflex and its respiratory modulation; (c) stretch receptors in the lung or the chest; (d) central interaction between mechanisms controlling the respiration and those controlling the heart rate.

In several studies different solitary phenomena were demonstrated which indicate, that the respiratory modulation of the heart rate (HR) can be characterized in terms of the interaction of two oscillatory systems (2,3,4). However, these studies were performed in small experimental groups and the range in which the respiratory frequency (RF) changed was partly different. That's why it is difficult to say whether these phenomena are consistent.

The purpose of the present study was
1) to define how different breathing patterns effect the heart rhythm, and
2) to examine whether the results are reproducible in a larger group of subjects.

METHODS

SUBJECTS

21 young healthy volunteers of both sexes (aged between 18 and 31 years) were included in the study. All subjects had a normal lung function and a normal ECG and were without a history of previous cardiopulmonary or other remarkable diseases. Studies were carried out in the 1st series in the evening (16.00-18.00 p.m.) and in the 2nd series and in the morning (8.00-10.00 a.m.).

EQUIPMENT

The R-wave of a bipolar lead from the thorax was detected automatically by the ECG apparatus and led to an interface which calculated the R-R interval. The data were continuously stored on the hard disc of a computer.

Respiration was measured in the 1st series with a pneumotachograph. In the 2nd series the changes in thoracic circumference were measured by a strain gauge placed around the middle part of the thorax.

DATA ANALYSIS

The instantaneous HR was calculated from the R-R interval. Spectral analysis of fluctuations in heart rate was performed using a fast-Fourier-transform based method. Before this the HR signal was sampled at 2 Hz and filtered with a cut-off frequency of 0.01 Hz.

EXPERIMENTAL PROCEDURE

In both series the subjects first were encouraged to breath quietly and spontaneously. Series 1: Breathing at different given breathing frequencies and free chosen tidal volume (17 subjects): A metronome (60 beats/min) assisted the subject in controlling breathing frequency. The respiration cycles had a duration of 2, 4, 6, 8, 10, 12, 14 and 16 sec, which resemble frequencies of 0.5, 0.25, 0.167, 0.125, 0.1, 0.083, 0.071 and 0.063 Hz, respectively. The subject breathed for 5 min at each frequency. Series 2 : Breathing at different frequencies and constant tidal volume (7 subjects): Each subject monitored his respiration signal (thoracic circumference) on an oscilloscope. The subjects were instructed to keep their tidal volumes sinusoidal. The amplitude was calibrated at 15% of the vital capacity. The respiratory frequencies were those of series

1 and additionally 0.3 and 0.4 Hz. The duration of each experimental period was 150 sec. Between the phases of controlled breathing the subjects rested for a few minutes to normalize pCO_2.

RESULTS

Table 1: Percentage of testpersons with peaks in the low (LF), middle (MF), and high frequency (HF) range of the heart rate spectra and of harmonics and sidebands.

** respiratory frequency (Hz)	LF	MF	HF	harmonics	sidebands
			(%)		
res	100	76	100		
0.5	100	94	100		
0.4	100	100	100		56*
0.3	100	100	100		100*
0.25	100	94	100		
0.167	100	65	94	12	
0.125	100	100		88	
0.1	94	100		82	
0.083	100	100		94	
0.071	100	100		100	
0.063	65	100		94	

* only for 7 subjects of the 2nd series

Spontaneous quiet breathing: There are interindividual differences in breathing frequency, the latter varying between 0.11 and 0.32 Hz. Power spectral analysis shows high variability in total power and more or less distinct peaks at the low (0.01-0.05 Hz), middle (0.05-0.15 Hz), and high frequency range (> 0.15 Hz, Fig.1).

Voluntarily controlled breathing: The power spectra show more uniform patterns. The respiratory related peak at the respiration frequency and a peak at 0.1 Hz are clearly visible in all spectra. The patterns change depending on RF. If the frequency of breathing is between 0.063 and 0.125 Hz a distinct 0.1 Hz-peak cannot be seen. However, the spectrum presents additional peaks at whole number multiples of the main frequency (Fig. 1 and 2). If the respiration frequency is 0.167 Hz and greater the 0.1 Hz component and the respiration related component are visible (Fig. 1 and 2). At breathing frequencies of 0.3 Hz there are additional peaks at 0.2 and 0.4 Hz, and breathing at 0.4 Hz additional peaks lies in most cases at 0.3 and 0.5 Hz. Very high respiration frequencies of 0.5 Hz produce more variable spectra with clear interindividual differences (Fig. 2). All of the effects are highly reproducible (see Tab. 1). The frequency responses are the same for the subjects of the 1st and 2nd series.

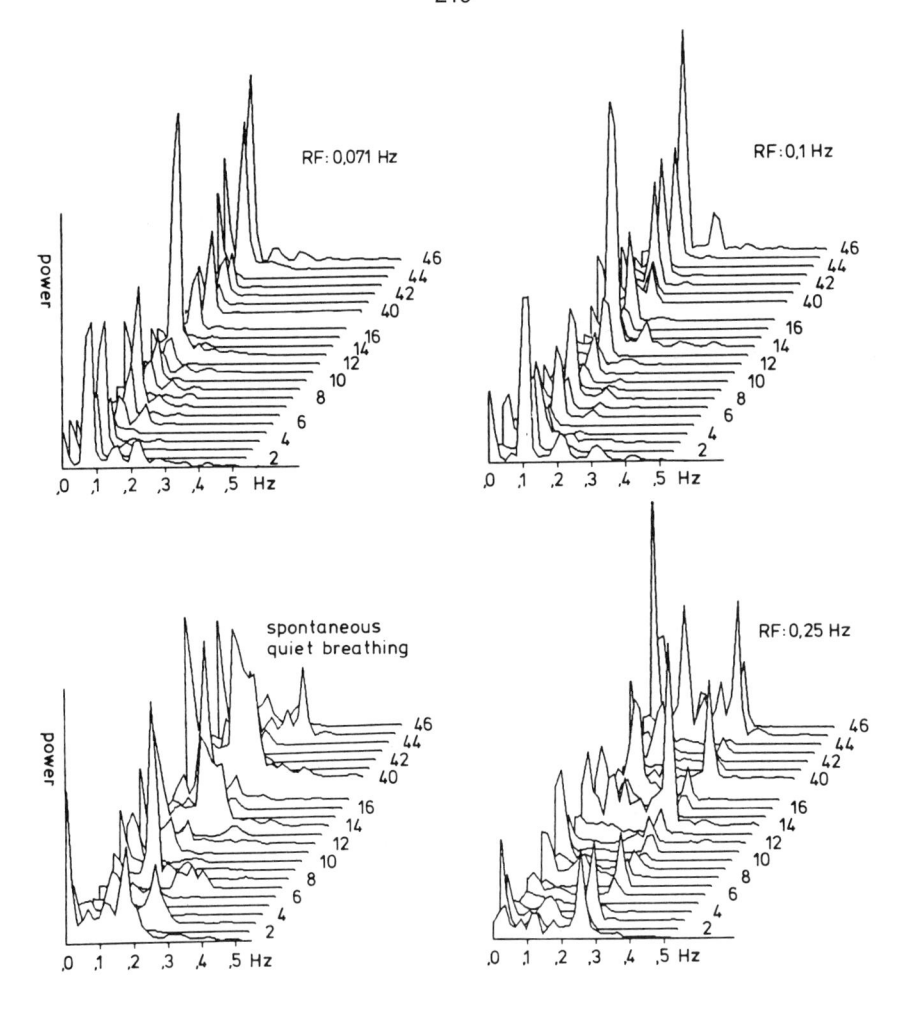

Fig.1: Power spectra for all subjects of series 1 and 2 showing the effect of quiet breathing and voluntarily controlled breathing at frequencies of 0.071, 0.1, and 0.25 Hz on the heart rhythm.

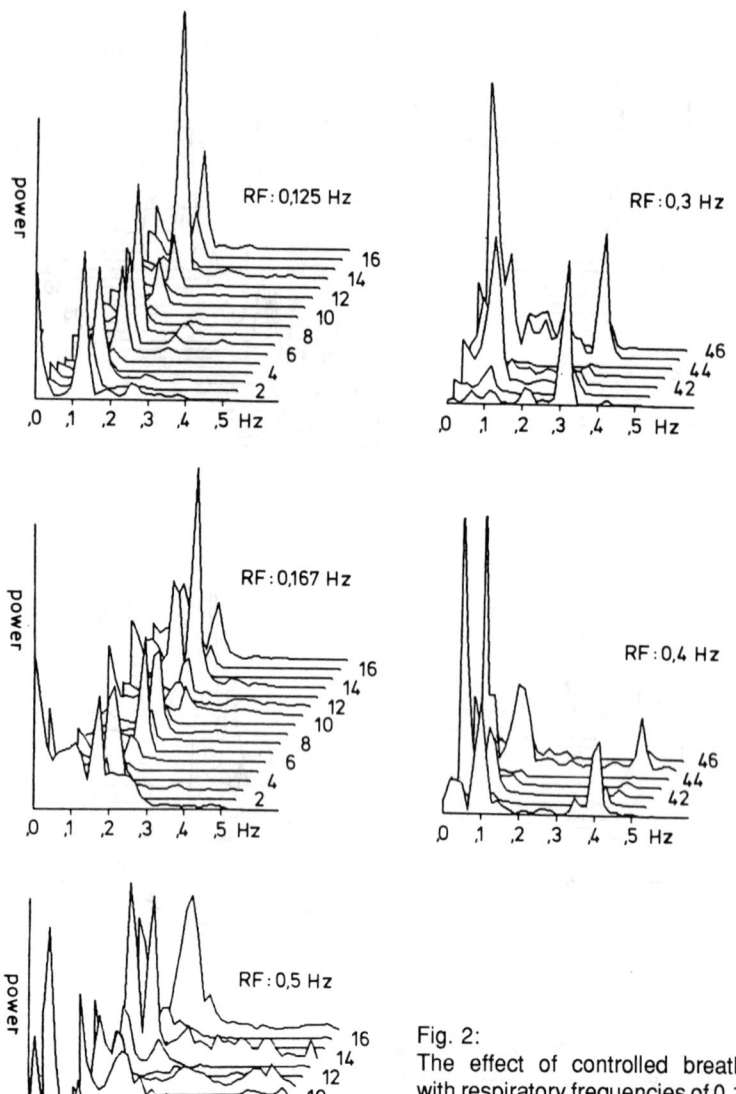

Fig. 2:
The effect of controlled breathing with respiratory frequencies of 0.125, 0.167, and 0.5 Hz (left, 1st series) and 0.3 and 0.4 Hz (right, 2nd series) on heart rate power spectra.

DISCUSSION

The power spectral analysis shows three regions of activity in the HR-spectrum. These are a low-frequency region (LF) at about 0.03 Hz, a region about 0.1 Hz (middle-frequency, MF), and a high-frequency component (HF) at the appropriate RF (1). The LF component is associated with the thermoregulatory system. The 0.1-Hz peak represents the oscillatory rhythm of the blood pressure control system (5) and the HF-peak is due to RSA.

Our study shows clear differences in the total power and the magnitude of the LF, MF, and HF-component according to the subjects breathing spontaneously. However, LF and HF-components are seen in all subjects and a MF-peak is present in 76% of them. The absence of the MF-component is likely caused by a great LF-peak, which masks the MF-peak in these cases.

If the subjects breath controlled the spectra become more uniform and the respiratory effect is clearly visible. It is shown that at low RF the respiration related peak and the whole number multiples dominate in the spectrum. These additional peaks are not an artificial product of FFT but a part of the HR signal, as we could discover by filtering of the main frequency and subsequent FFT-analysis. The mechanisms underlying this influence of respiration on the HR are not yet clear. But among others there are theories considering RSA as a result of the interaction of two nonlinear systems exhibiting frequency entrainment, whereby the main oscillator consists of the vasomotor response of the baroreceptor loop, and the respiration is the second oscillator (3). The effect of respiration becomes apparent in the range between total entrainment and non-entrainment of the heart rhythm and is dependent on the amplitude and the frequency of breathing. The power spectra of the HR demonstrated in this study show clear entrainment at low RF with patterns of harmonic entrainment. At higher respiratory frequencies (> 0.125 Hz) the rhythm of the cardiovascular system (0.1 Hz peak) as well as a clear HF-peak become evident. Moreover, there are additional peaks around the HF- peak. Now only a relative entrainment becomes effective and the phenomenon of modulation can be recognized in some cases. If a further increase of the RF occurs the spectral patterns develop to be multiform, the system tends to be non-entrained.

Comparing the results of the first with the second series we could not discover an influence of the tidal volume on the frequency response. The reason may be that the breathing volumes do not differ enough between both groups. Further experiments are required in which the tidal volume is changed systematically to describe the system completely.

CONCLUSIONS

Depending on the respiratory frequency the heart rate shows patterns of total, relative, and harmonic entrainment as well as modulation effects. The phenomena are well reproducible in a group of 21 young volunteers.

ACKNOWLEDGEMENT

The authors thank Dr. Kate P. Leiterer for her assistance in preparing and translating the manuscript.

REFERENCES

1. AKSELROD, S., D. GORDON, F.A. UBEL et al. (1981): Power spectrum analysis of heart rate fluctuations: A quantitative probe of beat-to-beat cardiovascular control. Science 213, 220-222
2. ANGELONE, A. & N.A. COULTER (1964): Respiratory sinus arrhythmia: a frequency dependent phenomenon. J. Appl. Physiol. 19, 479-482
3. KITNEY, R., D.A. LINKENS, A.C. SELMAN & A. McDONALD (1982): The interaction between heart rate and respiration: part II-nonlinear analysis based on computer modelling. Automedica 4, 141-153
4. SELMAN, A., A. McDONALD, R. KITNEY & D. LINKENS (1982): The interaction between heart rate and respiration: Part I-Experimental studies in man. Automedica 4, 131-139
5. SAYERS, B.MCA. (1973): Analysis of heart rate variability. Ergonomics 16, 17-32

VEGETATIVE RHYTHMS AND PERCEPTION OF TIME

J.STEBEL

Universität Leipzig, Carl-Ludwig-Institut für Physiologie,
Liebigstr. -7, D-O 7010 Leipzig, FRG

Investigations of the frequency of cyclic fluctuations in vegetatively influenced functions in humans led to a discrete spectrum of the modulating rhythms (range of selected periods: 4 up to 180 seconds). A cycle of 9 sec and/or harmonics appear most frequently. This statement came out correspondingly from heart rate, breathing rate, and involuntary horizontal eye movements.

Reaction times in relation to interstimulus intervals exhibit an increasing process with a discontinuity at 9 sec. Any dependencies of the amount of load (simple or choice reactions, durations of the test series) do not exist: the shape of curves remains the same with all these variations; choice reactions cause only a parallel shift.

Estimations of time in 51 distinct unfilled intervals from 0.5 up to 360 sec duration offer steplike changes (especially at 9 sec). Standard deviations as a criterion of the estimations' width about individual means show a nearly linear increase, superimposed by logarithmic developments, with points of intersections lying at multiples of 9 sec, that means at periods which authoritatively contribute to determine the spectrum of cycles in physiological functions.

RHYTHMS IN VEGETATIVE FUNCTIONS

The purpose of this study was to examine simultaneously the frequencies of period durations of rhythms in different physiological functions to compare the obtained spectra whether an uniformity is present and, therefore, a general genesis of this cycles one can take into consideration. The data of ECG (R-R-intervals), respiration and eye movements were recorded in test series for a period of 16 minutes from 21 subjects on various conditions in 88 trials and submitted to a spectral analysis by a special approach dividing the entire test duration into parts (window method), drawing up partial time series and computation of correlation functions.

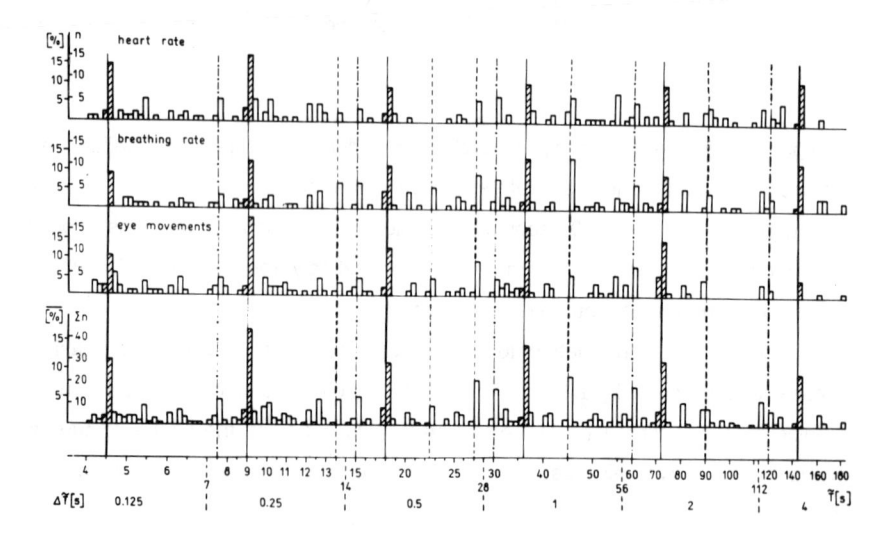

Fig.1: Frequencies of discovered periods of the modulating rhythms: Above: single addition for each variable Below: totals of all variables together n: absolute frequency of a period within 88 trials %: mean occurrence of a cycle The screen of abscissa approximately corresponds to the range of inaccuracy (T) within the determination of a period ΔT. Shaded columns: detected integer multiples of 9 sec, including a tolerance of about -.8 % in each case, e.g.: 9 sec: 8.75 - 9.-5 or 7- sec: 70 - 74 sec. NB: also pay attention to other multiples of 9 or 4.5 sec, e.g.: 13.5, -7, 45 sec, and to further cycles with their multiples, in the diagram directed to them by auxiliary lines of the same kind.

A correspondence of the spectral distributions is obvious (Fig. 1), although one and the same cycle need not appear simultaneously in the three variables. Integer multiples within a variable can occur simultaneously at a time or independently of another at any time. These are the results of several experiments: rest, eyes open, eyes closed, mental arithmetics, sitting and lying down position. The different conditions do not change the general size of the histogram, i.e., there are universally preferred periods.

REACTION TIMES

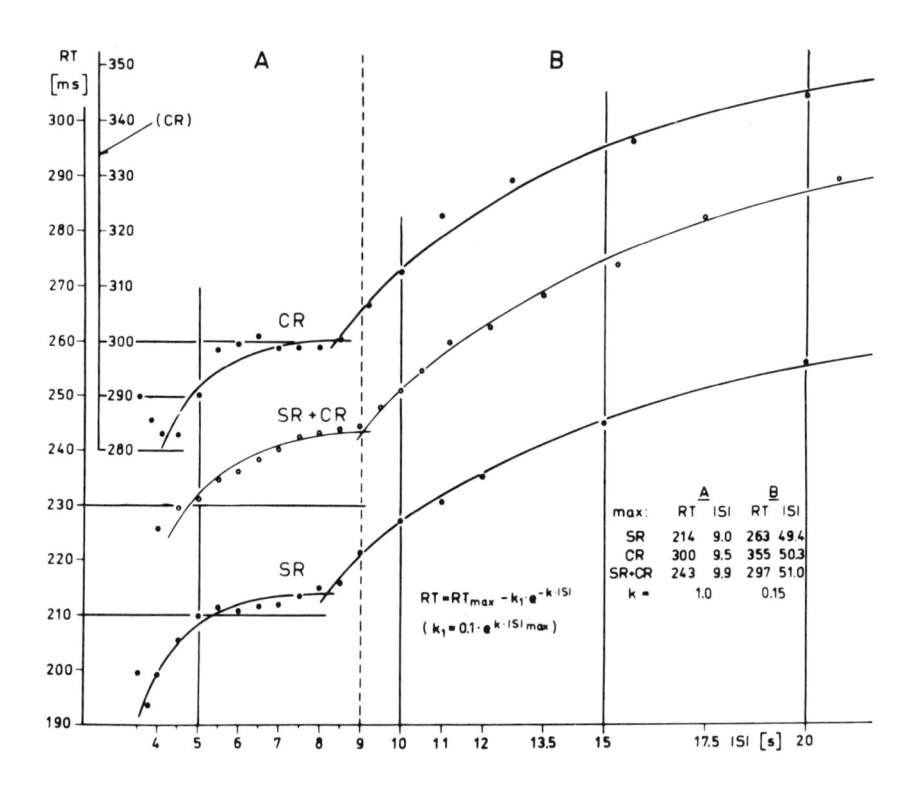

Fig.2: Reaction times as functions of the ISI (smoothed values of test series means). SR = simple reaction; CR = two-choice reaction; SR+CR = combined results of SR and CR test series, in which the duration of a trial was constant

The stimuli were presented visually with consistent intervals in each test series of ~5 presentations for simple reactions (SR) and 40 for choice reactions (CR), both with the instruction to press a response key as quickly as possible. SR means the reaction to white light, CR the two-choice reaction to red or green light stimuli in connection to the attached right-hand or left-hand key, respectively. Whereas the interstimulus interval (ISI) remained the same in SR and CR series, the colors varied stochastically in CR. The reaction time (RT) is the duration of the stimulus appearance until the switching off the light. 18 different ISIs were tested in the range from 3.5 up to -0 sec in healthy young adults.

Since the number of signals within a series was kept constant the duration of the series is longer with increasing ISI. A second trial should exclude any influences associated perhaps with the above established fact. By keeping a constant duration (6 min) one had to take into the bargain a diminuation of the signal's number with growing ISI. The combined results of SR and CR tasks on this condition (designed as "SR+CR") are only of importance for comparing the course of those RTs in dependence of the ISIs. Fig.2 - shows a parallel shift in the increasing developments of the RT, i.e., the CR requires only a fixed duration (about 90 msec) for stimulus categorization and response selection, independent of the time interval. The results reveal two separate courses with points of intersection near ISI 9 sec in each case. Exponential approximations, partially computed in the parts A and B (with boundaries at 9 sec), proved to be similar in all variants.

ESTIMATION OF TIME

Flashes of light marked the boundaries of constant time intervals repeatedly one after another. The volunteers were asked to estimate this period, i.e. the duration of the running time, without the possibility of counting or making use of any other mental aid. Corrected valuations should be obtained by the added repetitions of the present ISI. With a sufficient subjects exercise it was allowed to exceed the examined periods up to 6 minutes.

Since in perception of time periods a general subjective deviation from the real time is possible (over- or underestimation), the standard deviation (SD) in relation to the individual mean declaration can be used as a good criterion to determine the estimation range of any fixed interval, because these SDs hardly differ between individuals. In this manner the averaged SDs served to achieve an insight into the current temporal process.

The time lapse of the SD can be described in sections by logarithmic functions succeeding to previous ones with increasing steepness, in the course of a logarithmic period passes over into the next part with quadruples of 9 sec (Fig. 3). On the other hand these partial developments contribute to a proportional current time passing progression of the SDs, achieving this trend step by step.

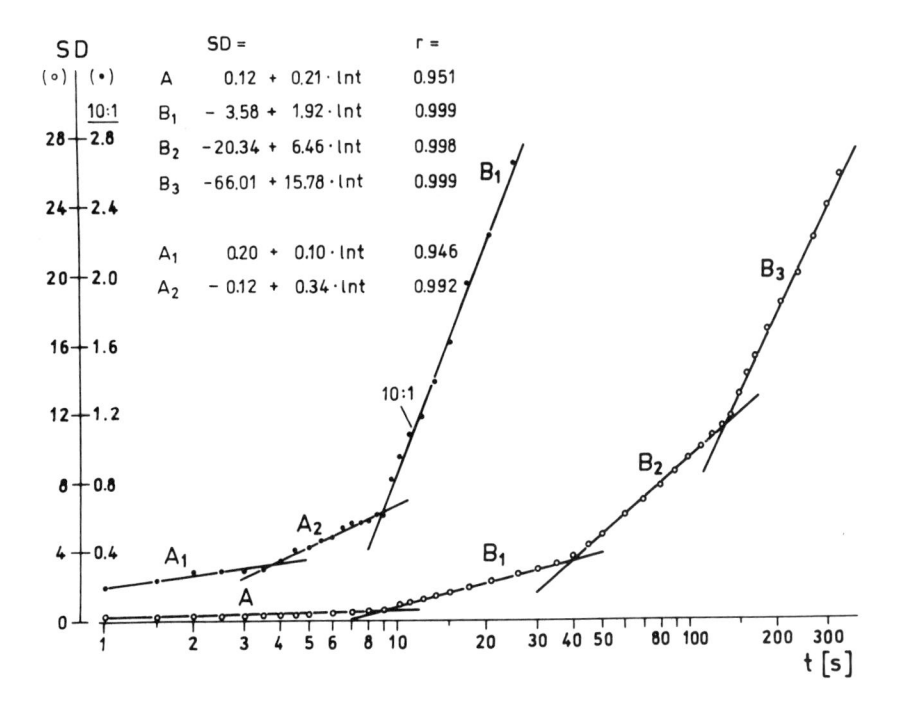

		SD =	r =
A		$0.12 + 0.21 \cdot \ln t$	0.951
B_1		$-3.58 + 1.92 \cdot \ln t$	0.999
B_2		$-20.34 + 6.46 \cdot \ln t$	0.998
B_3		$-66.01 + 15.78 \cdot \ln t$	0.999
A_1		$0.20 + 0.10 \cdot \ln t$	0.946
A_2		$-0.12 + 0.34 \cdot \ln t$	0.992

Fig.3: Standard deviations of time-estimation tests with logarithmic approximations in sections (smoothed values).
- abscissa logarithmic - upper curve and right ordinate scale: fine structure of the short intervals results; herein it puts in an appearance of a further splitting part A into two parts - t means the duration of an estimated time interval

CONCLUSIONS

Preferred cycles occurring as integer multiples (related to their universal spectral frequency) correspond again within several variables. These facts can point to ubiquitous uniform origins or an assumption of a common central regulation, whereas the results of the time-estimation tests may be based only on central nervous mechanisms, underlying on an integer graduated time structure. The concordant courses of reaction times offer an existence of different basic processes in different ranges of time.

It is to be expected that the findings support the long existing hypothesis of a discrete time organization, being inherent in mental lapses, and its relation to cyclic regulating functions of the brain. Furthermore, if one sums up all these non-stationary occurrences, a presence of a general time quantum structure can hardly be denied. But, the 9-sec break appears as the most prominent event, which was also found in other investigations of our own, concerning the examined range of time. (Details of these statements are not yet published.)

COUPLING DURING BIMANUAL ALTERNATING HAND MOVEMENTS IN NORMAL SUBJECTS AND HEMI-PARKINSONIAN PATIENTS

H. HEFTER, B. BÖSE, M. PIOTRKIEWCZ & H.-J. FREUND

Department of Neurology University of Düsseldorf, Moorenstraβe 5,
D-W 4000 Düsseldorf 1, FRG

INTRODUCTION

The human motor system does not allow the independent performance of two different voluntary motor tasks at the same time. KELSO et al. (1) testing simultaneously performed voluntary alternating finger and speech movements in normals observed a considerable coupling between these different movements. Also analysis of simultaneously performed alternating index finger movements of both hands revealed a high degree of coupling (2). This coupling was frequency dependent. Using pendulums of different length on the right and left side, and instructing normal subjects to swing the pendulum simultaneously at the most comfortable rate, TURVEY et al. (3) succeeded to show a highly complex interaction of the oscillations of both sides. The most comfortable rates during simultaneous swinging were different on both sides compared to the frequencies when each pendulum had to be swung alone. Recently KELSO & De GUZMAN (4) tested the interaction of voluntary alternating movements of one arm while driving the other arm alternating with a torque motor over a large amplitude and frequency range. Analyzing the Arnold-tongues describing the zones of fixed coupling in comparison to zones of chaotic behavior, he found a highly complex pattern in the frequency-amplitude-plane.

We were interested in the functional role of different parts of the central nervous system in these coupling processes. In patients, investigations comparable to those in normals are rare. BABINSKY (5) describing the diadochokinesia test mentioned that the paralysed side influences the unaffected side. COHN (6) measuring bilateral pro- and supination movements in patients with one paretic hand extended Babinski's observation reporting a reduction of movement rate and amplitude in the unaffected side by movements of the paretic hand. HAUSMANOWA (7) analyses repetitive ball pressing in hemiparetic patients. Maximal pressing frequency was much higher in the unaffected compared to the affected side. Onset of ball pressing of the affected side during pressing of the unaffected hand reduced pressing rate in the unaffected side considerably. When the unaffected side started to move during movements of the affected side also reduction of movement rate was observed in both hands. All these experiments describe a matching of both sides and a reduction of the level of performance dominated by the most disturbed side.

The lesions of the patients in the previously mentioned studies were not characterized in detail. Therefore we analyses bi-manual interactions in patients with an ideopathic hemi-parkinsonian syndrome being a clearly defined disease entity. Difficulties of Parkinsonian patients in performing different motor tasks simultaneously were recently reported by BENECKE et al. (8,9). A significant prolongation of motor execution times during bilateral performance of motor tasks compared to the performance with only one hand or arm, was found. The present study was designed to analyze coupling of simultaneous hand movements in hemi-parkinsonian patients during a modified version of Hausmanowa's approach.

METHODS

Normal Subjects and Patients
In 8 normal subjects (20-35 years; 6 right handed, 1 ambidextrous, 1 left handed; 4 male, 4 female) and in 7 Parkinsonian patients (mean age: 56 ± 13; six being mainly affected only on one side, stage 1-2 according to HOEHN & YAHR (10)) bi-manual alternating movements were analysed. All patients were only mildly affected and on medication during the recordings.

Data Recording
Patients and normal subjects set in a comfortable chair with both forearms being supported and fixed to two supports in a semi-pronated position. The hands were free to move in the wrist joint in the horizontal plane. The fingers were taped together with

adhesive tape so that only movements in the wrist joint were possible. Movements were recorded using a Selspot II-two camera system with three LEDs for each arm. Sample frequency per LED was 100 Hz. Frequency resolution of the spectra was 0.02 Hz.

The normal subjects were analysed under 6 different conditions but only two will be described here. These two conditions will be called the DLND- and the LNDLN-condition. They result from attaching a weight of 1 kg to the non-dominant (LN)-hand. In the DLND-condition the dominant (D)-hand had to be moved at the highest possible rate for about 2 seconds. Then, for about 4 seconds, both hands had to be moved simultaneously at the maximal possible rate. Then the loaded, non-dominant hand had to be stopped while the dominant hand continued to move at the highest possible rate. In the LNDLN-condition, the non-dominant hand started first. Then the dominant hand to be moved additionally. Then the movements of the dominant hand had to be stopped after 4 seconds while the non-dominant hand continued to move for a period of 2 further seconds. This is illustrated in Fig. 1. For each condition 5 trials were recorded.

In the hemi-parkinsonian patients two similar tasks were analysed which will be called the LAL- and the ALA-condition with the more affected hand (A-hand) corresponding to the loaded (LN)-hand in the normal subjects and the less affected hand (L-hand) corresponding to the non-loaded (D)-hand.

Data Analysis

Corresponding to the first 2, the next 4 and the last 2 seconds of the trials slightly shorter time segments were selected by cursor interaction from the recordings of the movements in horizontal direction and fourieranalysed. The peak frequency of each resulting power spectrum density function was used as the mean movement rate during the underlying interval.

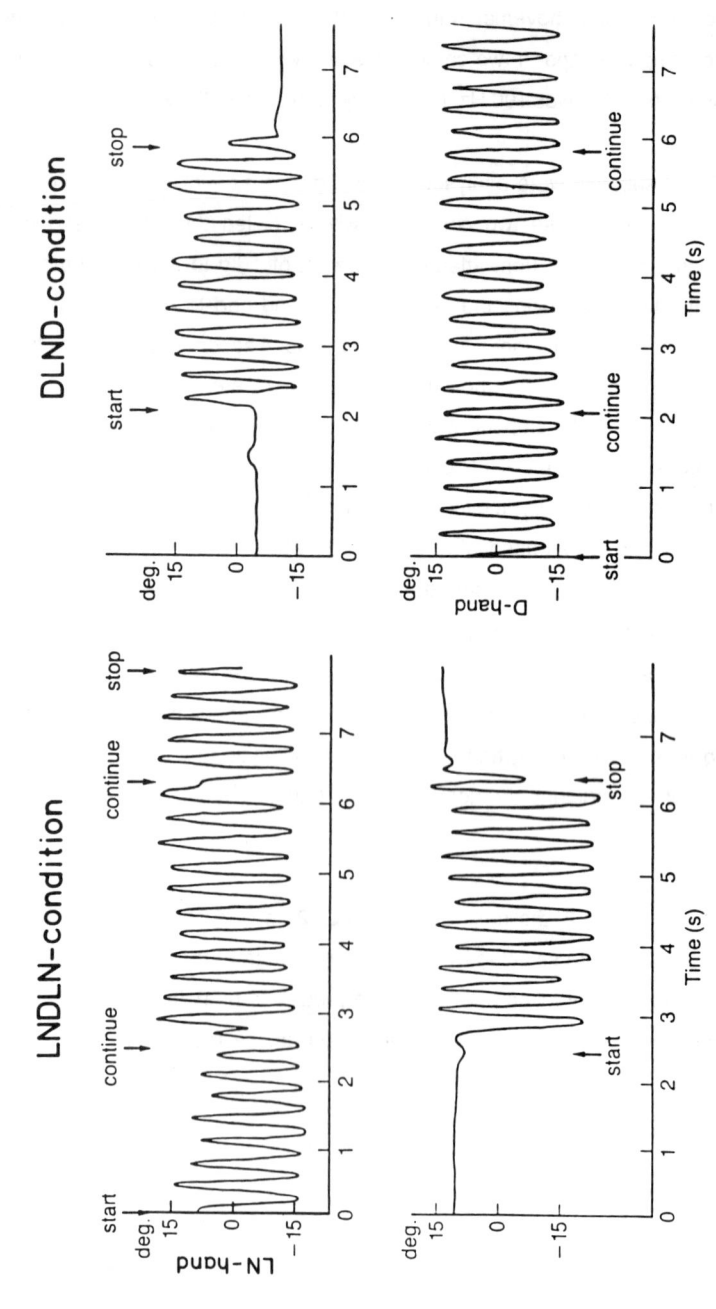

Fig.1: Explanation of the DLND- and the LNDLN-condition. In the DLND-condition the D-hand starts moving. While the D-hand continues to move the LN-hand starts to move for about 4 seconds. Then the LN-hand stops moving while the D-hand keeps on moving. The LNDLN-condition is completely analogues.

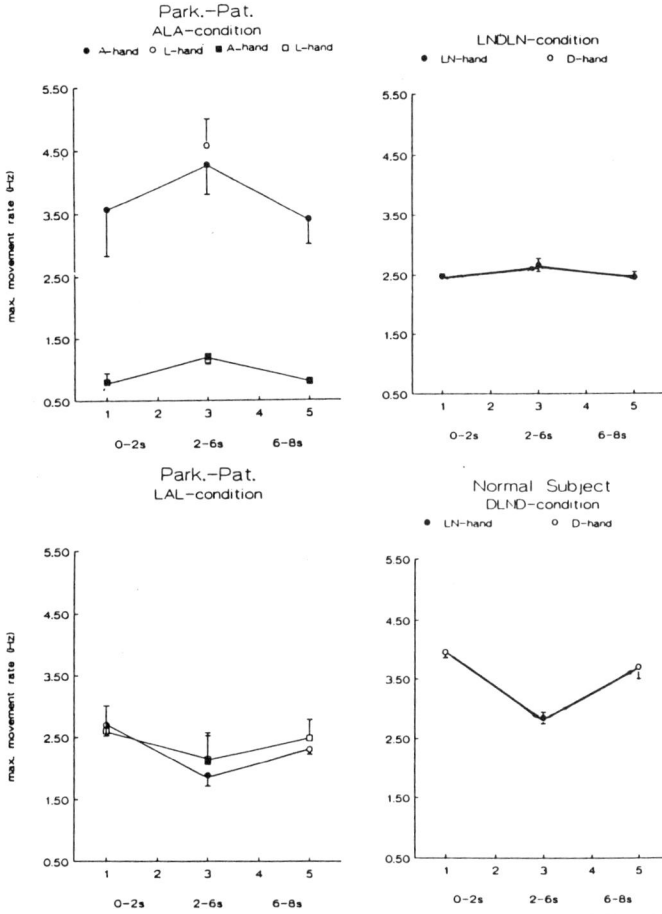

Fig.2: (left side; upper part) Two examples for the increase of MMR of the A-hand when the L-hand starts moving and for the decrease of MMR of the A-hand when the L-hand stops moving in the ALA-condition in 2 hemi-parkinsonian patients.
(left side; lower part) Two examples for the decrease of MMR in the L-hand when the A-hand starts moving and for the increase of MMR of the L-hand when the A-hand stops moving in the LAL-condition in 2 hemi-parkinsonian patients.
(right side; upper part) One example for the increase of MMR in the LN-hand when the D-hand starts moving and for the decrease of MMR of the LN-hand when the D-hand stops moving in the LNDLN-condition in a normal subject.
(right side; lower part) One example for the decrease of MMR in the D-hand when the LN-hand starts moving and for the increase of MMR of the D-hand when the LN-hand stops moving in the DLND-condition in a normal subject.

RESULTS

A) In Hemi-parkinsonian patients

In the Parkinsonian patients maximum movement rate (MMR) was significantly ($p<0.05$) lower in the more affected hand (A-hand: 2.94 ± 1.51 Hz) compared to the less affected hand (L-hand: 3.29 ± 1.17 Hz). Only in two patients the difference between both hands was less than 0.3 Hz.

In the LAL-condition, in which the less affected hand started to move, a significant reduction of its MMR was found in 4 out of 7 patients, when the more affected hand started to move. Two examples are presented in Fig. 2 (left side). Five out of 7 patients showed an increase of MMR rate of the L-hand when the A-hand stopped moving, but this effect was significant only in one patient and the group mean value did not reveal a significant effect.

In the ALA-condition, when the more affected side started to move, an increase of MMR was observed in 4 patients, when the less affected side was moved additionally. This effect was significant in 2 out of 7 patients (Fig. 2). These two patients also showed a significant decrease of MMR of the A-hand after the L-hand stopped moving (Fig. 2, left side).

In 64% of all ALA-trials regardless whether the initial frequency was different between both sides, the same frequency was found during simultaneous movements. In the LAL-condition, even 79% of all trials revealed the same frequency on both sides.

B) In normal subjects

One kg additional weight on the non-dominant hand reduced MMR significantly from 4.12 Hz to 2.89 Hz. Thus after loading, the difference in MMR between both hands was more pronounced in the normal subjects than in the patients.

Both a significant decrease of mean MMR and a significant increase of mean MMR of the D-hand was found in the DLND condition, when the LN-hand started and stopped moving. A less clear, but on the basis of single comparisons significant increase of the mean MMR of the LN-hand as well as a significant decrease of the LN-hand was found when the D-hand started and stopped moving. In 3 out of 8 subjects, a significant decrease and increase of MMR was found in the DLND-condition (Fig. 2, right side). In 2 out of 8 subjects, a significant increase of MMR was found in the LNDLN-condition (Fig. 2, right side).

In 62% of all trials the same frequency was found on either side during simultaneous movements in the DLND-condition and in 59% of the trials in the LNDLN-condition.

DISCUSSION

During the bimanual movements tested in this study both hemi-parkinsonian patients and normal subjects showed a high degree of coupling.

Coupling was evident: when the movement rate of one hand was significantly changed by movements of the other hand, the first time when the second hand was moved in addition and both hands were then moved simultaneously and a second time when the second hand stopped moving. Such coupling was observed in normals and patients. The degree of coupling showed a considerable variation from subject to subject and patient to patient. In the patients a more pronounced coupling could be observed compared to the normal subjects probably because of the lower mean MMR since the coupling effect was strongest in those subjects and patients who had a low movement rate and a clear difference in the MMR between both sides. This may be a hint that the sensorimotor integration process needs time. If the movement rate is too fast, the integration process takes too long to guarantee a precise coupling. Frequency dependent coupling has been described for simultaneous alternating index finger movements (2) and eye and hand movements (11).

Since we were able to demonstrate that the same coupling phenomena could be induced in normal subjects by changes of the peripheral mechanical conditions as observed in hemi-parkinsonian patients suffering from an intrinsic transmitter deficit and hemi-paretic patients (7) it is not true that the observed coupling in these patients results from their special lesions of the CNS. Impairment of single structures in the CNS may cause tight coupling as for example in patients with mirror movements who show the highest degree of coupling of both hands probably because of a bilateral projection of one motor cortex to the spinal cord (12). These patients are forced to make the same movement on either side. Patients with a hypertrophic degeneration of the inferior olive leading to palatal myoclonus may show a 1:1 coupling between oscillations in different motor subsystems (13). But these are special lesions with special changes in the neuroanatomy.

We think that coupling of both hands always occurs when clear differences between MMR on both sides is present and results from a complex interaction and sensorimotor integration of a variety of different brain structures with the spinal cord and brainstem structures known as central pattern generators playing a major role.

ACKNOWLEDGEMENT

This study was supplied by grants from the Deutsche Forschungsgemeinschaft (SFB 194, A5)

REFERENCES

1. KELSO, J.A.S., B. TULLER & K.S. HARRIS (1981): A "dynamic pattern" perspective on the control and coordination of movement. In: P.F. MAC NEILAGE (ed.): The production of Speech. Springer, New York, pp. 137-173
2. HAKEN, H., J.A.S. KELSO & H. BUNZ (1985): A theoretical model of phase transitions in human hand movements. Biol. Cybern. 51, 347-356
3. TURVEY, M.T., L.D. ROSENBLUM, P.N. KUGLER & R.C. SCHMIDT (1986): Fluctuatious and phase symmetry in coordinated rhythmic movements. J. Exp. Psychology: Human Percep and Perform 124, 564-583
4. KELSO, J.A.S. & G.C. DE GUZMAN (1989): Phase attractive dynamics and pattern selection in complex multi-frequency behaviours. Soc. Neur. Abstr. Vol. 15, A4. 14
5. BABINSKI, J. (1902): Sur le role du cervelet dans les actes volitionnels neccessitant une succession rapide des mouvements (diadococinesie). Rev. neurol. 10, 1013-1017
6. COHN, R. (1951): Interaction in bilaterally simultaneous voluntary motor function. Am. Arch. Neurol. & Psychiatry 65, 472-478
7. HAUSMANOVA-PETRUSEOVICZ, I. (1959): Interaction in simultaneous motor functions. Am. Arch. Neurol. & Psychiatry 81, 173-181
8. BENECKE, R., J.C. ROTHWELL, J.P.R. DICK et al. (1986a): Performance of simultaneous movements with Parkinson's disease. Brain 109, 739-757
9. BENECKE, R., J.C. ROTHWELL, B.L. DAY et al. (1986): Motor strategies involved in the performance of sequential movements. Exp. Brain Res. 63, 585-595
10. HOEHN, M.M. & M.D. YAHR (1967): Parkisonism: onset, progression, and mortality. Neurology 17, 427-442
11. FREUND, H.-J. (1986): Time control in hand movements. In: H.-J. FREUND, U. BÜTTNER, B. COHEN & J. NOTH (eds.): The oculomotor and skeletalmotor system. Prog. Brain Res. 64, 287-294
12. CHAN, J.-L. & E.D. ROSS (1988): Left-handed mirror writing following right anterior cerebral artery infarction: Evidence for nonmirror transformation of motor programs by right supplementary motor area. Neurology 38, 59-65
13. HEFTER, H., E. LOGIGIAN, O.W. WITTE et al. (1991): Mechanisms underlying the pathogenesis of oscillatory activity in palatal myoclonus. (this volume)

CLINICAL ASPECTS

CIRCADIAN RHYTHMS IN CARDIOVASCULAR DISEASES

N.L. ASLANIAN

Institute of Cardiology, P.Sevak Street 5, Yerevan-375044, Armenia

For a physiologic function to be considered rhythmic it needs not be precisely repetitive, but it must show some regularity of repetition. It must be shown that its occurrence as a matter of chance is unlikely. With the use of computers it has become possible to reveal and quantify rhythmic, predictable variability of biologic functions.

The changes in the organism in time are very complex. We can consider the following changes: slow but progressive increase or decrease of the investigated index without returning to its initial value during the observation span. Such a change is called a trend. The second type of change can be considered when the index deviates from the initial value but soon after it returns to it, i.e. the index fluctuates. When fluctuations are repeated we can speak about undulating process. When the cycles of an undulating process are nearly the same, we can use the term rhythm, rhythmicity. But besides these determined changes (useful signal) there also may be casual changes (noise). When we record a trend during the observation span it does not mean yet that the process is not rhythmic. The process may be rhythmic but the observation span will be insufficient to detect the rhythm. On the other hand, when we have revealed a cycle, we cannot speak about rhythmicity yet. The cycle must be repeated.

Whether the data are collected by selfmeasurement alone or in combination with automatic measurement, as is readily feasible for several variables, the time series thus obtained are the sum of our responses to varied cyclic and other environmental factors. Some of the frequencies, validated as statistically significant changes, usually are defined as rhythms. These rhythms are synchronized by environmental schedules, primary social rather than physical.

The frequency of rhythms varies within large limits. But revealing rhythms is a difficult task, because rhythms with different frequencies are superimposed in the organism. And this brings to distortion of waveform. The revealing of rhythms is possible when mathematical modelling of the rhythmic function is used. Rhythms change with development, maturation and aging.

For the most part I shall speak about circadian rhythms and about disorders of the circadian timing system. However I shall make some attempts to go into other frequencies such as circaseptan and circadiseptan.

METHODS

Constraints imposed by finance, time and ethics may limit the number of experiments that one wishes to perform. However in the case of chronobiological studies, a restriction in the frequency of sampling can influence the result obtained considerably. The less frequent the sampling interval is, the more the details of the rhythm are lost, though they are convenient but are misleading. To reveal circadian rhythms, the observations on each subject or organism should be made for at least 48 h. In my opinion, it is advisable to measure for 3-5 days and to measure the index investigated at intervals of not more than 4 h. When using such an approach I encountered a problem and sought to develop different mathematical approaches. For the estimation of the period of the rhythm we recommend two methods (1): an approximation method and a method estimating the repetition of the following one after another fragments of investigated process. By means of the approximation method sinusoidal rhythms with known (cosinor) and unknown (nonlinear least-squares) periods are revealed.

Quite often, in the case of prominent components the linear least squares method (cosinor) yields satisfactory results (8). Sometimes, however, the linear least squares method may detect pseudoperiods or fail to detect true periods close to each other. Thus, the output of the single cosinor is neither a complete nor the best approach. The problem is partly resolved by combining the linear least squares and nonlinear least squares procedures. This approach may directly estimate the period along with its confidence interval (4,5,9). An analysis based upon several mathematical methods is preferable. Similar results obtained by more than one technique can give the researcher more confidence in their validity.

Thus, there are several mathematical methods for modelling of rhythmic processes. They are based on the principle of approximation of real rhythmic processes by trigonometric polynomials with unknown periods. But these methods may loose their meaning if the quantity of harmonics is nearer and nearer to the number of measurements of the index investigated. With a little change the above mentioned methods may be used when the measurements are done with unequidistant intervals. In this case we are faced with the phenomenon of revealing of spurious periods too.

In that case when by the nonlinear least squares statistically significant rhythms are not revealed we used the method estimating the repetition of the following one after another fragments of the process which is based on the method of dispersion analysis (1). Finally, years of biologic rhythms investigation dictated to us that it may be advisable to carry out manual measurements at nonequidistant intervals especially in the absence of automatic recording devices. The aim of such an approach is to keep the natural and ordinary regimen of the subject to be studied without disturbance during the hours of sleep at night. For this latter purpose we developed (2) a new mathematical method, which may be used to reveal all kinds of rhythms whether the points measured are equidistant or non-equidistant. In this method, based on the algorithm of minimization of the area of the displacement of the diagram of the index investigated, the a priori assumption about the waveform of the process investigated is not accepted. This program affords the opportunity to verify the supposed periodicity, its statistical significance, to find independently the best fitted period, with the given significance, if the investigated process is really rhythmic. The mean value, its mean deviation, the maximum and the minimum values and the time of their occurrence are also determined by this program. The calculated values and the diagram of the index time series are printed and the limits of each cycle are noted.

Subjects

The subjects studied by us were patients with cardiovascular diseases, healthy persons, healthy pregnants and neonates. We studied ultradian (with 3-20 h periods), circadian (20-28 h) and infradian (28-96 h) rhythms of cardiovascular and renal excretory functions. We have taken measurements with 1 or 4 h intervals during 48-120 h.

RESULTS

Statistically significant rhythms (repeated cycles) in heart rate (HR), systolic (SBP), diastolic (DBP) blood pressures and ECG indices were found in 95 % of healthy subjects studied as well in 91 % of the indices of their renal mineral excretory function. Among statistically significant rhythms, 92 % were circadian, 5 % ultradian and 3 % infradian. The mean circadian period was 23.8 +/-0.1 h. The acrophases of the circadian rhythms of healthy subjects were not alike for the group as a whole.

We studied 28 healthy pregnants (38-40 weeks), 30 puerperas (first delivery) and 33 healthy full-term neonates who have been monitored with a Nippon Colin sphygmomanometer to 72 h at 4 h intervals during the three days after birth. Significant and especially circadian rhythms were fewer in pregnants and puerperas than in healthy non-pregnants and were very rare in neonates. Our data of BP and HR rhythms in pregnants and neonates are very similar to that of HALBERG et al. (12).

At present, neonates are being given screening tests for relatively rare diseases, such as phenylketonuria. The time has come to give emphasis to assessing risk for major cardiovascular diseases, such as coronary heart diseases, hypertensive disease, etc. Previous studies have been based mostly on casual blood pressure measurements; a chronobiologic approach is the better alternative. HALBERG et al. (10) have used this chronobiologic approach to predict for neonates, from blood pressure readings, the likelihood of high blood pressure in later life. As expected from earlier studies by HELLBRÜGGE (13) the extent of development of approximate 24-h cardiovascular rhythmicity shows great interindividual differences in terms of circadian amplitude. Circadian amplitude was more prominent in the group with a family history of high blood pressure than in the group without such a family history (6,7,11).

Patients

The results of our investigations showed that in patients with exertional angina, the circadian organization of hemodynamics and ECG indices are disturbed. However the degree of disturbance is not the same for different indices. Thus, it is evident that the percentage of circadian rhythms diminishes from 95 % found in the controls as noted in HR, SBP and PQ interval rhythm. The lowest percentage of circadian rhythm is noted for the height of the T wave of the ECG. The rhythms of T wave are mainly desynchronized and a state of infradian neorhythmostasis is occurred. Probably this is due to the lability of the T wave and to chronic coronary circulatory insufficiency, which is a stress factor promoting time structure disturbance of the myocardium and its bioelectricity (3).

The role of several electrolytes and trace elements in the mechanism of vascular tone changes is well recognized. Particularly, the intra- and extracellular gradients of sodium, potassium, and calcium play a definite role in the spasm of peripheral arterioles, coronary arteries, and in heart contractions. The content of copper increases in the organism in the case of atherosclerosis and the content of iron and zinc decreases. Vanadium inhibits the synthesis of cholesterol at the point of mevalonic acid formation. The increase of cadmium enhances the development of arterial hypertension.

The results of our investigation showed that in patients with exertional angina, in most cases, infradian rhythms are observed. The percentage of circadian rhythms of Mg, P, Ca, and V decreases and the percentage of infradian rhythms of Cu and Zn increases. The infradian rhythms of Na, Cl, Mg, Cd, and Cr prevail over the circadian ones. However, an increase in the percentage of ultradian rhythms of Cl, Mg, Fe, and Cd may also be observed. At the same time ultradian rhythms of the excretion of Fe are more prominent than the circadian or infradian rhythms.

Circadian variations in the cardiovascular and urinary indices in men have been reported by KANABROCKI et al. (14). These authors investigated circadian changes by observing each subject during 27 h and using cosinor analysis of HALBERG (8). Nevertheless, our data of minerals excretion and cardiovascular indices rhythms concerning mesor and amplitude in healthy subjects were almost similar to the data reported by KANABROCKI et al.(14).

Today we cannot finally confirm if disrhythmostasis or disrhythmias are due only to ischemic heart disease. We must note that mean age is higher in patients with ischemic heart disease than in healthy subjects, and age may influence the results.

Thus, the chronobiologic approach provides an opportunity to reveal early stages of disease at a time at which there are no marked symptoms, but there are nonspecific disturbances manifested as desynchronization. This approach gives the possibility of scrutinizing the pathogenesis of disease from a new point of view.

Circaseptan and circadiseptan rhythms

In another investigation we studied circaseptan and circadiseptan rhythms of ECG indices in patients with acute myocardial infarction. For this purpose the ECG indices were recorded in 12 leads every day at 10.00 beginning from the 4th day after infarction during 30 days. The data were analyzed by linear least squares in combination with autocorrelation analysis. In healthy subjects circaseptan (with 4-10 days periods) and circadiseptan (with 11-17 days periods) rhythms were revealed. In patients with acute myocardial infarction the percentage of statistically significant circaseptan and circadiseptan rhythms was lower than in healthy subjects (60 % in patients versus 90 % in healthy subjects). In patients there were significant changes of mesors and amplitudes. But in patients with infarction in 60-90 % circaseptan and circadiseptan rhythms of ST segment were revealed. The changes of rhythm parameters of ECG indices in patients depended on the complications of infarction and on the accompanied diseases. These rhythmological data may help in the prophylaxis of complications and prediction of the severity of the course of acute myocardial infarction.

ACKNOWLEDGEMENT

The author gratefully acknowledges Miss Isabell Aslanian for the preparation of the manuscript.

REFERENCES

1. ASLANIAN, N.L., E.M. KRISHCHIAN & D.G. ASSATRIAN (1987): On the methodology of chronobiological investigations in clinical medicine. In: B. TARQUINI (ed.): Social diseases and chronobiology, Esculapio Pub., Bologna, pp.155-161
2. ASLANIAN, N.L. & G.G. HAGOPIAN (1990): A method of revealing and determination of biological rhythm parameters. Problems of Chronobiology (Yerevan) 1, 167-170

3. ASLANIAN, N.L. (1990): Chronobiological aspects of ischemic heart disease. In: D.K. HAYES, J.E. PAULY & R.J. REITER (eds.): Chronobiology: Its role in clinical medicine, general biology, and agriculture. Part B, Wilney-Liss, Inc., New York, pp. 583-592
4. AYALA, D.E. & R.C. HERMIDA (1989): Combined linear-nonlinear approach for biologic signal processing. Proc. 7th IASTED Int. Symp. Applied Informatics, Grindelwald, Switzerland, pp. 67-70
5. AYALA, D.E. & R.C. HERMIDA (1990): Infradian variation in the incidence of giardiasis assessed by linear-nonlinear rhythmometry. In: D.K. HAYES, J.E. PAULY,R. & J.REITER: Chronobiology: Its role in clinical medicine, general biology, and agriculture. Part B, Wilney-Liss, Inc., New York, pp. 202-220
6. CARANDENTE,F., A.ANGELI & S. VENTURA (1988): Autorhythmometry and automatic rhythmometry for the diagnosis of amplitude-hypertension and of mesor-hypertension: the hyperbaric impact as an index of vascular damage. Chronobiologia. 15, 246
7. CORNÉLISSEN, G., F. HALBERG, B. TARQUINI et al. (1988): Neonatal blood pressure monitoring for early cardiovascular risk assessment. Chronobiologia. 15 (Suppl. 1), 264
8. HALBERG, F., F. CARANDENTE, G. CORNÉLISSEN & G.S. KATINAS (1977): Glossary of Chronobiology. Chronobiologia, 4 (Suppl. 1), 189 pp.
9. HALBERG, F., E. HALBERG, W. NELSON et al. (1982): Chronobiology and laboratory medicine in developing areas. Proc. 1st African and Mediterranean Congress for Clinical Chemistry. Milan, Nov. 11-15, 1980, pp. 113-156
10. HALBERG, F., G. CORNÉLISSEN, C. BINGHAM et al. (1986): Neonatal monitoring to assess risk for hypertension. Postgraduate medicine. 79, No. 1, 44-46
11. HALBERG, F., E. BAKKEN, G. CORNÉLISSEN et al. (1989): Blood pressure assessment in a broad chronobiologic perspective. In: H. REFSUM, J.A. SULG & K. RASMUSSEN (eds.): Heart & Brain, Brain & Heart. Springer-Verlag, Berlin, pp. 142-162
12. HALBERG, F., G. CORNÉLISSEN & E. BAKKEN (1990): Caregiving merged with chronobiologic outcome assessment, research, and education in health maintenance organizations. In: D.K. HAYES, J.E. PAULY & R.J. REITER (eds.): Chronobiology: Its role in clinical medicine, general biology, and agriculture. Part B. Wiley-Liss, Inc., New York, pp. 491-549
13. HELLBRÜGGE, T. (1960): The development of circadian rhythms in infants. Cold Spr. Harb. Symp. quant. Biol. 25, 311-324
14. KANABROCKI, E.L., R.B. SOTHERN, L.E. SCHEVING et al. (1988): Ten-year-replicated circadian profiles for 36 physiological, serological, and urinary variables in healthy men. Chronobiology International. 5, 237-284

EVALUATION OF THE SPONTANEOUS VARIATIONS OF CARDIOVASCULAR VARIABLES.

K.UEZONO[1], M.BOTHMANN[2], G.HILDEBRANDT[2], R. MOOG[2] & T.KAWASAKI[1]

[1] Institute of Health Sciences, Kyushu University, Kasuga 816, Japan
[2] Institut für Arbeitsphysiologie und Rehabilitationsforschung, Philipps Universität Marburg, Robert-Koch-Str 7a, D-W-3550 Marburg, FRG

INTRODUCTION

Automatic devices enable us to measure several cardiovascular variables frequently and automatically for more than 24 hours. Such measurements were one presupposition to ascertain that variables like blood pressure, pulse rate, and even body temperature covariate substantially with activity (5,10,13), nutrition (1), mental stress (11), and even position (4) - etc (comp. 6,8). These effects do not only result in comparable rapid changes, they can cause "baseline displacements" over several hours too. Moreover, these long lasting changes are reported to depend on the circadian phase (9). Therefore, the usual procedures of measurements may only give little insight into the spontaneous rhythmicity.

Several routines and mathematical methods have been already developed in order to minimize disturbing masking effects or to dissociate spontaneous circadian courses and masking effects (e.g. constant routines; 7,15).

These methods to dissociate masking effects by mathematical means assume additivity between masking effects and the spontaneous circadian rhythm. For example, if masking effects are phasedependend this assumption is missed (e.g.7).

On the other hand, relations between circadian parameters of physiological variables like the circadian amplitude and cardiovascular diseases were found when subjects followed their usual living routines (2,14). However, the readings of these parameters do in fact represent a superimposition of the spontaneous behavior of the circadian system, and masking effects depending on the circadian variation of reagibility too.

Therefore, the aims of this study were to determine the effect of tranquilization and mild activities on the daily courses of cardiovascular variables and to determine the ability of low level activities to hide spontaneous circadian rhythms.

MATERIAL AND METHODS

In order to reduce masking effects measurements were performed during constant routines (control days (11)). On these standard control days subjects were kept in a constant climate on continuous bedrest for 24 hours while on a low-protein diet. Food intake was equally distributed over the period by providing a meal every three hours; on these occasions urine samples were taken, as well as a measure of subjective vigilance. The latter was also taken at intervening hours whenever EEG readings indicated that the subjects were awake. Apart from these disturbances subjects were allowed to sleep at their convenience. A personal light (40W) was available to them to read by, if they chose.

In order to further minimize masking effects on the one hand, and to study the effects of stepwise increased physical activity, for each subject this standard control day was modified twice.

1. On "tranquilization days" (lowest activity level) subjects took 4 doses of 30 mg phenobarbital each. Tranquilization started at 9.00 hrs (5h's before the usual start of the control day) and were repeated 3 times every 3 hrs.

2. On "strain days". Every 3 hrs, after food intake etc. and after at least 5 min of immobility, physical activity was interpolated for 30 min in the control day. Each load consisted of 10 min standing, 5 min sitting on a bicycle ergometer, 5 min cycling (50 cycles/min) without extra load, 5 min cycling with 25 Watt and 5-min quiet sitting on the ergometer, and 5 min of recumbency.

Twelve apparently healthy male students between 21 and 37 years of age served as subjects. Each subject underwent all three routines, with intervals of 4 to 10 days. Measurements were started always at 12.00 hrs and ended after 14.00 hrs. In respect to the nightly troughs of rectal temperature the individual circadian phase position ranged between 3.00 and 6.00 hrs, varying between indifferent circadian phase types and evening types.

Systolic (SYS BP) and diastolic (DIA BP) blood pressure were recorded automatically by an oscillometric method (BP-8800, Nippon Colin Co., Japan) from the left upper arm every 15 minutes throughout the control day, and on the strain days every 1 min during the 40 min periods of the stepwise physical interventions. Heart rate (HR), ECG as well as rectal temperature were recorded continuously.

As the entry to the chamber by personnel might affect cardiovascular variables as well as eating etc., the respective data were excluded and interpolated (15 min, 45 min on strain days). In order to reduce variance due to interindividual differences in circadian phase type, the daily profiles were shifted to adjust the timing of rectal temperature trough to a common minimum (3.00 hrs) on standard control days.

RESULTS AND DISCUSSION

Fig.1 compares the average daily courses of SYS BP, DIA BP, and HR. All curves are more or less superimposed by periodic changes of about 3 h period duration, presumably evoked by the experimental schedule, Therefore, these variation will, in the first step not be considered. We filtered all individual curves (moving weighted average, 5 points on either side). The blood pressure levels on tranquilization days were significantly lower than those on the standard control- and on the strain- days (undisturbed intervals). Whereas no significant differences in shapes, phases, or amplitudes could be detected.

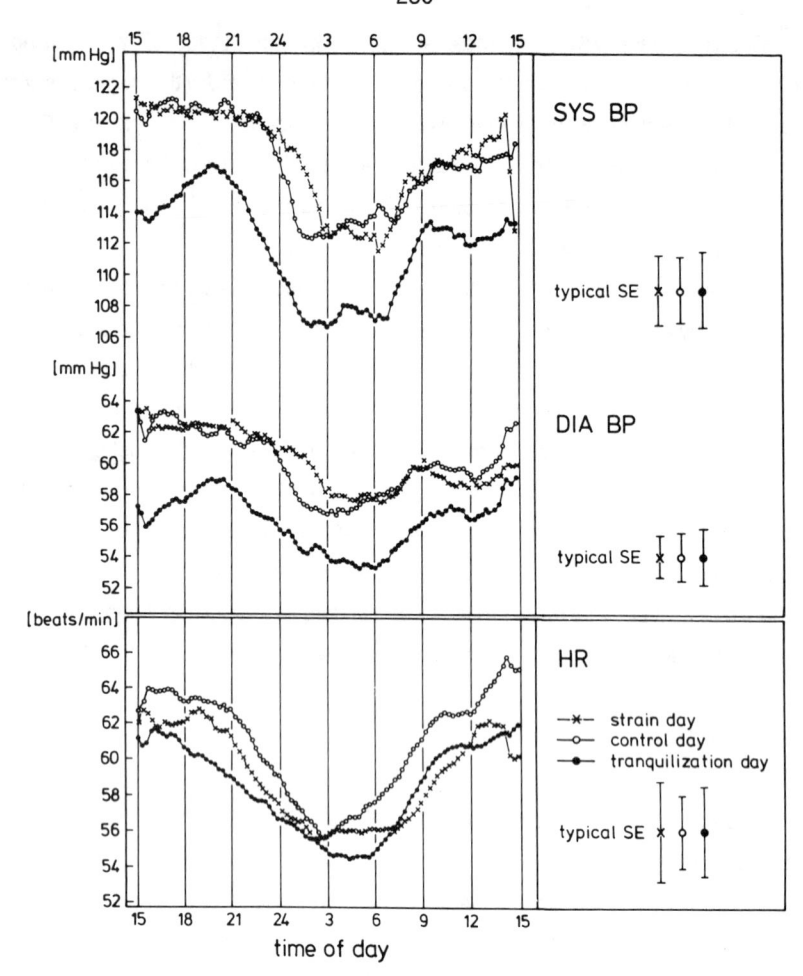

Fig.1: Comparisons of the average daily courses of SYS BP, DIA BP, and HR on tranquilization- standard control- and on strain- days. (moving weighted average, 5 points on either side).

Concerning the blood pressure values and HR on strain days the disturbing influences of the physical load periods are short lasting in such a way that concerning these variables analogous conditions exist as compared to the standard control days in the undisturbed intervals .

In order to prove possible influences of the circadian phase on the reactions of the circulatory parameters, Fig.2 shows the daily courses of the mean 3-hourly calculated deviations of the blood pressure values and heart rate from the respective levels during tranquilization day.

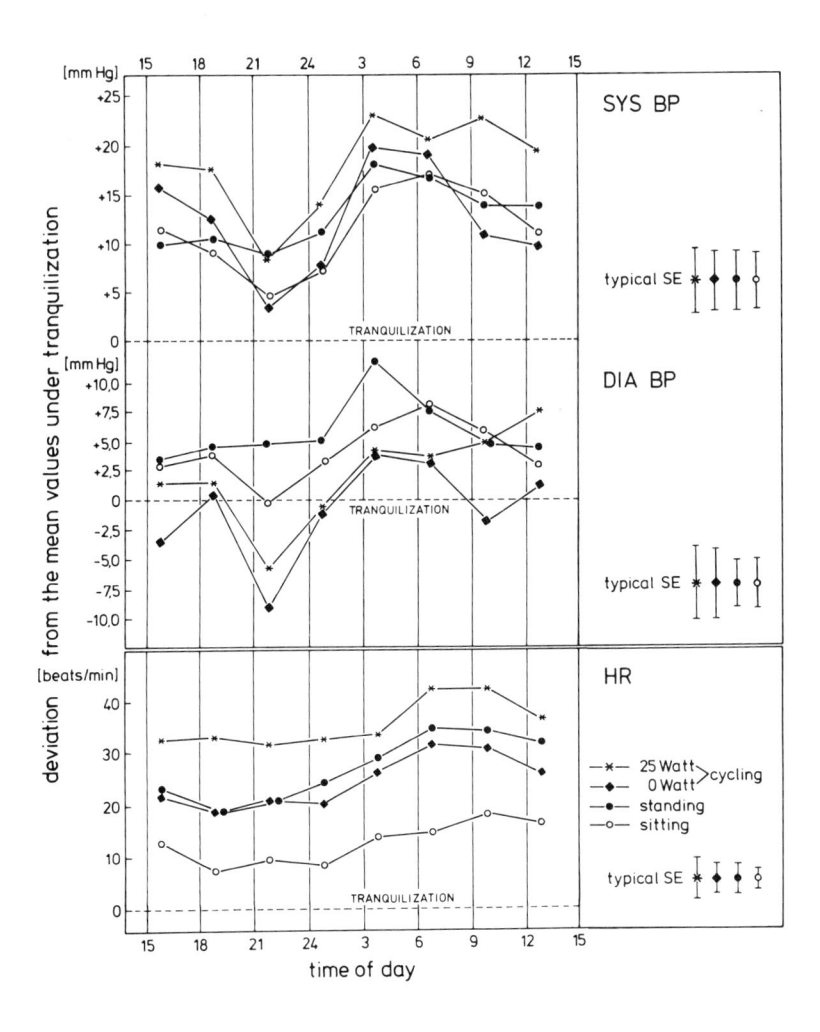

Fig.2: Daily courses of the mean 3-hourly calculated deviations of the blood pressure values and heart rate from the respective levels during tranquilization day.

There is a clear-cut circadian influence on the responses of blood pressure and heart rate to the various steps of physical load. The maximum of responsiveness is passed during the first half of the day, whereas the minimum occurs in the evening hours. This is true also for the DIA BP, which tends to negative responses during the afternoon and positive ones in the forenoon. Of course, the extends of the reactions depend also on the amount of physical load.

Hence, there is no doubt that the responsiveness of circulatory parameters studied depend on the circadian phase. Moreover, even under relatively small physical loads at certain circadian phases the (masking) reactions can exceed the spontaneous amplitude of the circadian variations. This is in line with older results (3), which already demonstrated a circadian cycle of non-specific responsiveness in man, the maximum of which passing in the forenoon.

From the point of view of masking of circadian rhythms it can be stated, that the circadian variation of blood pressure and heart rate do not result from masking influences of heart rate do not result from masking influences of varying states of activity. Masking effects depend on the spontaneous circadian changes of reagibility, and can, depending on the extend of stimulation, exceed the spontaneous amplitude of the circadian variations of the respective functions. All these facts do not allow to calculate masking effects on the basis of standardized assumptions.

REFERENCES

1. FOERTSCH, H.-U. (1968): Untersuchungen zur Wärmebilanz des Menschen nach Nahrungsaufnahme in thermisch neutraler und warmer Umgebung., Dissertation aus dem Max-Planck-Institut für Arbeitsphysiologie, Dortmund
2. HALBERG, F., J. DRAYER, G. CORNÉLISSON, M.A. WEBER (1984): Cardiovascular reference data base for recognizing circadian MESOR- and amplitude- hypertension in apparently healthy men. Chronobiologia, 11, 275-298
3. HILDEBRANDT, G. & G. BESTEHORN (1977): Studies on the circadian rhythm of lacrimation. Chronobiologia, 4, 1-6 & 210-244
4. JAMIESON, M.J., S. WEBSTER, S. PHILIPS et al.(1990): The measurement of blood pressure: sitting or supine, once or twice? J. Hypertens., 8, 635-640
5. MILLAR-CRAIG, M.W., C.N. BISHOP & E.B. RAFTERY (1978): Circadian variation of blood pressure. Lancet, 1, 798-797
6. MINORS, D.S. & J.M. WATHERHOUSE (eds; 1989): Chronobiology International, 6,
7. MINORS, D.S. & J.M. WATHERHOUSE (1992): Investigating the endogenous component of human circadian rhythms: a review of some simple alternatives to constant routines. Chronobiology International, 9, 55-78
8. MOOG, R., G. HILDEBRANDT & P. ZEZULA (1987): Comparison between different agents of masking. In: G. HILDEBRANDT, R. MOOG & F. RASCHKE (eds.): Chronobiology and Chronomedicine. Peter Lang, Frankfurt am Main, pp.136-139
9. MOOG, R. (1986): Disturbances of the circadian system due to masking effects. In: L. Rensing, U. an der HEIDEN & M.C.MACKEY (Eds.): Springer, Berlin-Heidelberg-NewYork-London. pp. 186-188
10. MOOG, R., & G. HILDEBRANDT (1989): Adaptation to shift work - experimental approaches with reduced masking effects. Chronobiology International, 6, 65-75
11. PARATI, G., G. POMIDOSSI, R. CASADEI et al. (1988): Comparisons of the cardiovascular effects of different laboratory stressors and their relationship with blood pressure variability. J. Hypertens., 6, 481-488
12. PERLOFF, D., M. SOKOLOW & R. COWAN (1983): The prognostic value of ambulatory blood pressures. YAMA, 249, 2792-2798
13. ROWLANDS, B.D., T.J. STALLARD, R.D.S. WATSON & W.A. LITTER (1980): The influence of physical activity on arterial pressure during ambulatory recordings in man. Clin.Sci., 58, 115-117
14. SMIRK, F.H., A.M.O. VEALE & K. ALSTAD (1959): Basal and supplemental blood pressure in relationship to life expectancy and hypertension symptomatology. NZ Med. J., 58,711-735
15. SPENCER, M.B. (1989): Regression models for the estimation of circadian rhythms in the presence of affects due to masking. Chronobiology International, 6, 77-91

CIRCADIAN BLOOD PRESSURE RHYTHM IN THE THIRD TRIMESTER OF PREGNANCY

C. MAGGIONI[1], B. BRODBECK[1], L. DETTI[2], G. MELLO[2] & G. BENZI[2]

[1]Clinica Ostetrico-Ginecologica dell' Universita' di Milano
[2]II Clinica Ostetrico-Ginecologica dell' Universita' di Firenze

INTRODUCTION

During pregnancy Blood Pressure (BP) has been reported to show a circadian variability, but little is known about the relation between rhythm characteristics and pathological conditions such as preeclampsia (PIH), intrauterine growth retardation (IUGR) or diabetes. In this study we investigated if a BP rhythm profile is specific for these pathological conditions and we analyze which BP rhythm parameter is more useful in the BP description for each pathology and more correlated to fetal outcome.

PATIENTS AND METHOD

63 subjects have been automatically monitored during the III trimester of pregnancy. All the subjects were hospitalized and submitted to the same regimen of life. 24-hour blood pressure profiles have been obtained at 15 min intervals using an automatic devise (Spacelab). None of the patients was receiving any drugs or medication.
    The patients include different groups:

- CONTROL (healthy controls)   n=26
- RISK (before clinical diagnosis of PIH)   n=7
- PIH (pregnancy induced hypertension)   n=5
- IUGR (intrauterine growth retardation)   n=12
- DIABETES (classes A -B -C)   n=10
- CHRONIC HYPERTENSION (personal history of hypertension and/or diagnosis before 20th week of pregnancy)   n=3

The rhythm analysis and statistical analysis were performed by: means and standard deviation, single cosinor and population cosinor, parameter test (Hotelling's t-squares and Student T-test). The rhythm parameters were defined as:
Mesor: the rhythm adjusted mean,
Amplitude: measure of predictable extent of change,
Acrophase: measure of timing of overall high values.

RESULTS

The results are shown in Fig. 1 to 4. The SBP is characterized by a Mesor increase in the PIH and hypertensive groups and a phase revers al in the PIH group. The DBP is characterized by an amplitude increase in the IUGR group. The best correlating parameter to the fetal outcome is the diastolic amplitude (Fig. 6).

DISCUSSION

A circadian periodicity of BP during pregnancy has been established by MALEK et al. (14), MEISS & HALBERG (16), and others. A decrease in BP during sleep was seen in all pregnant women except in case of PIH. This reversal of diurnal pattern in PIH has been reported by SELIGNAM (26) using the oscillographic automatic cuff type recorder, by REDMAN & BEILIN (24) using a Roche arteriosond monitor, by SAWYER & HALPERIN (25) using a Dinamap device, by MUNAGHAM (19) with an intraarterial catheter as well as by BEILIN & DEACON (3), VALENTIN & LAFFARGUE (28) and MIYAMOTO & SHIMOKAWA (17).

In our study, the PIH group also shows a nocturnal acrophase; but not the RISK group. This may be important with regard to the diagnosis of hypertension, with regard to the timing and dose of medication, but it is not a PIH predictive sign. In this study the rhythm parameter which is specific to PIH and RISK group is the increase of SBP Mesor first of all, then DBP Mesor when the disease is clinically evident. It has been recently suggested (8) that in the preeclamptic group the cardiac output is elevated very early in pregnancy where a compensatory vasodilatation maintains normotension. The rise of BP would then start a series of reactions in the vessel

CIRCADIAN RHYTHM SBP

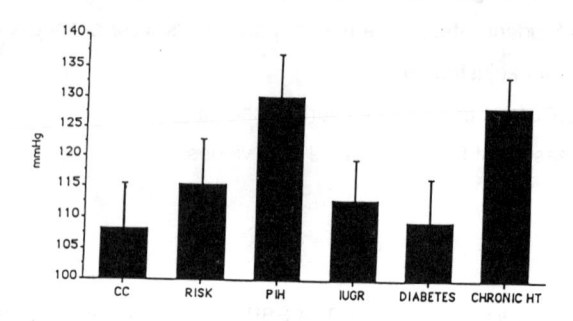

Means (±SD) of Systolic Blood Pressure MESOR

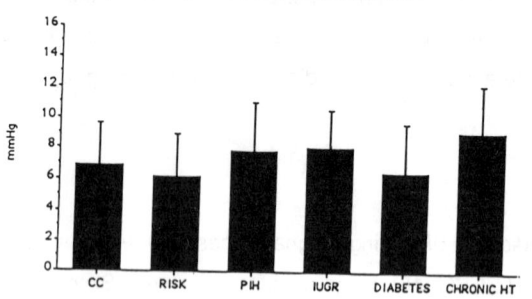

Means (±SD) of Systolic Blood Pressure AMPLITUDE

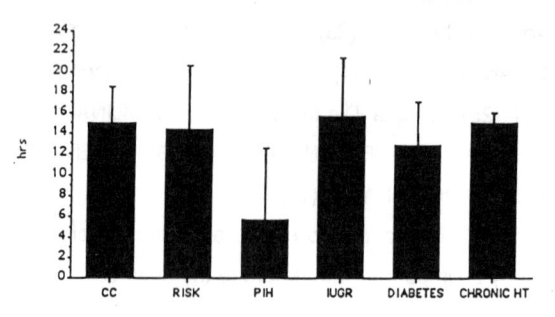

Means (±SD) of Systolic Blood Pressure ACROPHASE

Fig. 1: CC(healthy control) ,RISK (pre-clinic PIH), PIH (pregnancy induced hypertension) ,IUGR (intrauterine growth retardation),DIABETES ,CHRONIC HYPERTENSION , during the III trimester of pregnancy

CIRCADIAN RHYTHM DBP

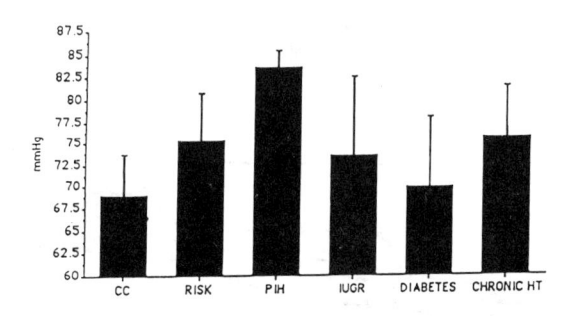

Means (±SD) of Diastolic Blood Pressure MESOR

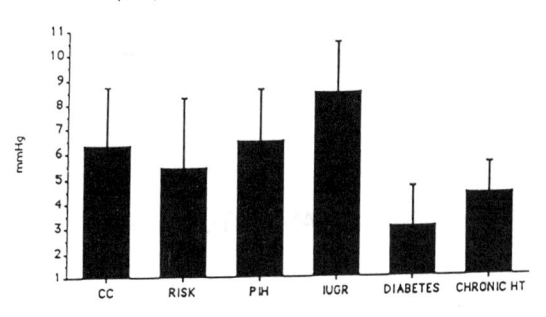

Means (±SD) of Diastolic Blood Pressure AMPLITUDE

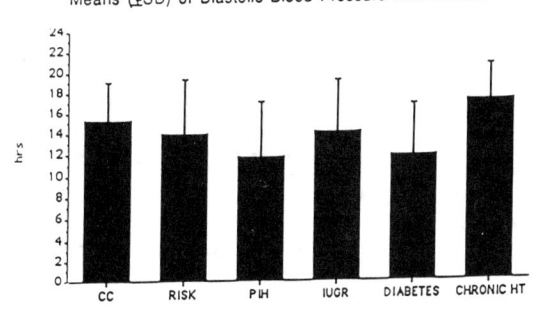

Means (±SD) of Diastolic Blood Pressure ACROPHASE

Fig. 2: CC(healthy control) ,RISK (pre-clinic PIH), PIH (pregnancy induced hypertension) ,IUGR (intrauterine growth retardation),DIABETES ,CHRONIC HYPERTENSION , during the III trimester of pregnancy

Percentiles comparaison for MESOR DBP

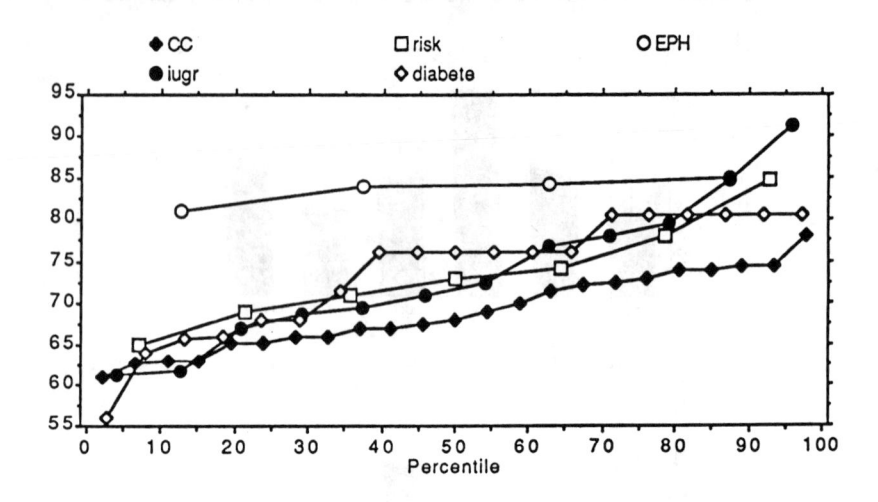

Percentiles comparaison for AMPLITUDE DBP

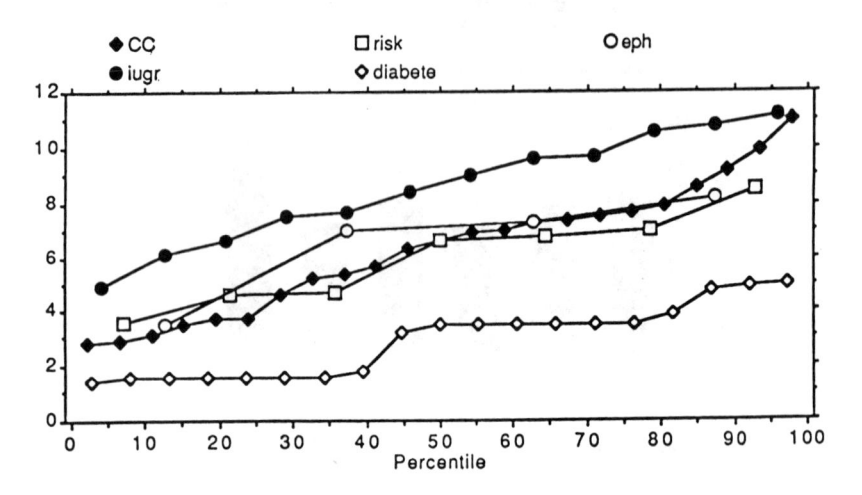

Fig. 3:

resistance, which gradually increases as the manifestations of the disease become severe. Such a cross over of hemodynamics from high cardiac output state to high resistance state with progression of the disease would explain the heterogenicity of hemodynamics observed in many studies. The high resistance state in PIH seems to be more frequently associated to the IUGR complicated pregnancy.

It has been reported that in the PIH complicated pregnancy MBP, cardiac output and stroke volume are lower in mothers with SGA rather than in those with AGA fetus (20). The underlying mechanism for IUGR could at least initially be different and in our study BP profile of the IUGR group is different from the PIH. The BP is a hemodynamic parameter directly dependent on cardiac output and systemic vascular resistance. In pregnancy, utero-placental blood flow increases dramatically to support the nutritional demands of the rapidly growing fetus. This is made possible by an increase in maternal plasma volume and cardiac output, a decrease in vascular resistance, and an increase in the fractional distribution of cardiac output to the uterus. At the beginning of normal pregnancy the trophoblast invades the myometrial cells of the placental vascular bed, reaching, finally, the endothelium of the spiral arteries at the 18th week.

In the IUGR complicated pregnancy peripheric resistances increase with a normal cardiac output and a low blood volume. In fact, reduction of normal maternal blood volume expansion and increase in venous tone are associated with hypovolemia and IUGR. GALLERY observed an inverse correlation between diastolic BP and plasma volume and, at least, to a certain point, a direct correlation between DBP and baby weight. The increase in BP in case of pure IUGR may be a secondary and compensatory mechanism and it may be destroyed by the antihypertensive therapy (WARKENTEN). Restoring the blood volume could exacerbate this reflex, increasing the cardiac output and finally the maternal BP.

We previously demonstrated by a Atrial Natriuretic Peptide study that there is a compensatory increase in blood volume also in normotensive IUGR, so increasing ANP levels to levels comparable to the PIH. In fact our actual study demonstrates that the BP rhythm parameter which better characterizes the IUGR group is the diastolic amplitude increase.

In conclusion, the IUGR group is characterized by an increase in diastolic amplitude, the PIH group by a systolic Mesor increase, the diabetic group by a normotension. These observations suggest that growth retardation can induce a superimposed PIH when the compensatory mechanism (amplitude increase) is not able to preserve fetal placental perfusion: an increase in the Mesor values in this pathology should be looked as a warning sign of PIH risk. On the other hand, pregnancy hypertension may produce growth retardation when the reverse phase and the Mesor increase are not sufficient to assure the fetal placental perfusion. An increase in diastolic Mesor is a predictive sign of hypertension, an increase in diastolic amplitude may be a predictive sign of IUGR.

If we look at the fetal outcome (considering "good" and "bad" babies depending on gestational age and sex correlated weight at birth, Apgar score) we see that, although there is not a direct correlation between maternal BP and baby weight, the fetal outcome is significantly different according to diastolic maternal amplitude.

REFERENCES

1. AHOKAS, R.A. & B.M. SIBAI (1990): The relationship between experimentally determined litter size and maternal blood pressure in spontaneously hypertensive rats. Am. J. Obst. Gynec. 162 (3), 841-847
2. BEAUFILS, M. (1990): Prise en charge de l'hipertension chez la femme enceinte diabetique. Diabete metabol. 16/2 bis, 144-148
3. BEILIN, L.J. & J. DEACON (1983): Diurnal rhythms of blood pressure, plasma renin activity, angiotensin II, and catecholamines in normotensive and hypertensive pregnancies. Clin. Exp. Hypertension (B: hypertension in pregnancy) 2, 271-293
4. BROWSENS, I. & H.G. DIXON (1977): Fetal growth retardation and arteries of the placental bed. Br. J. Obstet. Gynaecol. 84, 656-653
5. CLAPP III, J.F. (1985): Maternal heart in pregnancy. Am. J. Obst. Gynec. 152, 659-660
6. CORNÉLISSEN, G. & R. KOPHER (1989): Chronobiologic ambulatory cardiovascular monitoring during pregnancy in Group Health of Minnesota. I. E. E. E. 225-233
7. DANIEL, S.S. & P.J. TROPPER (1989): Prevention of the normal expansion of maternal plasma volume: a model for chronic fetal hypoxaemia. J. Dev. Physiol. 11/4, 225-233
8. EASTERLING, T.R. & T.J. BENEDETTI (1989): Preeclampsia: A hyperdynamic disease model. Am. J. Obstet. Gynec. 160, 1447-1453
9. HOHMAN, M. & M.K. McLAUGHLIN (1990): Maternal kardiovascular adaptation während der Schwangerschaft. Geburtshilfe Frauenheilkd 50/4, 255-262
10. KHONG, T.Y. & I. BRONSENS (1986): Inadequate maternal vascular response to placentation in pregnancies complicated by pre-eclampsia and by small for gestational age infants. Brit. J. of Obstet. Gynecol. 93, 1049-1059
11. KIRSHON, B. & K.J. MOISE (1988): Role of volume expansion in severe preeclampsia. Surg. Gynecol. Obstet. 1675, 367-371
13. MAGGIONI, C. & A. MANFREDI (1989): ANF: una chiave interpretiva all'iposviluppo fetale? Atti Congresso Nazionale Societá Italiana Ostetricia e Ginecologia CIC Edizioni Internazionali. pp. 777-783
14. MALEK, J. & K. SULK (1962): Daily rhythm in leukocytes, blood pressure, pulse rate and temperature during pregnancy. Ann. N. Y. Acad. Sc. 8, 1018-1041
15. McQUEEN, J. & A. JARDINE (1990): Interaction of angiotensin II and atrial natriuretic peptide in the human feto-placental unit. Am. J. Hypertens. 3/8 I, 641-644
16. MEISS, P.J. & F. HALBERG (1981): Circadian rhythms in blood pressure, breast surface temperature and blood dehydroepiandrosterone sulfate in human pregnancy - potential chronotherapeutic markers. In: C.A. WALKER, C.M. WINGET & K.F.A. SOLEMAN (eds.): Chronopharmacology and Chronotherapeutics. Florida A & M University Foundation, Tallahassee, Florida, pp 159-176

17. MIYAMOTO, S. & H. SHIMOKAWA (1989): A possible explanation for nocturnal hypertension in preeclamptics. Clinic. Exp. Hyperten. part B Hypert. Pregnancy 8/3, 495-506
18. MOUTQUIN, J.M. & C. RAINVILLE (1985): A prospective study of blood pressure in pregnancy: prediction of preeclampsia. Am J. Obst. Gynec. 151, 191-196
19. MURNAGHAN, G.A. (1976): Hypertension in pregnancy. Postgrad. Med. J. 52 (Suppl. 7), 123-196
20. NISSEL, H. & N. LUNELL (1988): Maternal hemodynamics and impaired fetal growth in pregnancy induced hypertension. Obst. Gynec. 71/2, 163-166
21. O'HAVE, J.A. (1988): The enigman of insulin resistance and hypertension. Insulin resistance, blood pressure and the circulation. Am. J. Med. 84, 505-510
22. PIPKIN, F.B. (1989): The changing pressor response to angiotensine II in early human pregnancy. Clin. Exp. Hypert. Part B Hypert. Pregn. 8/3, 551-559
23. RATH, W. & J. SCHRADER (1990): 24 Stunden-Blutdruckmessungen im Verlauf der normalen Schwangerschaft und bei hypertensiven Schwangeren. Klin. Wochenschr. 68/15, 768-773
24. REDMAN, C.W.G. & L.J. BEILIN (1976): Variability of blood pressure in normal and abnormal pregnancy. M.D. LINDHEIMER, A. KATS & F.P. ZUSPAN (eds.): Hypertension in pregnancy. W. Ley, New York, pp. 53-60
25. SAWYER, M.M. & L. HALPERIN (1981): Diurnal and short-term variation of blood pressure: comparison of preeclamptic, chronic hypertensive, and normotensive patient. Obst. Gynecol. 58/3, 291-294
26. SELIGMAN, S.A. (1971): Diurnal blood pressure variation in pregnancy. The J. of Obst. and Gynec. of the British Commonwealth 78, 417-422
27. SOHN, C. & H. FENDEL (1988): The renal artery and uterine circulation in normal and gestotischer Schwangerschaften. Geburtshilfe perinatal 192/2, 43-88
28. VALENTIN, B. & F. LAFFARGUE (1985): Evaluation of the clinical and biologic means of assessing fetal prognosis in pregnancy toxiemias. A propose of 223 cases. Source J. Gynecol. Obstet. Biol. Reprod. 14/4, 499-505.

PREDICTABLE PROFILE OF BLOOD PRESSURE VARIABILITY IN THE COURSE OF A HEALTHY HUMAN PREGNANCY

D.E. AYALA & R.C. HERMIDA

Bioengineering & Chronobiology Laboratories, E.T.S.I. Telecomunicación, University of Vigo, Apartado 62, Vigo, Spain.

INTRODUCTION

Ambulatory and non-invasive blood pressure (BP) monitoring is increasingly used in the diagnosis and treatment of "hypertension". The use of rhythmometric methods for analysis of time series with sparse and non-equidistant observations in combination with automatic and portable recording systems has already proved its value in assessing the antihypertensive efficacy of a prescribed drug not only for groups but also on an individualized basis (7). The major interest in BP readings comes as a need to better understand and to more accurately diagnose clinical high BP states. "Hypertension" is a common chronic condition that - as mesor (midline estimating statistic of rhythm, M) or circadian amplitude (A) hypertension - may affect up to 25% of human adults (12). This condition is an important risk factor for strokes, heart attacks and other vascular diseases (9). Treatment of high BP by pharmacologic methods lowers the incidence of these catastrophic complications and prolongs life (7). Accordingly, there has been a strong incentive to identify and treat individuals with high BP.

Of particular interest are pregnant women: pregnancies complicated by an elevated BP and/or preeclampsia contribute markedly to perinatal mortality (4). It also appears that a history of preeclampsia or at least of gestational hypertension in a prior pregnancy places the pregnant woman and her offspring at a high risk for a later development of high BP 7 to 12 years later (11). So far, the BP and heart rate (HR) assessment in pregnant women has relied mostly on a few measurements taken in the physician's office. Such conventional time-unspecified single measurements may be misleading because BP and HR vary according to a spectrum of rhythms (the circadian in particular) and because measurements may be influenced, among other factors, by the patient's emotional state, position, diet, and external stimuli (7). Self-measurement, if done systematically, offers an alternative, but it interferes with daytime activities and is not feasible during sleep. Moreover, the variability of BP, even among healthy individuals, is such that the identification and the proper definition of high BP is highly ambiguous, mainly when relying on single time-unspecified measurements. Even when based, not on one or two, but on a mean of several casual rather than systematic measurements, a BP found to be "high" or "low" is often unreliable. This is due both to the large variability of BP and the circumstance that unusually high or low values may occur only at certain times that may not be covered by casual sampling, as in the case of nightly hypertension. The development of automatic instrumentation for indirect non-invasive ambulatory BP monitoring makes possible to follow the time course of BP variation around the clock in large groups (10). The use of these monitors has provided a method of BP assessment that may compensate for some of the limitations of office and even self-measurements.

The evaluation of predictable variability in BP and HR by 1) the use of fully ambulatory devices, and 2) chronobiologic data processing, assesses early cardiovascular disease risk, e.g. in pregnancy. We have used this approach to quantify changes in 24-hour synchronized (circadian) characteristics of BP and HR in the second pregnancy of a clinically healthy woman (2).

SUBJECT AND METHODS

A clinically healthy pregnant woman (DEA) used an ABPM-630 (manufactured by Colin Medical Instruments Inc., Komaki City, Japan) to monitor her BP and HR at ~1-hour intervals with few interruptions. Starting on the first gestational week, she used the monitor for 2 out of each 6 days during the whole pregnancy, for a total of 39 48-h profiles.

The instrument is a small (165 x 36 x 89 mm), lightweight (830 g, including battery and gas cartridge) fully ambulatory monitor carried in a special shoulder pouch designed for pregnancy monitoring. This monitor is powered by CO_2 cartridges for cuff inflation and is battery-operated for the management of data collection, scheduling and storage in a solid-state memory that can hold over 600 sets of measurements. The instrument uses both an oscillometric and auscultatory (Riva Rocci-Korotkoff) method of measuring BP. Both approaches also provide a measure of HR. Additionally, four control keys allow the subject to enter information with a previously agreed code to identify the times of food intake, activity and position (walking, running, sitting, sleeping), drug administration (if any), etc. Data can be easily transfered from the memory cassette to a computer by the use of an AA-200 Colin Analyzer and software specifically designed by us for that purpose. After coding, data were analyzed on a Macintosh IIfx by rhythmometric procedures (8) and by polynomial regression analysis. In particular, circadian parameters of BP and HR were computed for each single 48-h profile by the least-squares fit of a 24-h cosine curve. Regression analysis of parameters thus obtained revealed patterns of variation of circadian Ms and As with gestational age.

RESULTS

The predictable variability of circadian BP M can be approximated by a second order polynomial model on gestational age: a steadily linear decrease in systolic (Fig.1), diastolic (Fig.2), and mean arterial BP up to the 22nd week of pregnancy is followed by an increase in BP up to the day of delivery. Circadian BP A is, however, slightly decreasing throughout pregnancy (linear correlation coefficient r=-.365, -.379, and -.497 for systolic, diastolic, and mean arterial BP, with P=.022, .017, and .001, respectively). For HR, results indicate a small linear increase in M (r=.344, P=.032), and no predictable change in A (r=-.030, P=.858).

These results corroborate those found for the first pregnancy on the same woman (2). In that case, BP and HR were monitored at ~1-h intervals for 47 consecutive days between the 9th and the 16th week of pregnancy, and for about 7 consecutive days each month thereafter until the day of delivery. Regression analysis on the circadian BP Ms indicated also a second-order model of variation for gestational age for systolic, diastolic, and mean arterial BP (2). Moreover, the second-order models approximating BP variability in both pregnancies have similar coefficients for gesta-

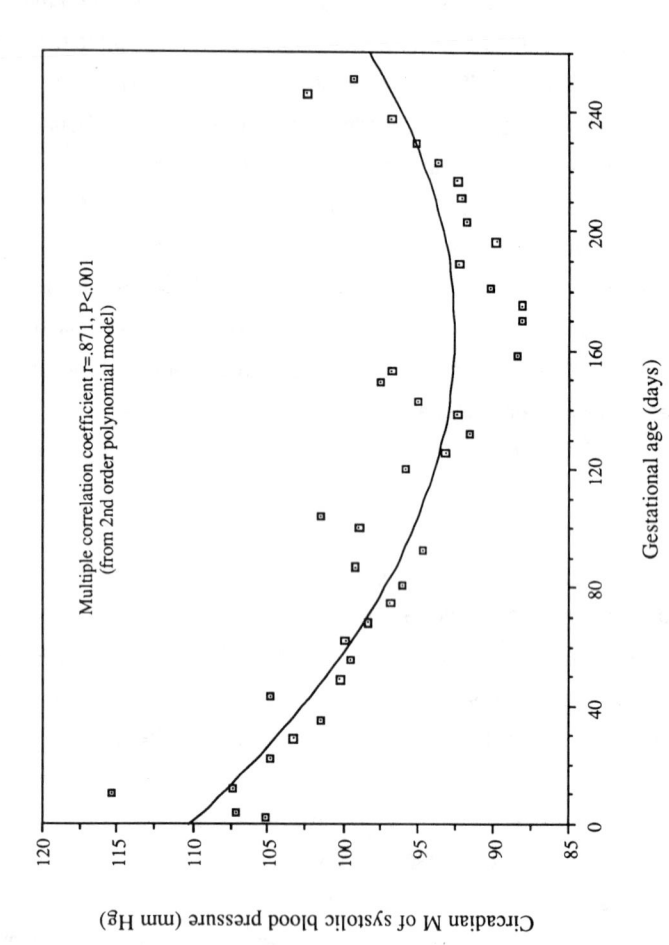

Fig.1: Variation of circadian mesor (M) of systolic blood pressure along gestational age in the second pregnancy of a clinically healthy woman.

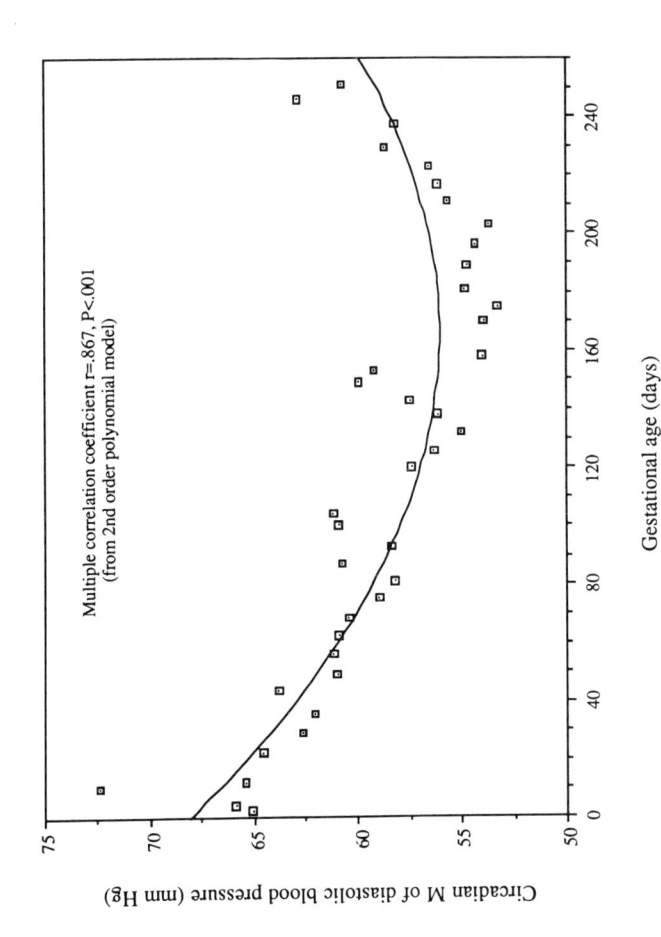

Fig.2: Variation of circadian mesor (M) of diastolic blood pressure along gestational age in the second pregnancy of a clinically healthy woman.

tional age (p=.668 and .669 for first and second order coefficients, respectively, for systolic BP; P=.339 and .397 for first and second order coefficients, respectively, fordiastolic BP), but lower BP values throughout the second pregnancy. This consistent and statistically significant decrease of about 8 mm Hg (p<.001) for systolic BP and 4 mm Hg (p<.001) for diastolic BP in the course of the second pregnancy could be due not only to the expected decrease in BP for subsequent pregnancies as compared to the first, but also to the average added intake of 625 mg/day of calcium during most of the second pregnancy.

It is important to note that the total range of variation observed in the circadian M is of 23 and 21 mm Hg for systolic and diastolic BP, respectively, in the first pregnancy, and of 27 and 21 mm Hg for systolic and diastolic BP in the second pregnancy. This range is about three times larger if we considered individual observations instead of circadian Ms.

DISCUSSION

Earlier transverse studies of BP during human pregnancy and an earlier quasi-longitudinal study of a pregnancy associated with diabetes (3) are here complemented by the longitudinal monitoring of a clinically healthy pregnancy. These longitudinal results can be compared with those of an earlier hybrid (transverse and longitudinal) study (4) on 22 women studied in at least two different stages of their pregnancy. Since not all women provided profiles at the same gestational age, a normalization procedure was used: as a first approximation, the rate of change per week was calculated assuming linearity, for each woman separately, and assigned to each gestational week between the dates of monitoring. These rates of change per week were further averaged for each week across all women (4). This approach clearly showed, for systolic and diastolic BP, statistically significant changes: a decrease between the 12th and the 15th week, and an increase between the 30th and 32nd week. In the case of HR, the increase could not be shown to be of statistically significance in any particular gestational week (4). The trends obtained on the basis of this hybrid study are also apparent in the longitudinal monitoring investigated herein (2).

In dealing with apparently healthy individuals, one important factor usually ignored when studying biologic variables within conventional physiologic normal ranges is the timing of a clinical measurement in relation to biologic rhythms. Time-varying reference limits that adjust for the rhythmic behavior of BP and HR have accordingly been suggested (6). These reference standards are derived from data provided by healthy peer groups taking into consideration changes in mean and variance as a function of rhythm stage (1,5). From this point of view, the study here presented represents a double example of longitudinal monitoring in a healthy pregnancy, keeping the everyday life conditions by the use of a fully ambulatory and non-invasive BP monitor. Non-invasive BP monitoring combined with the proper rhythmometric analysis on dense and long data series offers a complementary and early cardiovascular risk assessment in pregnant women.

ACKNOWLEDGEMENTS

This research was supported in part by Dirección General de Investigación Científica y Técnica, DGICYT, Ministerio de Educación y Ciencia (PA86-0229 & PB88-0546); Consellería de Educación e Ordenación Universitaria, Xunta de Galicia (XUGA-709-CO388) and Comisión Interministerial de Ciencia y Tecnología, Plan Nacional de Salud (SAL90-0500).

REFERENCES

1. AYALA,D.E., R.C. HERMIDA, J.R. FERNÁNDEZ & S. SÁNCHEZ (1989): Chronobiologic analysis of blood pressure variability in healthy subjects. Proc. 11th Ann. Int. Conf. IEEE Engineering in Medicine and Biology Society. Seattle, WA, Nov. 9-12, pp. 1622-1623
2. AYALA, D.E., R.C. HERMIDA, S. SÁNCHEZ & F. HALBERG (1990): Analysis of blood pressure assessed by ambulatory noninvasive monitoring in a healthy pregnancy. Proc. 40th ISMM Int. Conf. Mini and Microcomputers and Their Applications. Lugano, Switzerland, pp. 40-43
3. CHRISTENSEN, K., G. CORNÉLISSEN, F. HALBERG et al. (1988): "Circadian blood pressure (BP) and heart rate (HR) rhythms in pregnancy, complicated by diabetes. J. Minn. Acad. Sci. 53, 16-17
4. CORNÉLISSEN, G., R. KOPHER, P. BRAT et al. (1989): Chronobiologic ambulatory cardiovascular monitoring during pregnancy in Group Health of Minnesota. Proc. 2nd Ann. IEEE Symp. Computer-Based Medical Systems. Minneapolis, MN, June 26-27, pp. 226-237

5. FERNÁNDEZ, J.R., K. OTSUKA, E. HALBERG et al. (in press): Circadian chronodesm of systolic and diastolic blood pressure and heart rate for clinically healthy young adult women in Japan, North America and Spain. Chronobiology International
6. HALBERG,F.,J.I.M.DRAYER,G.CORNÉLISSEN&M.A.WEBER(1984):Cardiovascular reference data base for recognizing circadian mesor-and amplitude-hypertension in apparently healthy men. Chronobiologia 11, 275-298
7. HALBERG, F., G. CORNÉLISSEN, W. JINYI & P.K. ZACHARIAH (1989): Chronopharmacologic individualized and group assessment of outcomes in antihypertensive drug trials. Proc. 2nd Ann. IEEE Symp. Computer-Based Medical Systems. Minneapolis, MN, June 26-27, pp. 253-259
8. HERMIDA, R.C. (1987): Chronobiologic data analysis systems with emphasis in chronotherapeutic marker rhythmometry and chronoepidemiologic risk assessment. In: L.E. SCHEVING, F. HALBERG & C.F. EHRET (eds.): Chronobiotechnology and Chronobiological Engineering. Martinus Niijhoff, NATO ASI Series E, No. 120, Dordrecht, The Netherlands, pp. 88-119
9. KANNEL, W.B., W.P. CASTELLI, P.M. MCNAMARA & P. SORLIE (1969): Some factors affecting morbidity and mortality in hypertension. The Flamingham Study. Milbank Memorial Fund Q. 47, 116-142
10. OTSUKA, K., G. CORNÉLISSEN, N. ASLANIAN et al. (1989): Circadian period of human blood pressure and heart rate in clinical health under ordinary conditions. Proc. 2nd Ann. IEEE Symp. Computer-Based Medical Systems. Minneapolis, MN, June 26-27, 206-213
11. SVENSSON, A., L. SIGSTRÖM, B. ANDERSCH & L. HANSSON (1983): Maternal hypertension during pregnancy and high blood pressure in children. Clin. exp. Hypertension (B: Hypertension in Pregnancy) 2, 203-209
12. WEBER, M.A., J.I.M. DRAYER & D.D. BREWER (1987): Repetitive blood pressure measurements: Clinical issues, techniques, and data analysis. In: L.E. SCHEVING, F. HALBERG & C.F. EHRET (eds.): Chronobiotechnology and Chronobiological Engineering. Martinus Niijhoff, NATO ASI Series E, No. 120, Dordrecht, The Netherlands, pp. 270-277

A CIRCADIAN NONINVASIVE MONITORING OF HEMODYNAMIC PARAMETERS IN PATIENTS WITH CORONARY ARTERY DISEASE

J. LIETAVA, A. DUKAT, I. BALAZOVJECH & Z. MIKES

II. Department of Internal Medicine, Medical Faculty, Hospital of Comenius University, Mickiewiczova 13, Bratislava 813 69, Czechoslovakia

INTRODUCTION

An analysis of circadian rhythms could be helpful in the diagnostics of coronary artery disease (CAD) patients. Over the past 20 years, the unequivocal periodicity pattern has been established in such clinical manifestation of CAD as the onset of myocardial infarction (MI) (12), of distribution of ischemic episodes (5), of sudden death or related diseases as stroke (8). The majority of studies was based on an epidemiological approach.

The studies concerning the evaluation of individual hemodynamic status mainly deal with the ECG and blood pressure (BP) chronograms, and only exceptionally they refer to circadian changes of cardiac output (CO) and contractility parameters. The lack of CO orientated studies could be due to methodological obstacles.

We try to examine the hemodynamic CO circadian changes using the method of thoracic electric bioimpendance (TEB). Numerous authors have reported a good correlation in CO measured by TEB and by thermodilution, by indirect Fick CO_2 rebreathing method and by Doppler method (3,4,7), but this method has not been used until now for determination of circadian changes of CO in CAD patients.

METHOD

The TEB is based on measuring of steady-state and pulsatile components of thoracic resistance to the flow of high-frequency and low amplitude alternating current. On the basis of the rate of bioimpendace changes (first approximation of changes to time)

are measured or calculated following volumic, contractile, and performance parameters indexed to the body surface area, where it is applicable: 1. cardiac index (CI) [l/min/m^2], 2. stroke index (SI) [ml/m^2], 3. end-diastolic index (EDI) [ml/m^2], 4. ejection fraction (EF) [%], 5. heart rate (HR) [bpm], 6. peak flow index (PFI) [ml/sec/m^2], 7. index of contractility (IC) [1/sec], 8. acceleration index (ACI) [1/sec^2], 9. ejection ratio (ER) [%], 10. systolic time ratio (STR) [%].

We used the noninvasive cardiomonitor NCCOM3 R7 (BoMed Co.) measuring the TEB waveform beat-by-beat. Every 16 accepted beats were averaged for one record. For data recording and further processing was used a software developed in our laboratory. The records had been averaged to 10 minute intervals to suit the cosinor analysis program.

Blood pressure was measured indirectly by Pressurometer IV (Del Mar Avionics comp). We chose an usual 10 minute interval for data collecting.

Statistical evaluation was made using the cosinor analysis developed by MIKULECKY et al. (9) based on 3 harmonic periods - 24, 12 and 8 hours. The chronobiologic characteristics (mesor, amplitude, and acrophase) were evaluated by Mann-Whitney test (Alpha = 0.05).

We examined twelve patients with verified CAD. Selection criteria were typical angina pectoris (AP) with ST depression on Holter - ECG or MI history. Eight patients had AP, three patients had acute Q MI, one non-Q MI. The patients without clinical symptoms and signs of heart failure were monitored during the chronic stage of the disease, at least 4 weeks after MI. The control group consisted of six subjects admitted for vertebrogenic algic syndroma. Those subjects were monitored in the asymptomatic stage of the disease, shortly before discharge.

All subjects were men. A more detailed comparison of the characteristic of the groups were as follows: average age of patients was 52.4 vs. 40.2 years of those of control group (p<0.01); average height 170 vs. 169 cm (NS); average weight 74 vs. 63 kg (p<0.01), BP 125/74 vs. 121/72 mmHg (both NS).

Monitoring began in all subjects in the morning hours. Since recording depended on the connection with computer, the movement of the patients was room-restricted by a 6 meter electric cord. Paradoxically, this limitation could be judged as advantageous, because of standardization of the subjects' movement regimen.

RESULTS

We found significant presence of diurnal rhythm in all measured parameters in both groups, but not ultradian rhythms (12 hours and 8 hour rhythms), however, an unexpected finding was a similar macroscopical ultradian periodicity with two peaks in the majority of measured parameters in both groups. The first one occurred in the morning from 6.00 a.m. to noon, the lower one was present in the evening between 6.00 p.m and midnight. The acrophase analysis of the ultradian rhythms did not reveal any differences between patients and controls. These disproportions in the results and relatively small groups were reasons why we concentrated on the analysis of the circadian periodicity.

To the most pronounced findings belonged the periodicity pattern of contractility parameters with a decrease between midnight and 6.00 a.m with a succeeding increase to maximal values between 6.00 a.m. and noon. This pattern was similar in all contractile parameters, but became special importance in the mesor analysis of ACI chronodesm, as will be reasoned later.

Statistically significant differences were found in volumic parameter mesors between both groups. The important chronograms were those of CI, SI, and EDI. CI, SI, EDI mesors were significantly lower in CAD group ($p < 0.01$, $p < 0.05$, resp. $p < 0.05$) but the amplitudes were almost equal. Because of the same HR in both groups, differences in CI could be explained by the different SI alone. The amplitudes were similar in both groups. Similar findings were reported also by other authors (1).

We found decreased pump performance in CAD group - STR was increased, EF and ER decreased, but the differences did not reach statistically significant levels. Practically the same were the amplitude values.

As expected, the controls tended to have better contractility, but only differences in PFI mesors reached the statistically significant level ($p < 0.05$). The absence of significant differences in contractility parameters mesors might have been influenced by small numbers of subjects in both groups because ACI and IC mesors were close to the statistical significance (p<0.06, p<0.08 resp.).

Tab. 1: Comparison of chronobiologic characteristics in CAD group and controls (median values and range; Ampl.: Amplitude, Acrop.:Acrophase).

PATIENTS

	Mesor	range	Ampl.	range	Acrop.	range
CI	2.9	2.0 **	0.3	0.56	-210	85.4
SI	39.7	37.9 *	3.0	8.44	-199	313.9
EDI	74.4	54.9 *	4.7	10.52	-224	261.5
EF	55.6	21.2	1.5	2.84	-244	328.5
HR	71.6	31.6	7.3	8.49	-204	104.1
PFI	264.4	231.0 *	18.0	13.92	-217	227.3
IC	38.6	44.7	2.8	5.58	-216	235.4
ACI	110.6	102.2	9.2	11.56 *	-220	215.8
TFI	28.1	14.4	1.2	2.43	-198	194.9
STR	36.9	29.7	2.2	4.37	-141	196.0
ER	35.0	13.4	1.7	1.76	-231	111.2

CONTROLS

	Mesor	range	Ampl.	range	Acrop.	range
CI	3.7	2.4	0.1	0.37	-232	54.4
SI	53.8	38.9	3.0	6.83	-278	325.2
EDI	94.3	56.3	4.0	11.54	-254	288.2
EF	53.8	21.2	1.8	2.84	-155	304.1
HR	70.2	29.1	6.8	7.38	-210	53.3
PFI	373.5	212.0	18.8	19.47	-240	233.2
IC	43.8	36.6	2.1	4.18	-239	234.0
ACI	123.3	97.8	5.3	10.47	-174	214.8
TFI	28.0	7.2	1.4	0.84	-208	152.7
STR	32.5	10.3	2.6	2.65	-165	133.3
ER	34.2	9.3	2.1	1.98	-234	94.8

* = p<0.06;   ** = p<0.01

As mentioned above, interesting differences were those of ACI. ACI is relatively independent of the volumic part of contractility and reflects changes of inotropy. Although the mesor test showed a statistically insignificantly decreased ACI in CAD patients, the amplitude test revealed higher values ($p<0.05$) in these subjects. This fact suggests that CAD patients having impaired contractility exhibit also disadvantageous contractile adaptation to daily activity.

DISCUSSION

The morning increase of contractile and volumic parameters may be a point of causal relationship between hemodynamics and increased incidence of sudden death, MI and ischemia, and strokes as reported by numerous studies (5,6,12). Our results also correspond with the known periodicity of sympathetic activity, fluctuation of coronary vasoconstriction, and changes of blood viscosity and hypercoagulable state (10).

Decreased ACI mesor and increased ACI amplitude in CAD patients could indicate the worsened inotropic adaptation to a daily activity. Changed contractility may be especially important in the morning hours, during which are known increased risks of sudden cardiovascular events (12). Moreover, differences in contractility, as recorded using 24-hour TEB, monitoring could be useful in discriminating of CAD patients without manifest heart failure. However, these findings need more study and further verifications in more patients.

CONCLUSION

We examined twelve patients with CAD without signs and symptoms of heart failure using the circadian monitoring of thoracic electric bioimpedence (TEB). We revealed decreased mesor values of volumic parameters (cardiac index, stroke index and end-diastolic index) in CAD patients. There were no statistically significant differences in performance and contractility mesors between both groups, but we found an increased amplitude of acceleration index chronodesm in CAD patients.

We have concluded that the 24-hour TEB monitoring could be valuable in the study of the hemodynamic periodicity in patients with the CAD.

ACKNOWLEDGEMENT

Authors are gratefully indebted to Professor Miroslav Mikulecky and his coworkers for kind permission to use their program for our study. We are also deeply grateful to Dr. Juraj Bereznak from Mathematic-Physical Faculty of Comenius University in Bratislava for his excellent computer consultations and development of the software used.

REFERENCES

1. ADAMIAN, K.G., S.V. GRIGORIAN, V.M. SHUKHIAN & N.L. ASLANIAN (1983): Chronobiological aspects of ischemic heart disease. Chronobiology 1982-1983. Karger, Basel, pp. 558-565.
2. BEAMER, A.D., Th.H. LEE, E.F. COOK et al. (1987): Diagnostic implications for the circadian variation of the onset of chest pain. Am. J. Cardiol., 60, 13, 998-1002
3. BERNSTEIN, D.P. (1989): Noninvasive cardiac output measurement. In: Textbook of Critical Care Medicine. Saunders, New York, pp. 159-185
4. EDMUNDS, A.T., S. GODFREY & M. TOOLEY (1982): Cardiac output measured by transthoracic impedance cardiography at rest, during exercise and at various lung volumes. Clin. Sci., 63 (1), 107-113
5. GOTTLIEB, S.O. (1988): Circadian pattern of myocardial ischemia: Pathophysiologic and therapeutic considerations. J. Cardiovasc. Pharmacol., 12, S18-S21
6. HJALMARSON, A., E.A. GILPIN, P. NICOD et.al. (1989): Differing circadian pattern of symptom onset in subgroups of patients with acute myocardial infarction. Circulation, 80 (2), 267-275
7. KLOCKE, R.K., A. HEIN, G. MAGER et al. (1989): Kontinuierliches nichtinvasives real-time-monitoring des Herzzeitvolumens mittels der neuen beat-to-beat Bioimpedance (BoMed NCCOM3). Intensivmedizin, 26 (2), 161-167
8. MARLER, J.R., Th.R. PRICE, G.L. CLARK et al. (1989): Morning increase in onset of ischemic stroke. Stroke. 20 (6), 473-476
9. MIKULECKY, M., L. KUBACEK, A. VALACHOVA et al. (1990): Numerical and graphical solution of time series analysis. Manual. Teaching Hospital. Comenius University, Bratislava. p. 41
10. MULLER, J.E (1987): Circadian variation of cardiovascular disease and sympathetic activity. J. Cardiovasc. Pharmacol., 10, S104-S111
11. MULLER, J.E., G.H. TOFLER & P.H. STONE (1989): Circadian variation and trigger onset of acute cardiovascular disease. Circulation, 79 (4), 733-743
12. WILLICH, S.N., T. LINDERER, K. WEGSCHEIDER & R. SCHOEDER (1989), Zir- kadiane Variation in der Inzidenz des Myokardinfarkts. Dtsch. med. Wschr., 114 (16), 613-617

CHRONOBIOLOGICAL ASPECTS OF VALVULAR DISEASES OF THE HEART IN PREGNANT WOMEN

R.M. ZASLAVSKAYA, B.G. VARSHISKY & M.M. TEIBLOOM

Gorodskaya bolnitsa N 60, 111123, Shosse Enthusiastov 84/1, Moscow, RUSSIA

Functional system "mother-fetus" during pregnancy exemplifies a complicated spacetime organization, every organization level in living systems being provided with its own range of endogenous rhythms. Nevertheless the questions of circadian organization of cardiovascular system in normal pregnancy and in its pathology are not adequately worked out. At the same time the investigations in this field will allow us to estimate the functional state of the organism of a pregnant woman, and also provide an insight into the mechanisms of cardiovascular system disturbances in diseases of blood circulation organs in pregnant woman, and, finally will help reveal the regulations of time adaptation to this kind of pathology during pregnancy and improve the therapy.

MATERIAL AND METHODS

Biorhythmologic studies of hemodynamics were conducted in 40 healthy pregnant women from 18 to 33 years old in the second half of pregnancy and in 94 pregnant women with acquired and congenital valvular heart diseases from 18 to 40 years old. All the acquired valvular heart diseases (AVHD) were of rheumatic etiology. Diagnosis of valve lesion was based on the anamnestic data, patients complaints, data of objective investigations methods (auscultation, roentgenography, phonocardiography, electrocardiography). In 75% of pregnant women the diagnosis of valvular heart diseases (VHD) had been established before the present pregnancy and verified with the help of invasive methods of investigation and echocardiography. All the pregnant women with AVHD had the inactive stage of rheumatism and the sinus rhythm of heart electrical activity. Parameters of heart electromechanic function were studied by means of electrocardiographic (ECG) and polycardiographic (PCG) methods. All the parameters of arterial blood pressure (BP) and electromechanical function were examined in the course of 24 hours at 7.00, 11.00, 15.00, 19.00, 23.00 h. There were 26 pregnant women with mitral valve deficiency (MVD) 32, with mitral stenosis (MS) 13, with aortic ostium stenosis (AS) 9, with interatrial septum defect (IASD) 8, with interventricular septum defect (IVSD) and 6 with open ductus arteriosus (ODA). All women with VHD had the 2nd half of pregnancy. Daily dynamics of systolic, diastolic, and pulse BP (BPs, BPd, BPp), heart rate (HR), intraatrial (P), atria-ventricular (P-Q), intraventricular (QRS) conductivity duration, electric systole, diastole (Se, De), mechanic and total systole and diastolic (Sm, Dm, St, Dt) interval Q-Itone, isometric contraction (Ic), expulsion phase (E), tension period (T) duration were checked. The data obtained in the course of investigation of daily dynamics of studied blood circulation parameters were analyzed according to the program of individual and population "Cosinor-analysis" and with the use of χ^2 method. In cases where acrophases of investigated parameters are set up in the course of 7.00 till 23.00, they have been classified as "day" type rhythms. If acrophase are set up during the period from 23.00 till 7.00, the rhythm type has been defined as "night".

RESULTS AND DISCUSSION

The studies have shown 95% confidence of the presence of daily rhythm of BPs, BPp and HR in healthy pregnant women. BPd does not reveal a regular average group rhythm. The analysis of individual acrophases distribution has revealed the "day" type of BPs, BPp and HR circadian rhythm in 95%, 92,5% and 87,5% accordingly of pregnant healthy women (at 17.18, 16.18, 13.54 h accordingly). Thus, in 90% of these persons the inner and outer synchronization 'of BPs, BPp, HR circadian rhythms takes place. The acrophases of P-Q, Se, De, Sm, St, Dt, T duration were displayed in the night hours, but QRS ones - at 18.18 (Table 1). The analysis of our biorhythmologic investigations has shown that practically healthy women during the 2nd half of pregnancy are characterized with a regular circadian organization of blood circulation. It is manifested through a strictly regular circadian chronostructure of rhythms of heart electromechanical function parameters and BP. The acrophases of BPs, BPp, HR,

Table 1:
Circadian rhythms of blood pressure and heart electromechanic function parameters in practically healthy women during the second half of pregnancy (according to the data of average group "Cosinor-analysis" and χ^2-method).

Parameter	Mesor	Amplitude	Acrophase
	(95% limits of confidence)		
BPs, mm Hg	117,4 (114,1- 120,7)	6,7 (4,4 -9,1)	17.18 (16.12 -18.18)
BPd, mm Hg	77,1 (74,7 -79,5)	-	-
BPp, mm HG	40,3 (37,8 -42,8)	5,6 (3,1 -8,1)	16.48 (15.24 -18.12
HR [er 1 min	81 (71-84)	5 (3-7)	13.54 (12.00 -15.18)
P-Q, sec	0,136 (0,122-0,149)	0,009 (0,002-0,016)	02.36 (18.56-05.24)
QRs, sec	0,073 (0,07-0,076)	0,002 (0,001-0,005)	18.18 (15.12-01.24)
Se, sec	0,348 (0,34-0,355)	0,016 (0,01-0,02)	01.42 (23.48-02.36)
De, sec	0,41 (0379-0,430)	0,029 (0,014-0,045)	02.30 (23.48-04.36)
Q-I tone, sec	0,10 (0,097-0,105)	-	-
Ic, sec	0,037 (0,031-0,042)	-	-
T, sec	0,135 (0,129-0,141)	0,006 (0,002-0,011)	04.54 (02.36-8.00)
E, sec	0,237 (0,229-0,245)	-	-
St, sec	0,371 (0,363-0,379)	0,016 (0,009-0,023)	02.54 (01.54-03.54)
Dt, sec	0,388 (0,359-0,416)	0,03 (0,018-0,043)	01.18 (22.54-03.06)
Sm, sec	0,271 (0,264-0,279)	0,012 (0,005-0,019)	03.00 (01.24-04.30)

QRS are set up at day time (since 13.55 till 18.18) and the acrophases of P-Q, Se, De, Sm, St, Dt, Ic duration are set up during the night (from 01.40 till 05.00). The electromechanical activity of the heart increases during the period of daytime, which is testified by the rise of automatism function, acceleration of conductivity, and increase of heart contractile ability. During sleep the shifts with the opposite

Tabl.2:
Circadian rhythm of heart electromechanic function parameters and blood pressure in pregnant women with mitral valve insufficiency and cardiac failure (CF), stage I-II (according to the data of average - group "Cosinor-analysis").

Parameter	Mesor	Amplitude	Acrophase
	95% limits of confidence		
BPs, mm Hg	115 (111-119)	6,5 (4,2-8,8)	16.36 (15.24-19.06)
BPd, mm Hg	72,5 (70-76)	–	–
HR per 1 min	77,3 (77-81)	5 (3-7)	12.36 (11.00-14.48)
P, sec	0,09 (0,084-0,092)	–	
PQ, sec	0,145 (0,136-0,154)	–	–
QRS, sec	0,073 (0,069-0,078)	–	–
Se, sec	0,347 (0,337-0,357)	0,08 (0,003-0,012)	01.30 (23.36-03.54)
De, sec	0,446 (0,413-0,478)	0,04 (0,02-0,02-0,06)	00.48 (23.06-02.54)
Q-Itone, sec.	0,08 (0,07-0,085)	–	
Ic, sec.	0,037 (0,033-0,042)	0,006 (0,001-0,011)	05.12 (01.42-08.42)
T, sec	0,113 (0,015-0,122)	0,007 (0,003-0,011)	05.42 (03.48-09.00)
E, sec	0,236 (0,225-0,247)	0,009 (0,002-0,016)	05.42 (03.48-09.00)
E, sec	0,236 (0,225-0,247)	0,009 (0,002-0,016)	00.24 (21.45-03.06)
Sm, sec	0,273 (0,263-0,282)	0,013 (0,005-0,02)	02.18 (00.06-04.12)
St, sec	0,35 (0,338-0,36)	0,012 (0,005-0,019)	02.36 (00.48-04.54)
Dt, sec	0,35 (0,338-0,36)	0,038 (0,019-0,057)	00.24 (22.54-2.32)

sign are observed. The described hemodynamic circadian rhythm is physiologically expedient for pregnant women, as it provides a relatively high level of cardiovascular system adaptation at day time.

Pregnant women with MVD had the same chronostructure of hemodynamic parameters as healthy pregnant women with a small increase of frequency of inner and outer desynchronization (in 29% of cases in comparison to healthy pregnant women in 25% of cases). Acrophases of BPs, HR are set up in the day time. Se, De, Ic, E, St, Dt duration acrophases are revealed mainly at night (Table 2). Rhythm acrophase in MVD and in healthy pregnant women have no differences in excess of 3 hours. In contrast to practically healthy women, the circadian rhythms of P and QRS duration in pregnant women with MVD are absent. Pregnant women with MS had prominent disturbances of circadian organization of the electromechanical heart function and BP, displayed by daily rhythms inversion with prominent internal and external desynchronization of the given parameters (in 80% of cases). Electromechanical heart activity increases at night and decreases in the afternoon. BPs, BPd circadian rhythms are not reveal regular. Circadian rhythms of HR, P, P-Q, Se, De, Sm, St, Dt, E duration have been displayed in MS patients, though two absolutely different types of these parameters of circadian organization in healthy pregnant women and women with MS were observed: The "day" type of HR and "night" type of P, P-Q, Se, De, Sm, St, Dt, E duration are typical for the first, and the "night" type of HR, and "day" type of basic electromechanical parameters rhythms - for the MS patients.

Thus, in pregnant women with MS, electromechanical activity grows at night and decreases in the course of daytime.

Table 3:
Distribution of circadian rhythm acrophases of heart electromechanical function and blood pressure parameters in healthy women and in pregnant women with mitral stenosis and cardiac failure (CF), stage I-II.

Groups of observed women	"Day' type of circadian rhythm	"Night" type of circadian rhythm
	Parameters	
Healthy pregnant women	BPs, BPp, HR, QRS	P-Q, Se, De, Sm, St, Dt, T, E
Pregnant women with MS, CF, stage I-II	P, P-Q, Se, De Sm, St, Dt, E	HR

Biorhythmologic studies of hemodynamics in a group of 13 pregnant women with aortic ostium stenosis (AS), CF, stage I-II, aged from 29 to 36 years old, allowed us to establish with 95% limits of confidence the presence of BPs, P-Q, QRS duration rhythms (Table 4). It follows from this table that the "day" type of QRS duration and BPs rhythms and "night" type of P-Q duration rhythms are characteristics of HR, Se, De, Q-Itone, Ic, T, E, Sm, St, Dt. These data enable us to differentiate between pregnant women with AS and pregnant women of other groups. Disappearance of circadian rhythms of myocardial electromechanical function parameters in AS women should be considered as an unfavorable phenomenon.

Considerable inner and outer desynchronization of BP and rhythms of heart electromechanical function parameters were observed in 47% of AS cases.

In a group of 9 women with pregnancy in its second half, suffering from interatrial septum defect (IASD) the circadian rhythms of BPs, HR, De, E, Sm, St, Dt were revealed. Circadian rhythms of BPd, P, P-Q, QRS, Q-Itone, Ic, T duration were absent (Table 5). Thus, the chronostructure of circadian rhythms of heart electromechanical activity and BP parameters in IASD patients is characterized by the "day" type of HR and BPs rhythms and "night" type of De, E, Sm, St, and Dt duration rhythms. In pregnant women with IASD the circadian organization of cardiohemodynamic parameters are found, manifest by the absence of P-Q, QRS, Se, Ic, T duration rhythms and also by the phenomena of outer and inner desynchronization. These disorders may point at the functional overload of blood circulation system in pregnant women with IASD. The rhythmological investigations of cardiohemodynamics in a group of 8 pregnant women with interventricular septum defect (IVSD), aged from 18 to 27 years enabled us to reveal circadian rhythms of BPs, HR, P, Se, De, E, Sm, St, Dt duration (Table 6). The time organization of heart electromechanical function and BP parameters in pregnant women with IVSD is characterized by the "day" type of HR, BPs and P rhythm and the "night" type of Se, De, Sm, St, Dt, E duration rhythms, the disorder of cardiohemodynamic parameters chronostructure, revealing themselves by the absence of P-Q, QRS, T, Ic rhythms and by the phenomenon of outer and inner circadian rhythm desynchronization.

The analysis of biorhythmologic investigation of cardiohemodynamic parameters in a group of 6 pregnant women, aged from 17 to 38 years old, suffering from opened ductus (ODA) showed the presence of Hr, Se, De, Sm, St, Dt duration

Table 4:
Circadian rhythm of heart electromechanic function and BP parameters in women with pregnancy in its second half, suffering from aortic ostium stenosis (according to the data of average group "Cosinor-analysis")

Parameter	Mesor	Amplitude	Acrophase
	95% limits of confidence		
BPs, mm HG	111,4 (108–115)	5,0 (0,7–8,7)	13.42 (07.00 –16.42)
BPd, mm Hg	78 (74–83)	–	–
HR, per 1 min	78 (68–87)	–	–
P, sec	0,06 (0,02–0,1)	–	–
PQ, sec	0,14 (0,13–0,16)	0,006 (0,002–0,009)	02.18 (19.06–05.00)
QRS, sec	0,07 (0,06–0,08)	0,003 (0,0–0,006)	17.48 (04.30–01.36)
Se, sec	0,34 (0,33–0,35)	–	–
De, sec	0,46 (0,38–0,53)	–	–
Q-I tone, sec	0,06 (0,05–0,07)	–	–
Ic, sec	0,04 (0,03–0,05)	–	–
T, sec	0,10 (0,08–0,11)	–	–
E, sec	0,23 (0,21–0,25)	–	–
Sm, sec	0,27 (0,25–0,29)	–	–
St, sec	0,33 (0,31–0,35)	–	–
Dt, sec	0,47 (0,4–0,54)	–	–

Table 5:
Circadian rhythm of heart electromechanical activity and blood pressure parameters in pregnant women with interatrial septum defect /IASD/ (according to the data of average - group "Cosinor - analysis").

Parameter	Mesor	Amplitude	Acrophase
	95% limits of confidence		
BPs, mm Hg	110,3 (102-109)	6,4 (3-10)	15.42p.m. (13.24p.m.18.30p.m.)
BPd, mm Hg	74 (68-79)	–	–
HR, per 1 min	79 (63-95)	8,3 (3,3-13,3)	14.30p.m. (11.06a.m.17.06p.m.)
P, sec	0,09 (0,08-0,09)	–	–
P-Q, sec	0,14 (0,13-0,15)	–	–
QRS, sec	0,06 (0,058-0,07)	–	–
Se, sec	0,33 (0,32-0,37)	–	–
De, sec	0,47 (0,36-0,58)	0,08 (0,04-0,12)	02.12 a.m. (23.06p.m.-05.00a.m.)
Q-Itone, sec	0,08 (0,06-0,09)	–	–
Ic, sec	0,04 (0,03-0,05)	–	–
T, sec	0,12 (0,1-0,14)	–	–
E, sec	0,23 (0,21-0,26)	0,017 (0,008-0,03)	01.06 a.m. (22.00p.m.03.54a.m)
Sm, sec	0,28 (0,25-0,30)	0,018 (0,004-0,03)	01.54 a.m. (21.24p.m.-04.24a.m.)
St, sec	0,35 (0,31-0,39)	0,019 (0,004-0,035)	02.24 a.m. (22.30p.m.-05.12a.m.)
Dt, sec	0,46 (0,25-0,57)	0,06 (0,02-0,11)	02.12 a.m. (22.06p.m.-05.42a.m)

circadian rhythm (Table 7). It is seen from data presented in this table that the circadian rhythms of BPs, BPd, Q-Itone, Ic, T, E are absent. The chronostructure of circadian rhythms of heart electromechanicactivity and BP parameters in pregnant women

Table 6:
Circadian rhythm of heart electromechanical function and BP parameters in pregnant women with interventricular septum defect

Parameter	Mesor	Amplitude	Acrophase
	95% limits of confidence		
Bps, mm Hg	110,5 (102-119)	7,0 (1,1-12,8)	16.48 p.m. (13.18p.m.21.48p.m.)
BPd, mm Hg	74,5 (69-80)	-	-
HR per 1 min	82 (70-94)	7,64 (4-11)	13.00 p.m. (10.18a.m.15.00p.m.)
P, sec	0,08 (0,07-0,09)	0,003 (0,001-0,006)	18.12 p.m. (14.06 p.m.00.48p.m.)
PQ, sec	0,16 (0,14-0,19)	-	-
QRS, sec	0,08 (0,06-0,086)	-	-
Se, sec	0,34 (0,32-0,36)	0,013 (0,002-0,024)	02.00 a.m. (17.42p.m. 3.18a.m.)
De, sec	0,42 (0,33-0,51)	0,06 (0,03-0,09)	01.00 a.m. (22.24p.m. 02.30a.m)
Q-Itone, sec	0,09 (0,07-0,12)	-	-
Ic, sec	0,05 (0,03-0,06)	-	-
T, sec	0,13 (0,11-0,15)	-	-
E, sec	0,23 (0,21-0,26)	0,02 (0,008-0,03)	02.06 a.m. (22.18p.m. 04.06a.m.)
Sm, sec	0,27 (0,26-0,29)	0,02 (0,012-0,03)	04.06 a.m. (01.42a.m. 06.54a.m.)
St, sec	0,37 (0,35-0,38)	0,02 (0,009-0,03)	03.30 a.m. (1.18a.m. 06.06a.m.)
Dt, sec	0,4 (0,3-0,5)	0,05 (0,008-0,09)	00.36 a.m. (20.06p.m. 02.54a.m.)

with ODA is characterized by the "day" type of HRM the "night" type of Se, De, Sm, St and Dt rhythm. In contrast to healthy pregnant women, circadian rhythms of BPs, P-Q, QRS, T and Ic are absent in those with ODA, that may point to the disturbances in circadian organization of cardiohemodynamic parameters in ODA. The data thus obtained can be used as supplementary biorhythmologic criteria for cardiovascular system functional state estimation in pregnant women and for determination of a degree of adequacy of its adaptive reconstructions. The latter can be useful in an earlier diagnosis of functional pathology of the cardiovascular system in the course of pregnancy. The chonobiological investigations in different groups of pregnant women with VHD enable to clarify the pathogenesis of cardiohemodynamics disorders and to reveal the degree of adaptive possibilities of cardiovascular system during the pregnancy. In cases of MS, CF stage I-II, we had to administer cardiac glycosides (CG). A course of 7 days digoxin therapy in a dose of 0,75 mg intravenously brings about a tendency to normalization of circadian rhythm of cardiohemodynamic parameters. Acute clinico-pharmacological tests with digoxin (0,75 mg intravenously), conducted in 3 groups of 32 pregnant women with MS in the morning (at 8.00), day time (at 13.00) and in the evening (at 20.00), allowed to reveal the most extensive shifts of cardiovascular parameters in the evening.

This phenomenon may indicate rhythmicity of digoxin metabolism, of digoxin concentration in plasma and on the cardiac and vascular receptors and, finally, rhythms of these receptors sensitivity.

Table 7:
Circadian rhythm of heart electromechanical function and blood pressure parameters in pregnant women with opened ductus arteriosus (according to the data of average group "Cosinor -analysis")

Parameter	Messor	Amplitude	Acrophase
	95% limits of confidence		
BPs, mm Hg	137 (132-141)	-	-
BPd, mm Hg	70 (64-75)	-	-
HR, per 1 min	75 (68-81)	7,02 (2-11)	15.24 p.m. (6,00 a.m.-16.24 p.m.)
P, sec	0,09 (0,08-0,10)	-	-
P-Q, sec	0,15 (0,14-0,17)	-	-
QRS, sec	0,08 (0,06-0,09)	-	-
Se, sec	0,36 (0,33-0,39)	0,02 (0,008-0,03)	04.30 a.m. (01.54a.m.-07.06a.m.)
De, sec	0,46 (0,4-0,52)	0,07 (0,008-0,13)	03.24 a.m. (20.06p.m.-05.00a.m.)
Q-Itone, sec	0,08 (0,06-0,09)	-	-
Ic, sec	0,06 (0,05-0,07)	-	-
T, sec	0,16 (0,1-0,22)	-	-
E, sec	0,24 (0,22-0,26)	-	-
Sm, sec	0,30 (0,27-0,33)	0,02 (0,02-0,03)	04.36 a.m. (02.30a.m.-07.30a.m.)
St, sec	0,37 (0,34-0,41)	0,03 (0,02-0,04)	04.24 a.m. (02.12a.m.-06.06a.m.)
Dt, sec	044 (0,39-0,50)	0,06 (0,007-0,11)	03.24 a.m. (19.12p.m.-5.06a.m.)

LOSS OF THE CIRCADIAN RHYTHM OF PLASMA CONCENTRATIONS OF ATRIAL NATRIURETIC PEPTIDE IN CONGESTIVE HEART FAILURE

F. PORTALUPPI, L. MONTANARI, L. VERGNANI,
A. D'AMBROSI[1], M. FERLINI & E. DEGLI UBERTI

Chair of Internal Medicine and Endocrinology Section, University of Ferrara,
[1]Institute of Internal Medicine, University of Ferrara, via Savonarola 9, I-44100 Ferrara, Italy.

INTRODUCTION

The occurrence and extent of a circadian rhythm in the circulating concentration of the natriuretic hormone atrial natriuretic peptide (ANP) has been the subject of a number of investigations, both in normal and disease states. The interest raised by this matter is due to the fact that most of the circulating peptides are likely to exert a modulatory role on the cardiovascular system, either at the peripheral or the cardiac level, so that it seems important to look for the rhythmicity of these humoral factors.

Normal (1) and essential hypertensive (2) subjects studied in our hospital showed significant circadian variations in plasma concentrations of ANP. In essential hypertension, the circadian rhythm of ANP is set at higher circulating levels, but is otherwise similar to the circadian rhythm found in normals. Congestive heart failure (CHF) is a condition known to be associated with high plasma concentrations of ANP (3,4,5), but we could find only one study concerning the diurnal change of ANP in this disease (6). We investigated the circadian rhythm of ANP in a group of patients affected by CHF, by using a standardized chronobiological inferential statistical method.

MATERIAL AND METHODS

We studied 10 patients in CHF due to either ischemic heart disease in 8 or mitral regurgitation in 2. There were 6 men and 4 women ranging in age from 44 to 62 years (mean ±SE : 55.8 ±4.4). All were in normal sinus rhythm. They entered the study when clinically stabilized and optimally controlled on digoxin and frusemide alone. The day of the study, frusemide was not administered. The patients were in functional class II as classified by the New York Heart Association (NYHA) functional criteria. In addition, 10 normal volunteers (seven men and three women), 55.3 ±4.2 (range: 39-70) years of age, were studied as control group.

After informed consent was obtained, the patients and subjects were restricted in bed supported at 45° on pillows kept on a standard daily diet containing 7 g of salt per day for at least 3 days. The sleep span in darkness lasted from 22.00 to 06.30 h. Meals were given at precise times as follows: breakfast at 08.15 h, lunch at 12.15 h, small snack at 16.15 h, and dinner at 20.15 h. Beginning at 8.00 h, venous blood samples were drawn every 4 h for 24 h into precooled plastic tubes containing 1 mg/ml Lethylene-diamine- tetraacetic acid disodium salt and aprotinin (Trasylol, Bayer, Milan, Italy; 500 kIU/ml). They were promptly centrifuged at 3000 xg for 15 min at 0°C and then the plasma was frozen at -80°C until the assay was performed. All samples were processed in duplicate in the same assay.

ANP was measured by RIA combined with an extraction step, using reagents supplied by Immuno Technology Service Production BV (Wijchen, The Netherlands). The intra- and interassay coefficients of variation are 8.6 and 11.6%, respectively, and the sensitivity is 2.1 pmol/ml.

All 24-h data are expressed as the mean ±standard error. Analysis of variance was used for comparison of mean variable values between times of assessment. Duncan's multiple range test was used to determine the location of significant differences in mean values. Cosinor analysis was also performed using a computerized procedure that we developed (7). Intergroup differences of rhythm parameters were assessed by the Bingham test. In all tests, a $p<0.05$ was considered significant.

RESULTS

Fig. 1 shows the 24 h levels of ANP in CHF patients and normal controls. Tab. 1 reports the results of the cosinor analysis on the ANP data obtained in the groups studied.

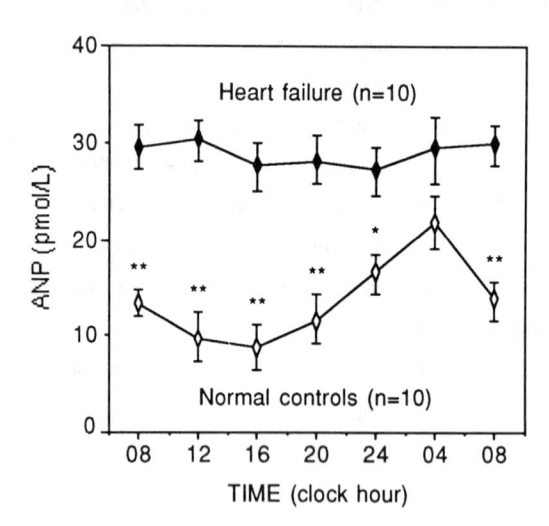

Fig. 1:
Circadian variation of plasma concentrations of atrial natriuretic peptide in controls and patients with congestive heart failure. * $p < 0.05$, ** $p < 0.01$, as compared with values at 04.00 h.

Tab. 1: Mean circadian rhythm parameters of atrial natriuretic peptide in normal controls and in congestive heart failure.

	P (Ftest)	Mesor +/-s.e.m. pmol/L	Ampl. +/-s.e.m. pmol/L	Acrophase (95% conf.limits) Degrees	Clock Hours
Normal controls (n=10)	<0.01	13.4 +/- 1.7	5.7 +/- 1.1	-42.5 (-332.4,-40.0)	02.50 (22.10, 0240)
Heart failure (n=10)	0.12	28.6 +/- 2.4**	1.2 +/- 0.5	-137.6 (--,--)	09.11 (--,--)

P (F Test) = significance level of the circadian rhythm; n.s. =not significant; Ampl. = amplitude. *P < 0.01, and **P < 0.001,as compared with the corresponding value in the control group.

The mean (±SE) circadian mesor of ANP plasma levels was 13.4 ±1.7 pmol/l in the control group and 28.6 ±2.4 pmol/l in the group of CHF patients. In the control group, the peak plasma ANP concentration of 21.5 ±2.7 pmol/l occurred at 04.00 h, while the lowest concentration of 8.8 ±2.4 pmol/l was measured at 16.00 h ($p<0.01$). Analysis of variance demonstrated that the peak value was significantly different from all other time points. In CHF patients, the comparison of mean ANP levels between times of assessment revealed no significant difference throughout the 24 h. Cosinor analysis confirmed the disappearance of the circadian rhythm of ANP in CHF patients.

DISCUSSION

Our CHF patients displayed mean plasma concentrations of ANP three times higher than normal. This is in keeping with previous reports of high plasma concentrations of ANP in relation to the clinical severity of CHF and with a significant inverse correlation with left ventricular function (3,4,5). The elevated plasma levels of ANP in CHF may be ascribed to the myocardial strain (8), secondary to increased intracavitary pressures, volume overload, and tachycardia. Augmented synthesis and secretion by both atrial and ventricular cardiocytes can be demonstrated (9), with increased turnover and depleted stores of ANP in proportion to the severity of the disease (10).

A circadian rhythm in the plasma concentrations of ANP, unrelated to postural changes and peaking in the early morning hours, is established in normal subjects (1,2) and was present in our control group. In CHF, the changes in the circadian rhythm of ANP, reported in one previous study (6), is confirmed by our data. The mechanism underlying the lack of a diurnal variability in the circulating concentrations of ANP in CHF can be postulated to be a saturation of the receptor sensing system of ANP, since both a limited transformation rate of its precursor (11) and an intracellular defect that prevents the mediation of its signal into biological action (12) have been demonstrated in failing hearts. The fact that a flat pattern already becomes evident when a patient is in NYHA class II supports the concept that ANP release is aimed to counteract the detrimental effects of stimulation of renin and the sympathetic nervous system even in the early phases of CHF. When CHF is severe, however, the counterbalancing effects of ANP may be offset by vasoconstriction and fluid and sodium retention.

In conclusion, our data demonstrate that the circadian pattern of ANP is altered in patients with CHF, due to the absence of the relative increase in nocturnal secretion.

REFERENCES

1. PORTALUPPI, F., L. MONTANARI, B. BAGNI et al. (1989): Circadian rhythms of atrial natriuretic peptide, blood pressure, and heart rate in normal subjects. Cardiology 76, 428-432
2. PORTALUPPI, F., B. BAGNI, E. DEGLI UBERTI et al. (1990): Circadian rhythms of atrial natriuretic peptide, renin, aldosterone, cortisol, blood pressure, and heart rate in normal and hypertensive subjects. J. Hypertens. 8, 85-95
3. ARENDT, R.M., A.L. GERBES, D. RITTER et al. (1986): Atrial natriuretic factor in plasma of patients with arterial hypertension, heart failure or cirrhosis of the liver. J. Hypertens. 4(Suppl), S131-S135
4. SCHIFFRIN, E.L. & R. TAILLEFER (1986): Correlation of left ventricular ejection fraction and plasma atrial natriuretic peptide in congestive heart failure (letter). N. Engl. J. Med. 315, 765-766
5. HARA, H., T. OGIHARA, J. SHIMA et al. (1987): Plasma atrial natriuretic peptide level as an index for the severity of congestive heart failure. Clin. Cardiol. 10, 437-442
6. YOSHINO, F., N. SAKUMA, T. DATE et al. (1989): Diurnal change of plasma atrial natriuretic peptide concentrations in patients with congestive heart failure. Am. Heart J. 117, 1316-1319
7. PORTALUPPI, F. & M. RIGHETTI (1987): Cosinor analysis of long time series on Apple microcomputers. In: A. SERIO, R. O'MOORER, A. TARDINI & F.H. ROGER (eds.): Medical Informatics Europe'87. Proceedings of the Seventh International Congress of the European Federation for Medical Informatics. Edi. Press, Roma, pp. 1267-1271
8. PORTALUPPI, F., A. PRADELLA, L. MONTANARI et al. (1990): Atrial strain is the main determinant of release of atrial natriuretic peptide. Int. J. Cardiol. 29(3), 297-303
9. ARBUSTINI, E., A. PUCCI, M. GRASSO et al. (1990): Expression of natriuretic peptide in ventricular myocardium of failing human hearts and its correlation with the severity of clinical and hemodynamic impairment. Am. J. Cardiol. 66, 973-980
10. SUGAWARA, A., K. NAKAO, N. MORII et al. (1988): Synthesis of atrial natriuretic polypeptide in human failing hearts. Evidence for altered processing of atrial natriuretic polypeptide precursor and augmented synthesis of beta-human ANP. J. Clin. Invest. 81, 1962-1970
11. CANTIN, M., G. THIBAULT, J.F. DING et al. (1988): ANF in experimental congestive heart failure. Am. J. Pathol. 130, 552-568
12. RIEGGER, G.A., D. ELSNER, E.P. KROMER et al. (1988): Atrial natriuretic peptide in congestive heart failure in the dog: plasma levels, cyclic guanosine monophosphate, ultrastructure of atrial myoendocrine cells, and hemodynamic, hormonal, and renal effects. Circulation 77, 398-406

CIRCADIAN VARIATION OF HEART RATE AND HEART RATE VARIABILITY IN SHORT TERM SURVIVORS AND NON-SURVIVORS AFTER ACUTE MYOCARDIAL INFARCTION

VAN LEEUWEN,P., C.HECKMANN, P.ENGELKE, G.KESTING & H.C.KÜMMELL

Department of Internal Medicine, Gemeinschaftskrankenhaus, University of Witten/Herdecke, D-W-5804 Herdecke, FRG

INTRODUCTION

A number of studies have shown that heart rate (HR) and HR variability can be used to identify high risk patients after acute myocardial infarction (AMI) (5,7,1). Furthermore, time of day also seems to play a role in the prognostic value of these parameters (8), implying that the presence or absence of circadian patterns may be associated with the severity of disease. However, these studies made their observations well after the acute event, usually weeks or months later, and related their results to long term mortality. The purpose of our study was to examine HR and HR variability as well as their circadian rhythms immediately following AMI and relate these results to survival within the following weeks.

PATIENTS AND METHODS

The study population consisted of 57 patients admitted to hospital with a diagnosis of AMI. This diagnosis was confirmed when two of the following were present: characteristic chest pain, electrocardiographic (ECG) changes and elevation of cardiac enzymes. On this basis 7 patients were excluded from the study, as well as further 6 patients as a result of incomplete data acquisition during hospital stay. The remaining 44 patients were divided into two groups retrospectively on the basis of survival. 39 patients recovered and were discharged. 5 patients died in hospital between 10 and 35 days after admission, 3 at approximately 3 weeks.

Using an Oxford Medilog II recording system, 24-hour ECG recordings were acquired within 36 hours after admission and around 3 weeks later. We were able to acquire data of only two of the non-surviving patients at three weeks; these data were excluded from further group comparisons. An Oxford Medilog ECG Analyser MA 14 was used to read cycle length (RR interval) into a Masscomp 5300 computer with a resolution of 1 ms for further analysis. Basis for the determination of the rhythmic parameters over 24 hours were 288 consecutive 5-minute periods from each ECG recording. Arrhythmic beats were excluded from the data processing. Mean HR was calculated from the RR-intervals of each five minute period. The standard deviation (SD) of the RR-intervals over this period was used to calculate HR variability. The five minute values for HR and HR variability were in turn used to calculate hourly means. The circadian patterns were approximated by 3 component harmonic fit on the basis of Fourier analysis. Values are given as means ± SD (standard error of the mean (SEM) in the figure). Hourly and 24-hour means were compared using paired within groups and unpaired t-tests between groups.

RESULTS

Within the first three days after admission the 24-hour HR mean of the survivors was significantly lower than that of the non-survivors (92±17 vs. 116±20 bpm, p<0.01). The hourly values of the two groups were significantly different throughout the complete 24-hour period, whereby the most obvious differences occurred during the daytime hours (Fig. 1). During the following three weeks the mean 24-hour HR of the survivor group fell significantly to 77±11 bpm (p<0.001). These overall changes in HR were not uniform over the 24 hour period: the greatest decrease occurred during the nighttime. Harmonic analysis demonstrated the differences in circadian rhythms between the groups and times.

Within the first 36 hours after admission the 24-hour mean HR variability of the survivors was significantly higher than that of the non-survivors (29±13 vs. 12±6 ms, p<0.001). Most of the difference lay between midnight and early afternoon, peaking around 07.00-08.00 (see Fig. 1). After three weeks the 24-hour mean HR variability of the survivors increased to 32±12 ms. Although this was non-significant, the hourly values between 01.00 and 09.00 did increase significantly.

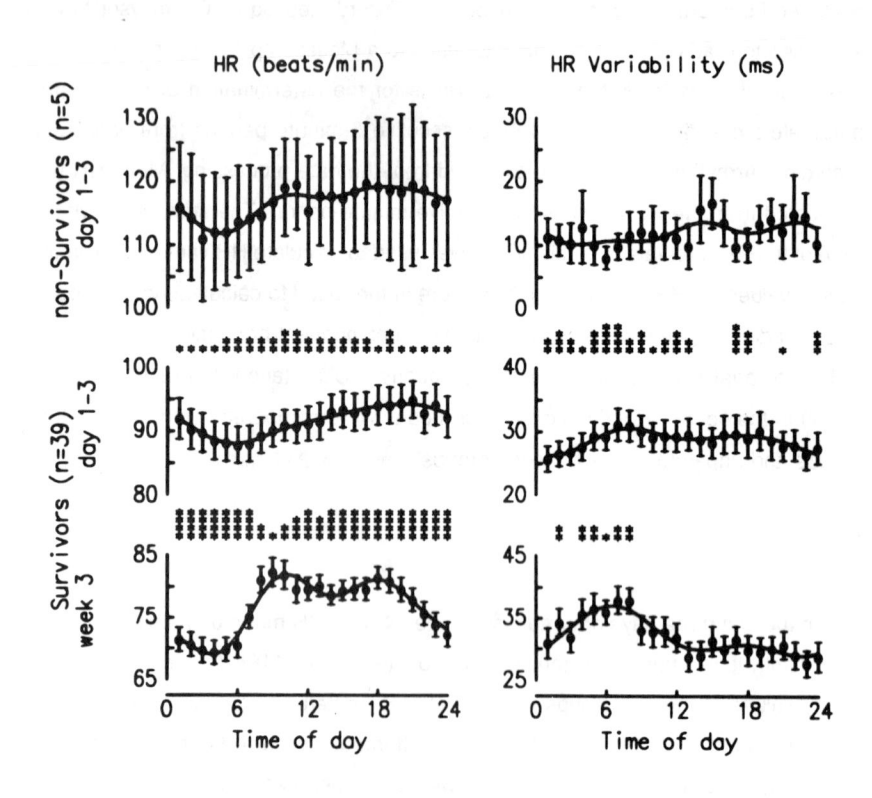

Fig.1: Mean hourly HR (bpm±SEM) and HR variability (ms±SEM) over 24 hours with corresponding 3-component harmonic fit of the non-survivors at day 1-3 and the survivors at day 1-3 and week 3.* p<0.05; ** p<0.01; *** p<0.001; **** p<0.0001; levels of significance given for adjoining plots (non-survivors vs. survivors: day 1-3, and survivors: day 1-3 vs. week 3).Circadian variation of heart rate and heart rate variability in short term survivors and non-survivors after acute myocardial infarction.

DISCUSSION

It has previously been shown that elevated HR post AMI can be associated with increased 1-year mortality (5). Our results indicate that this may also be true for mortality within the first few weeks after AMI and therefore HR may also help identify short term high risk patients. On the other hand circadian variation of HR cannot be associated with survival at this point. However, after three weeks the typical pattern with a minimum at night, a maximum during the daytime and a concomitant increase in amplitude was present in the survivors. Although this may in part have been due to increased daily activity, such circadian rhythms have also been observed in cardiovascularly healthy subjects confined to bed (3). Therefore the presence of circadian variation in these patients may also be regarded as a sign of recovery (4).

It has been reported in the literature that HR variability is lower in high risk patients (7,1), and that the greatest predictive value of HR variability is in the morning hours (6). In this study we have, however, been able to show 1) that it may be possible to identify high risk patients almost immediately after admission on the basis of HR variability, and 2) that the risk is acute, i.e. low HR variability may be associated to death within weeks. Furthermore, a comparison of the harmonic curves indicates a tendency of increasing HR variability between groups and over time, in particular with a peak in the morning hours. In fact, reports in the literature (2) and own unpublished data confirm that, in healthy subjects, measuring HR variability using the SD of the RR intervals produces values well over 50 ms and morning peaks up to 90 ms or more. The data obtained in this study showed that over time the values of the surviving patients were developing towards normal.

It may be concluded that circadian variation of HR parameters can be determined very early after acute events. Such time dependent variation in particular of HR but also of its variability can be associated with recovery after AMI.

REFERENCES

1. BIGGER, J.T., R.E. KLEIGER, J.L. FLEISS et al. (1988): The Multicenter Post-Infarction Research Group. Components of heart rate variability measured during healing of acute myocardial infarction. Am.J. Cardiol., 61, 208-215.
2. CORNÉLISSEN, G., E. BAKKEN, P. DELMORE et al. (1990): From various kinds of heart rate variability to chronobiology. Am.J.Cardiol, 66, 863-868.
3. EWING, D.J., J.M.M. NEILSON, C.M. SHAPIRO et al. (1991): Twenty four hour heart rate variability: effects of posture, sleep, and time of day in healthy controls and comparison with bedside tests of autonomic function in diabetic patients. Br.Heart.J. 65, 239-244.
4. HECKMANN, C. & M. BUSCH (1987): Changes in circadian pattern of heart rate in patients after acute myocardial infarction. In: G. HILDEBRANDT, R. MOOG, & F. RASCHKE (eds.): Chronobiology and Chronomedicine. Peter Lang, Frankfurt AM Bern New York Paris, pp. 299-306.
5. HJALMARSON, Å., E.A. GILPIN, J. KJEKSHUS et al. (1990): Influence of heart rate on mortality after acute myocardial infarction. Am.J.Cardiol., 65, 547-553.
6. HUIKURI, H.V., K.M. KESSLER, E. TERRACALL et al. (1990): Reproducibility and circadian rhythm of heart rate variability in healthy subjects. Am.J.Cardiol., 65, 391-393.
7. KLEIGER, R.E., J.P. MILLER, J.T. BIGGER & A.J. MOSS (1987): The Multicenter Post-infarction Research Group. Decreased heart rate variability and its association with increased mortality after acute myocardial infarction. Am.J.Cardiol., 59, 256-262.
8. MALIK, M., T. FARRELL & A.J. CAMM (1990): Circadian rhythm of heart rate variability after acute myocardial infarction and its influence on the prognostic value of heart rate variability. Am.J.Cardiol., 66, 1049-1054.

RHYTHMIC ABILITIES IN PATIENTS WITH FUNCTIONAL CARDIAC ARRHYTHMIAS

M. KAYSER & R. RICHTER

Universitätskrankenhaus Eppendorf, Abteilung für Psychosomatik und Psychotherapie, Martinistr.52, D-W-2000 Hamburg 20, FRG

INTRODUCTION

In his work "Musikpsychologie" the musicologist Ernst KURTH (1) described his observation, that "rhythmic performance fails when physical or nervous functions are weakened" and moreover "rhythmic performance declines sooner than the musical ear or the technique being once established". He mentioned that "sudden fluctuations in rhythm" are related to anxious affections in a way that "dysrhythmias" are caused by anxiety. It is striking that cardiac arrhythmias appear in a very similar way, whereby in particular the so-called "functional cardiac arrhythmias" are most likely to be interpreted as an anxiety equivalent.

Starting from this point dysrhythmias in different physiological functions should correlate with performance in musical rhythms. Thus we investigated the question, if patients with functional cardiac arrhythmias perform worse in "rhythm perception & production" when compared to healthy controls.

METHODS

31 patients suffering from functional cardiac arrhythmias were included in our study. Organic causes had been previously excluded by clinical, non-invasive cardiological diagnostics. Patients suffered from ventricular extrasystoles (n=11, 35.5%), paroxysmal supraventricular tachycardias (n=11, 35.5%), and the combination of both (n=9, 29%).

31 healthy control subjects (16 women, 15 men) were matched in regard to sex, age, school education, and last not least the musical education level. The last variable was defined and standardized by a "Musical Anamnesis Questionnaire" including the topics

1. Music listening customs (how often and on which occasions do you listen to music?)
2. practice (dexterity) on musical instrument (Did you learn a musical instrument? Did you have lessons or a higher education?)
3. knowledge in musical theory
4. vocal abilities
5. "Attractability" by rhythmical music (e.g.: Do you usually accompany music by foot tapping? Do you like dancing?)

To examine the "rhythmic ability" we used factorized tests developed by HIRIARTBORDE & FRAISSE (2). These authors studied the nature of rhythmic abilities among physical education teachers, using some of the well-known WING- and SEASHORE-Tests of musical ability, as well as eleven tests of their own. Validation of their results was attempted by comparison with dance examination scores and rating by instructors.

They concluded that three independent factors account for a 'sense of rhythm'. These include (1) the perception of rhythmical structures, (2) the ability to synchronize with them, and (3) "elaborating complex movements with one or several limbs when accompanying or reproducing rhythms".

The original test battery of HIRIARTBORDE & FRAISSE (2) seemed to be too voluminous for our practical application, so we selected the tests with the highest factor loadings for each of the three factors.

On the first two tests (belonging to factor 1) the subjects listened to rhythmic patterns played by a taperecorder; the task was to reproduce these patterns by marking them on a sheet of paper.

An example: *   * *   *   * *

Test 1 deals with temporal patterns, test 2 with intensity patterns. The tests 3 and 4 refer to the second factor, the ability of rhythmic synchronization; the subject has to tap synchronously with the repeating patterns from the tape.

Another example: * *** ** *   * *** ** * etc.

A polygraph-recording registrates the tap-impulses together with the impulses from the taperecorder, to analyze time-taps. Test 5 times the maximum speed of alternating hand/foot-taps: The subject shall tap by turns on a hand- and a foot-key, that is as fast as he can, but strictly one after the other. The speed is again recorded by the polygraph-recording.

To control for the effect of mood and emotional state we administered a standardized questionnaire, the Bf-S rating scale.

RESULTS

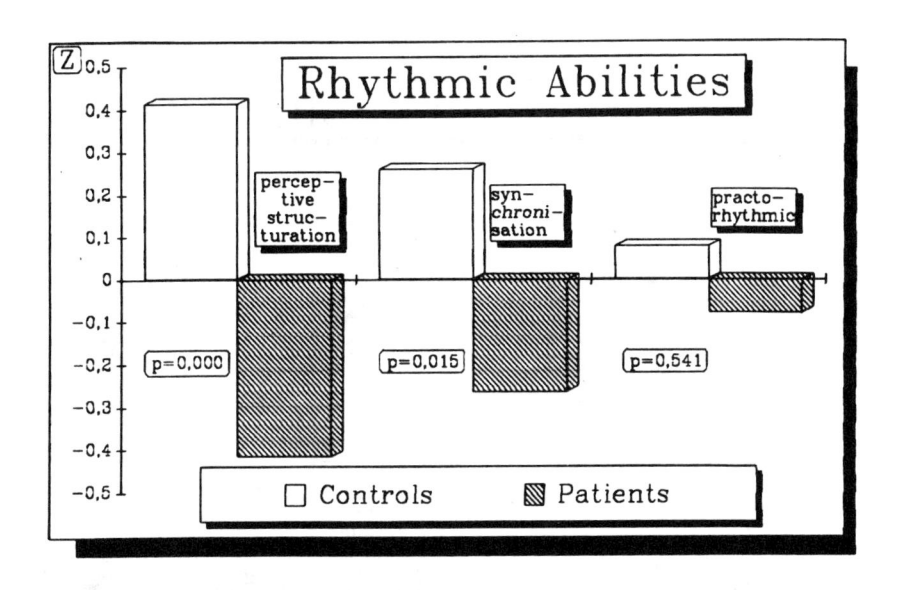

Fig.1: Test results in regard to "factors of rhythmicity"; t-test significance between groups

The results in the rhythm tests were z-transformed prior to statistical tests (Fig. 1).

Between patients and healthy controls we found highly significant differences in "perceptual structuration", the first of FRAISSE' rhythm factors, as well as significant differences in the second factor, the ability of "rhythmic anticipation". These differences demonstrated a clearly poorer performance of patients with dysrhythmias. In the third factor "practo-rhythmic" the patients still have lower scores, however, there were no significant differences.

Post-hoc splitting the patient group in patients with extrasystoles, tachycardia and the combination of both revealed some more interesting differences (Fig. 2): Patients with tachycardia are highly poor in perceptive structuration and rhythmic synchronization; the practo-rhythmic factor 3 has again a different distribution with the highest scores in the "tachycardia-group".

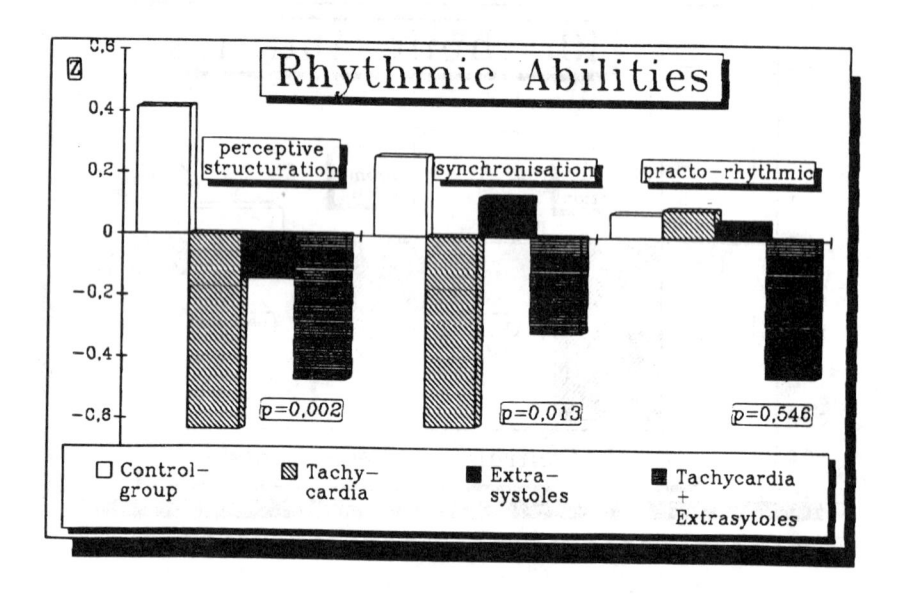

Fig.2: Test results / splitted patient group; Analysis of variance between groups

DISCUSSION

Statistical tests revealed a significant difference in emotional state (Bf-Score) between patients and controls. As would be expected, the patients reached higher scores - that is, they were in a worse emotional state than the controls (patients: mean=24.71, s=14.22; controls: mean=11.13, s=10.05; p=0.0001). But this difference does not explain the differences in rhythmic abilities as reported above: the correlations between Bf-Score on the one hand with the rhythmic perception and performance are not significant.

Several conclusions may be drawn: Our findings suggest a relationship between physiological arrhythmias and "mental" rhythmic abilities. Our design cannot answer the question whether a primarily disturbed cardiac rhythm is the reason for the worse musical rhythmicity or vice versa. Nevertheless the results provide further evidence for a generally disturbed rhythmicity in patients with functional arrhythmias.

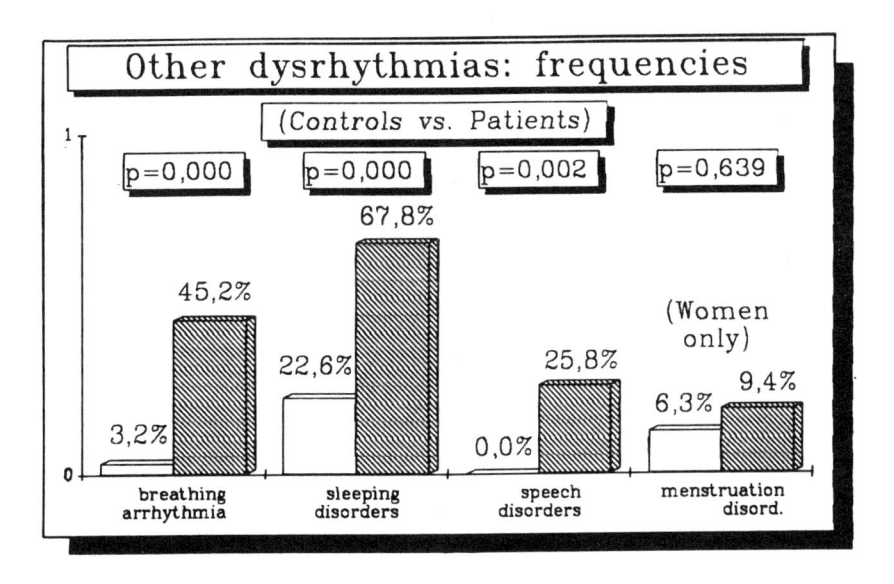

Fig.3: Frequencies of other dysrhythmias, t-test significance between groups

This suggestion is supported by the coincidence between dysrhythmias in other physiological functions (e.g. breathing arrhythmias, sleeping-, speech-, and menstruation disorders) and cardiac arrhythmias (Fig. 3).

REFERENCES

1. KURTH, E. 1947): Musikpsychologie: Bern
2. HIRIARTBORDE, E. & P. FRAISSE (1968): Les aptitudes rythmiques, Paris

CHANGES IN PULSE-RESPIRATION RATIO IN PATIENTS WITH IMMUNOGENIC HYPERTHYROIDISM (GRAVES' DISEASE)

C. HECKMANN, P. ENGELKE, G. KESTING, P. VAN LEEUWEN & KH. RUDORFF

Department of Internal Medicine, Gemeinschaftskrankenhaus,
University of Witten/Herdecke, D-W-5804 Herdecke, FRG

INTRODUCTION

Over the past decades the group of HILDEBRANDT has widely investigated the coordination of cardiovascular and respiratory rhythms (2,3,4). One of the main results of this work is that the pulse-respiration ratio (PRR) can be viewed as a marker of man's rhythmical functional order. The interaction of these rhythms leads to normalization at rest (autonomous order) and to adaptation under stress (heteronomous order). This corresponds to a phase coupling and frequency coordination between cardiac and respiratory rhythms in particular during night sleep and to continuous frequency changes with respect to the degree of workload during daytime activity. The former is expressed in a whole number PRR whereas this is usually not the case in the latter. Although it is well known that hyperthyroidism affects cardiac activity, however, it is unclear to what extent the coordination of the cardiac and respiratory rhythms is affected.

METHODS

Using an Oxford Medilog II recording system we obtained 24-hour ECG recordings from thirteen patients with immunogenic hyperthyroidism prior to thyroistatic therapy and after remission of clinical signs. Basis for the determination of the rhythmic paramaters over 24 hours were 288 consecutive 5-minute periods from each ECG recording. Respiration was estimated on the basis of the analysis of respiratory sinus arrhythmia (6). Heart rate and respiratory rate were calculated for each

5-minute period. These values were then used to calculate hourly means. For purposes of comparison we have included the data of 39 patients after acute myocardial infarction (AMI), 24 patients with anorexia nervosa as well as 18 healthy subjects.

RESULTS

PULSE RESPIRATION RATIO
Grave's Disease

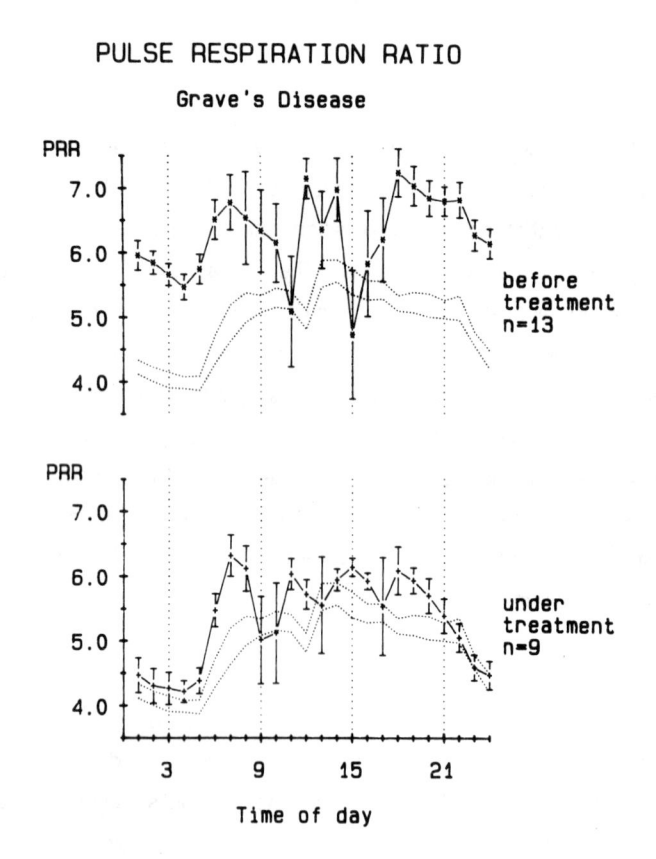

Fig. 1: Pulse-respiration ratio (PRR) over 24 hours of patients with Graves' Disease before treatment and under treatment. Dotted lines show ±1 standard deviation range of healthy subjects.

The patients suffering from hyperthyroidism in untreated Graves' Disease displayed an elevated PRR throughout the whole 24-hour period with values fluctuating around 6.5. There was no evidence of a normal circadian pattern as found in healthy subjects with a minimum during the nighttime, an increase in the morning hours and bimodal maxima during the daytime (see Fig. 1, top). After thyroistatic treatment there was a substantial decrease in the nighttime PRR values of the patients. This, together with a stabilisation of the daytime values, led to a 24-hour profile similar to that of healthy subjects (Fig. 1, bottom).

DISCUSSION

Our results suggest that hyperthyroidism leads not only to changes in cardiac activity but also to changes in the coordination of cardiac and respiratory rhythms. In particular the autonomous order (normalisation at night) is disturbed as indicated by the elevated PRR values. This indicates a loss of balance in autonomous functions. The normal healthy state is characterised by rhythmic alteration between heteronomous and autonomous order. Different disease states disturb this alternation in differing ways. Patients who have suffered an acute myocardial infarction display an elevated PRR with very little variation over 24 hours, indicating a predominance of sympathetic tone (see Fig. 2). It has been demonstrated that in normal recovery after AMI normalisation of the circardian pattern of PRR occurs (5). The respective paper also describes one case with a fatal outcome in which the PRR deviation increased during the clinical course. The behaviour of PRR of the survivors of AMI is similar to that of the patients with Graves' Disease with respect to the lack of night-time normalisation but it differs in that there is a complete lack of variation throughout the 24-hour period in the patients with AMI. Patients with Graves' Disease suffer from weight loss. This is also the case in anorexia nervosa. The latter, however, usually display normal circadian PRR variation (see Fig. 2). This may be due to the fact that autonomic balance is less disturbed in the latter. However, some of these patients initially displayed abnormal circadian PRR variation. In these cases normalisation of body weight went along with normalistion of PRR rhythms (1).

PULSE RESPIRATION RATIO
patients at admission

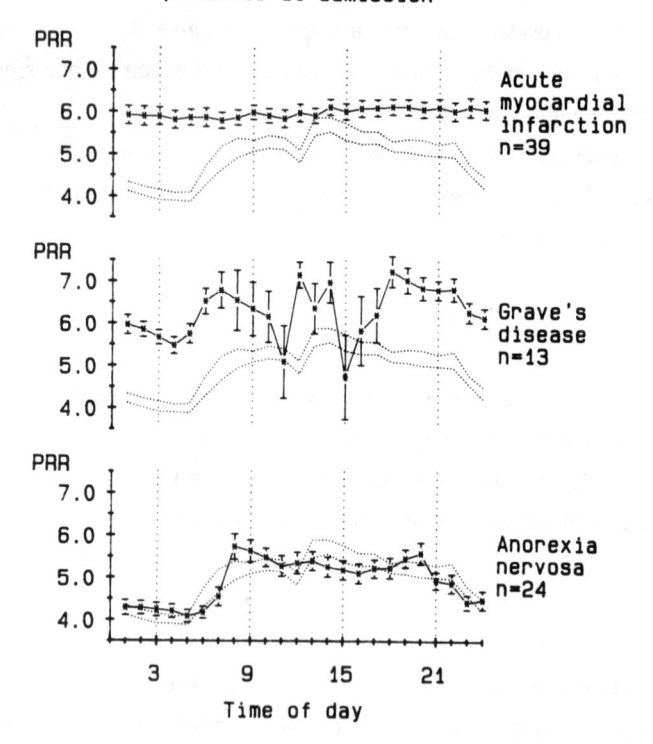

Fig.2: Pulse respiration ratio (PRR) over 24 hours of patients after acute myocardial infarction, with Graves' Disease (before treatment) and with aorexia nervosa. Shaded areas show ±1 standard deviation range of healthy subjects.

The 24-hour course of PRR in patients with Graves' Disease appear to lie between these two extremes.

In general, treatment success may be judged by the changes in circadian rhythm of PRR. In particular the success of thyroistatic therapy is reflected by the renewed coordination of respiratory and cardiac rhythms.

REFERENCES

1. HECKMANN C., P. MEINARDUS, P. MATTHIESSEN et al. (1990): Frequency coordination of pulse and respiration in adolescent patients with eating disorders. In: E. MORGAN (ed.), Chronobiology and Chronomedicine. Peter Lang Verlag, Frankfurt a.M., 135-142.
2. HILDEBRANDT, G. (1960): Die rhythmische Funktionsordnung von Puls und Atmung. Zeitschrift für angewandte Bäder- und Klimaheilkunde 7, 533-615.
3. HILDEBRANDT, G. (1986): Coordination of cardiac and respiratory rhythms and therapeutical effects on it. J. Autonomous Nervous System; Suppl., 253-263.
4. HILDEBRANDT, G. (1988): Temporal order of ultradian rhythms in man. In: W.T. HEKKENS, G.A. KERKHOF, W.J. RIETVELD (eds.), Trends in Chronobiology. Pergamon Press, Oxford New York, pp. 107-122.
5. KÜMMELL H.C. & C. HECKMANN (1987): Herzinfarkt und rhythmisches System. Beitr.erw.Heilk. 1987; Sonderheft: 15-27.
6. RASCHKE, F., W. BOCKELBRINK & G. HILDEBRANDT (1977): Spectral analysis of momentary heart rate for examination of recovery during night sleep. In: W.P. KOELLA, P. LEVIN (eds.), Sleep 1976. Proc.3rd Europ.Congr.Sleep Research. S.Karger, Basel: 298-301.

RESPIRATORY SINUS ARRHYTHMIA AFTER EXTERNAL COLD WATER APPLICATION

H.-J. RUDOLPH

Department for Holistic Medicine and Physiological Healing,
Free University Berlin, Krankenhaus Moabit, IV. Innere Abteilung,
Turmstr. 21, D-W-1000 Berlin 21, FRG

INTRODUCTION

Generally, respiratory sinus arrhythmia (RSA) is known to result from reciprocal sympathetic and parasympathetic heart innervation (1). During quiet nonstressed supine rest, however, cardiac sympathetic activity is neglectable in normal man (2). Consequently, in this situation, the degree of RSA can be used to estimate parasympathetic heart innervation (3).

Physical exercises as well as external cold water applications have been found to develop active and passive toughness (4). Additionally, with intensive physical exercise, long-term changes of parasympathetic heart innervation have been suggested to reflect sufficiency or lack of regenerative capacity (5). Similar effects of cold water treatments can be expected, but have not yet been described. In a first step, we therefore examined RSA before and after a single external cold water application.

MATERIAL AND METHODS

From August to October 1990 eight healthy male caucasians (age 24 to 47 years) were examined in a randomized cross-over trial. Night workers were excluded and tiring activities as well as coffee, alcoholic beverages, and tobacco products were forbidden for four hours before examination.

Data were recorded in supine position before and after application of thoracic cold wet (CWSP) or dry sheet packs (DSP) (application according to (6), water temperature 16-17 °C, starting time 14.00, after lunch). From the 15th minute before till the 60th minute after application of thoracic packs, every 2 to 10 minutes cardiac beat-to-beat intervals and nostril air temperatures (indicating the rhythm of respiration) were recorded simultaneously for one minute each, using a cardiotachometer and a thermosensor (Epicon, Munich, FRG).

Evaluation was done on an IBM-compatible personal computer. The degree of sinus arrhythmia was determined by the mean successive difference (MSD) of beat-to-beat intervals, earlier described by ECKOLDT (7), and the mean respiratory sinus arrhythmia amplitude (RSAA), calculated automatically by a modified peak-trough method (8). Statistics were computed with SPSS/PC+ Software (SPSS Inc., Chicago, USA).

RESULTS

Time courses of respiratory frequency (RF) and RSAA are depicted in Fig. 1. After application of DSP/CWSP, initially, RF decreased and RSAA increased significantly. In the following minutes both parameters returned to base line values. No significant group differences (DSP vs. CWSP) were found in RF, RSAA, MSD, and heart rate (HR) up to the 15th minute. But 20 minutes after application of DSP/CWSP, RSAA and MSD increased significantly with CWSP only. RF and HR remained unchanged at this time. Subsequently, RSAA and MSD showed base line values without significant group differences. Tab. 1 summarizes the results.

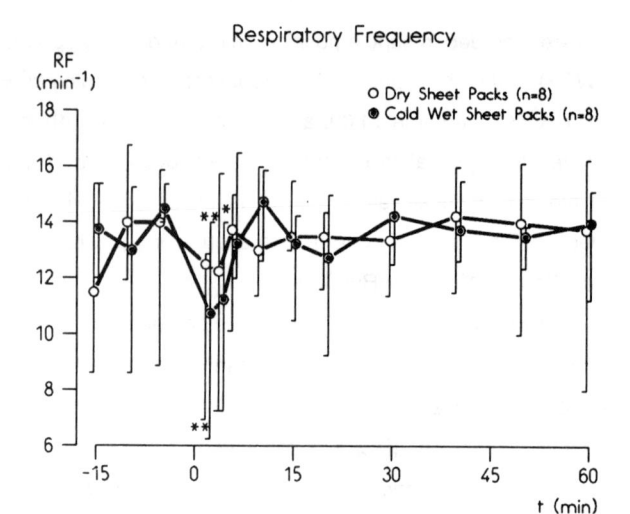

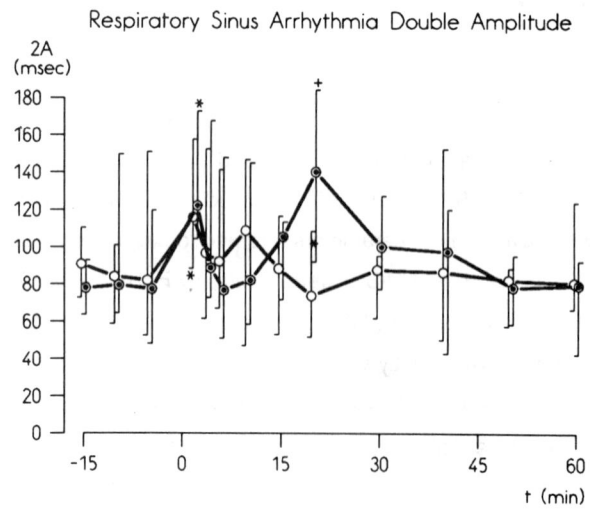

Fig. 1:
Time courses of respiratory frequency (RF) and the double amplitude (2A) of respiratory sinus arrhythmia before and after application of dry (DSP) or cold wet sheet packs (CWSP). Medians, quartiles, Wilcoxon's matched-pairs signed-ranks test. Asterisks at the lower quartiles indicate significant differences vs. the 5th minute before DSP, at the upper quartiles vs. the 5th minute before CWSP; between medians group differences (DSP vs. CWSP) are indicated ($^+$ $2p \leq 0.1$, * $2p \leq 0.05$, ** $2p \leq 0.025$).

Tab.1: Respiratory frequency (RF), heart rate (HR), double amplitude (2A) of respiratory sinus arrhythmia and mean successive difference (MSD) 5 minutes before and 2 as well as 20 minutes after application of dry (DSP) or cold wet sheet packs (CWSP). Medians, Wilcoxon's matched-pairs signed-ranks test (⁺ 2p≤0.1, * 2p≤0.05, ** 2p≤0.025). Within 'wet' and 'dry' subgroups asterisks indicate significant differences vs. the 5 th minute before DSP/CWSP.

	5th min before DSP/CWSP			2nd min after DSP/CWSP			20th min after DSP/CWSP		
	wet	dry	w/d	wet	dry	w/d	wet	dry	w/d
RF (beats/min)	14.5	14.0	ns	10.8**	12−.5**	ns	12.8ns	13.5ns	ns
HR (beats/min)	59	60	ns	58+	61	ns	58+	57*	ns
2A (ms)	77	82	ns	122*	116*	ns	140+	74 ns	*
MSD (ms)	32	41	ns	44ns	43+	ns	56*	33 ns	+

DISCUSSION

Increases of respiratory sinus arrhythmia amplitude (RSAA) and mean successive difference (MSD) have been found twice: Once initially with dry (DSP) as well as cold wet sheet packs (CWSP), and once again after 20 minutes with CWSP only. The first increase coincided with a decrease of respiratory frequency (RF), but 20 minutes after application of CWSP, RF (and heart rate) remained unchanged.

Breathing patterns modulate RSA via reflex responses, released by thoracic stretch receptor discharges (9). HIRSCH & BISHOP (10) described decreases of RSAA at RFs above a characteristic corner frequency (7.2 ± 1.5 cycles/min); additionally RSAA increases with respiratory tidal volume (10), and, via baroreceptor reflex, after arterial blood pressure elevation (11). During supine rest, however, elevations of blood pressure are unlikely to occur in normal man, and increases of respiratory tidal volume are understood to be compensated spontaneously by a decrease of RF. Therefore, the second increase of RSAA and MSD can be expected to result not from one of the above-mentioned reflex responses, but instead from a transient increase of tonic parasympathetic heart innervation (12).

ACKNOWLEDGEMENT

This work was supported by the Rut- & Klaus-Bahlsen-Foundation, and Kneipp-Werke, FRG.

REFERENCES

1. KOLLAI, M. & K. KOIZUMI (1979): Reciprocal and non-reciprocal action of the vagal and sympathetic nerves innervating the heart. J. Auto. Nerv. Syst. 1, 33-52
2. PFEIFER, M.A., D. COOK, J. BRODSKY et al. (1982): Quantitative evaluation of cardiac parasympathetic activity in normal and diabetic man. Diabetes 31, 339-345
3. GENOVELY, H. & M.A. PFEIFER (1988): RR-variation: the autonomic test of choice in diabetes. Diabetes/Metabolism Reviews 4, 255-271
4. DIENSTBIER, R.A. (1989): Arousal and physiological toughness: implications for mental and physical health. Psychological Review 96, 84-100
5. ABEL, H.-H., R. KRAUSE, R. BERGER et al. (1991): Differentes Langzeitverhalten der chronotropen Herzkontrolle während körperlicher Ruhe. Berichtsband Dt. Sportärztekongreβ, München, Oktober 1990. Zugschwerdt Verlag, München, (in press)
6. BRÜGGEMANN, W. (1980): Technik der Kneipp-Hydrotherapie. In: W. BRÜGGEMANN (ed.): Kneipptherapie. Springer, Berlin, pp. 84-99
7. ECKOLDT, K. (1984): Verfahren und Ergebnisse der quantitativen automatischen Analyse der Herzfrequenz und deren Spontanvariabilität. Dt. Gesundh. Wesen 39, 856-863
8. FROSTIG, Z. & R.D. FROSTIG (1987): Analysis of frequency components in time series data. Journal of Neuroscience Methods 22, 79-87
9. CLYNES, M. (1960): Respiratory sinus arrhythmia: laws derived from computer simulation. J. Appl. Physiol. 15, 863-874
10. HIRSCH, J.A. & B. BISHOP (1981): Respiratory sinus arrhythmia in humans: how breathing pattern modulates heart rate. Am. J. Physiol. 241, H620-H629
11. KATONA, P.G. & J.I.H. FELIX (1975): Respiratory sinus arrhythmia: noninvasive measure of parasympathetic cardiac control. J. Appl. Physiol. 39, 801-805
12. KOLLAI, M. & K. KOIZUMI (1981): Cardiovascular reflexes and interrelationships between sympathetic and parasympathetic activity. J. Auto. Nerv. Syst. 4, 135-148

ON SEMILUNAR CHANGES IN CARDIOVASCULAR MORTALITY

J. SITAR

Department of Internal Medicine, Policlinic, 66436 Kurim u Brna
Czechoslovakia

INTRODUCTION

During recent decades the concept of our living environment has also extended into space. Although some pathological conditions have a complicated genesis, it may be possible to find certain cosmic influences which have a practical importance.

The question was asked, whether any correlation exists between the position of the moon and sudden cardiovascular mortality.

METHODS

During the years 1975 - 1983 the frequency of sudden cardiovascular mortality in the industrial town of Brno, Czechoslovakia, was followed. Since all of the deceased died suddenly, none of them could be delivered to a hospital in time to receive medical treatment adequate for an acute emergency condition. In all cases an autopsy was carried out and the diagnoses of cardiovascular disorder were confirmed. The number of cases occurring during these nine years was 1437.

In our research diagnoses of cardiac or cerebral ischaemia and infarction of atherosclerotic, thrombotic or embolic origin and even cases with pulmonary embolism were included; hemorrhages of the central nervous system were not.

We arranged the mortality data by means of transferring of epochs, having regard for the various time differences between successive moon phases. These time differences are caused by the eccentricity of the lunar orbit and the different velocity of the moon's revolution, according to KEPLER's laws.

RESULTS

The results can be seen in Fig. 1. There are two maxima and two minima of sudden cardiovascular mortality during lunar phases in the data accumulated over a period of all nine years. Mortality maxima are shortly before both moon quarters; the first minimum is one day before the new moon. The differences between maximal and minimal mean mortality frequencies are statistically highly significant. The mean percentual difference is 20.5 %.

We subsequently divided all cases into three separate groups according to different periods of solar activity, characterized by the occurrence of sunspots. That

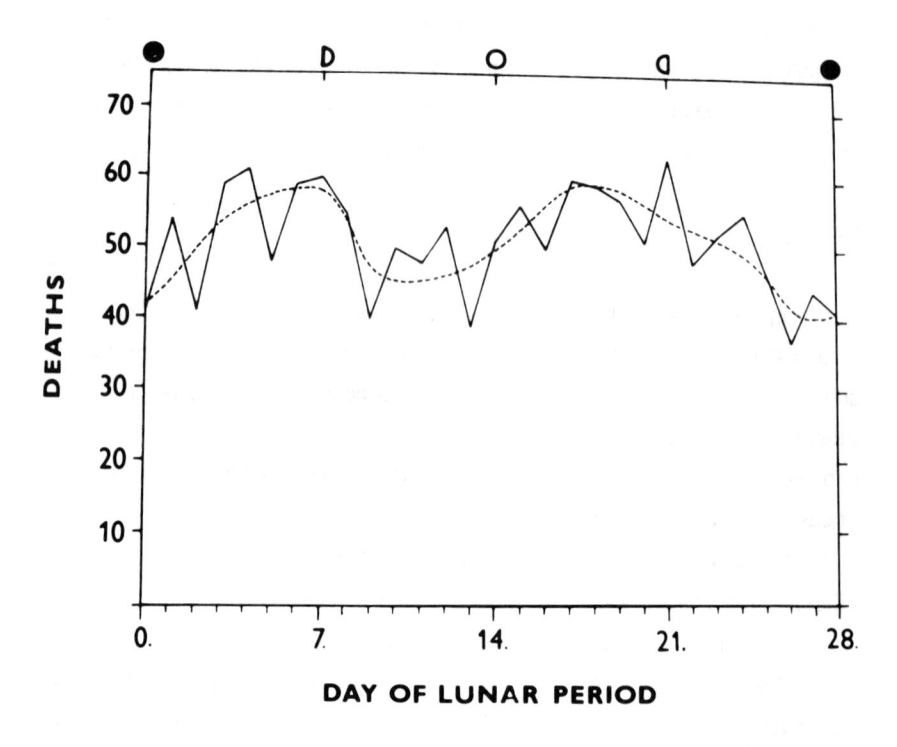

Fig.1: Two maxima and two minima of sudden cardiovascular mortality in the course of lunar phases. Follow-up study of 1437 fatal cases during nine years 1975-1983.

way we could see changes in the cardiovascular mortality in the years of maximal, intermediate, and minimal solar activity (see Fig. 2). In all these three graphs we see two maxima and two minima in sudden cardiovascular mortality as well.

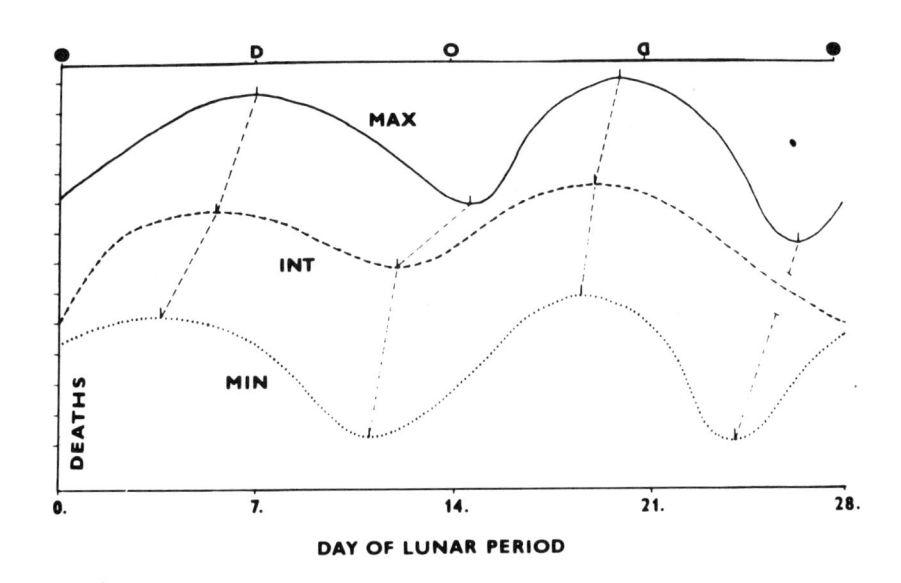

Fig.2: All cases of sudden cardiovascular deaths, divided into three separate groups, according to maximal (MAX), intermediate (INT) and minimal (MIN) solar activity (after the occurrence of sunspots). The phase shift of all excursions over time is evident.

Moreover, we can see the phase shift of all excursions over time, so that during high solar activity the maxima and minima of mortality related closely to the lunar phases, while during periods of intermediate and minimum solar activity mortality curves shift to reach their maxima and minima from 1.5 to 3.5 days earlier than in years of maximum solar activity.

DISCUSSION

In our opinion, among the causes of semilunar change of mortality may be the combination of lunar and solar gravitation forces, which vary in the same semilunar period and may exert changes in the various layers of the earth's atmosphere and even in the ionosphere, similar to sudden tides in the ocean. But, the phase shift of mortality curves according to the period of solar activity indicates that the influence of solar corpuscular radiation may also be involved. This radiation, especially its velocity, causes geomagnetic disturbances (1,2,4), corresponding to our cardiovascular mortality maxima.

It is well known that after great solar flares, with intensive corpuscular radiation there is a higher frequency of geomagnetic storms and of aurorae. More cardiovascular disorders are also observed. If the correlation described above really does exists, then close to lunar quarters aurorae would appear more frequently as well. In our own study (3) we have proved that in lunar periodicity aurorae really occur more frequently close to lunar quarters. Moreover, close to the last quarter of the moon aurorae are relatively more frequent than close to the first quarter. The mean increase of the frequency of aurorae close to lunar quarters is +27.7 % and is statistically highly significant. It corresponds well to the mortality increase in these periods, which according to our investigation is +20.5%.

It is therefore very likely that the increased cardiovascular mortality rate close to lunar quarters correlates with increased geomagnetic activity. The relationship is certainly not direct and we must search for the specific causes which exert an unfavorable effect on our cardiovascular system.

REFERENCES

1. BIGG, E.K. (1963): The influence of the moon on geomagnetic disturbances. J. Geophys. Res. 68 (5), 1409-1413
2. RÉTHLY, A. & Z. BERKES (1963): Nordlichtbeobachtungen in Ungarn (1523-1960). Budapest, pp. 145-146
3. SITAR, J. (1990): K příčinám lunárich změn kardiovaskulární úmrtnosti. Cas. lék. čes. 129 (45), 1425-1430
4. STOLOV, H.L. & A.G.W. CAMERON (1964): Variations of geomagnetic activity with lunar phase. J. Geophys. Res. 69 (23), 4976-4982

… 320

DO CHANGES OF MICROVASCULAR BLOOD FLOW OF NASAL MUCOSA PLAY A ROLE IN OCCURENCE OF THE LATERALITY RHYTHM OF NASAL BREATHING?

M. TAFIL-KLAWE* & G. HILDEBRANDT

Institut für Arbeitsphysiologie u.Rehabilitationsforschung,
Philipps Universität Marburg, D-3550 Marburg/Lahn, Robert-Koch-Str.7a, FRG,
*Dept.of Physiology, Medical.Academy. 00-927 Warsaw, Krak.
Przedmiescie. 26/28, Poland.

INTRODUCTION

Rhythmicity is an ubiquitous biological phenomenon. Already old hindu "mystery sciences" showed a laterality of the nasal thoroughfare and its ultradian rhythmical variations (10). Later on, in many human subjects and experimental animals was observed, that the airflow resistance regularly shifts from one side of the nose to the other. As the resistance on the one side increases, the resistance on the other side decreases proportionately. After 3-4 h this inequality reverses (3,4,5,8,12,16). Blood vessels, in the turbinates especially, are arranged in such a manner as to provide an erectile capacity (corpora cavernosa), whereas the degree of nasal vascular congestion is primarily related to autonomic control (14). Many authors suggested, that rhythmical changes of capillary, arterial, and venous congestion and of flow in the turbinates are responsible for the functional laterality of nasal resistance (1,7,11). The aim of the present study was to show, how the microcirculatory vascular bed of nasal mucosa does participate in occurence of the laterality rhythm of nasal breathing.

METHODS

7 healthy volunteers, aging 25-36 years, were included in this study.

Measurement and calculation of pure nasal resistance:

Every 2h during 24h the airway resistance was measured using the oscillatory method (Siregnost FD5) (8) during oral (OR) and nasal breathing, separately for the right and left side of the nose (RR, LR respectively). Pure nasal resistance was calculated also separately for the both nose sides: nRR for the right, nLR for the left side, from

eqation: nRR (nLR) = RR (LR) - OR (16).

Measurement and calculation of the nasal microcirculatory mucosal blood flow:

The Laser Doppler Flowmeter (PERIFLUX, Perimed, Sweden) was used to measure microcirculatory blood flow in mucosa of the anterior part of the inferior turbinate (2,13). The recordings were performed with a probe PF 103. In order to achieve the reproducibility of measurements, performed also every 2h, the probe was fixed always in the same position using a nasal stamp, prepared for each subject individually. The results of microcirculatory blood flow measurements were expressed in arbitrary units (perfusion units).

RESULTS

1. Two pulsatile fluctuations of the blood flow were observed in the right- und left-side of the nasal mucosa (Fig.1): a slow one (0.8-1.2 min^{-1}) and a second in the range of the cardiac cycle.
2. Variations of nRR and nLR showed an ultradian rhythm, in the course of which a maximum/minimum of resistance occured in an average every 4h for each nose side (16).
3. The ultradian variations of microcirculatory blood flow in the nasal mucosa (separately for each side of nose), paralleled the ultradian variations of nRR and nLR (Fig.1).

A positive significant correlation between pure nasal resistance and nasal mucosa microcirculation was found separately for the right and left nose side (Fig.2).

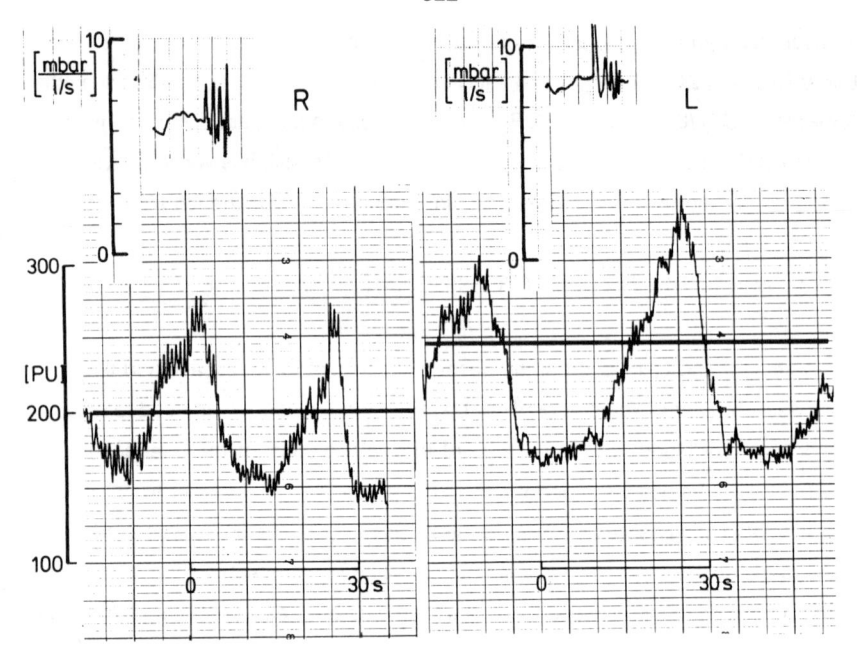

Fig.1: Recordings of the nasal mucosa blood flow, together with right/left-sided separate measurements of the nasal resistance.

DISCUSSION

The structure and function of the cardiovascular system in human subjects have been studied by several techniques during the last century, but most data showed a detailed picture of the hemodynamics of blood flow in the major blood vessels. The Laser Doppler Flowmeter is a new und non-invasive method for measuring microcirculation. We have used the Laser Doppler in order to measure microcirculation in the nasal mucosa of inferior turbinate. The effective penetration depth of the laser photon for nasal mucosa is not known (13). Like OLSSON (13), we presume that our results showed changes in superficial blood flow in the nasal mucosa such as in skin (6). The output signal from the flowmeter corresponds to the integrated motion of red cells in

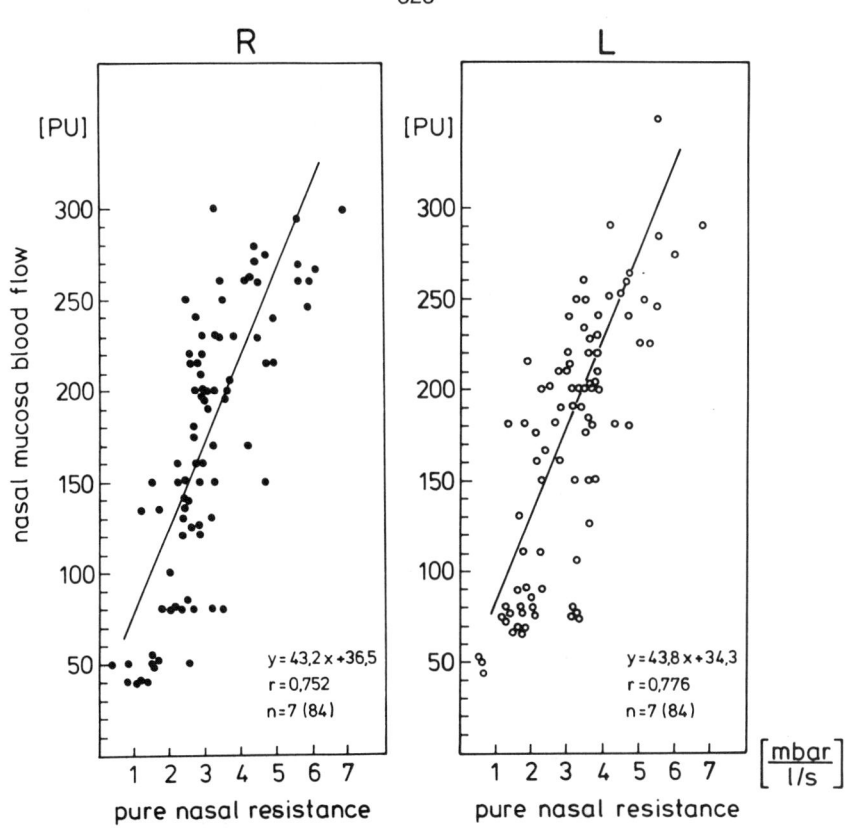

Fig.2: Linear relationship between the nasal mucosa microcirculation and pure nasal resistance calculated for the right/left side of the nose separately.

the measuring volume. Therefore the contributions to this signal from small arterioles, capillaries, arteriovenous anastomoses and venules cannot easily be separated. Rhythmic oscillations of the blood flow, observed in the present study, could have been due to spontaneous variations in sympathetic nerve activity (13,15). Our results indicate that the rhythmic ultradian changes of microcirculatory blood flow in the nasal mucosa might be one of the factors, producing ultradian fluctuations of nasal resistance and responsible for the nasal functional laterality.

REFERENCES

1. BEICKERT,P. (1951): Halbseitenrhythmus und vegetative Innervation. Arch.Ohr.-Nas.-Kehlk.Heil. 157, 404-411
2. DRUCE,H., C.PATOW, R.BONNER et al.(1983): Nasal blood flow responses to nasal provocation. J.Allergy Clin.Immun., 71 :80 (Abstract)
3. ECCLES,R. (1977): Proceedings: cyclic changes in human nasal resistance to air flow (Abstr.) J.Physiol. (London), 272, 75P-76P
4. ECCLES,R. (1978): The central rhythm of the nasal cycle. Acta Oto-Laryngol., 86, 464-468
5. ECCLES,R. & R.L.MAYNARD (1975): Proceedings: studies on the nasal cycle in the immobilized pig (Abstr) J.Physiol.London, 247
6. ENGELHART,M. & J.K.KRISTENSEN (1983): Evaluation of cutaneous 113 blood flow response by Xenon washout and a laser-Doppler flowmeter. J.Invest. Dermatol., 80, 12-15
7. HEETDERKS,D.R. (1927): Observations on the reaction of normal nasal mucous membrane. Amer.J.Sci., 174, 231-244
8. JÄGER, R.I. (1970): Untersuchungen über den Seitigkeitsrhythmus der Nasenatmung. Med.Inaug.Dissert.Marburg/Lahn
9. KORN,P. v. (1971): Die Bestimmung des Atemwegswiderstandes mit oszillatorischen Meßbprinzip. In NOLTE & V.KORN (eds). Oszillatorische Messung des Atemwiderstandes. Feistle. München
10. KASYAPA, P.R.P. (1895): Die Wissenschaft des Atems. Leipzig: Wilhelm Friedrich
11. KAYSER,R. (1895): Die exacte Messung der Luftdurchgängigkeit der Nase. Arch.f.Laryng., 3, 101-120
12. KEUNING,J. (1968): On the nasal cycle. Rhinology Rhin., 6, 99-136
13. OLSSON,P., M.BENDE & P.OHLIN (1985): The Laser Doppler Flowmeter for Measuring Microcirculation in Human Nasal Mucosa. Acta Otol., 99, 133-9
14. PROCTOR,D.F. (1977): Nasal physiology and defence of the lungs Am.Rev.of Respir.Dis., 115, 97-129
15. TENLAND,T. (1982): On Laser Doppler Flowmetry. Linköping Studies and Technology Dissertat.No. 83.
16. THIELE,A.E. (1989): Tagesrhythmische Untersuchungen über den oralen und nasalen Atemwegswiderstand unter Berücksichtigung der Nasenseitigkeit Med.Inaug.Dissert.Marburg/Lahn

RESPIRATORY PATTERNS IN PATIENTS WITH SLEEP APNEA EVALUATED BY AUTOMATIC TECHNIQUES

T. PENZEL, U. MARX, J.H. PETER, H. SCHNEIDER & P. VON WICHERT

Medical Poliklinik of Philipps-University, Baldingerstr. 1
D-W-3550 Marburg, FRG

INTRODUCTION

Sleep related breathing disorders (SRBD) consist of many different breathing abnormalities. They can be grouped in disorders with sleep apnea and in hypoventilation disorders (1). Sleep apnea appears in three different forms which are central, mixed, and obstructive. The hypoventilation syndrome can be distinguished in primary and secondary hypoventilation. In patients aged between 20 and 65 years obstructive sleep apnea is the most prevalent finding. Obstructive sleep apnea is characterized by cessations of respiratory flow with increased respiratory effort due to upper airway obstruction. Central apneas with a lack of respiratory flow and without respiratory effort are very rare. Periods of hypoventilation are often found in patients with obstructive snoring. A quantitative evaluation of nocturnal breathing is necessary to assess a diagnosis which reflects the details of respiratory disorders being the basis of appropriate therapy. This evaluation is very time consuming if performed visually and therefore it is often reduced to a counting of the occurance of apneas only. We developed an automatic technique to analyze respiration which enabled us to keep information on all breaths for a better diagnosis. This new technique was applied to 17 patients with sleep apnea and eight healthy volunteers. Respiratory patterns were investigated in respect of their relation to sleep stages.

METHODS

All patients underwent polysomnography for two consecutive nights. In this study the second night was used for further analysis. To evaluate sleep stages, EEG, EOG, and EMG were recorded according to standard criteria of RECHTSCHAFFEN & KALES (7). ECG was recorded to determine heart rate. Respiratory flow was assessed by means of nasal thermistors (8) and respiratory effort was obtained by inductive plethysmography (Respitrace) for thoracic and abdominal effort separately. Oxygen saturation was assessed by pulse oximetry (Nellcor N-100). All signals were digitized by computer with a sampling rate of 64 Hz (Intertechnique IN 1200). The digital recordings of respiration and oxygen saturation were used to analyze disturbed respiration. Statistical analysis was done on an IBM compatible personal computer.

Analysis technique

A new method to analyze respiratory waveform was developed to reduce respiratory signals to their essentials (4). This method is based on earlier investigations by RASCHKE (5,6) and KORTEN & HADDAD (2). The analysis method uses the first derivation of the signal dy/dt to estimate preliminary points of inspiration and expiration. The two estimated points are the starting points for a search of the final location of inspiration and expiration using the original signal of respiration. Thus inspiratory time T_I, expiratory time T_E, and a relative respiratory volume V_T were determined for each breath detected in the signal. These values were stored for later processing together with the time at which the breath was found. The values are calculated three times independently for the three different respiratory signals recorded, which are nasal airflow, thoracic and abdominal respiratory effort.

Using the reduced data the intervals between successive inspirations and expirations were calculated. If the interval between two successive inspirations in the airflow signal exceeded 10 seconds this interval was defined as an apnea (Fig. 1). Distinction whether an obstructive or central apnea was found, depended on the respiratory effort signals. In the case there was no respiratory effort during the apnea, it was classified as central, otherwise as obstructive apnea. Using the three signals it is also possible to determine paradoxical breathing. Paradoxical breathing reflects upper airway obstruction. Partial upper airway obstruction often indicates a later development of sleep apnea.

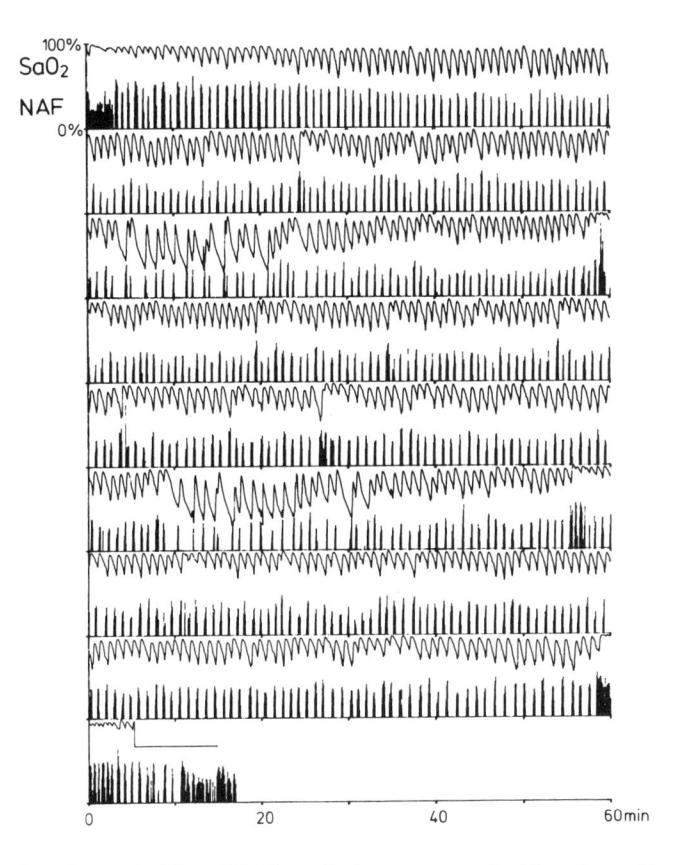

Fig.1: Results of nasal airflow (NAF) analysis are presented together with oxygen saturation (SaO_2) for an entire night recording (duration: 8 hours 17 minutes). Each horizontal line presents one hour of compressed data. Each vertical bar indicates a breath detected. The vertical length of each bar was drawn according to the relative volume of the breath. Thus narrow bars, as can be seen during the first three minutes and during the last seven minutes, reflect regular breathing. All breath to breath intervals which are longer than 10 seconds indicate an apnea. This patient had apneas during the entire recording. In the third hour and in the sixth hour irregular apnea duration with severe desaturations can be observed. Here two REM sleep episodes of 20 minutes duration were found.

Oxygen saturation is evaluated in terms of desaturation and duration of desaturation. Evaluation is performed strictly linked to each apnea detected. As the lowest value of oxygen saturation is reached a few seconds after the end of each apnea, the search for minimal saturation starts at the end of each apnea. A search backwards for maximal saturation looks for the starting point of the decline. Thus it is possible to determine for each apnea the concomittant desaturation and duration of desaturation (Fig. 1).

Subjects

Recordings of 17 patients were analyzed using the described method. Patients were selected on the basis of the results of an ambulatory apnea recording (3). The ambulatory recording had to indicate a clear presence of sleep apnea (more than 20 apneas per hour of sleep) without having other complaints or severe cardiovascular disorders. Mean age of the patients was 48 years (35-60), mean Broca index was 130 (104-163), and mean apnea index according to ambulatory recording was 43 apneas per hour of sleep (21-72). All patients complained of daytime sleepiness and loud snoring. They performed a lung function test and daytime blood gas values were determined. These values did not exceed normal limits.

RESULTS

A total of 6925 apneas were analyzed in the 17 patients selected. Most apneas were found in sleep stage 1 and 2 (76%). Only few apneas were found during slow wave sleep (stage 3 and 4) (2%). 17% of all apneas were found during REM sleep. The remaining apneas were found during stage "awake". Analysis of apnea duration proved that the longest apneas were found during REM sleep. The overall mean was 37.6 sec (individual means ranged from 22.0 to 67.7 sec). The shortest apneas were found during stage "awake" (mean 22.6 sec). Apneas during stage 3 and 4 (24.8 and 17.0 sec resp.) were also shorter than apneas during stage 1 and 2 (25.0 and 31.4 sec resp.).

Crosscorrelation between apnea duration and oxygen desaturation was carried out. The longer the apnea was the more saturation did fall. Correlation was highest

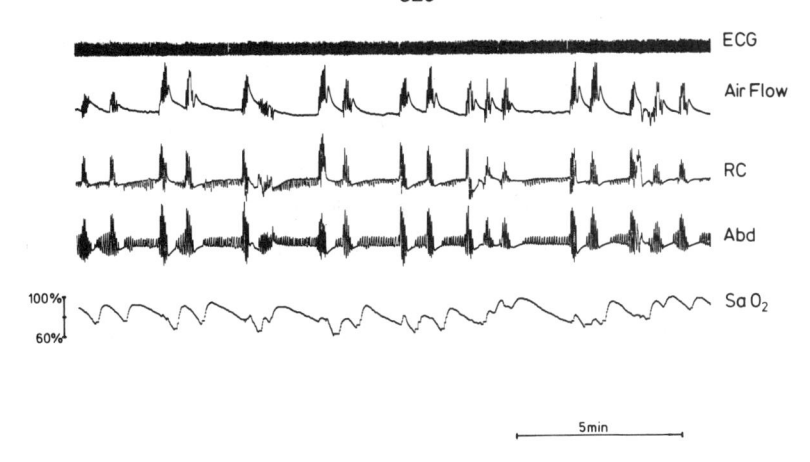

Fig.2: A sample recording of respiration and oxygen saturation during REM sleep shows the irregularity of apnea duration and desaturation. From top to bottom ECG, nasal airflow (NAF), thoracic respiratory effort (RC), abdominal respiratory effort (Abd), and oxygen saturation (SaO_2) were recorded.

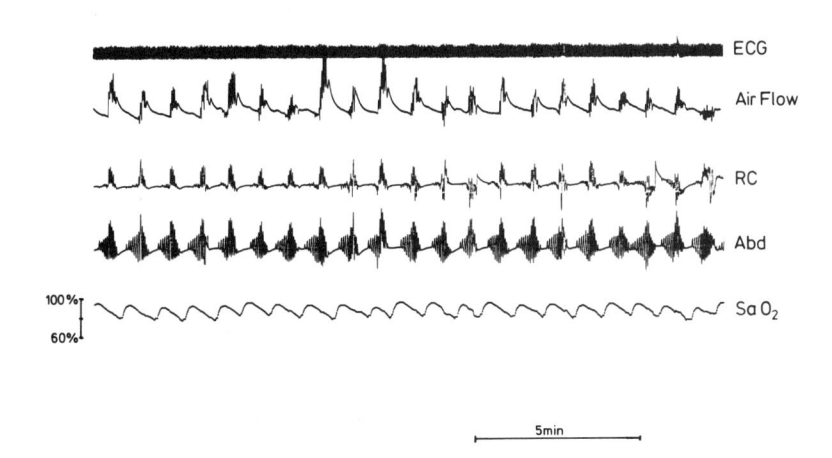

Fig.3: A sample recording of respiration and oxygen saturation of the same patient as shown in Fig. 2. Here the patient is in sleep stage 2 and has mixed apneas (each apnea starts without obstructive efforts) of very regular duration. The recorded channels are the same as in Fig. 2.

in REM sleep (r=0.60) and lowest in stages 3 and 4 (r=0.46 and r=0.21). Correlation in stages 1 and 2 was in between (r=0.57 and r=0.51).

We analyzed the speed of desaturation over time for all desaturations $dSaO_2/dt$. This speed did vary very much between individual patients. In all patients we found an increased speed of desaturation during REM sleep (overall mean: 0.64 %/sec) compared with light sleep stages (mean: 0.50 %/sec) and compared with apneas found during stage "awake" (mean: 0.38 %/sec). A control study performed with eight normal students which were asked to produce voluntary apneas when being awake, reached a speed of desaturation $dSaO_2/dt$ of 0.20 %/sec.

DISCUSSION

The new analysis proved to be able to recognize sleep apnea and to determine their characteristics in terms of duration and concomittant oxygen desaturation. The method also allows to distinguish between the different types of sleep apnea (i.e. central, mixed, and obstructive). Beside that the respiratory details obtained about all apneas allow to investigate the underlying mechanisms of sleep apnea. Evaluation of the relation between the stage of sleep and apnea parameters revealed that the regulation of respiration even in pathologies is different in three different states of sleep, which are light sleep (stage 1 and 2), slow wave sleep (stage 3 and 4), and REM sleep. It may be astonishing that a considerable number of apneas was found during stage awake. But this is a result of the rules for sleep staging which were set up by RECHTSCHAFFEN & KALES (7). The rules do not take account of drowsiness before sleep stage 1. In patients with severe sleep apnea the first apneas were already recorded during drowsiness.

The difference in respiratory regulation was also found in the difference of the relation between apnea duration and oxygen desaturation, in the speed of desaturation $dSaO_2/dt$, and in the regularity of apnea duration. The longest apneas and the severest oxygen desaturations were found during REM sleep (Fig. 2). During REM sleep apneas were not only the longest, but duration was also very irregular. Typically patients had a characteristic mean apnea duration which did not vary during non-REM sleep (Fig. 3). But during REM sleep, apneas occurred which were dramatically longer as well as shorter than the individual mean duration.

CONCLUSION

The computerized analysis of nocturnal respiration and oxygen saturation proved to be helpful in the diagnosis of sleep related breathing disorders. The investigation showed that it is necessary to record and analyze at least four respiration related signals which are airflow, thoracic and abdominal respiratory effort, and oxygen saturation. A detailed analysis with the help of computerized techniques may not only speed up routine clinical evaluation of polysomnograms but may also serve as a methodology to clarify interactions between sleep and respiration. We found that there are distinct differences of respiratory patterns during non-REM and REM sleep in patients with sleep related breathing disorders concerning the occurance of sleep apnea. This indicates that patients with sleep apnea are more endangered during REM sleep than during non-REM sleep. This has to be considered for therapy: during therapeutic nights periods of REM sleep must be supervised very carefully.

REFERENCES

1. ICSD INTERNATIONAL CLASSIFICATION OF SLEEP DISORDERS (1990): Diagnostic and Coding Manual. Diagnostic Classification Steering Committee, THORPY M.J., Chairman. Rochester, Minnesota: American Sleep Disorders Association
2. KORTEN J.B. & G.G. HADDAD (1989): Respiratory waveform pattern recognition using digital techniques. Comput. Biol. Med. 19, 207-217
3. PENZEL T., G. AMEND, K. MEINZER et al. (1990): MESAM: a Heart Rate and Snoring Recorder for Detection of Obstructive Sleep Apnea. Sleep 13, 175-182
4. PENZEL T., J.H. PETER, H. SCHNEIDER & P. VON WICHERT (1991): Computeranalyse der gestörten Atmung bei Patienten mit Schlafapnoe. Pneumologie 45, 213-216
5. RASCHKE, F. (1981): Die Kopplung zwischen Herzschlag und Atmung beim Menschen - Untersuchungen zur Frequenz- und Phasenkoordination mit neuen Verfahren der automatischen Analyse. Inaug. Diss. Fachbereich Medizin Univ. Marburg
6. RASCHKE, F., J. MAYER, T. PENZEL et al. (1987): Assessment of the time structure of sleep apnea. In: PETER, J.H., T. PODSZUS & P. VON WICHERT (eds.): Sleep Related Disorders and Internal Diseases. Springer, Berlin Heidelberg, pp. 135-139
7. RECHTSCHAFFEN, A. & A. KALES (1968): A Manual of Standardized Terminology, Techniques and Scoring System for Sleep Stages of Human Subjects. Publ. Health Service, U.S. Government Printing Office, Washington D.C.
8. STUTTE, K.H. (1967): Untersuchungen über die Phasenkopplung zwischen Herzschlag und Atmung beim Menschen. Inaug. Diss. Fachbereich Medizin, Univ. Marburg

CIRCADIAN REACTIONS TO NCPAP-TREATMENT

M. VOGEL[1], G. HILDEBRANDT[1], R. MOOG[1] & J.H. PETER[2]

[1]Institut für Arbeitsphysiologie und Rehabilitationsforschung,
Robert Koch Str. 7a, D-W-3550 Marburg, FRG
[2]Medizinische Poliklinik Marburg, Baldingerstr., D-W-3550 Marburg, FRG

INTRODUCTION

Obstructive Sleep Apnea (OSA) means cessation of breathing during sleep, mainly due to obstruction of the upper airway. The critical changes of blood gases as life-threatening consequence to apneas are corrected by arousal of the central nervous system, which is followed by an increase of the level of vigilance and causes disturbances of the sleep structure (2). This chain reaction leads to Excessive Daytime Sleepiness (EDS), one of the main symptoms of OSA (1). Although close connections between sleep/sleep disturbances and circadian system are well known, the circadian system did not get great attention in investigation and therapy of sleep apnea up to now.

The study described underneath investigated the circadian system by controlling the rhythms of vigilance and core temperature in patients with OSA before and with treatment with Nasal Continuous Positiv Airway Pressure (NCPAP).

METHODS

Eleven male patients (aged 38-69, mean 51.6 ±7.8 years) with OSA (mean 40 ±15 apneas per hour sleep, range 18-69) took part at a 24-hour lasting constant routine, which started at 15.00 h. During the examination, masking effects were minimized. To reduce activity, patients had to stay in bed. Every 3 hours a diet was given and patients answered the Thayer-list (7) for subjective vigilance rating. Although masking effects of sleep upon core temperature are discussed (5), sleep was allowed ad libitum, because the interactions between OSA and the resulting sleep fragmentation play an important role in this investigation. EEG, EOG, EMG and core temperature were measured continuously. The constant routine was repeated after 3 nights with nCPAP-therapy. The night before the constant routines, a polysomnography was performed.

RESULTS

NCPAP-therapy prevents apneas during sleep. As a consequence, sleep structure tends to normalize, as is demonstrated by the distribution of sleep stages during the twenty-four hours of the constant routine in Fig. 1 (vigilance was scored according to 6). Patients show an extreme duration of sleep before therapy - the average lies

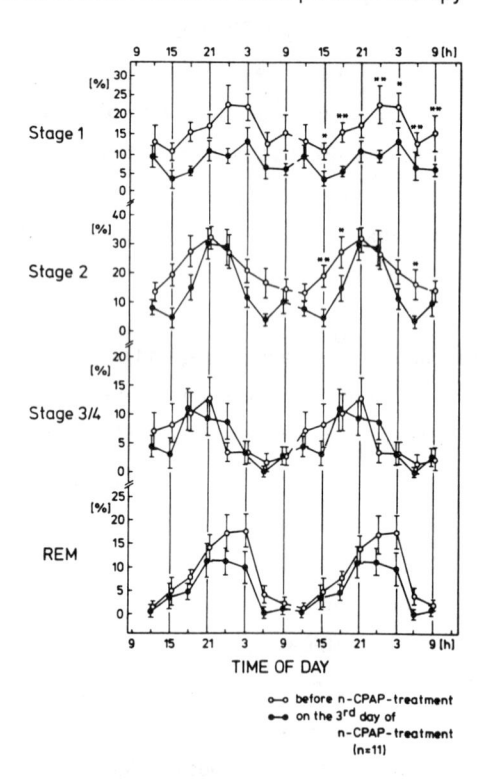

Fig.1: Distribution of sleep stages during the twenty-four hours of the constant routine. Mean over all patients ±standard error. (Wilcoxon test *: $p \leq 0.05$; **: $p \leq 0.01$)

above 12 hours sleep. This means a long sleep duration even under the relative monotonous conditions of the examination, which is reduced significantly with nCPAP-therapy. Over all subjects we recognize differences mainly in sleep stages I and II. Fig. 2 shows the daily distribution of sleep. Under treatment, sleep is better concentrated to night hours. There seems to be a rehabilitation of the of the before therapy. At night time the subjects exhibited the same duration of sleep stage 3 and 4 after therapy as they did before therapy over the hole day (compare Fig.1).

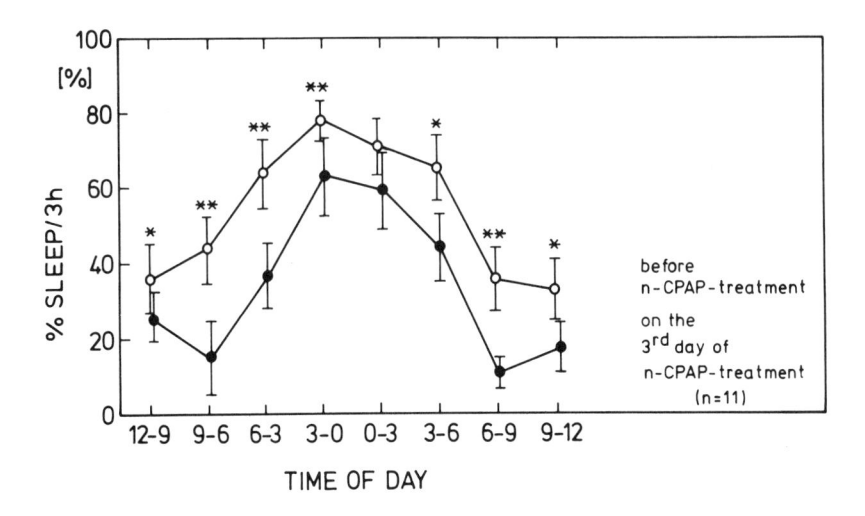

Fig.2: Circadian variation of sleep (mean over all patients ±standard error). The ordinate shows the percentual part of total sleep of the 3-hour intervals between the vigilance ratings.
(Wilcoxon test *: $p \leq 0.05$; **: $p \leq 0.01$)

The circadian variation of core temperature is shown in Fig. 3 (part a) in consideration of the individuell circadian phase, which was determined visually. Sleep fragmentation manifests itself by an unusual small daily variation of the amplitude (mean=0.64°) because of only poor decrease of temperature during the trophotropic phase. Already shorttime-application of nCPAP-therapy (3 nights) effected a significantly greater average temperature decline to the minimum with resulting higher daily variation of core temperature (mean=0.79°). In individual cases superpositions of the circadian rhythms with twelve-hour or shorter ultradian periods - which has been described for patients with sleep disturbances first by MENZEL (4)- were observed also. These ultradian rhythms are reduced under treatment.

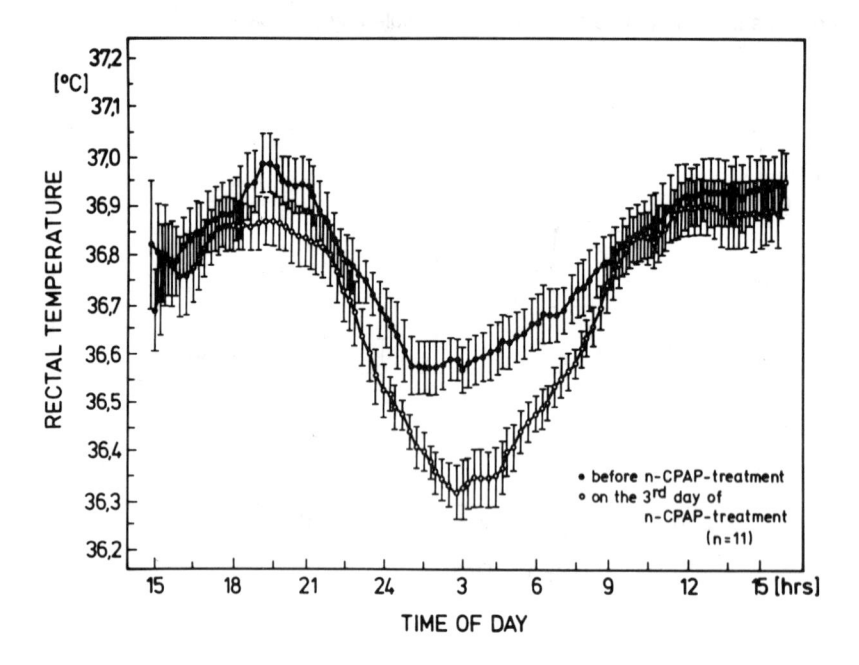

Fig.3: Circadian variation of core temperature as measured constantly during the 24 hour constant routine. Mean over all patients ±standard error. To balance individuell differences, the individuell circadian phase was determined visually, and core temperature is figured refering to the minimum, which -for the whole group- lies about 3 a.m. and therefore corresponds with the generally as standard taken value. The bars at the right indicate average temperature ±standard error of all patients over the whole day.

The subjective ratings of wakefulness (Fig. 4) showed before and with therapy the well-known circadian course. The minimum of vigilance corresponds to the temperature minimum. Therapy results in a significantly higher rating of vigilance over the whole day.

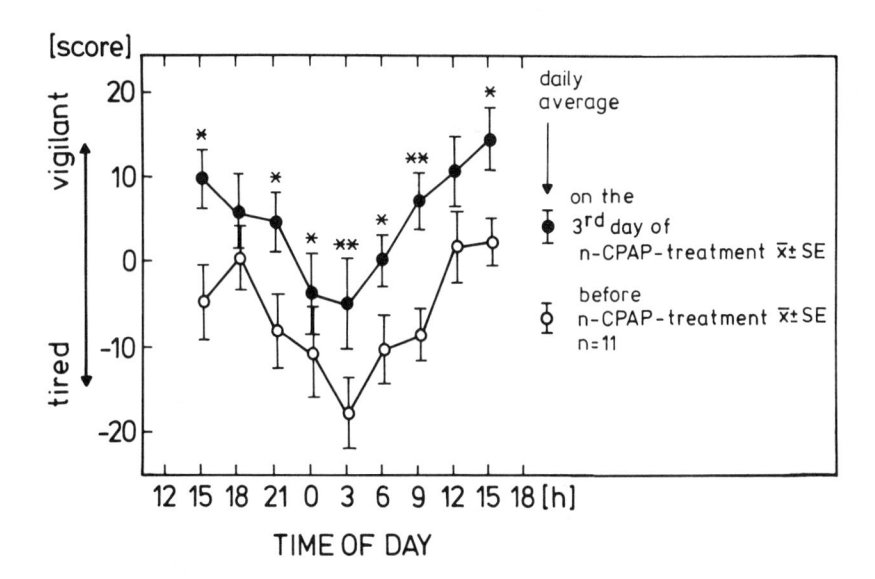

Fig.4: Subjective rating of wakefulness according to the evaluation of the Thayer-list (mean over all patients ±standard error, Wilcoxon test, *: $p \leq 0.05$; **: $p \leq 0.01$)

DISCUSSION

These results show for various variables that sleep apnea causes a disturbance of the circadian system, which particularly concerns the usual trophotropic phase, in which the organism cannot get sufficient recovery due to permanent sleep disturbances. The distribution of sleep stages during the constant routine confirm data of night-polysomnography (for example 3). NCPAP-therapy causes a significant decrease of the previously increased duration of total sleep, especially stages 1 and 2. The rhythm of core temperature -as a represantative of the circadian system- shows a noticeable amplitude-flattening due to only small temperature decrease during the trophotropic phase, compared to healthy people or with nCPAP-therapy. As a consequence of close reciprocal effects of sleep-/wake-behavior and core temperature within the circadian system this result had to been expected, because a high variability of circadian rhythms usually is expressed by a small daily amplitude (8).

This study was a first step to involve the circadian system closer in investigation and therapy of obstructive sleep apnea, and the results demonstrate it's importantant role. If these findings are exclusively an expression of a change of spontaneous behaviour - for which the reduction of ultradian rhythms may be a hint- or if they may be due to for example a demasking because of a reduction of the sleep deficit has to be shown by further evaluation and research.

REFERENCES

1. DEMENT, W.C., M.A. CARDASKON & G. RICHARDSON (1978):Excessiv Daytime Sleepiness in the Sleep Apnea Syndrome.In: C. GUILLMINAULT & W.C. DEMENT (eds.), Sleep Apnea Syndromes. AR Liss, New York, pp. 23-46
2. GUILLMINAULT, C., J. VAN DE HOED & M.M. MITLER (1978): Clinical overview of the Sleep Apnea Syndromes. In: C. GUILLMINAULT & W.C. DEMENT (eds.), Sleep Apnea Syndromes. AR Liss, New York, pp. 1-12
3. ISSA, F.G. & C.E. SULLIVAN (1985): The immediate effects of nasal continuous positive airpressure treatment on sleep pattern in patients with obstructive sleep apnea syndrome. Electoencephalography and clin. Neurophysiology 63: 10-17
4. MENZEL, W. (1955): Spontane Leistungsschwankungen im menschlichen Organismus. Verh. Dtsch. Ges. f. Arbeitsschutz 3, 232-240
5. MINORS, D.S. & J.M. WATERHOUSE (1989): Masking in humans: the problem and some attempts to solve it. Chronobiology International 6, 29-53
6. RECHTSCHAFFEN, A. & A. KALES (eds.) (1968): A manual of standardized terminology, techniques and scoring system for sleep stages of human subjects Publ. Health Service, US Government Printing Office, Washigton D.C.
7. THAYER, R.E. (1967): Measurement of activation through self-report. Psychological Reports, 20: 663-678
8. WEVER, R. A. (1985): Characteristics of circadian rhythms in human functions. In: R.J. WURTMAN & F. WALDHAUSER (eds.), Melatonin in humans.Center for Brain Sciences and Metabolism Charitable Trust, Cambridge, MA USA, pp. 291-352

CIRCADIAN RHYTHM OF ERYTHROCYTE SEDIMENTATION RATE, C-REACTIVE PROTEIN AND SOLUBLE INTERLEUKIN-2 RECEPTOR IN PATIENTS WITH ACTIVE RHEUMATOID ARTHRITIS

M. HEROLD[1] & R. GÜNTHER[2]

University of Innsbruck, [1]Dept. of Internal Medicine, [2]Dept. of Physical Medicine, Anichstrasse 35, A-6020 Innsbruck, Austria

INTRODUCTION

Rheumatoid arthritis is a chronic disease of unknown etiology with nonsuppurative inflammation of joints. The onset of disease may be at any age but most often occurs between 25 and 50 years. The course of the disease varies greatly and is characterized by striking tendency toward spontaneous remission and exacerbation. No specific diagnostic criteria are known, and diagnosis is based on clinical and laboratory findings which are found in RA with high incidence (1). One of the main diagnostic criteria is stiffness of joints in the morning that improves with progression of daytime.

Clinical symptoms of RA show significant circadian rhythms with maximum pain in the morning. Well measureable functions like grip strength or walking time show the best values according to a better feeling in the afternoon (3, 5, 8, 12, 13). Parallel to improvement of clinical symptoms mood also changes with daytime and shows best feeling in the afternoon (7).

We had described a significant circadian rhythm of CRP in patients with active RA and supposed that changes of clinical symptoms might be the consequence of a circadian rhythm of inflammation in joints as it is reflected by CRP-serum concentrations (6).

Autoimmune diseases like RA activate cellular immune mechanisms. T-cell activation is mainly regulated by interleukin-2 which exerts its growth promoting activities via a specific receptor. The interleukin-2 receptor is a 55-65 kD transmembrane glycoprotein located in large numbers on the surface of T-cells. As a consequence of T-cell activation, a smaller 10 kD protein is secreted into biological fluids as so called soluble interleukin-2 receptor (sIL-2R; 14).

We measured sIL-2R concentrations in patients with RA to find out whether this specific factor of T-cell activation also shows day time dependent changes as it was seen with CRP, an unspecific acute phase reactant.

MATERIAL AND METHODS

Patients: 19 men with rheumatoid arthritis were studied in two different groups. Blood was collected by venapuncture in the first group at the times 09.00, 12.00, 14.00, 18.00, and 22.00 and in the second group at 09.00, 14.00, 18.00, and 20.00. Erythrocyte sedimentation rate (ESR) was measured immediately after blood collection by Westergreen's method. Serum was collected, portioned, and frozen at -20 °C until determination of sIL-2R and CRP. We took care that samples of one patient were assayed within the same run to keep variation caused by chemical analysis as low as possible.

Reagents: Commercially available reagents were used. sIL-2R was measured using the enzyme immunoassay from T-Cell Science, Cambridge, MA. CRP was detected turbidimetrically after reaction with specific antibodies on a Cobas Mira analyzer. The reagents were prepared by Orion company from Espoo, Finland.

Statistical analysis: Circadian rhythms were calculated by Halberg's cosinor method (2). Statistical comparisons and correlations were performed with non-parametric tests using the StatView- software package for Apple MacIntosh computers. The Wilcoxon-Wilcox signed rank test and Spearman's rank correlation test were used.

Abbreviations: CL, confidence limits; CRP, C-reactive protein; ESR, erythrocyte sedimentation rate; n, number; p, probability of statistical error; phi, acrophase referred to local midnight, -360° equals 24 hours; RA, rheumatoid arthritis; SE, standard error of the mean; sIL-2R, soluble interleukin-2 receptor;

RESULTS

9 patients had elevated (> 500 U/ml) serum concentrations of sIL-2R and 8 patients showed CRP-concentrations higher than normal (> 8 mg/l). ESR was only detected in the second group. 8 of those 10 men with RA showed increased ESR (>15 mm after 1 hour). The concentration of sIL-2R was significantly correlated with CRP-concentration (p = 0.004) and with ESR (p = 0.066).

Population mean cosinor calculation with sIL-2R concentrations of all 19 patients revealed no statistical significant rhythm (n = 19, p = 0.42).

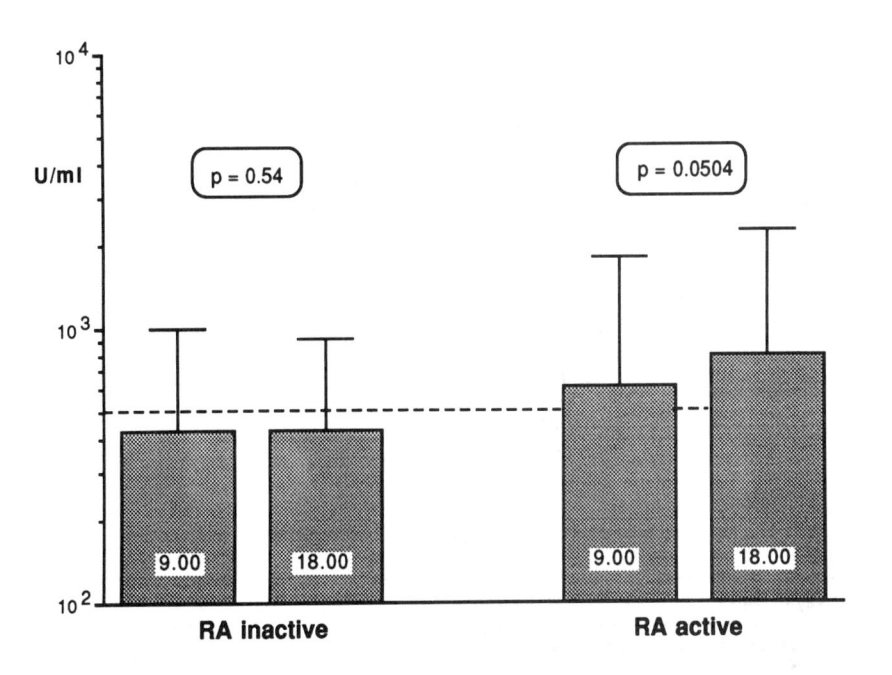

Fig.1: Serum concentartions (median values) of sIL-2R at 9.00 and 18.00 in 10 patients with inactive and 9 patients with active RA. The broken line shows the upper limit of the normal range. 8 out of 9 patients with active RA showed an increase of sIL-2R concentration

The patients were divided into two groups, 10 patients with mean sIL-2R concentrations in the normal range (< 500 U/ml) and 9 patients with elevated sIL-2R values. Elevated sIL-2R values correspond to increased disease activity. The concentrations of blood samples collected at the times 8.00 and 18.00, those two times when blood was sampled in both groups, were compared. Patients with inactive arthritis showed no difference of sIL-2R concentrations at 8.00 and 18.00, but a significant ($p = 0.05$) difference was found for patients with active arthritis. Concentrations at 18.00 were higher than at 8.00 (Fig. 1).

Population mean cosinor results are shown in Fig. 2. Significant circadian rhythms were detected in patients with active RA for sIL-2R, CRP and ESR. The acrophases of sIL-2R and of ESR were located in the afternoon, acrophases of CRP close to midnight.

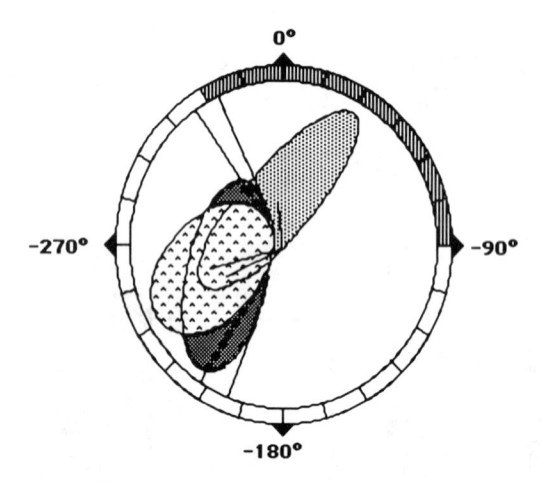

key to ellipses	n	p	PR	mesor ± SE	amplitude (95% CL)	acrophase (95% CL)
sIL-2R	9	0.0171	54	746 ± 80	59 (11 , 131)	-242 (-201, -335)
CRP	8	0.0538	78	43.6 ± 12.5	3.8	-353
ESR	8	0.0231	81	32.2 ± 6.1	4.9 (0.8, 9.3)	-254 (-214, -323)

Fig. 2: Results of population mean cosinor analysis of serum concentrations of sIL-2R, CRP, and ESR in patients with active RA.

DISCUSSION

Several studies confirmed the circadian variation of signs and symptoms in patients with RA. The reason is unknown but may be caused by a circadian rhythm of the immune system and its associated inflammatory responses. Also our findings with a circadian rhythm of ESR and the serum concentrations of sIL-2R and CRP indicate a rhythm of inflammation and its immune stimulation.

sIL-2R is secreted from activated T-cells. The circadian rhythm of sIL-2R may be the direct consequence of a circadian rhythm of circulating T-cell numbers or of T-cell functions. In patients with RA as well as in healthy people a circadian rhythm in the number of white blood cells, lymphocytes and lymphocyte subtypes has been documented (9, 10). Also reactions of lymphocytes after stimulation in vitro were seen to be time dependent.

We have also found a circadian rhythm of ESR, which is the most widely employed test to assess the degree of inflammation in disease. ESR reflects both the acute phase response in inflammatory and tissue-damaging disorders and the increased immunglobulin production after immune stimulation. It has been described that diurnal variation of ESR is related to feeding. 5 nonfasting patients with RA showed lowest ESR in the early afternoon but no circadian rhythm was found in 4 fasting patients (11). In contrast we found in our patients who ate daily at 8.00, 12.00, and 19.00 the highest ESR in the afternoon.

CRP is a fast reacting acute phase protein. Increases of CRP concentrations are seen within a few hours after tissue damaging. The biological half-life of serum CRP is about 8 hours. Changes of CRP serum concentrations may also be caused by the diurnal varying inflammatory stimulation.

Our results suggest that disease activity and inflammation vary with the time of the day in RA and may cause the circadian rhythm of the clinical symptoms.

REFERENCES

1. ARNETT, F.C., S.M. EDWORTHY, D.A. BLOCH et.al. (1988): The American Rheumatism Association 1987 revised criteria for the classification of rheumatoid arthritis. Arthritis Rheum. 3, 315-324
2. HALBERG, F., E.A. JOHNSON, W. NELSON & R. SOTHERN (1972): Autorhythmometry procedures for physiologic self-measurements and their analysis. Physiol. Teach 1, 1-11
3. HARKNESS, J.A.L., M.B. RICHTER, G.S. PANAYI et al. (1982): Circadian variation in disease activity in rheumatoid arthritis. Br. Med. J. 284, 51-554
4. HEROLD, M., R. GÜNTHER & F. HALBERG (1984): Synchronization of circadian rhythms in patients with rheumatoid arthritis by a three-day course of ACTH. E. HAUS & H.F. KABAT (eds.), Chronobiology 1982-1983, S. Karger, New York, ii, 500-504
5. HEROLD, M. & R. GÜNTHER (1984): Chronobiologische Untersuchungen während Radon-Balneotherapie . Z Phys Med Baln Med Klim (Sonderheft 1) 13, 68-76
6. HEROLD, M. & R. GÜNTHER (1987): Circadian rhythm of C-reactive protein (CRP) in patients with rheumatoid arthritis (RA). In: E.J. PAULY, L.E. SCHEVING (eds.), Advances in Chronobiology, Part B. Alan R. Liss. Inc. New York, ii, 271-279
7. HEROLD, M., V. GÜNTHER & R. GÜNTHER (1990): Circadian Changes of Mood and Disease Activity in Patients with Rheumatoid Arthritis. J. interdiscipl. Cycle Res. 21, 203-205
8. KIRKHAM, B.W. & G.S. PANAYI (1989): Diurnal periodicity of cortisol secretion, immune reactivity, and disease activity in rheumatoid arthritis: implications for steroid treatment. Br. J. Rheumatol. 28, 154-157
9. KNAPP, M.S. & R. POWNALL (1984): Lymphocytes are rhythmic: Is this important?. Br. Med. J. 289, 1328-1330
10. LEVI, F., Ch. CANON, J. BLUM et al. (1983): Large amplitude circadian rhythm in helper: suppressor ratio of peripheral blood lymphocytes. Lancet, ii, 462-463
11. MALLYA, R.K., H. BERRY, B.E.W. MACE et al. (1982): Diurnal variation of erythrocyte sedimentation rate related to feeding. Lancet, ii, 389-390
12. PETERSEN, I., G. BAATRUP, I. BRANDSLUND et al. (1986): Circadian and diurnal variation of circulating immune complexes, complement-mediated solubilization, and the complement split product C3d in rheumatoid arthritis. Scand. J. Rheumatol. 15 113-118
13. ROBERTSON, J.C., M.G. HELLIWELL, E.G. CANTRELL et al. (1982): Circadian variation in disease activity in rheumatoid arthritis. Br. Med. J. 284, 1114-1115
14. RUBIN, L.A., C.C. KURMAN, M.E. FRITZ et al. (1985): Soluble interleukin-2 receptors are released from activated human lymphoid cells in vitro. J. Immunol. 135, 3172-3177

CIRCADIAN VARIATIONS OF CELLULAR IMMUNE PARAMETERS IN HEALTHY SUBJECTS

CHR. GUTENBRUNNER & E. HEINDRICHS

Institut für Kurmedizinische Forschung Bad Wildungen, Langemarckstr.2, D-W 3590 Bad Wildungen, FRG

INTRODUCTION

The circadian variations of physiological functions not only concern resting conditions but especially the reagibility to different stressors. This is e.g. well known in the fields of thermoregulation, water and electrolyte balance, physical training and others (cf. 7). The resistance against infections depends on the activity of the immune system, the activity of which underlies circadian variations (3, 5, 8, 9; and others, see also 4). In the last few years several investigations on the circadian rhythmicity of different immune parameters were carried out, however, up to now, no studies were made under exclusion of masking effects.

MATRIAL AND METHODS

In the present study, 13 healthy male subjects, aged 20-33 years, were investigated under 24-hour resting conditions without any external light stimulation. Blood samples were taken in 4-hour intervals around the 24-h day. In these samples the numbers of leucocytes, granulocytes and total lymphocytes were counted. Using immunofluorescent techniques the numbers of B-lymphocytes, T-lymphocytes as well as the helper- and supressor-cells were determined. For the differentiation of the T-cell subsets the T-cell markers T3, T4, and T8 were used. The phagocyte function of leucocytes was measured with the Nitroblue Tetrazolium Reduction test. This test was carried out under basic conditions and after in vitro stimulation. The concentrations of plasma immunoglobulins (IgA, IgM, and IgG) were determinded by means of nephelometry.

RESULTS

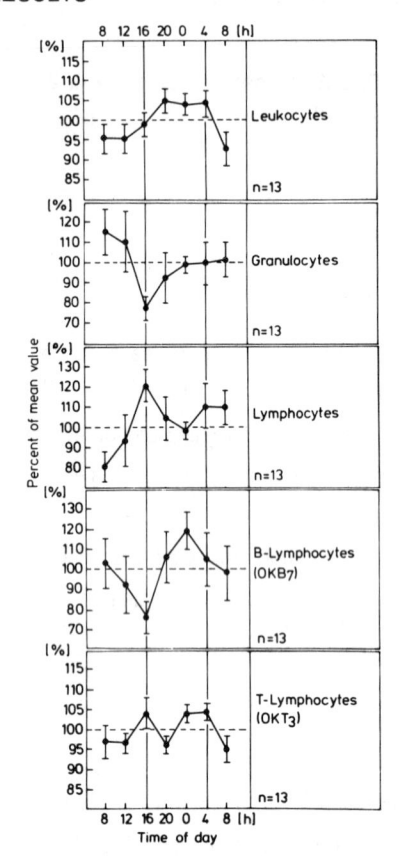

Fig. 1:
Mean circadian courses of the number of leucocytes, granulocytes, lymphocytes as well as the B- and T-lymphocytes in peripheral blood of healthy subjects under constant resting conditions and equally distributed food and fluid intake. Brackets indicate standard errors.

In Fig. 1 the number of circulating leucocytes, granulocytes, and lymphocytes are shown throughout the 24-hour period. The number of leucocytes exhibits a clear circadian rhythm with maximum values in the night. The acrophase is located around midnight. The number of granulocytes and lymphocytes showed inverse circadian variations with a relative lymphocytosis during daytime and granulocytosis at night. However, these curves show a decreasing tendency of granulocytes and an increasing trend of lymphocytes in the course of the 24 hour resting period. The mean number of B-cells exhibit a synchronous circadian variation with an acrophase at 4.00 h and a significant amplitude of about 40%. The course of the number of total T-cells is less clear and the amplitude is smaller as compared to the B-cells. Two maxima occur indicating a frequency multiplication.

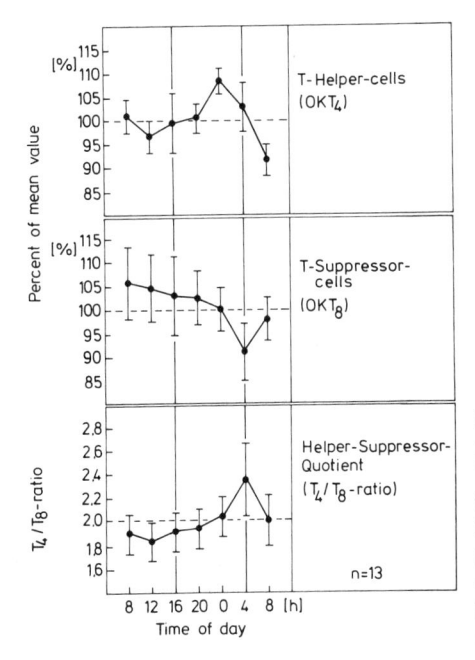

Fig. 2: Mean circadian courses of the number of the T-helper (OKT_4) and T-suppressor (OKT_8) cells as well as the T_4/T_8-ratio in peripheral blood of healthy subjects under constant resting conditions and equally distributed food and fluid intake. Brackets indicate standard errors (from 4).

However, the circadian variation of T-cells is caused mainly by the variability of T-helper cells (Fig. 2). The 24-hour follow-up of this parameter obviousely shows maximum values around midnight. However the mean number of T-supressor cells exhibits a nightly minimum. Therefore, the helper-supressor ratio reaches maximum values during nighttime.

The in vitro phagocyte activity of the leucocytes did not exhibit a clear circadian variation, however, the non-stimulated phagocytosis was sligthly smaller at midnight. These variations were not significant. Looking at the individual curves of phagocytic activity, two types of circadian variability seem to exist: about half of the test persons exhibited minimum values around 12.00 h, the others around midnight. However, this observation must be subject to further investigations.

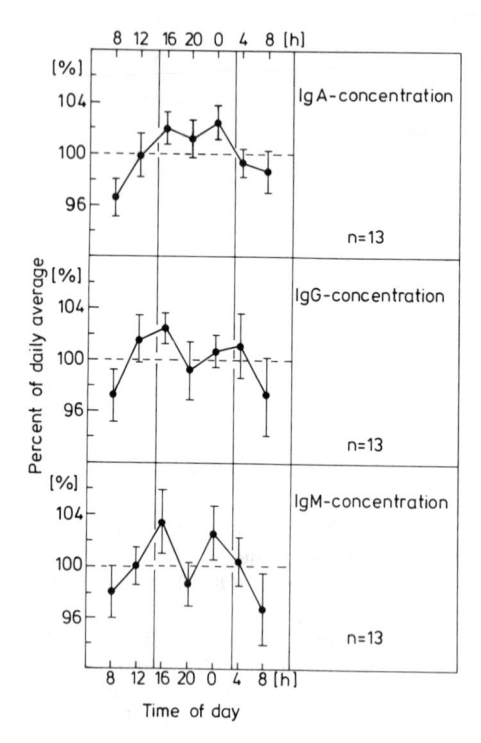

Fig. 3:
Mean circadian courses of the plasma concentrations of the immunoglobulins A, G, and M of healthy subjects under constant resting conditions and equally distributed food and fluid intake. Brackets indicate standard errors.

In Fig. 3 the mean plasma concentrations of the immunoglobulines A, G, and M are shown. In all cases, the mean immunoglobulin concentrations exhibit a circadian variability with increased values in the afternoon. Minimum values were found in the early morning. In addition the mean concentrations of IgG and IgM are slightly lowered in the 20.00-hour sample, indicating a superimposed 12-hour periodicity. All in all, the plasma immunoglobuline concentrations have their maximum values at day time.

DISCUSSION

Our results show, that the number of immune cells in peripheral blood undergo circadian variations even under constant resting conditions. Therefore these rhythms can be considered to be endogenous circadian rhythms. Different types of cells show different acrophases: The numbers of total leucocytes, granulocytes as well as total B- and T-cells and T-helper cells are higher in the night whereas lymphocytes and T-suppressor cells exhibit maximum values during daytime. These results coincide with other investigations (1, 6, 8, 9, 10, 11, and others). However, in the cited studies, masking effects were not excluded. An influence of the water balance of the organism on the described variations can be excluded, because of the fact that a decrease of cell numbers during nighttime caused by the nightly hydraemia had to be expected. All in all, our results can be interpreted as a maximum of proliferation of immune cells during the trophophase of the circadian system. However, the rhythms of immune cells in peripheral blood may alternate with the number of immune cells in the tissues.

The serum concentrations of immunoglobulines exhibited maximum values in the late afternoon. HALBERG et al. (5) found acrophases of these parameters between 11.00 and 14.00 hours. Therefore an alternation of proliferation and metabolic activity of immunologic cells can be supposed. This seems to be likely because the DNA-synthesis of circulating lymphocytes have maximum values during forenoon (2). Corresponding to the second circadian peak of the IgG and IgM concentrations another circadian maximum of DNA-synthesis was found in the evening. Comparable findings proving with an alternation of proliferation and metabolic activity are reported from other tissues, e.g. the mucosa of the intestine (12).

REFERENCES

1. BERTOUCH, J.V., P.J. ROBERT-THOMSON & J. BRADLEY (1983): Diurnal variation of lymphocyte subsets identified by monoclonal antibodies. Brit. med. J. 286: 1171-1172
2. CARTER, J.B., G.D. BARR, A.S. LEVIN et al. (1975): Standardization of tissue culture conditions for spontaneous thymidine-2 ^{14}C incorporation by unstimulated normal human peripheral lymphocytes: circadian rhythm of DNA synthesis. J. Allergy Clin. Immunol. 56: 191- 205
3. COVE-SMITH, M.S., P.A. KABLER, R. POWNALL & M.S. KNAPP (1987): Circadian variation in an immune response in man. Brit. med. J. II: 253
4. GUTENBRUNNER, Chr. & B. FOROUTAN (1990): Tagesrhythmische Schwankungen im Immunsystem. Z. Phys. Med. 19: 339-345.
5. HALBERG, F., D. DUFFERT & H. v. MAYERSBACH (1977): Circadian rhythm in serum immunoglobulins of clinically healthy young men. Chronobiologia 4: 114.
6. HAUS, E., D.J. LAKATUA, J. SWOYER & L. SACKETT-LUNDEEN (1983): Chronobiology in hematology and immunology. Am. J. Anat. 186: 467-517
7. HILDEBRANDT, G. (1985):Therapeutische Physiologie, Grundlagen der Kurortbehandlung. In: W. AMELUNG & G. HILDEBRANDT (Hrsg.): Balneologie und medizinische Klimatologie. Band 1. Springer-Verlag, Berlin-Heidelberg-New York-Tokyo
8. LÉVI, F., C. CANON, J.P. BLUM et al. (1985): Large amplitude circadian rhythm in helper:suppressor ratio of peripheral blood lymphocytes. Lancet II: 462-463
9. LÉVI, F., C. CANON, J.P. BLUM et al. (1985): Circadian and/or circahemidian rhythms in nine lymphocyte-related variables from peripheral blood in healthy subjects. J. Immunol. 134: 217-225
10. LÉVI, F., A. REINBERG & C. CANON (1989): Clinical immunology and allergy. In: J. ARENDT, D.S. MINORS & J.M. WATERHOUSE (Eds.): Biological Rhythms in Clinical Practice. Wright, London-Boston-Singapore-Sydney-Toronto-Wellington, pp. 99-135.
11. RITCHIE, W.S., I. OSWALD, H.S. MICKLEM et al. (1983): Circadian variation of lymphocyte subpopulations: a study with monoclonal antibodies. Brit. med. J. 268: 1773-1775
12. SCHEVING, L.E., J.E. PAULY, T.H. TSAI & L.A. SCHEVING (1983): Chronobiology of cell proliferation, implications for cancer chemotherapy. In: A. REINBERG & M.H. SMOLENSKY (Eds.): Biological rhythms and medicine. Cellular, metabolic, physiopathologic and pharmacologic aspects. Springer-Verlag, New York-Berlin- Heidelberg-Tokyo, pp. 79-130.

CIRCADIAN AND CIRCASEPTAN VARIATIONS IN TUBERCULIN SKIN REACTION

L. PÖLLMANN

Institut für Arbeitsphysiologie und Rehabilitationsforschung
der Philipps-Universität, Marburg, Robert-Koch-Str. 7 a, D-W-3550 Marburg, FRG

INTRODUCTION

In occupational medicine, the tuberculin test is necessary for all employees working in infections diseases wards. The tuberculin test should indicate the susceptibility of their personnel to tuberculosis. It is recommended to read the result qualitatively and quantitatively (5).

In previous studies it was found, that the skin reaction to tuberculin is a function of application time. The reaction was reported to be maximal during the night hours and minimal during the day-time on day 4 after application (2, 3). In contrary to these findings, we found maximal skin reactions in the morning seven days after application (11).

A reason for these difference could be a reactive circaseptan process modulating the respective circadian courses (cf. 7), another cause for differences between the tests in use.

METHODS

Therefore, we controlled the skin reaction of two different tuberculin tests over more than 21 days every morning at 8.00 h in occupational practice, and measured the reaction to the below listed tests on day seven at six equally spaced times of day.

- TUBERGEN ^{R} (application of tuberculin by plastic pins; n=47; manufactured by Behring, D-3550 Marburg),

- MERIEUX ^{R} (application of tuberculin by a row of plastic pins; n=7; manufactured by Institut MERIEUX, D-6906 Leimen),

- MENDEL-MANTOUX [R] (intradermal injection; n=4; manufactured by Behring, D-3550 Marburg),
- HEAF [R] (intradermal "multiple puncture gun" injection; n=4)
- TINE [R] (application of tuberculin by four rough and sharp metal pins; n=12; manufactured by Lederle, D-8190 Wolfratshausen; for details see 5)

RESULTS

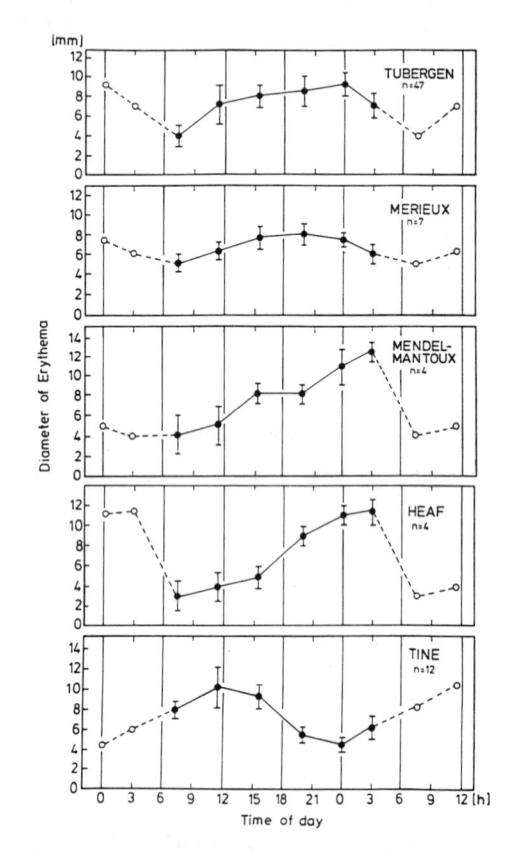

Fig.1: Diameter of the erythema at day 7 after application of tuberculin tests: TUBERGEN [R], n=17; MERIEUX [R], n=7; MENDEL-MANTOUX [R], n=4; HEAF [R], n=4; TINE [R], n=12. The brackets indicate standard errors.

Fig. 1 shows in general the size of the area of skin alteration, registered around day 7 after application. TUBERGEN [R] and MERIEUX [R] show a peak of expansion in the late evening and minimum in the morning (upper chart of Fig. 1). MENDEL-MANTOUX [R] and HEAF [R] test represent the circadian maximum in the morning (around 12.00) (lower part).

It may thus be assumed, that the extent of the tuberculin skin test depends on the size and depth of the skin wound, because the wound after TINE [R] is more profound than after TUBERGEN [R] and MERIEUX [R]. Moreover, the time of day of the alteration of the skin influences the result; the different layers of the skin react differently according to the time of day of application.

The follow-up of the skin alterations over more than 21 days (TUBERGEN [R], n=47; TINE [R], n=12) demonstrates differences of the skin reaction with maximal swelling around day 5 - 9 (Fig. 2). The alterations reveal the mode of a circaseptan reaction (6,7).

DISCUSSION

The circadian variations of the results, obtained with HEAF [R] or MENDEL-MANTOUX [R] test coincide with those of COVE-SMITH et al. (2, 3) with an intradermal injection of purified protein derivative of tuberculin (PPD) to a standard depth (c.f. 1, 9). The observed circadian variation coincides with the circadian variation of the histamin susceptibily of skin (4, 10, 12). It may be assumed that different mechanisms could be involved in the results of the diverse tuberculin tests.

Nevertheless, the highest reaction, measured by means of the largest erythema, is found around day 7 and 8 for all the systems tested. The best time of reading is not the generally recommended time of 72 hours after administration, but some days later around day 7 or 8. For practical medicine, this result is of great importance, because about 15 % of the employees did not react at the 72 hours readings, but clearly showed positive infiltrations around day 7 (TUBERGEN [R], TINE [R]) (8).

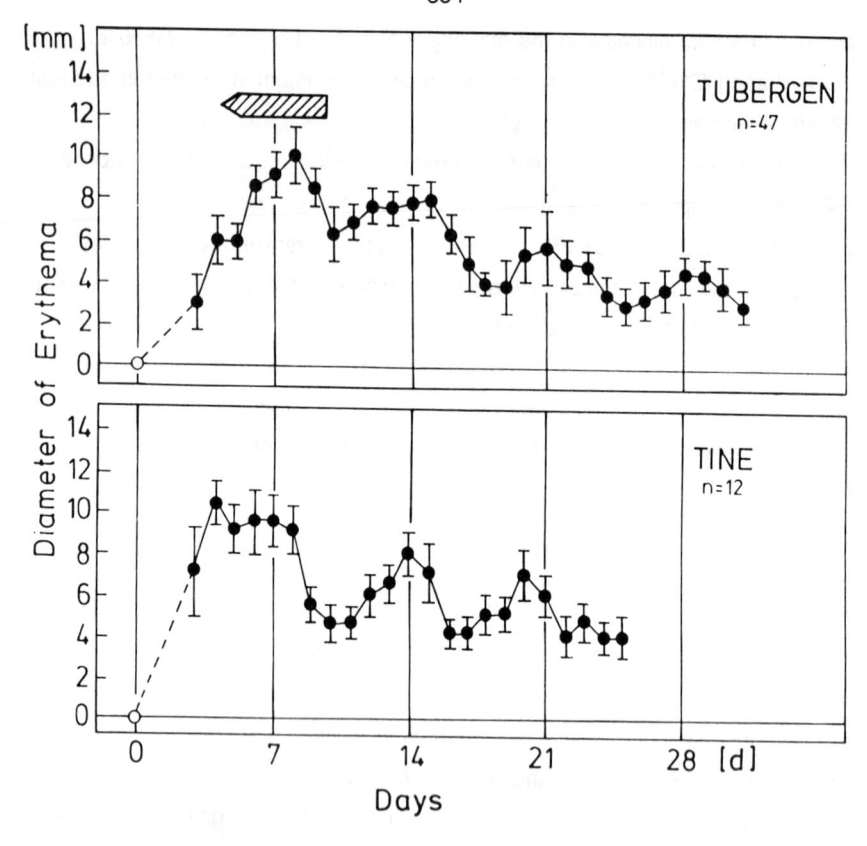

Fig.2: Follow-up of the diameter of the erythema after application of TUBERGEN [R] (n=42) and TINE [R] (n=12). The diameter is read at precisely 8.00 a.m. on consecutive days over a longer period. The recommended time of the reaction is given as 5-9 days (arrow). Brackets indicate standard errors.

REFERENCES

1. BUREAU, J.-P., G. LABREQUE, M. COUPE & L. GARELLY (1986): Influence of BCG administration time on the in-vivo migration of leucocytes. Chronobiology International 3, 23-28
2. COVE-SMITH, J.R., P. KABLER, R. POWNALL & M.S. KNAPP (1978): Circadian variation in an immune response in man. Brit. med. J., II, 253-254
3. COVE-SMITH, J.R., R. POWNALL, T.A. KABLER & M.S. KNAPP (1979): Circadian variations in the cell-mediated immune response in man and their response to prednisolone. In: A. REINBERG & F. HALBERG (eds.): Chronopharmacology. Pergamon Press, Oxford-New York-Toronto-Sydney-Paris-Frankfurt. pp. 369-373
4. GUTENBRUNNER, Chr. & H. GEIS (1986): Vegetative Einflüsse auf die Histamin-Empfindlichkeit der Haut und deren adaptive Modifikationen bei wiederholten intracutanen Testungen. Z. Phys. Med. Baln. Med. Klim. 15, 108-115
5. HERTL, M. (1985): Tuberkulindiagnostik. Dtsch. Krankenpflegezeitschrift 38, 185-193
6. HILDEBRANDT, G. (1962): Biologische Rhythmen und ihre Bedeutung für die Bäder- und Klimaheilkunde. In: W. AMELUNG & A. EVERS (eds.): Handbuch der Bäder- und Klimaheilkunde. Schattauer, Stuttgart. pp. 730-785
7. HILDEBRANDT, G. (1985): Therapeutische Physiologie, Grundlagen der Kurortbehandlung. In: W. AMELUNG & G. HILDEBRANDT (eds.): Balneologie und medizinische Klimatologie. Springer, Berlin-Heidelberg-New York-Tokyo. Bd. 1, pp. 1-271
8. HOFMANN, F. & W. GROTZ (1986): Probleme der Tuberkulintestung bei arbeitsmedizinischen Vorsorgeuntersuchungen im Gesundheitsdienst. In: F. HOFMANN & U. STÖSSEL (eds.): Arbeitsmedizin im Gesundheitsdienst. Gentner, Stuttgart. pp. 165-169
9. KNAPP, M.S. & R. POWNALL (1980): Biological rhythms in cell-mediated immunity findings from rats and man and their potential clinical relevance. In: M.H. SMOLENSKY, A. REINBERG & J.P. McGOVERN (eds.): Recent advances in the chronobiology of allergy and immunology. Pergamon Press, Oxford-New York-Toronto-Sydney-Paris-Frankfurt. pp. 323-331
10. LEE, R.E., M.H. SMOLENSKY, C.S. LEACH & J.P. McGOVERN (1977): Circadian rhythms in the cutaneous reactivity to histamine and selected antigens, including phase relationship to urinary cortisol. Annals of allergy 38, 231-236
11. PÖLLMANN, L. & B. PÖLLMANN (1988): Untersuchungen zum Tuberkulin-Stempeltest. In: F. HOFMANN & U. STÖSSEL (eds.): Arbeitsmedizin im Gesundheitsdienst. Gentner, Stuttgart. Bd.2, pp. 183-188
12. REINBERG, A., E. SIDI & J. GHATA (1965): Circadian reactivity rhythms of human skin to histamine or allergen and adrenal cycle. J. allergy 36, 273-283

MOON CYCLE AND ACUTE DIARRHEAL INFECTIONS IN BRATISLAVA 1988-1990

M. MIKULECKY Jr.[1] & P. ONDREJKA[2]

[1]Clinic of Infectious and Parasitic Diseases, Faculty of Medicine,
Comenius University, Limbova 5, 833 05 Bratislava
[2]1st Medical Clinic, Teaching Hospital, Comenius Slovak Republic, CSFR

INTRODUCTION

One of chronomedical criteria claimed for differentiation between common acute bacterial and viral infections in general medical practice was the time relation to the separate phases of synodic lunar cycle (5).

The aim of the present paper is to pursue an analogical study on acute diarrheal infections.

MATERIAL AND METHODS

All hospital admissions of adult patients for an acute diarrheal gastrointestinal infection (59 with shigellosis, 134 with salmonellosis, 646 with other etiology) between January 1988 and December 1990 in Bratislava have been registered. The source data file is open to any kind of verification.

Each day of admission has been expressed in terms of the synodic lunar cycle, with the new moon as day zero. Moving averages of daily numbers of admissions, from three consecutive days of lunar cycle each, were used for construction of plexogram for the total of 839 admissions.

The plexogram was processed by R.A. FISHER's periodogram (1) and HALBERG's cosinor method (2) modified by MIKULECKY et al. (6). Irrespective of the result obtained by periodogram, the estimated plexogram was constructed using the regression with 6 periodic components. Their period lengths were fixed at one synodic lunar cycle (SLC, 29.53 days), SLC/2 (14.77 days), SLC/3 (9.84 days), SLC/4 (7.38 days), SLC/5 (5.91 days) and SLC/6 (4.92 days).

RESULTS AND DISCUSSION

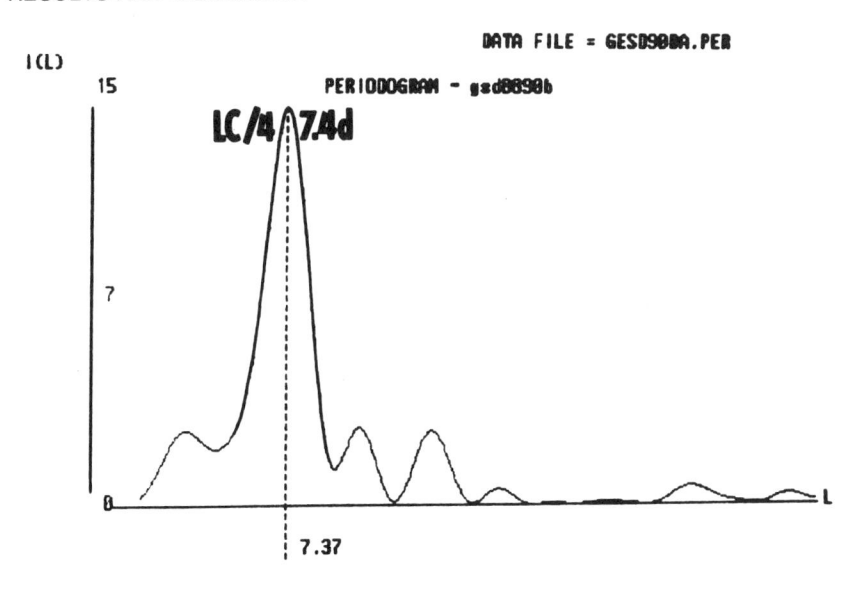

Fig. 1: Periodogram

The resulting periodogram (Fig. 1) displays a clear dominance of a rhythm with an incredibly exact agreement with the theoretical assumption (6) that the DERER's and HALBERG's circaseptans (3,4) represent, at least in some cases, the quarter of synodic lunar cycle, i.e. 7.38 days, rather than the social week.

Fig. 2 shows the data with the fitted plexogram, p values for separate periodic components and coefficient of determination. The latter value says that almost 86 % of the variance has been explained by the given regression.

The cosinor graph on Fig. 3 illustrates the statistical significance of separate periodic components as well as multivariate relations of their amplitudes and acrophases.

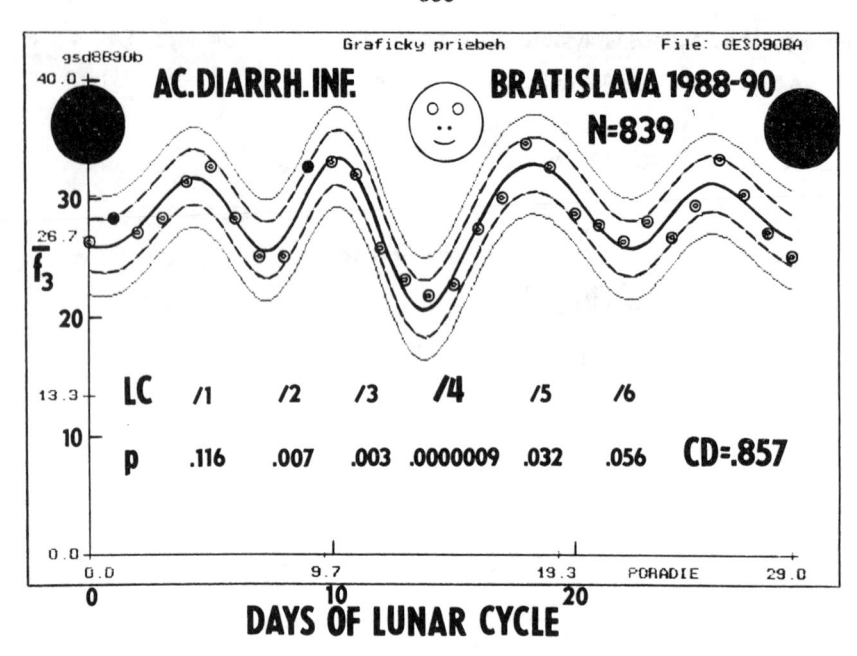

Fig.2: Moving averages of numbers of admissions (f3) per day of lunar cycle (dots). The regression approximating function is shown with its 95 % confidence (narrower) and 95 % tolerance corridor. The p values and that of the coefficient of determination (CD) are given, too.

In conclusion, an approximately circaseptan (exactly 7.4 - day) rhythmicity, bound on lunar cycle, surprisingly dominates in the daily numbers of hospital admissions of patients with an acute diarrheal infection in Bratislava during the last three years. The conspicuous peaks are located at the 4th - 5th day, 10th - 11th day, 18th - 19th day and 25th - 26th day of the cycle. We have, so far, no idea about the pathophysiological background of this phenomenon. Both, the host as well as the pathogenic microorganism might be influenced by periodic changes in the lunisolar system.

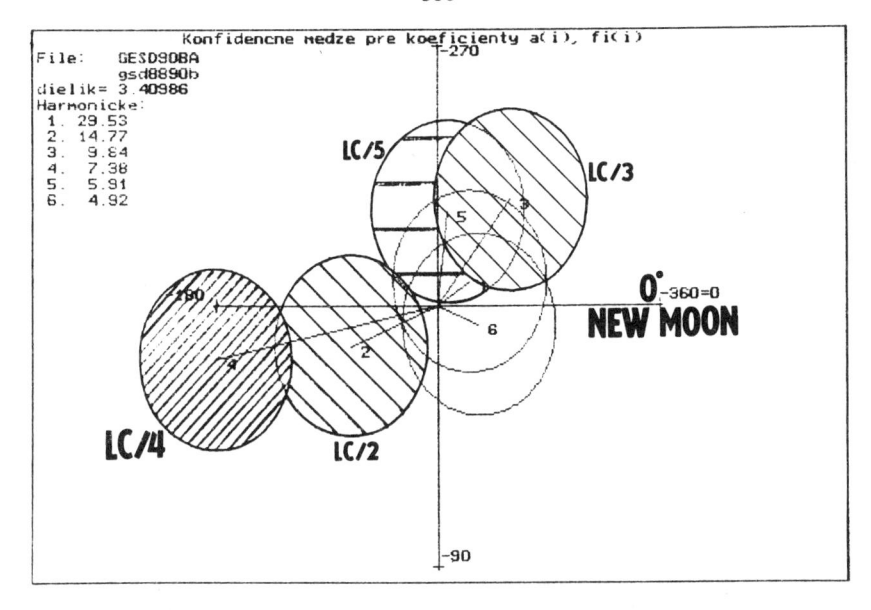

Fig.3: Halberg's polar plot of amplitudes against acrophases, showing corresponding 95 % confidence ellipses, for the 6 periodic components.

REFERENCES

1. ANDEL, J. (1976): Statistical analysis of time series. (Czech.) Statni nakladatelstvi technicke literatury. Prague
2. BINGHAM, C., B. ARBOGAST, G. CORNELISSEN et al. (1982): Inferential statistical methods for estimating and comparing cosinor parameters. Chronobiologia 9, 397
3. HALBERG, F., E. HALBERG, F. HALBERG & J. HALBERG (1985): Circaseptan (about 7-day) and circasemiseptan (about 3.5-day) rhythms and contributions by Ladislav Derer. 1. General methodological approach and bilogical aspects. Biologia 41, 233. Bratislava
4. HALBERG, F., E. HALBERG, F. HALBERG & J. HALBERG (1986): Dtto. 2. Examples from botany, zoology, and medicine. Biologia 40, 1119. Bratislava
5. HEJL, Z. & J. POCHOBRADSKY (1990): Chronobiological criteria for diagnosis of acute infectious diseases. In: E. MORGAN (ed.): Chronobiology and chronomedicine. Basic research and applications. P. Lang, Frankfurt am Main-Bern-New York-Paris, pp. 143-150
6. MIKULECKY, M. (1991): Enright's criticism of Cosinor analysis: the solution? J. interdiscipl. Cycle Res. 22, 157
7. MIKULECKY, M., L. KUBACEK & A. VALACH (1991): Time series analysis with periodic components. Manual for TSA-PC program. Comenius University, Bratislava

CIRCANNUAL PATTERN OF SALMONELLA INCIDENCE PREDICTED BY LINEAR-NONLINEAR RHYTHMOMETRY

R.C. HERMIDA & D.E. AYALA

Bioengineering & Chronobiology Laboratories, E.T.S.I. Telecomunicación, University of Vigo, Apartado 62, Vigo (Pontevedra) 36280, Spain,

INTRODUCTION

Salmonellosis (SA) is a bacterial disease commonly manifested by an acute enterocolitis, with sudden onset of headache, abdominal pain, diarrhea, nausea and sometimes vomiting. Dehydration, specially among infants, may be severe. Fever is nearly always present. Numerous serotypes of Salmonella are pathogenic for both animals and man. Worldwide in distribution, SA has been more extensively reported in North America and Europe. Food is the predominant vehicle of infection and, as such, SA is classified as a foodborne disease. The incidence rate for infection is highest for infants and young children. Epidemiologically, *Salmonella enterocolitis* may occur in small outbreaks in the general population. Large outbreaks in hospitals, day care centers, restaurants, and nursing homes are common, and usually arise from food contaminated at its source, or, less often, during handling by an ill patient or a carrier, but may be due to person-to-person spread. Fecal-oral transmission from person to person is important specially when diarrhea is present (5). The organisms can multiply in a variety of foods, especially milk, to attain a very high infective level. Hospital epidemics tend to be protracted, with organisms persisting in the environment; they often start with contaminated food and continue by person-to-person transmission via the hands of personnel or contaminated instruments. Ordinarily, deaths are uncommon except in the very young, the very old, or the debilitated. However, morbidity and associated costs of SA may be high (3,5). Early recognition and treatment may be necessary to prevent its spread in the community. Intervention by preventive health education is indicated. The timing of such educational endeavors may be adjusted to any rhythmic and to that extent predictable changes in the incidence of SA (3).

In the search for rhythmicity, least squares techniques are useful for curve fitting when it is desirable to obtain a functional form that best fits a given set of measurements. The procedure consists of fitting a (cosine) curve to the data at trial periods in a given domain. The goodness of fit is indicated by minimizing the sum of squares of differences between the actual measurements and the estimated functional form or best fitting curve. The least squares methods of fitting relatively simple models are based on regression techniques and, as such, are applicable to the analysis of nonequidistant observations (6,8). Depending on whether the periods are assumed to be known or unknown, linear least squares (LLS; (8)) or nonlinear least squares (NLLS; (2,4)) rhythmometry can be applied. The LLS method of fitting a set of fixed-length trial periods, one at a time, should only be used in awareness of the possibility of 1) fitting several components concomitantly and of 2) allowing the periods be adjusted, as additional parameters to be estimated. In considering the relative merits of these two approaches, it seemed reasonable to combine both instead of dealing with them as alternatives (4,7). This is done by introducing the procedure of first fitting one period at a time as a LLS spectrum in regions of biologic interest. In a second step, the linearly obtained results are used as input for a NLLS analysis. This latter procedure provides interval as well as point estimations for a single period or for several periods characterizing the data. We thus arrive to a sequential procedure that was used in the search for circannual variability in the incidence of SA in Minnesota (USA).

MATERIALS AND METHODS

Monthly totals of detected cases of SA in Minnesota (USA) between 1980 and 1986 were first computed from the reported cases for each of the 87 counties to the Division of Disease Prevention and Control of the Minnesota Department of Health (3). Data were assigned to the 15th of the corresponding month, and then analyzed on a microcomputer Macintosh IIfx by LLS-NLLS rhythmometric methods written in standard Fortran. Additionally, data from 1987 to 1989 were later available and used to test the degree of agreement between the predicted and the real incidence of SA for those three years.

The combined LLS-NLLS procedure works as follows: First, a so-called linear-in-time analysis is performed by the sequential LLS fit of trial periods, with the analyst choosing the domain of periods to be analyzed (by setting the length of the first period and the number of periods to be analyzed) and the distance between consecutive trial periods. In our particular case, the domain of interest covered from 3 months to 5 years, with an increment of 3 months between consecutive trial periods. Physiologically pertinent and statistically significant components in the time series are thus detected and their A and ϕ objectively quantified (8,10).

The point estimates thus obtained for all statistically significant components are used as initial values in a subsequent NLLS analysis. In this case, the use of the Marquardt algorithm (2,9) allows us to validate multiple periodicities statistically and to develop their confidence limits, whether or not the data are equally spaced. The last step in the sequential LLS-NLLS approach consists on performing a stepwise selection of candidate components. The procedure is as follows: a) A component is conservatively excluded from further analysis if the 95% confidence limits for its amplitude (A) overlaps zero. b) All components with periods associated with overlapping 95% confidence limits have to be combined into one. c) The selection of candidate periods follows similar criteria than those used for model building in multiple regression analysis, with proper provisions for multiple testing (4).

RESULTS

A linear harmonic analysis of the data on the incidence of SA between 1980 and 1986 shows the anticipated 1-year synchronized circannual rhythm with a period of 8766 hours (P<0.001). By the fit of the cosine curve with the given period, characteristics of this significant component are also described.These circannual parameters are indicated in the table at the bottom of Fig.1. This figure shows, on the left, a circannual chronogram, with bimonthly means and standard errors (S.E.) of the total number of detected cases per month of SA for the years considered. The cosine curve represented in the chronogram corresponds to the best fitted model obtained

Fig. 1: Circannual variation in number of detected cases per month of salmonellosis in Minnesota, USA, between 1980 and 1986 (Left: Chronogram with mean ± s.e. at 2-months intervals; Right: Polar plot).

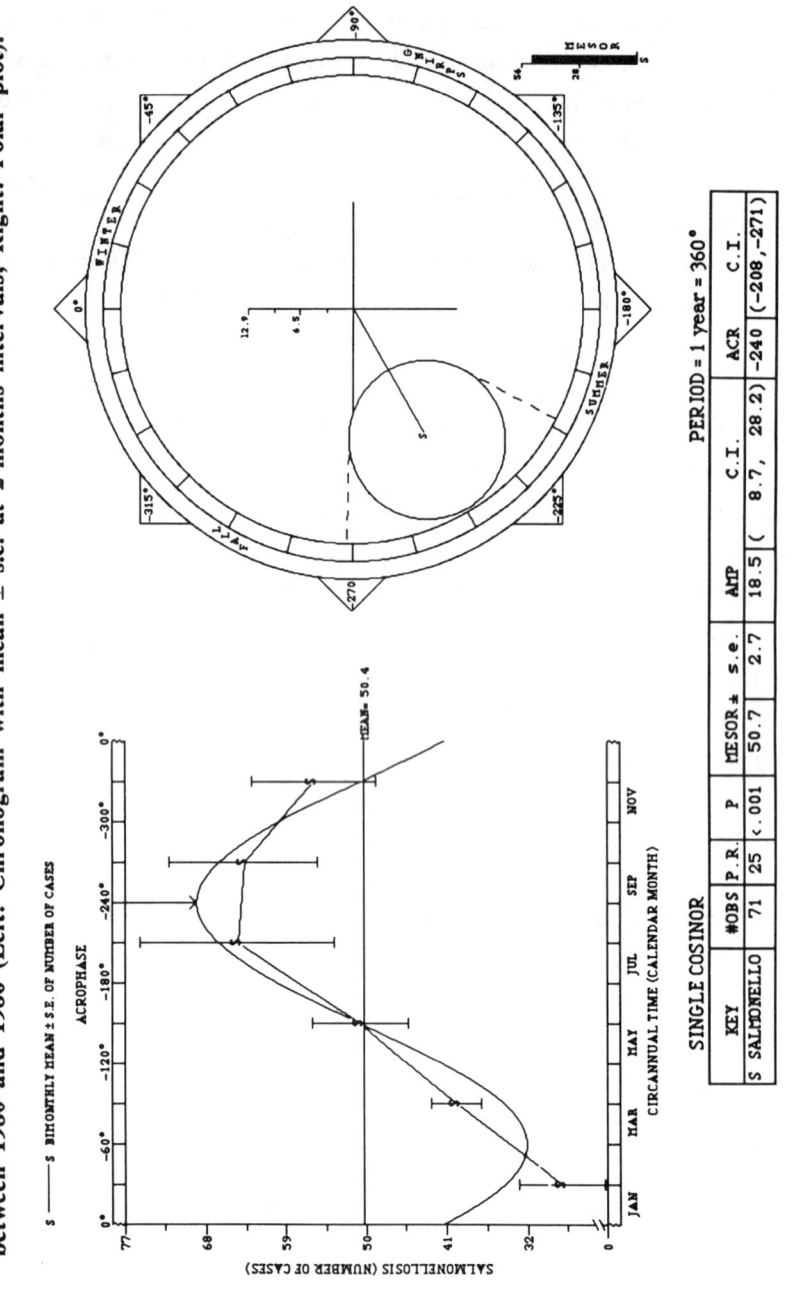

Figure 2: Predictable circannual pattern in the incidence of salmonellosis for consecutive spans of time (1980-86 and 1987-89).

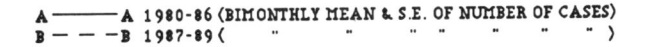

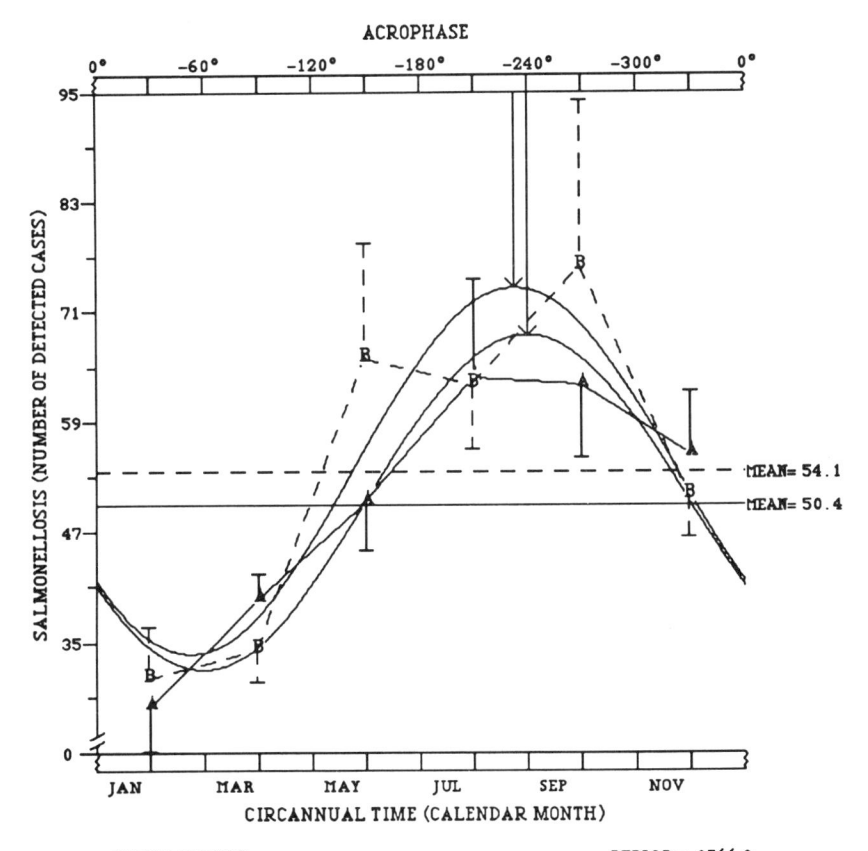

A ———— A 1980-86 (BIMONTHLY MEAN & S.E. OF NUMBER OF CASES)
B — — —B 1987-89 ("   "   " "   "   "   ")

SINGLE COSINOR   PERIOD = 8766.0

KEY	#OBS	P.R.	P	MESOR ± s.e.		AMP ± C.I.		ACR ± C.I.	
A——A	71	25	<.001	50.7	2.7	18.5	7.8	-240	25
B— -B	36	26	.007	54.2	4.2	20.2	12.1	-233	37

by LLS. The arrow from the upper horizontal axis indicates the circannual acrophase (ϕ; -240° from January 1, 1980, with 360°=365.25 days). The chronogram also shows that the monthly means are above the yearly mean (50 cases per month for the whole span considered) between June and December. Moreover, circannual rhythm parameters computed separately for consecutive years are similar in terms of ϕ. A comparison of circannual parameters indicates their consistency by the lack of differences among years (p=0.844; (3)). Fig. 1 also includes, on the right, a polar plot. The 95% conservative confidence limits for A and ϕ separately, computed as the projections of the confidence ellipse, are indicated in the table at the bottom of Fig. 1.

The rhythm-adjusted mean or mesor (midline estimating statistic of rhythm, M), A, ϕ and period of the circannual component found as statistically significant in the LLS method were then considered as initial values for the NLLS analysis. Results from the nonlinear approach give estimates and 95% confidence intervals (CI) for the period (8629.4 (8354.2-8904.6)), the M (50.9 (45.4-56.4) cases per month), the A (18.9 (11.0-26.7)), and the ϕ (-259° (-235°-283°)). This anticipated circannual component is presumably 1-year synchronized since its 95% CI covers the year and the 95% CI of the A does not overlap zero.

The linear function obtained with this combined approach was then used to predict the circannual variability in the incidence of SA for subsequent years (1987 to 1989). The degree of agreement between the predicted and the real incidence for those three years was statistically evaluated by the use of a parameter test for comparison of rhythmometric characteristics (6). Results indicate the validity of the prediction by the lack of differences with the real outcome (P=0.495, 0.811, 0.732, 0.910, and 0.877 for comparisons of Ms, As, ϕs, (A,ϕ) pairs, and complete waveform, respectively). This agreement can be visualized in Fig. 2. This figure represents a circannual chronogram with bimonthly means and s.e. of the total number of detected cases per month of SA for two consecutive spans, 1980-1986 and 1987-1989. The cosine curve represented in the chronogram for the first span corresponds to the best fitted model obtained by LLS for those seven consecutive years. This model was used to predict the incidence of SA for the second span. The cosine curve represented for the second span corresponds to the best model also obtained by LLS for the real incidence in those three years. The circannual parameters of the best fitted models for both spans are indicated in the table at the bottom of Fig. 2.

DISCUSSION

We have used a set of sequential procedures to show the predictable circannual variability in the incidence of SA. We first analyzed time series on monthly totals of detected cases of the disease for the quantification of bioperiodicities with known periods by the so-called single cosinor procedure (8,10) applied in a LLS spectrum over a spectral region of interest. If thus only one component is observed, results from subsequent nonlinear approaches would correspond to the linear result, as it is the case herein. The output of the single cosinor, however, is neither a complete nor the best approach in the case when multiple components with uncertain periods characterize the series examined. In such a general case, a NLLS approach is indicated. Quite often, in the case of prominent components, the LLS method yields satisfactory results. Sometimes, however, the LLS method may detect pseudoperiods or fail to detect true periods close to each other. These problems are partly resolved by combining the LLS and NLLS procedures (2,4,7). The sequential approach here used allows the detection of multiple periodicities in short, noisy, and nonequidistant data and, as such, can be generally applied in the modelling and simulation of biologic time series.

Foodborne diseases are associated with seasonal increases in the late summer and fall that may represent the influence of the ambient temperature on the ability of organisms to multiply in or on their reservoirs and sources, resulting in an increased concentration of organisms for potential contact with susceptible hosts. Additionally, the frequency of picnics and the lack of refrigeration add to the increased opportunity for small doses of agents to multiply to infectious doses. Salmonella surveillance data reflect a circannual variation of this sort (3). Similar circannual patterns were also found earlier for the incidence in Minnesota of giardiasis (2) and amebiasis (1). Even when further studies are required to identify factors responsible for the predictable circannual pattern in the incidence of SA here shown, awareness of this circannual variability may be important in the prevention, diagnosis, and treatment of this infectious disorder. Moreover, the attention of health professionals should now be focused on the prevention of SA and increased epidemiological surveillance. Time-specified health education programs with special attention to personal hygiene (3,5) and developed from the results here obtained could represent an important factor in decreasing the incidence of SA. Evaluation of this point is warranted from the results here found, indicating a large total predictable change in the incidence of SA (by a double A of 74% the average monthly incidence) and a highly stable circannual ϕ for consecutive years.

ACKNOWLEDGEMENTS

This research was supported in part by Consellería de Educación e Ordenación Universitaria, Xunta de Galicia (XUGA-709-0289) and Dirección General de Investigación Científica y Técnica, DGICYT, Ministerio de Educación y Ciencia (PB88-0546).

REFERENCES

1. AYALA, D.E. & R.C. HERMIDA (1987): Yearly variation in the incidence of amebiasis assessed by rhythmometric signal processing. Proc. 34th ISMM Int. Conf. on Mini- and Microcomputers and Their Applications. Lugano, Switzerland, June 29-July 2, pp.151-155
2. AYALA, D.E. & R.C. HERMIDA (1989): Combined linear-nonlinear approach for biologic signal processing. Proc. 7th IASTED Int. Symp. Applied Informatics. Grindelwald, Switzerland, Feb. 8-10, pp.67-70
3. AYALA, D.E. & R.C. HERMIDA (1990): Circannual variation in the incidence of salmonellosis. J. Interdiscipl. Cycle Res., $\underline{21}$, 165-167
4. AYALA, D.E., R.C. HERMIDA & R.J. ARROYAVE (1990): Sequential approach for analysis of sparse biologic time series, illustrated for the incidence of giardiasis. Proc. IASTED Int. Conf. Signal Processing and Digital Filtering. Lugano, Switzerland, June 18-21, pp.241-244
5. BENENSON, A.S. (ed.) (1985): Control of communicable diseases in man. The American Public Health Association, Washington, D.C., 14th edic.
6. BINGHAM, C., B. ARBOGAST, G. CORNELISSEN et al. (1982): Inferential statistical methods for estimating and comparing cosinor parameters. Chronobiologia $\underline{9}$, 397-439
7. HALBERG, F., E. HALBERG, W. NELSON et al. (1982): Chronobiology and laboratory medicine in developing areas. Proc. 1st African and Mediterranean Congress for Clinical Chemistry, Milan, Italy, 1980, pp.113-156
8. HERMIDA, R.C. (1987): Chronobiologic data analysis systems with emphasis in chronotherapeutic marker rhythmometry and chronoepidemiologic risk assessment. In: L.E. SCHEVING, F. HALBERG & C.F. EHRET (eds.): Chronobiotechnology and Chronobiological Engineering. Martinus Niijhoff, NATO ASI Series E, No. 120, Dordrecht, pp. 88-119
9. MARQUARDT, D.W. (1979): An algorithm for least squares estimation of nonlinear parameters. J. SIAM $\underline{11(2)}$, 431-441
10. NELSON, W., Y.L. TONG, J.K. LEE & F. HALBERG (1979): Methods for cosinor rhythmometry. Chronobiologia $\underline{6}$, 305-323

MULTIPLE COMPONENT ANALYSIS OF PLASMA INSULIN IN CHILDREN WITH STANDARD AND SHORT STATURE

L. GARCIA[1], R.C. HERMIDA[2], A. MOJÓN[2], D.E. AYALA[2], J.R. FERNÁNDEZ[2], C. LODEIRO[3] & T. IGLESIAS[1]

[1] Hospital Materno-Infantil 'Teresa Herrera', La Coruña, Spain.
[2] Bioengineering & Chronobiology Laboratories, E.T.S.I. Telecomunicación, University of Vigo, Apartado 62, Vigo 36280, Spain.
[3] Lab. RIA, Centro Oncológico de Galicia, La Coruña, Spain.

INTRODUCTION

Biologic time series obtained by time-specified sampling allow to recognize variability as a novel source of information to be extracted by appropriate hardware and software for use in prediction, prevention, diagnosis, and treatment of disease (7,8). Several decades of research worldwide have established that medical diagnosis can be subject to a much higher proportion of false positives and false negatives when only single samples are taken at arbitrary times of the day instead of taking periodicity into account (7). Special attention is given to the derivation of signal processing methods for the analysis of so-called hybrid data: time series collected from a group of subjects. Along these lines, methods of linear least-squares (LLS) rhythmometry have been designed for the detection of periodic components in short and noisy single time series (as they are generally presented in biologic applications) anticipated to be rhythmic with a given period (2,7,8).

Most variables of interest to laboratory medicine show indeed rhythmic and, as such, predictable changes with several frequencies (6). Given series of sufficient density and length, rhythms with different periods (from the ultradian to the infradian range) can be validated in the same data, in conjunction with trends associated with growth, development, and aging (6,7,9). In particular, insulin (IN) is characterized in children by circadian and ultradian variability (5,10). A comparison of circadian (the most prominent component) parameters obtained by LLS methods had not revealed, however, statistically significant differences between prepuberty children with standard and short stature (10).

In the analysis of IN variability, transformations or alternate models to the cosine curve should be considered since the tests of assumptions underlying the use of cosinor methods usually do not indicate sinusoidality and normality of residuals (10). In such a case, the use of a multiple component analysis to fit a model consisting of several cosine functions is also an alternative (1,7,9). This method yields an orthophase as a better estimate than the acrophase for the timing of high values in non-sinusoidal rhythms (1,11).

With the aim to study possible differences in rhythmic characteristics of IN secretion with stature, groups of children with short and standard stature were extensively sampled, extending earlier work in the description of circadian variability of plasma IN children with short stature (5). Blood was drawn for hormone determinations and several other physiological systems (including temperature, blood pressure, and heart rate) were monitored around the clock. Individual and population rhythm parameters were computed and compared from the time series thus obtained by the LLS fit of multiple statistically significant components.

SUBJECTS AND METHODS

We here analyzed data from 16 boys & 8 girls of short stature (2 to 4 standard deviations below their peer group mean), and 4 boys & 5 girls of standard stature, before any treatment were administered to the former. All children were in apparent good health at the time of the study. Subjects had 10.32 ± 0.49 years of age, and were living on a diurnal waking (ca. 07:30 to 22:30), nocturnal resting routine, consuming the usual hospital diet. Blood was drawn at about 3-hr intervals during most of the day and denser (at about half-hour intervals) 4 hrs around midnight, for a maximum of 16 blood samples in the span of up to 26 hrs. Serum was stored frozen at -60°C until radioimmunoassay for IN concentration (in µIU/ml). Data were then coded and analyzed on a minicomputer Macintosh IIfx by analysis of variance and by LLS procedures (2,7), including the fit of multiple components and the estimation of orthophase (11). Rhythm characteristics for any given period were also compared between groups according to stature and gender by the use of a parameter test (2).

LLS rhythmometry is based on regression techniques and, as such, it is applicable to the analysis of unequidistant observations. The procedure consists of fitting, one at a time, a set of (cosine) curves to the data, with the analyst choosing the domain of trial periods to be analyzed and the distance between consecutive trial periods. The goodness of fit is indicated by minimizing the sum of squares of the residuals from the analysis, computed as differences between the actual measurements and the estimated functional form or best fitting curve. Thus, one obtains, for each period considered, estimates of parameters characterizing the data, namely, the rhythm-adjusted mean or mesor (midline estimating statistic of rhythm, M), the amplitude (A), and the acrophase (ϕ) (2,7).

Methods of LLS rhythmometry include the single and the population-mean cosinor. The single cosinor is a method applicable to single biologic time series anticipated to be rhythmic with a given period (7). This procedure amounts to fit a cosine function of fixed (and anticipated) period to the data by least squares. In order to summarize results obtained for different individuals belonging to a same population, the rhythm parameters obtained by the single cosinor procedure may serve as input for a population-mean cosinor for further quantification (2,7,8). The rhythm characteristics obtained by the single cosinor are then considered as imputations or first order statistics. The population-mean cosinor, in turn, constitutes a second order statistic, applied to derive confidence intervals for rhythm parameters pertaining to the whole population. The parameter estimates are based on the means of estimates obtained from individuals in the sample, and their confidence intervals depend on the variability among individual parameter estimates. The rejection of the zero A assumption by single cosinor ("rhythm detection") refers to the given data set and does not allow extrapolation to the whole population. If the sample is to be characterized without further inference to others in the population, a single cosinor is much more efficient; however, when inference is to be drawn on the basis of the sample for the entire population, the population-mean cosinor is indicated (6,7).

The multiple LLS procedure, as a complement to the LLS methods explained above, provides point and interval estimates of the A and the ϕ for each fitted period.

A partial F-statistic is used to test the zero-A hypothesis for each fitted component. The method also provides, for the overall fit of multiple components, point and interval estimates of the M, the overall A, and the orthophase (1,11).

RESULTS

Circadian rhythm parameters were first computed for each individual series by the single cosinor fit of a 24-hour cosine curve. The parameter estimates thus obtained were used as imputations for assessing the circadian rhythm characteristics of each group by gender and stature separately with the population-mean cosinor method. The same procedure was then applied to obtain group characteristics for other significant components. A comparison of circadian parameters indicates similar characteristics between subjects of short and standard stature, whether one compares boys (p=.976, .565, .997, and .845 for comparisons of Ms, As, ϕs, and (A,ϕ) pairs, respectively), girls (p=.987, .740, .113, and .221) or all subjects irrespectively of gender (p=.791, .745, .097, and .118; (10)). Larger differences are found in circadian ϕ when subjects are compared by gender (p=.101, .093, and .008, for ϕ comparisons between boys and girls of standard stature, short stature, and independently of stature, respectively). The possible difference in circadian ϕ should be further checked as more data accumulate.

The LLS spectral analysis of the data reveals the anticipated statistically significant circadian rhythm as the principal spectral feature for all groups. Prominent components with periods of 12, 8, 4.8 and 2.67 hours can also be independently documented for some groups of children divided according to stature or gender. Accordingly, and given the non-sinusoidal waveform for IN, a multiple component analysis was undertaken for data of each group separately, as well as for all subjects. Results in Table 1 indicate concomitant periods that are statistically significant for each group of stature with their corresponding As and ϕs, as well as the overall M, A, and orthophase for each group considered. Both the multiple component analysis and the population-mean cosinor method fail to detect a circasemidian (12-hr) component for standard children. Moreover, a comparison of parameters from the 8-hr component indicates a statistically significant difference in φ (p=.003) between short and standard children (Fig. 1).

POPULATION MEAN COSINOR   PERIOD = 8 HOURS = 360°

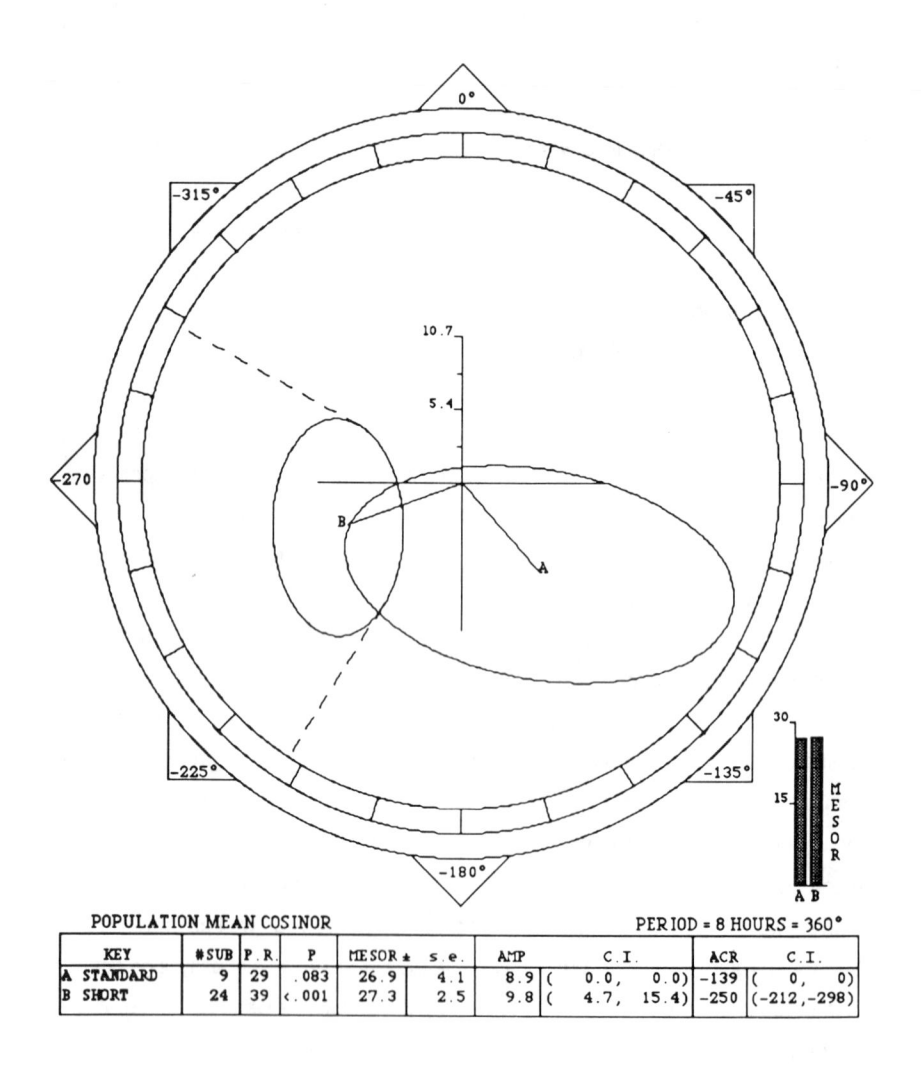

KEY	#SUB	P.R.	P	MESOR ± s.e.		AMP	C.I.		ACR	C.I.	
A STANDARD	9	29	.083	26.9	4.1	8.9	(0.0,	0.0)	-139	(0,	0)
B SHORT	24	39	<.001	27.3	2.5	9.8	(4.7,	15.4)	-250	(-212,	-298)

Fig. 1:Eight-hourly variation of plasma insulin in Spanish children with standard and short stature

Table 1: Multiple components analysis of plasma insulin in standard and short children.

Group	#of sbj.	Period	P	MSR* ± s.e.	Ampltd ±s.e.	Acrphsis.e.
Short Chldrn	24	24.00	<.001		9.80 ± 1.92	-260° ± 11°
		12.00	<.001		10.14 ± 1.87	-351° ± 10°
		8.00	<.001		9.86 ± 2.43	-235° ± 10°
		2.67	<.008		6.89 ± 2.02	-58° ± 18°
		Overall	<.001	27.01 ± 1.36	24.58	-205°
Stand. Chldrn	9	24.00	.023		10.19 ± 3.29	-254° ± 18°
		8.00	.033		9.12 ± 3.15	-115° ± 20°
		Overall	.005	25.81 ± 2.14	18.50	-276°
All Chldrn	33	24.00	<.001		8.85 ± 1.94	-277° ± 11°
		8.00	<.001		7.53 ± 2.27	-222° ± 12°
		4.80	.004		6.25 ± 1.85	-284° ± 14°
		4.00	<.001		7.66 ± 1.59	-201° ± 16°
		3.00	.017		5.46 ± 1.60	-124° ± 20°
		2.67	<.001		7.77 ± 1.86	-67° ± 13°
		Overall	<.001	28.25 ± 1.32	24.23	-330°

MSR: mesor (midline estimating statistic of rhythm, a rhythm-adjusted mean)
Ampltd: Amplitude (amplitude in μIU/ml), Acrphs: Acrophase, Chldrn: Children

DISCUSSION

The LLS analysis requires that the data obtained can be reasonably well represented by a cosine curve and non-sinusoidality limits the applicability of the method (7). For a meaningful LLS rhythmometric analysis, it is important, therefore, to determine the approximate sinusoidality of the data. This requires at least the inspection of the chronogram (display of data as a function of time) before application of the LLS procedure and/or a mathematical test for sinusoidality. Tests for normality of residuals and homogeneity of variance are also indicated and thus incorporated into our LLS programs (4). The fit of multiple components that are statistically and biologically significant accounts for non-sinusoidal waveforms and provides parameters characterizing them. When the data are non-sinusoidal, the LLS fit of a cosine curve may be used for rhythm detection, although this approach may not be as powerful as the simultaneous fit of all statistically significant components. The p-value obtained in testing the zero-A assumption should thus be regarded as reflecting whether the data are better approximated by a cosine curve than by a horizontal line (4,7,8).

Harmonic interpolation has also been suggested to obtain estimates of timing of high values on non-sinusoidal rhythms (3). This method provides the detailed shape of the waveform (by the fit of harmonics) and provides an estimation of the rhythm's high values, called the paraphase (3,7). The method, however, is only applicable to equidistant data covering an integral number of cycles with the period assumed to be known. The limitations of this alternative approach are, therefore, evident.

In considering the therapeutic effects of growth hormone, children in whom study of its efficacy seems appropriate include not only those who have growth hormone deficiency, but also those of short stature who are not growth hormone deficient (including children with the designations of familial short stature and constitutional growth delay). However, the most effective dose and mode of administration of growth hormone have yet to be defined (9). Information on marker rhythms (7) can be used to indicate the best time for treatment for any given child and to establish reference standards that are time-specified, for the interpretation of individual profiles or even single samples (8,9). In this sense, the computation of an orthophase by the concomitant fit of several components serves to estimate the best treatment-time. The ultradian components here documented, mostly for short children, should be taken into consideration in the design and later evaluation of a time-specified treatment of children of short stature.

Circadian (and ultradian) parameters of the IN rhythm may also serve to monitor any possible progression of a condition into a disease. Evaluation of risk by appropriately designed questionnaires should then next complement blood sampling for IN or other hormonal determination in groups of children differing in age, gender or ethnicity, among other factors, with the purpose to derive prediction regions for the individualized assessment of health, risk or disease (8).

ACKNOWLEDGEMENTS
This research was supported in part by Consellería de Educación e Ordenación Universitaria, Xunta de Galicia (DOGA 16-12-87 and XUGA-709-0289); Dirección General de Investigación Científica y Técnica, DGICYT, Ministerio de Educación y Ciencia (PB88-0546); and Fondo de Investigaciones Sanitarias de la Seguridad Social (FISss 89-0407).

REFERENCES

1. AYALA, D.E., R.C. HERMIDA, L. GARCÍA et al. (1990): Multiple component analysis of plasma growth hormone in children with standard and short stature, Chronobiology International 7(3), 217-220
2. BINGHAM, C., B. ARBOGAST, G. CORNELISSEN et al. (1982): Inferential statistical methods for estimating and comparing cosinor parameters, Chronobiologia 9, 397-439
3. DE PRINS, J., G. CORNELISSEN, D. HILLMAN et al. (1977): Harmonic interpolation yields paraphases and orthophases for biologic rhythms, Proc. XIII Int. Soc. Chronobiol. Pavia, Italy, Sept. 4-7, 333-344
4. FERNÁNDEZ, J.R. & R.C. HERMIDA (1990): A software package for linear least-squares rhythmometry written for the Macintosh computer, Journal of Interdisciplinary Cycle Research, 21(3), 186-189
5. GARCÍA, L., R.C. HERMIDA, T. IGLESIAS et al. (1990): Circadian rhythm of plasma insulin in children of short stature. In: E. MORGAN (ed.): Chronobiology and Chronomedicine: Basic Research and Applications. Verlag Peter Lang GmbH, Frankfurt am Main, pp. 171-180
6. HAUS, E., G.Y. NICOLAU, D. LAKATUA & L. SACKETT LUNDEEN (1988): Reference values for chronopharmacology. In: A. REINBERG, M. SMOLENSKY & G. LABRECQUE (eds.): Annual Review of Chronopharmacology, Vol. 4. Pergamon Press, Oxford, pp. 333-424
7. HERMIDA, R.C. (1987): Chronobiologic data analysis systems with emphasis in chronotherapeutic marker rhythmometry and chronoepidemiologic risk assessment. In: Chronobiotechnology and Chronobiological Engineering. L.E. SCHEVING, F. HALBERG & C.F. EHRET (eds.): Martinus Nijhoff, NATO ASI Series E, No. 120, Dordrecht, The Netherlands, pp. 88-119
8. HERMIDA, R.C., L. GARCÍA, J.R. FERNÁNDEZ & D.E. AYALA (1989): Software system for marker rhythmometry in pediatrics, Proc. 7th IASTED Int. Symp. Applied Informatics. Grindelwald, Switzerland, Febr. 8-10, 238-242
9. HERMIDA, R.C., L. GARCÍA, C. LODEIRO et al. (1989): Circadian and ultradian characteristics of plasma growth hormone in children with normal and short stature, J. Endocrinol. Invest. 12 (Suppl. 3), 69-73
10. LODEIRO, C., R. HERMIDA, L. GARCÍA et al. (1990): Comparison of circadian characteristics of plasma insulin in children with standard and short stature, J. Interdiscipl. Cycle Res. 21, 211-213
11. TONG, Y.L., W. NELSON, R.B. SOTHERN & F. HALBERG (1977): Estimation of the orthophase (timing of high values) on a non-sinusoidal rhythm-illustrated by the best timing for experimental cancer chronotherapy, Proc. XII Int. Conf. Int. Soc. Chronobiol., Il Ponte, Milán, 765-769,

SEX-DEPENDENT CIRCADIAN AND PULSATILE SECRETION PATTERNS OF ACTH AND CORTISOL IN NORMAL MAN.

F. ROELFSEMA & M. FRÖLICH

Department of Endocrinology, University Hospital, Rijnsburgerweg 10,
2333 AL Leiden, The Netherlands

INTRODUCTION

The episodic and diurnal secretions of ACTH and cortisol are established since the pioneering work of KRIEGER (1). Early studies on the secretion pattern of ACTH were hampered by rather insensitive assays, requiring large amounts of plasma, precluding frequent sampling (1,2,3). The introduction of specific and sensitive ACTH assays made it possible to study its secretion in a more accurate way (4,5).

The aim of this study was to investigate the precise relationship between ACTH and cortisol pulses as well as to establish eventual differences in ACTH and cortisol secretion patterns between males and females.

METHODS

Five healthy males (mean age 43, range 37-55 yrs) and 5 healthy female volunteers (mean age 35, range 32-41 yrs) were studied. Female subjects were investigated in the follicular phase of the menstrual cycle. After an overnight hospital stay, an indwelling i.v. cannula was inserted in a large vein of the forearm and blood samples were drawn at 10-min intervals starting at 08.00 h for the next 24 h. Meals were served at 08.00, 12.30, and 17.30 h. Lights were turned off between 22.00-24.00 h.

Plasma samples for ACTH were collected on ice in chilled EDTA-containing siliconized glass tubes and samples for cortisol in heparinized tubes, centrifuged at 4 °C, frozen and stored at -20 °C until assayed.

Plasma ACTH was measured by IRMA (Nichols Institute, San Juan, Capristrano, Ca, USA). The detection limit of this assay is 3 ng/l, the intra-assay precision varied from 7.4 to 2.8%. Plasma cortisol was measured by RIA (Sorin Biomedica, Milan, Italy). The intra-assay precision varied from 2 to 4%, detection limit 30 nmol/l. Each sample was assayed in duplicate, and all samples of any subject were assayed in one run.

Discrete peak detection was undertaken using Cluster, a computerized pulse analysis algorithm which identifies all significant increases and decreases within the ACTH and cortisol data series in relation to a dose-dependent measurement error obtained from a power function fit of all intrasample variances, plotted against mean sample concentrations (6). A 2 x 1 cluster configuration was used together with a t statistics of 2.0 for significant upstrokes and down-strokes for both the ACTH and cortisol series. Under these circumstances, the false positive rate is constrained to less than 5%. The locations and widths of all significant peaks were identified and the total number counted. In addition, the following pulse parameters were determined: maximal peak height and amplitude, area under the peak and interpulse valley concentration.

The circadian rhythm was analyzed by the Cosinor method. Significant cross-correlations between ACTH and cortisol pulses with the exact coincident probability program developed by VELDHUIS (7) and with ARIMA modeling using Autobox vs 2 (Automatic Forcasting Systems, Hatboro, PE).

RESULTS

The results for the cluster analysis are shown in Tab. 1. The number of ACTH pulses was significantly larger in males than in females, but otherwise the (mean) pulse characteristics were not different between males and females. The integrated area under the curve was larger in males than in females (20.443±895 ng.min/l vs 15.116±1.782, p=0.028) and the mean ACTH level (mean of 144 samples per subject) was higher in males than in females (14.2±0.6 ng/l vs 10.7±1.3, p=0.043). A similar analysis carried out for the cortisol data did not reveal any differences between male and female subjects.

ACTH and cortisol pulse properties in males and females.

	ACTH				Cortisol		
	males	females			males	females	
means 24 h. conc.	14.2±0.6	10.7±1.3[a]	ng/l		206±26	210±20	nmol/l
area	348±52	284±52	ng.min/l		4229±1067	3347±470	nmol.min/l
integrated area	5128±524	3287±491[b]	ng.min/l		81442±1718	71246±9255	nmol.min/l
height	17.8±0.9	15.5±0.9	ng/l		234±23	245±29	nmol/l
increment	7.63±0.9	7.13±1.0	ng/l		90±13	89±8	nmol/l
width	59±7.4	62±5.8	min		47.7±4.4	37.6±3.0	min
nadir	10.0±0.5	8.6±0.9	ng/l		140±15	143±17	nmol/l
pulse number	15.4±1.3	11.6±0.7[c]	no/24h		19±0.8	21.4±0.9	no/24h

Values are expressed as the mean±SEM. Statistical significance: [a] $p=0.043$ [b] $p=0.034$ [c] $p=0.037$ (two-sided t-test).

Both ACTH and cortisol secretion had a highly significant circadian rhythm. For ACTH the mesor was higher in males than in females (14.2 ± 0.25 vs 10.81 ± 0.24, $p=0.001$). In addition, the amplitude was larger in males than in females (5.5 ± 0.35 vs 3.3 ± 0.33, $p=0.001$) but the time of acrophase did not differ (8.40 ± 15 min vs 934 ± 35 min). The mesor for cortisol in males was 206 ± 5 and in females 204 ± 4 (NS). However, the amplitude was significantly larger in males than in females (148 ± 7 vs 104 ± 7, $p=0.001$). The acrophases did not differ (09.52 ± 10 min vs 09.49 ± 14).

The temporal distribution of the pulse characteristics of ACTH and cortisol was investigated by a 2-way ANOVA. To this end, the 24-h sampling period was divided into four equal parts, starting at 04.00h. The height of ACTH pulses was higher in males than in females ($p=0.007$, see Fig. 1), and as expected, highest values were found in the first period (i.e. from 04.00-10.00 h). Nadir values differed between sexes ($p=0.02$) and the period ($p<0.001$). No sex differences between pulse amplitude and pulse area could be demonstrated, but both parameters had a strong period effect ($p<0.001$). The interpulse interval had no period effect, indicating an equal distribution of pulses across the 24-h span. A similar analysis carried out for cortisol pulses did not show a sex difference of any of the pulse characteristics, but a highly significant period effect was present for pulse height
($p<0.01$), amplitude ($p<0.001$), area ($p<0.001$), and nadir level ($p <0.001$). The cortisol pulses were equally distributed along the day-night cycle.

For the analysis of the coincidence of pulses the data series of the individual subjects were appended. In these data series a total number of 135 ACTH pulses and 202 cortisol pulses were present. The cortisol pulses lagged 10 min after the ACTH pulse in 37 cases ($p<0.0001$), 20 min in 25 cases ($p<0.007$) and were exactly coincident in 24 cases ($p=0.07$). The cross-correlation analysis of the prewhitened ACTH and cortisol series revealed a significant correlation with no time lag in 3 subjects, a 10-min lag of the cortisol series in 5 subjects, and a 20-min lag in another subject. In one subject no correlation could be established within this time window.

Since the ACTH pulse parameters, especially pulse height, were larger in males than in females, whilst the cortisol did not show a sex dependency, a higher

ACTH Pulse Parameters

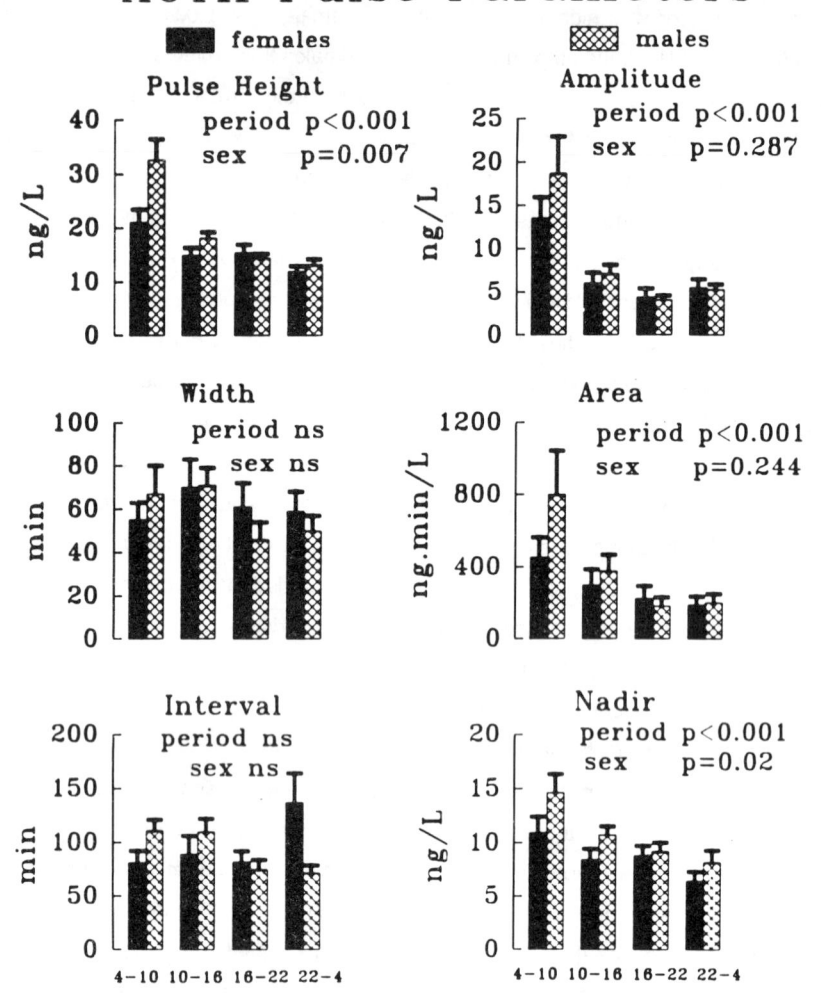

Fig.1: Cluster analysis of ACTH pulses in normal subjects. The 24-hour period was divided into four parts and the data analyzed by a 2-way ANOVA.

sensitivity of the female adrenal cortex towards ACTH than the male adrenal cortex is suggested. Indeed, the ratio of the total cortisol area and ACTH area was larger in females than in males (20.373±1.418 nmol/ng vs 14.672±2.251, p=0.028). Therefore, the analysis was extended by the calculation of the regression of the height of cortisol pulses and the corresponding ACTH pulses for both sexes. The results are depicted in Fig. 2. The regression slopes did not differ significantly (13.37±1.51 in males, and 15.21±2.33 in females, t=0.60). The mean distance between the parallel lines was 78.1 nmol/l (t=3.52, p<0.001).

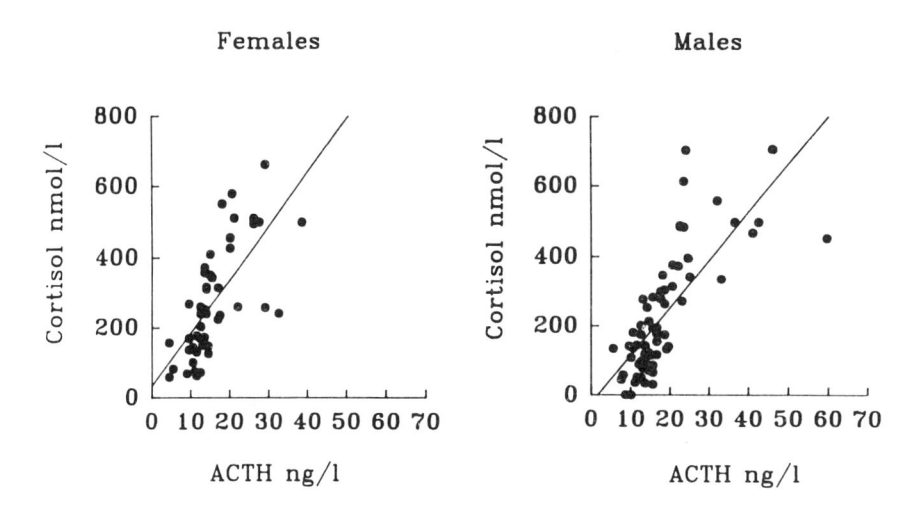

Fig.2: Regression analysis for cortisol pulse height vs ACTH pulse height in normal males and females.

DISCUSSION

The study confirms the well known pulsatile and circadian secretion patterns for ACTH and cortisol. Interestingly, the analysis revealed amplitude modulation rather than frequency modulation of the circadian secretion of both ACTH and cortisol, confirming recent data obtained in normal males by VELDHUIS et al. (6). In addition, cortisol pulses lagged behind ACTH pulses by 10 minutes, a result to be expected given the used time window. Similar results were obtained by VELDHUIS. However, in the study of VELDHUIS using a deconvolution analysis, but also in a comparable study of IRIMANESH et al.(8), a larger number of ACTH peaks was found than in our study. The results of the latter study are especially interesting, because the same ACTH assay was used with a comparable intra-assay coefficient of variation. In addition, the mean age of the male subjects in the two studies were comparable, and the same pulse analysis, with similar cluster sizes and thresholds was used. At present, we have no explanation for this discrepancy in findings. We did analyze however, the ACTH data published by HORROCKS et al. (4) with the cluster program and found that the number of pulses matched more closely our data than that of the group of VELDHUIS.

The total number of cortisol pulses was larger in our subjects than the number of ACTH pulses. As mentioned above, we cannot fully exclude that our analysis missed some ACTH pulses, but recent data suggest that cortisol secretion peaks may also be generated via a paracrine mechanism involving stimulation of chromaffin cells through the sympatho-adrenal system (9). The difference in ACTH pulse frequency between sexes was also found by HORROCKS et al.(4), but not by WALLACE et al. (10). The latter group studied 14 normal children, 9 male, 5 female, 9 prepubertal and 5 pubertal. By using Fourier analysis they found a dominant periodicity of 0.7-1.0h, representing 24-34 secretory episodes of ACTH. No differences in periodicity was found between males and females and between pubertal and prepubertal status.

In view of the findings discussed above, the present data suggest that the sex difference in ACTH pulse number is a phenomenon which appears late in puberty or in young adults, probably under the influence of sex steroid hormones on the hypothalamus.

Another finding in our study was the apparent higher sensitivity of the female adrenal cortex towards ACTH in terms of circulating cortisol levels. This result could not be explained by a difference in plasma clearance rates. The estimated plasma half life for ACTH was 34.7 ± 14.8 min in males, and 27.6 ± 6.8 min in females, the corresponding half lifes for cortisol were 38.5 ± 8.8 min and 39.8 ± 6.8 min. In rats, gonadal hormones have a profound effect on the pituitary-adrenal axis (both at the pituitary and at the adrenal level), but also on the hepatic clearance of glucocorticoids (11). This study in humans suggests that the main action of sex steroids is at the hypothalamic-pituitary level, as well as at the adrenal level.

REFERENCES

1. KRIEGER, D.T., W. ALLEN, F. RIZZO & H.P. KRIEGER (1971): Characterisation of the normal pattern of plasma corticosteroid levels. Journal of Clinical Endocrinology and Metabolism 32, 266-284
2. GALLAGHER, T.F., K. YOSHIDA, H.D. ROFFWARG et al. (1973): ACTH and cortisol secretory patterns in man. Journal of Clinical Endocrinology and Metabolism 53, 1022-1032
3. LINKOWSKI, P., J. MENDLEWICZ, R. LECLERCQ et al. (1985): The 24-hour profile of adrenocorticotropin and cortisol in major depressive illness. Journal of Clinical Endocrinology and Metabolism 61, 429-438
4. HORROCKS, P.M., A.F. JONES, W.A. RATCLIFFE et al. (1990): Patterns of ACTH and cortisol pulsatility over twenty-four hours in normal males and females. Clinical Endocrinology 32, 127-134
5. VELDHUIS, J.D., A. IRANMANESH, M.L. JOHNSON & G. LIZARRALDE (1990): Amplitude but not frequency modulation of adreno-corticotropin secretory bursts give rise to the nyctohemeral rhythm of corticotropic axis in man. Journal of Clinical Endocrinology and Metabolism 71, 452-463
6. VELDHUIS, J.D. & M.L. JOHNSON (1986): Cluster analysis: a simple, versatile and robust algorithm. American Journal of Physiology 250, E 486-E 493
7. VELDHUIS, J.D., A. IRANMANESH, I. CLARKE et al. (1989): Random and non-random coincidence between LH peaks and FSH, alpha subunit, prolactin, and GnRH pulsations. Journal of Neuroendocrinology 1, 1-10
8. IRINMANESH, A., G. LIZARRALDE, D. SHORT & J.D. VELDHUIS (1990): Intensive venous sampling paradigms disclose high frequency adrenocorticotropin release episodes in normal men. Journal of Clinical Endocrinology and Metabolism 71, 1276-1283
9. BORNSTEIN, S.R., M. EHRHART-BORNSTEIN, W.A. SCHERBAUM et al. (1990): Effects of splanchnic nerve stimulation on the adrenal cortex may be mediated by chromaffin cells in a paracrine manner. Endocrinology 127, 900-906
10. WALLACE, W.H.B., E.C. CROWNE, S.M. SHALET et al. (1991): Episodic ACTH and cortisol secretion in normal children. Clinical Endocrinology 34, 215-221
11. KITAY, J.I. (1963): Pituitary-adrenal function in the rat after gonadectomy and gonadal hormone replacement. Endocrinology 73, 253-260

THE CLINICAL USE OF NA/K OSCILLATIONS IN URINE

D. HAJEK

IInd Clinic of Internal Diseases, Faculty Hospital, Videnska 100, 657 15 Brno, Czechoslovakia

INTRODUCTION

The existence of physiological reactions common to various kinds of stressors e.g. the activation of steroids secretion was proved (5). The stress resistance is influenced by the regulation quality of steroids secretion, that can be estimated by means of sodium and potassium urine excretion. The experiments carried out on mice have proved that both the extreme values of sodium and potassium urine excretion and its extreme oscillations have an unfavorable effect on the individual stress resistance resulting in cortical-visceral disorders (4,7).

As man reveals markedly lower frequency of periodical processes and intensity of metabolism compared to mice due to allometry principle (1,2, 6) the method of studying sodium and potassium urine oscillations is not useful for clinical praxis without being modified.

The study was objected to develop a simple to use test of individual stress resistance that could be used in clinical praxis e.g. for the prevention of gastrointestinal stress ulcers and therapy effect estimation.

METHOD

21 probands were selected for the study: 11 men with polytrauma, age 42 ±9 years, 7 (4 men, 3 women) with duodenal ulcer, age 29 ±4 years, 3 women with severe endogenous depressing disorders, age 34, 47, 52 years. No modification of the therapy was done. Patients with polytrauma received even diuretics and steroids.

The amplitude and frequency of function F=D[Na] / D[K] where D[Na], D[K] are the change of sodium and potassium respectively urine concentration in 24 hours were evaluated. The samples were taken at 7 o'clock in the morning.

RESULTS

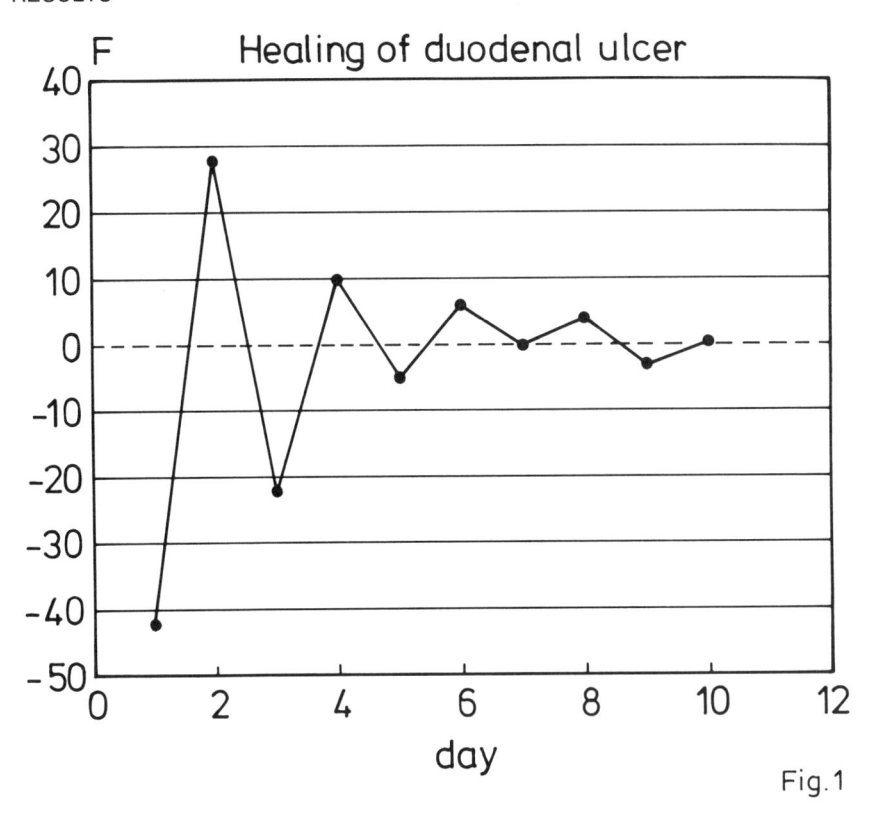

Healing of duodenal ulcer

Fig.1

The increasing amplitude and frequency of the function F indicated the occurrence of gastroduodenal bleeding in about 3 to 5 days. The gradual decrease was observed during the favorable treatment of duodenal ulcer (Fig.1). There were great interindividual differences among the patients of the same general health estimated by means of routine examination and polytrauma severity. High, only slowly decreasing amplitude and frequency of the function F indicated poor prognosis, gastroduodenal bleeding and circulation disturbances were most frequent complications.

The patients with cerebral commotion revealed high amplitude and frequency even after their complaints disappeared and no other pathological findings were present. The period increased approximately from 2 days to a week during the favorable treatment.

On the other hand the patients with endogenous depression revealed very slight oscillation that were gradually increasing during the successful treatment (Fig.2).

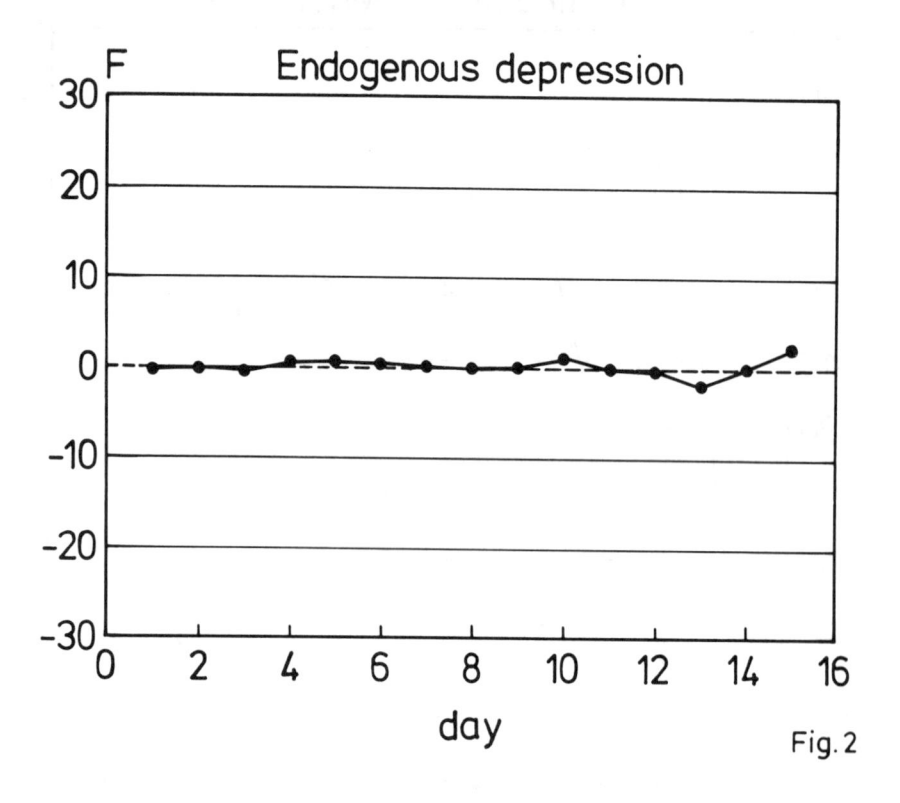

Fig. 2

DISCUSSION

The changes of frequency and amplitude of the pursued function were compatible with the clinical status. According to the allometry principles (1,2,3,6) and experimental results (4,7) the period of human Na/K urine oscillations was expected to be about 3 times longer compared to mouse, approximately 5-7 days at rest. The author's results proved this assumption.

As the approach to the problem was modified by the point of view of clinical praxis e.g. by the possibility to use the described method bed side by staff having only limited command of mathematics and low costs the exact evaluation by means of more sophisticated chronobiological methods was not performed.

CONCLUSIONS

The method proved to be useful in clinical praxis first of all in the prevention and treatment of stress gastroduodenal ulcer first of all for its simplicity and noninvasivity. The pathology of physiological regulations and its interindividual differences are probably of a great importance for clinical medicine. An interesting problem to be discussed is, whether endogenous depression and gastroduodenal stress ulcer represent an opposite type of pathology according to chronobiological point of view.

REFERENCES

1. CLARC, A.J. (1927): Comparative physiology of the heart. Cambridge
2. KLEIBER, M. (1932): Body size and metabolism. Hilgardia 6, 315-353
3. KRUTA, V. (1965): Nekolik pohledu na srovnavaci fysiologii srdce a obehu krevniho. SZdN, Prague
4. POSPISIL, M., J. SIKULOVA & F. SEVCIK (1965): The fluctuation of urine electrolyte excretion and its relation to mortality of multiple irradiated mice. Z. ges. exp. Med. 139, 112-121
5. SELYE, H. (1976): Stress in health and disease. Butterworts, Boston and London
6. SMITH, R.J. (1980): Rethinking allometry. J. theor. Biol. 87, 97-111
7. VACHA, J. & M. POSPISIL (1980): Changes in ratio of sodium and potassium in urine in stress situation. Scr. med. Fac. Med. Brun. 53, 27-32

ANTI-TUMOR ACTIVITY IN PINEAL GLANDS AND IN URINE DISPLAY SIMILAR CIRCANNUAL RHYTHMICITY

H. BARTSCH & C. BARTSCH

Section of Clinical Pharmacology, Department of Gynaecology,
University of Tübingen, D-W 7400 Tübingen, FRG

INTRODUCTION

The tumor-inhibiting activity of the pineal hormone melatonin in vivo is well established now (1). However, melatonin-free pineal extracts have also been shown to exert tumor-inhibiting activity, in vitro as well as in vivo (2,3). A number of findings indicate that these substances may be present in urine as well (for review see (2)). Attempts to detect the anti-tumor activity in human urine turned out to be very frustrating: strong tumor inhibition was observed as well as complete lack of effect. Only the application of chronobiological methods to analyze the obtained data shed light upon the contradictory findings: the presence of anti-tumor substance(s) in human urine seemed to be dependent on the season. Similar circannual rhythms were found for the presence of tumor-inhibiting activity in crude extracts prepared from rat pineal glands. These experiments are described in this paper and the question is raised whether there may be a correlation between these findings and the seasonal occurrence of human cancers.

TUMOR-INHIBITING ACTIVITY IN HUMAN URINE

The tumor-inhibiting effect of human urine was tested on three transplantable murine tumors with the aim to find a suitable model for the detection of the active substance in the course of a future isolation procedure.

(i) A spontaneous adenocarcinoma of the breast in a female Holtzman rat was serially transplanted subcutaneously and rats bearing such tumors were given human urine mixed with drinking water (50%) one week after inoculation, when tumors had become palpable, till death. Tumor-inhibiting activity was defined as increased number of surviving animals compared to the control group. The left panel of Fig. 1 shows the results of the most successful experiment with this tumor model depicting the average tumor size of living animals in the treated group as well as in the untreated controls. In the treated group, 73% of the animals survived and their tumors regressed whereas in the control group only 25% of the animals survived.

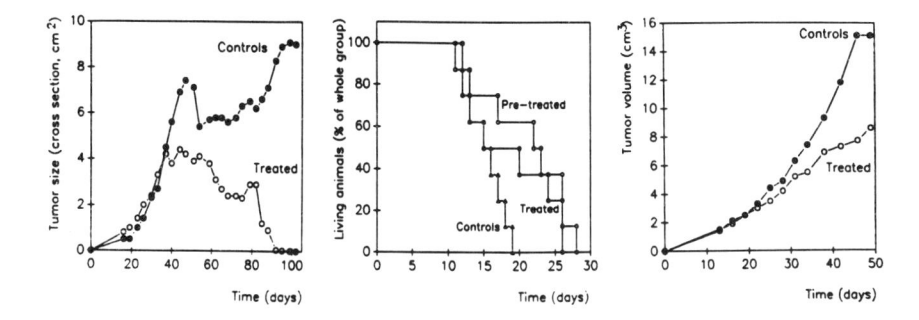

Fig. 1: left panel: Tumor growth of transplantable adenocarcinoma in rats (average tumor size of living animals in each group);
central panel: survival time of Swiss mice inoculated with fibrosarcoma ascites cells;
right panel: growth of Ehrlich solid tumors in mice (development of average tumor sizes of treated animals and untreated controls).

(ii) Methylcholanthrene-induced fibrosarcoma ascites in inbred Swiss mice, maintained over many years by serial intraperitoneal transplantation, was injected intraperitoneally into outbred Swiss mice. Urine treatment was started three weeks before tumor inoculation (pre-treated) or on the day of inoculation (treated) and continued until death. In this model, tumor-inhibiting activity was defined as increased survival time compared to the controls. Fig. 1, central panel, demonstrates the results of one experiment with this tumor model showing the number of surviving animals on each day of the experiment. In this experiment, the pre-treated group showed a 35% increase in survival time which was statistically significant ($p<0.05$) according to the Mann-Whitney test.

(iii) Ehrlich tumor cells were maintained by serial intraperitoneal transplantation in inbred Swiss mice. For the experiments solid tumors were induced by intramuscular injection of tumor cells into one hindleg of outbred animals. Urine was given in drinking water beginning four weeks before tumor inoculation till the end of the experiments when the animals were sacrificed and the tumor-inhibiting activity was assessed as decreased tumor volume compared to that of controls. The right panel of Fig. 1 shows the tumor-development of treated and control animals in one experiment with this tumor model. In the treated group the tumor volume at the end of the experiments was decreased by 36% (p<0.01) as compared to untreated controls.

More details about these experiments are given in (4).

CIRCANNUAL RHYTHM OF ANTI-TUMOR ACTIVITY IN HUMAN URINE

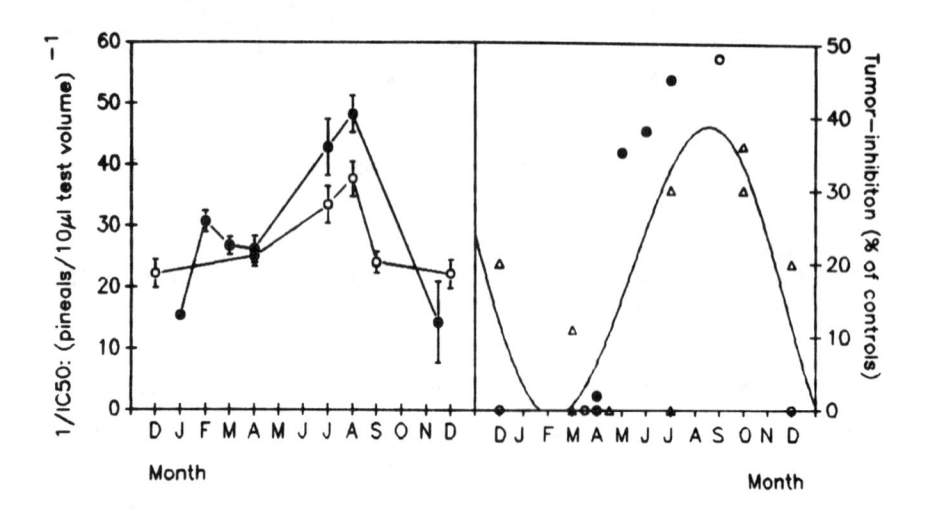

Fig. 2: right panel: Circannual rhythm of tumor-inhibiting activity in human urine as assessed in three tumor models (open circles: adenocarcinoma of the breast in female Holtzman rats; filled circles: solid Ehrlich tumors in Swiss mice; open triangles: fibrosarcoma ascites in Swiss mice,
left panel: circannual rhythm of pineal anti-tumor activity as assessed in two different experiments (1986/87: filled circles, data re-drawn from (6); 1989/90: open circles).

On the whole, the results obtained with the three tumor systems were contradictory: strong tumor inhibition was observed as well as complete lack of effect. When the results were plotted against the month in which the experiments were performed the seasonality of the activity became apparent.

Fig.2, right panel, shows the circannual rhythm of the tumor-inhibiting activity of human urine which, according to the cosinor analysis (5) is significant ($p<0.005$) and has its acrophase in August and the amplitude as % of the MESOR is 100%.

CIRCANNUAL RHYTHM OF PINEAL ANTI-TUMOR ACTIVITY

The tumor-inhibiting activity in individual rat pineal glands was analyzed over a complete year in two different experiments (6,7). The first one was carried out in 1986/87 with 167 pineal glands from rats of both sexes, aged between 20 and 120 days, killed at different times of the day. Ethanol extracts were prepared and applied to human erythroleukemia cells (6). Anti-tumor activity was defined as the reciprocal IC_{50}-value (i.e. the concentration necessary to obtain 50% inhibition of cell growth as compared to untreated controls) and expressed as (pineals/10µl test volume)$^{-1}$. A significant circannual rhythm was observed with all animals ($p<0.0001$, acrophase in July, amplitude as % of MESOR: 43%; see Fig. 2, left panel, filled circles) as well as with a sub-group consisting of only female rats, aged 100-120 days, killed at 9 a.m. ($p<0.001$, acrophase at the end of May, amplitude as % of MESOR: 60%) (6). The second experiment was carried out in 1989/90 with 52 nine months old male rats killed at 11 a.m. Again, a significant circannual rhythm was obtained ($p<0.001$, acrophase in July, amplitude as % of MESOR: 27%; see Fig. 2, left panel, open circles) (7). The decreased peak values and lower amplitude and MESOR may be due to the advanced age of these animals as compared to the first experiment.

DISCUSSION

This paper shows that an unidentified anti-tumor activity in the rat pineal gland shows a similar circannual rhythmicity as the tumor-inhibiting activity present in human urine. Although there is substantial indication in the literature that the urinary activity may have its origin in the pineal gland (2), it needs still to be proven. A common feature of both activities seems to be that they are not species-specific: human urine was active on tumors in rats and mice, while the pineal extracts inhibited human tumor cells; moreover, in the first half of the present century extracts from bovine pineal glands were used in the treatment of human cancer patients and beneficial effects were reported (for review see (8)). The question arises whether there is a correlation with the seasonality in the occurrence of human cancers which is reported by a number of authors (9-11). Since also various parameters of the immune function undergo circannual rhythms, e.g. the stimulation of rat lymphocytes was found to be higher in July than in January (12), it seems that the endogenous defence mechanisms against cancer may be higher in summer than in winter leading to a seasonality of various manifestations of tumorous growth, e.g. appearance of first symptoms and metastases, maximal tumor growth etc. The acrophases of these parameters may differ, depending on the type of tumor, its localization, and detection methods, but the underlying mechanisms of seasonality may still be the same. Further elucidation of these phenomena can have important implications for the prevention and treatment of human cancer.

ACKNOWLEDGMENTS

Hella and Christian Bartsch are research scholars of the Society of Servants of God, New Delhi, India, working in the Cancer Research Programme of the Society.
The experiments with the tumor-inhibiting activity in human urine were carried out by H.B. and C.B. during their stay at the Department of Biochemistry, All-India Institute of Medical Sciences, New Delhi, India. The authors thank Prof. T.N. Chapekar and Prof. G.P. Talwar for their support of this work.

REFERENCES

1. BLASK, D.E. & S.M. HILL (1988): Melatonin and cancer: basic and clinical aspects. In: A. MILES, D.R.S. PHILBRICK & C. THOMPSON (eds.): Melatonin - Clinical Perspectives. Oxford Medical Publications, Oxford, pp.128-173
2. BARTSCH, H. & C. BARTSCH (1988): Unidentified pineal substances with antitumor activity. In: D. GUPTA, A. ATTANASIO & R.J. REITER (eds.): The Pineal Gland and Cancer, Brain Research Promotion, London/Tübingen, pp.369-376.
3. BARTSCH, H., C. BARTSCH, W.E SIMON et al. (1991): Antitumor activity of the pineal gland: Effect of unidentified substances versus the effect of melatonin. Oncology, in press
4. BARTSCH, H. (1988): Untersuchungen zur Antitumorwirkung der Zirbeldrüse. Dissertation, Fakultät für Chemie und Pharmazie, Universität Tübingen.
5. BINGHAM, C., B. ARBOGAST, G. CORNELISSEN et al. (1982): Inferential statistical methods for estimating and comparing cosinor parameters. Chronobiologia 9, 397-439.
6. BARTSCH, H., C. BARTSCH & D. GUPTA (1990): Tumor-inhibiting activity in the rat pineal gland displays a circannual rhythm. J. Pineal. Res. 9, 171-178.
7. BARTSCH, H., C. BARTSCH & T. LIPPERT (1991): Melatonin und chronobiologische Aspekte von Krebs. Münch. Med. Wschr. 133, 49-54
8. LAPIN, V. (1976): Pineal gland and malignancy. Öst. Zschr. Onkol. 3, 51-60
9. COHEN, P., Y. WAX & B. MODAN (1983): Seasonality in the occurrence of breast cancer. Cancer Res 43, 892-896
10. JACOBSON, H., D.T. JANERICHT, P. NASCA et al. (1983): Circannual rhythmicity in the incidence of endocrine malignancy: evidence for neurohumoral control of cancer development and growth. Chronobiologia 10, 135
11. OWNBY, H.E., J. FREDERICK, R.F. MORTENSEN et al. (1986): Seasonal variation in tumor size at diagnosis and immunologic responses in human breast cancer. Invasion Metastasis 6, 246-256.
12. BARTSCH, H., C. BARTSCH & D. GUPTA (1990): Seasonal variations of endogenous defence mechanisms against cancer. In: D. GUPTA, H.A. WOLLMANN & M.B. RANKE (eds.): Neuroendocrinology: New Frontiers. Brain Research Promotion, London/Tübingen, pp.333-339

CIRCANNUAL PATTERN IN UTERINE CERVIX CANCER SCREENING CAMPAIGNS

R.C. HERMIDA[1], J.J. LÓPEZ-FRANCO[2], D.E. AYALA[1] & R.J. ARRÓYAVE[3]

[1] Bioengineering & Chronobiology Laboratories, E.T.S.I. Telecomunicación, University of Vigo, Apartado 62, Vigo, Spain.
[2] Laboratorio Regional de Citología Exfoliativa, HGZCMF #17, I.M.S.S., Monterrey, Nuevo Léon, México.
[3] Unidad de Medicina Familiar #27, I.M.S.S., Ciudad Guadalupe, Nuevo León, México.

INTRODUCTION

In a technical report (12), an expert committee of the World Health Organization stated 1964 that the potential scope of cancer prevention is only limited by the proportion of human cancers in which extrinsic factors are responsible. The categories of cancer that are influenced, directly or indirectly, by extrinsic factors include many tumors of the skin and mouth, the respiratory, gastrointestinal and urinary tracts, hormone dependent organs (such as the breast, thyroid, and uterus), hematopoietic and lymphopoietic systems, which, collectively, account for more than three-quarters of human cancers. Among those extrinsic factors one should consider, in addition to man-made or natural carcinogens, viral infections, nutritional deficiencies or excess, and other factors determined by sexual, reproductive, and personal behavior. It would seem, therefore, that the majority of human cancer is potentially preventable.

In particular, the impact of uterine cervix cancer (UCC) can be greatly reduced by regular vaginal examination and other preventive measures. With this aim, UCC screening programs had been developed and applied for several years in Mexico and elsewhere. One point still to be considered in such preventive programs is the possible circannual pattern in the morbidity or mortality of UCC. In this sense, DOLL & PETO (3) stated that changes in the incidence of particular types of cancer with the passage of time provide conclusive evidence that extrinsic factors affect those types of cancer. Therefore, variations with time in the incidence of cancer within particular communities are an evidence for the possibility of decreasing the morbi-mortality of cancer.

In fact, mortality from cancer reveals a circannual rhythmic pattern, with a high peak consistently observed in January from different studies and for different types of cancer (7). Earlier studies showed that mortality from cancer (nonspecific in relation to type of tumor) along 45 consecutive years at a San Francisco hospital followed a circannual rhythmic pattern, with a larger incidence observed in January and a secondary peak in August (5). In a different geographic setting (Edinburgh, Scotland), a difference as a function of season was also seen in the frequency of all deaths or of non-cancer deaths, with a peak in each case in January (10). When only deaths ascribed to breast cancer are analyzed, separately for women diagnosed before and after menopause, it was found that among those diagnosed premenopausally, deaths from breast cancer were more frequent in the summer, whereas among those diagnosed postmenopausally the deaths peaked in the winter (10). In relation to the morbidity of UCC, earlier reports indicate a higher incidence between the end of the fall and the beginning of the winter (7).

With respect to UCC screening campaigns, we earlier reported a highly statistically significant circannual rhythm in the number of preventive check-ups done in Mexico between 1979 and 1981 (7). This circannual variation was not related, however, to the reported UCC mortality or morbidity rates. In order to extend this study on the cost-effectiveness of UCC screening campaigns, we here analyzed by linear (LLS) and nonlinear least-squares (NLLS) methods the monthly totals of UCC screening check-ups done in the state of Nuevo León (México) between 1978 and 1987.

MATERIALS AND METHODS

Data analyzed here correspond to the monthly totals of UCC preventive check-ups done in the state of Nuevo León (México) between 1978 and 1987, as reported by the Regional Cytology Laboratory at Monterrey. This central laboratory was at that time one of the eight laboratories covering all the cytology needs from the Mexican Institute of Social Security (Instituto Mexicano del Seguro Social, IMSS). The IMSS provides health care for Mexicans with a regular job as well as for their family members, with fees usually being paid in shares by the government, the employer and the employee. The population under IMSS care in June 1985 was 30,908,092 people. Health care is provided in IMSS hospitals covering the whole country. Data reflect the number of check-ups done as a result of the screening campaign established by the IMSS in 1962 to detect UCC in women of 25 years of age and older. Data were coded and analyzed on a microcomputer Macintosh IIfx by the use of a software package written in standard Fortran 77 for LLS-NLLS rhythmometry and for graphical representation of results (4).

In the search for circannual variability in the number of preventive check-ups, data were first fitted by LLS according to the single cosinor method (2,6), a procedure applicable to single biologic time series anticipated to be rhythmic with a given period. This procedure amounts to fit a cosine function of fixed period to the data by least-squares. Thus, one obtains, for the period considered, an estimate of (a) the midline estimating statistic mean of rhythm or mesor (M), a rhythm-adjusted mean, defined as the average value of the rhythmic function (e.g. cosine curve) fitted to the data; (b) the amplitude (A), half the extent of rhythmic change in a cycle approximated by the fitted cosine curve (difference between the maximum and the M of the best fitted curve); and (c) the acrophase (ϕ), lag from a defined reference time point of the crest time in the cosine curve fitted to the data (6).

By the use of the single cosinor, physiologically pertinent and statistically significant components in the time series are detected and their A and ϕ objectively quantified. The point estimates thus obtained for all statistically significant components are used as initial values in a subsequent nonlinear analysis. In this case, the use of the Marquardt algorithm (11) allows us to validate multiple periodicities statistically and to develop their confidence limits, whether or not the data are equally spaced (1). The Marquardt algorithm is a maximum neighborhood method which performs an optimal interpolation between the gradient method and the Taylor series method. It combines the best features of these two approaches while avoiding their most serious limitations (11). Quite often, in the case of prominent components, the linear method yields satisfactory results. Sometimes, however, this linear approach may detect pseudoperiods or fail to detect true periods close to each other. These problems are mostly resolved by the use of the LLS-NLLS combined approach (1).

In order to study the time course of circannual characteristics for the span considered (1978-1987), rhythm parameters were also computed separately by single cosinor for data on each consecutive year. The characteristics thus obtained were subsequently compared with the so-called Bingham test for comparison of cosinor parameters (2).

RESULTS

The monthly totals of UCC preventive check-ups in Nuevo León do not vary at random. This is established in inferential statistical terms by the results of the fit of a 1-year cosine curve to the data. The non-circannual rhythm assumption is rejected ($p<0.001$). By the same fit, characteristics of the circannual rhythm are also described. These parameters are indicated in the table at the bottom of Fig. 1. This figure shows, on the left, a circannual chronogram (display of data as a function of time), with bimonthly means and standard errors (S.E.) of the total number of check-ups for the years considered. The cosine curve represented in the chronogram corresponds to the best fitted model obtained by single cosinor. The arrow from the upper horizontal axis indicates the circannual ϕ (-172° from midnight, 1 January, with 360°=365.25 days). The chronogram also shows that the monthly means are above the yearly mean (4933 check-ups/month) between April and August. In summary, lower-than-average incidences are found in fall and winter, whereas higher incidences follow in the spring, and even increase throughout the summer.

Fig. 1 also includes, on the right, a polar plot. In this presentation, the A and ϕ are represented as a directed line (vector). The length of that line indicates the A of the rhythm. The orientation of the line, i.e. its direction with respect to the circular scale, indicates the ϕ of the rhythm. The circular scale covers one period (or 360°). The scale from the pole (center of the circle) is graduated in the same units as the A (i.e. number of check-ups per month). A 95% confidence region for the pair (A,ϕ) is shown by an error ellipse around the tip of the vector. An ellipse not overlapping the center of the circle, as it is the case herein, indicates that the A differs from zero and

Fig. 1: Circannual variation in number of uterine cervix cancer check-ups done in Nuevo León (México) between 1978 and 1987 (Left: Chronogram with mean ± s.e. at 2-months intervals; Right: Polar plot).

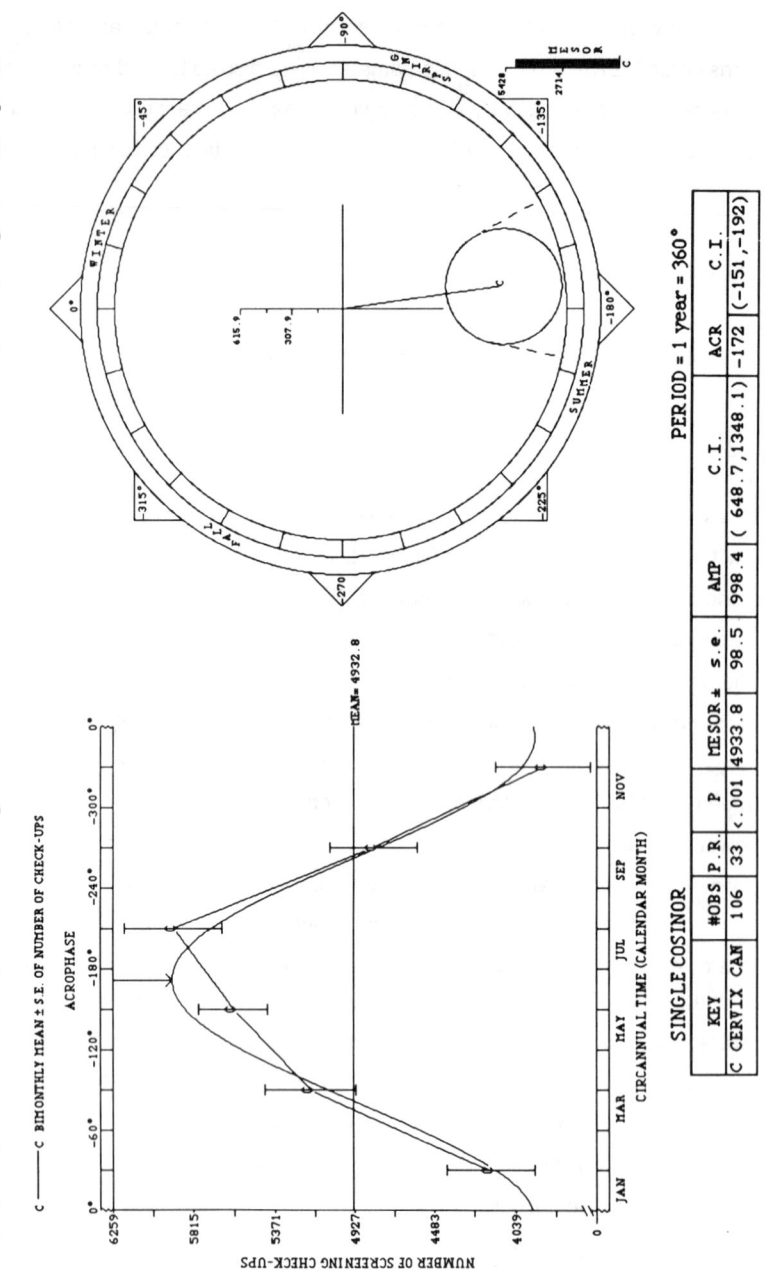

Figure 2: Circannual variation in the monthly number of preventive check-ups for deteccion of uterine cervix cancer in Nuevo León, México, for consecutive years (1984-1986).

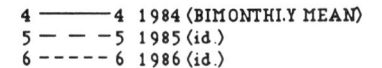

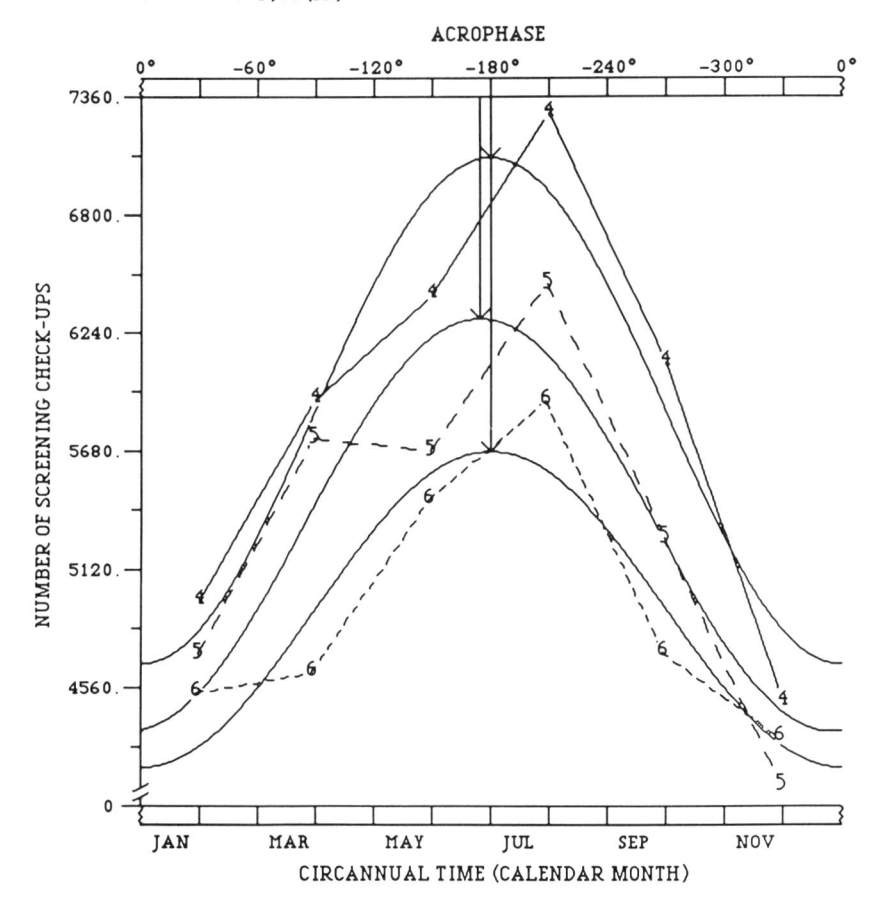

4 ———— 4 1984 (BIMONTHLY MEAN)
5 — — — 5 1985 (id.)
6 – – – – 6 1986 (id.)

SINGLE COSINOR   PERIOD = 8766.0

KEY	#OBS	P.R.	P	MESOR ± s.e.		AMP ± C.I.		ACR ± C.I.	
4——4	12	65	.009	5870.9	207.7	1199.9	665.2	-180	34
5— -5	12	54	.029	5332.4	209.8	976.0	673.3	-174	43
6– - -6	12	52	.037	4931.0	168.6	746.8	541.1	-180	46

that the rhythm is thus statistically significant (6). The 95% conservative confidence limits for A and φ separately, computed as the projections of the confidence ellipse, are also indicated in the table at the bottom of Fig. 1. The polar plot is also complemented by a histogram on the right, used for visual comparison of Ms when several series are represented concomitantly.

A linear harmonic analysis of the data in an infradian window covering from three months to five years (with an increment of three months between consecutive trial periods studied) shows the anticipated 1-year synchronized circannual rhythm as the only statistically significant component. When only one component is observed in the LLS analysis, as it is the case herein, results from the subsequent NLLS approach would correspond to the linear result. Accordingly, results from the nonlinear method give estimates and 95% confidence intervals for the period (8766.5 (8562.9-8970.1)), the M (4933.7, (4622.0-5245.5) check-ups/month), the A (998.4 (554.4-1442.5)), and the φ (-172°, (-147° - -197°)). This anticipated circannual component is presumably 1-year synchronized since its 95% confidence interval covers the year and the 95% confidence interval of the A does not overlap zero.

Circannual rhythm parameters were also computed separately for consecutive years. A parameter test for comparison of circannual characteristics (2) shows the consistency in circannual φ for consecutive years by the lack of "among-year" differences (p=0.971). Fig. 2 shows this similarity in circannual characteristics for 1984-1986, as an example. The figure represents, for each year, the bimonthly means of UCC preventive check-ups. Circannual parameters obtained by single cosinor for each year, with their respective S.E., are indicated in the table at the bottom of Fig.2.

DISCUSSION

The evidence that much human cancer is avoidable can be summarized under four heads: differences in the incidence of cancer among different settled communities, differences between migrants from a community and those who remain behind, variations with time in the incidence of cancer within particular communities, and the actual identification of many specific causes or preventive factors (3). Genetic factors and age also affect cancer onset rates, of course, but this does not affect the conclusion that much human cancer is avoidable.

With respect to the possible circannual pattern in the morbidity of cancer, changes in the incidence of particular types of cancer with the passage of time provide conclusive evidence that extrinsic factors affect those types of cancer. The simplest evidence of the preventability of cancer would be the demonstration by scientific experiment that a particular action actually leads to a reduction in the incidence of the disease. In this sense, the index of suspected cases from the IMSS screening campaign for UCC detection was 11.5 (per 10,000 women studied) on the early years of the campaign, descending to 4.9 in the late 80's (9). In any event, UCC still is the major cause of death in Mexico among women over 30 years with a history of multiparity, start of sexual activities before 18 years of age, multiple sexual partners, and chronic cervix infections (9).

The results here found indicate a circannual pattern in the UCC screening campaign, with a higher effort for detection done systematically during the summer. This timing seems to be related to the incidence of most common diseases in Mexico, such as giardiasis, amebiasis, and salmonellosis (8), but not to the reported mortality and morbidity UCC rates (7). In fact, the monthly totals of positive detected cases of UCC (expressed in percentage of the number of check-ups) on the same location and for the same years considered show a circannual pattern of variability ($p=0.013$), with a higher UCC incidence during the winter ($\phi=-49°$).

That is, screening is mostly done in Mexico when women visit the clinic for other reasons (not precisely to ask for the free check-up), and not when a higher rate of detection could be expected. Health educational and screening campaigns for prevention of UCC and other major conditions should be timed along the year according to morbidity and/or mortality statistics, for which a circannual pattern of variation has been already proved.

ACKNOWLEDGEMENTS

This research was supported in part by Consellería de Educación e Ordenación Universitaria, Xunta de Galicia (XUGA-32201B90) and Dirección General de Investigación Científica y Técnica, DGICYT, Ministerio de Educación y Ciencia (PB88-0546).

REFERENCES

1. AYALA, D.E. & R.C. HERMIDA (1989): Combined linear-nonlinear approach for biologic signal processing. Proc. 7th IASTED Int. Symp. Applied Informatics. Grindelwald, Switzerland, Feb. 8-10, pp. 67-70
2. BINGHAM, C., B. ARBOGAST, G. CORNELISSEN et al. (1982): Inferential statistical methods for estimating and comparing cosinor parameters. Chronobiologia 9, 397-439
3. DOL, R. & R. PETO (1981): The causes of cancer. J. Nat. Can. Inst. 66, 1191-1308
4. FERNÁNDEZ, J.R. & R.C. HERMIDA (1990): A software package for linear least-squares rhythmometry written for the Macintosh TM computer. Journal of Interdisciplinary Cycle Research 21 (3), 186-189
5. HALBERG, F.E., G. CORNÉLISSEN, T. KESSLER et al. (1988): About-yearly (circannual) pattern of cancer deaths, 1942-87, in patients of a San Francisco Hospital. Proc. 1st World Conf. Clin. Chronobiol., pg. 60
6. HERMIDA, R.C. (1987): Chronobiologic data analysis systems with emphasis in chronotherapeutic marker rhythmometry and chronoepidemiologic risk assessment. In: L.E. SCHEVING, F. HALBERG & C.F. EHRET (eds.): Chronobiotechnology and Chronobiological Engineering. Martinus Niijhoff, NATO ASI Series E, No. 120, Dordrecht, The Netherlands, pp. 88-119
7. HERMIDA, R.C., D.E. AYALA & R.J. ARRÓYAVE (1990): Do cancer screening programs have the right circannual timing?. J. Interdiscipl. Cycle Res. 21, 201-203
8. HERMIDA, R.C., D.E. AYALA & R.J. ARRÓYAVE (1990): Circannual incidence of Giardia lamblia in Mexico. Chronobiology International 7(4), 329-340
9. Instituto Mexicano del Seguro Social (1990): Normas Para la Atención y Control del Cáncer Cérvicouterino. Subdirección General Médica, I.M.S.S., 2nd edic.
10. LANGLANDS, A.O., H. SIMPSON, R.B. SOTHERN & F. HALBERG (1977): Different timing of circannual rhythm in mortality of women with breast cancer diagnosed before and after menopause. Proc. 8th Int. Sci. Mtg. Int. Epidemiol. Ass., San Juan, Puerto Rico, September 17-23
11. MARQUARDT, D.W. (1963): An algorithm for least squares estimation of nonlinear parameters. J. SIAM 11(2), 431-441
12. World Health Organization (1964): Prevention of Cancer. (Technical Report Series 276). WHO, Geneva

ENDOCRINE RHYTHMS IN PATIENTS WITH BREAST AND PROSTATE CANCER

C. BARTSCH & H. BARTSCH

Section of Clinical Pharmacology, Department of Gynaecology, University of Tübingen, D-W-7400 Tübingen, FRG

Breast and prostate cancer show a high incidence with increasing tendency in Western countries representing a serious health threat to men and women. Breast and prostate gland are accessory sex organs being part of the endocrine system and their tumors show hormone dependency during development and growth. Endogenous factors therefore play an important role in breast (BC) and prostate cancer (PC) as opposed to those malignancies which are mainly due to exogenous agents, e.g. lung cancer and occupational neoplasias. The incidence of these types of malignancies can drastically be reduced by preventing exposition to the carcinogens concerned. Cancer of breast and prostate should, on the other hand, be preventable if more is known about the detailed endocrine mechanisms involved.

Various endocrine therapies have been devised for the treatment of BC and PC to reduce the stimulating influence of sex steroids as well as prolactin on tumor growth but the endogenous hormonal mechanisms involved in malignant transformation and growth remain poorly understood. Epidemiological data point to a key role of the hormonal patterns to which an individual is exposed during life. In BC the age at menarche, of first pregnancy as well as of menopause are determinants for tumor risk. These events mark changes in the endocrine milieu that control development, growth and function of the mammary gland. A detailed recording of the actual hormonal changes during the different phases of life can hardly be achieved even by longitudinal studies. Therefore investigations are mainly confined to the acute endocrine situation present at the time of detection of malignancy. Earlier studies did not show endocrine changes when only single blood samples were used neglecting chronobiological aspects. In our investigations circadian blood sampling was performed to study the nocturnal surge of melatonin as well the diurnal variations of circulating adenohypophyseal and peripheral hormones. The rhythm of pineal melatonin is controlled by the circadian clock within the suprachiasmatic nuclei (SCN) tuned by retinal photic input and serves as chemical signal to convey temporal information to the body so that the endocrine system is adjusted to the time of day as well as to the season.

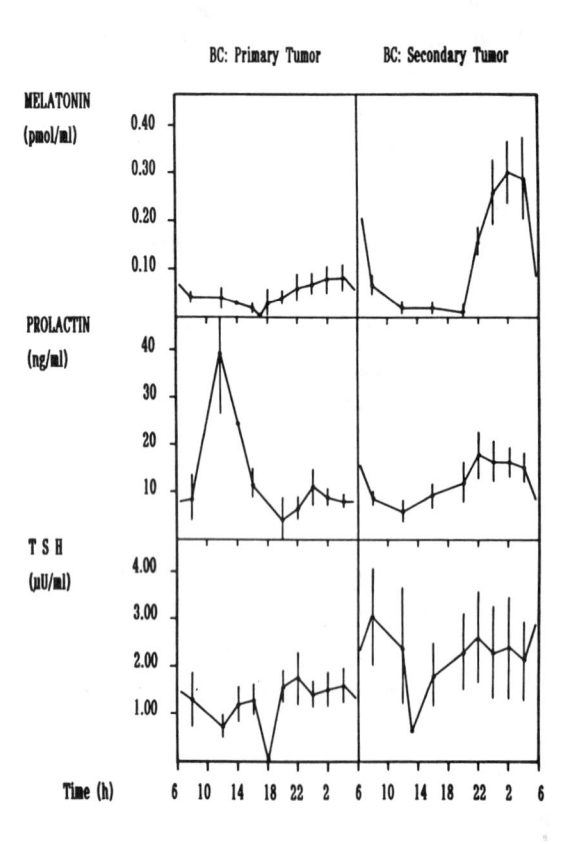

Fig. 1: The corresponding average concentrations ±SEM of serum melatonin, prolactin and thyroid-stimulating hormone (TSH) at the different time points over 24 hours are given for patients with primary breast cancer (BC: Primary Tumor, n=10) and for patients with secondary breast cancer (BC: Secondary Tumor, n=7-8).
(Reproduced from (2) with permission of J.B. Lippincott Company, Philadelphia, PA, U.S.A.)

A striking finding of studies on patients with BC (12,2) and PC (3,4) is that a significant and marked depletion of the amplitude and MESOR of serum melatonin occurs in patients with primary unoperated and untreated malignant tumors. The depletion of melatonin progresses with increasing size of the tumor (BC: (2); PC: BARTSCH et al., unpublished results) and seems to be the result of a depressed production and secretion by the pineal gland because changes in the peripheral metabolism of the hormone by the liver were ruled out (4,5). Depression of melatonin in cancer patients appears to be mainly limited to the phase of primary growth since patients with secondaries (local recidives, tumors of the contra-lateral breast, distant metastases) again show normal levels (2). Since melatonin acts as a chemical signal of the circadian oscillator of the SCN it can be presumed that disturbances in the endocrine balance may occur in patients with depressed melatonin. The results show in fact clear changes in the circadian secretion of adenohypophyseal hormones in BC (2,6) and PC (3,6,7), particularly for prolactin. In patients with PC no circadian rhythm of prolactin exists due to an ultradian pattern of secretion with elevated levels in the afternoon in addition to the usual night surge. In patients with BC highly elevated values are observed at noon which even exceed the nocturnal values leading to a drastic acrophase shift by eleven hours (see Fig.1). Ultradian rhythmicity can also be found for growth hormone and, to a lesser extent, for the gonadotropins, LH and FSH, in case of PC (see Fig.2). TSH secretion shows neither circadian nor ultradian patterns in BC and PC and their age-matched controls but a depressed MESOR in BC (see Fig.1) and PC. The overall balance of the central hormones, including melatonin and the adenohypophyseal hormones, seems to be marked by a pronounced loss of circadian secretory activity in the cancer patients. Elderly patients without malignant tumors still have circadian rhythms, however, with lower night-time synchronization among the different central hormones than young subjects (see Fig.2). In contrast, the hormones of peripheral glands, i.e. thyroxine, cortisol, and testosterone, show a high degree of stability of MESOR, amplitude, and acrophase in patients with PC (6). In primary BC a 1.5 h delay of the cortisol acrophase is found which leads to a coincidence with the unusual acrophase of prolactin at noon in these patients. The difference in cortisol secretion in BC in contrast to PC may indicate a possible sexual dimorphism in the control mechanisms of adrenal hormone secretion (BARTSCH et al., unpublished results).

MELATONIN
PROLACTIN
L H

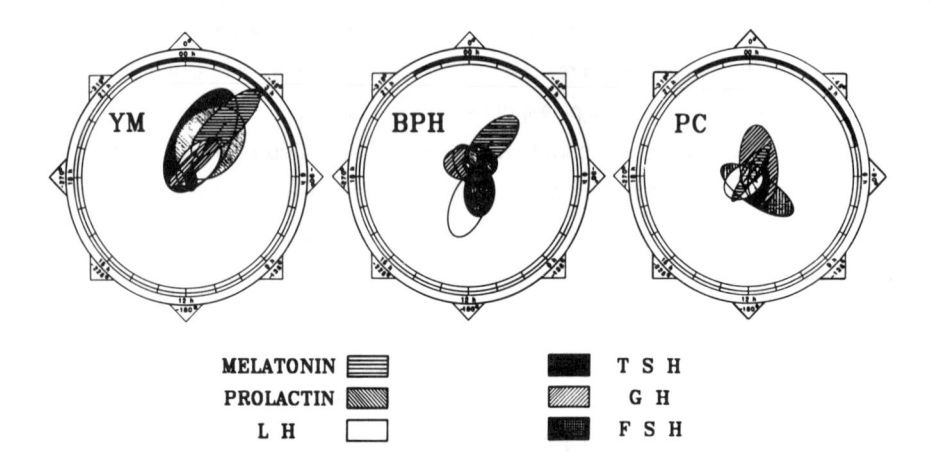

T S H
G H
F S H

Fig. 2: Cosinor plots of the endocrine rhytms in young men (YM: n=10), patients with benign prostatic hyperplasia (BPH: n=13) and patients with primary prostate cancer (PC: n=9) showing the cosinors of the different endocrine parameters as well as corresponding 95% confidence regions. (LH: luteinizing hormone, FSH: follicle-stimulating hormone, TSH: thyroid-stimulating hormone, GH: growth hormone).

Summarizing our findings it may be concluded that endocrine disturbances in patients with primary BC and PC are mainly confined to the central part of the endocrine system. A relative stability of the peripheral hormonal secretion indicates that a modified feedback control occurs at the level of the endocrine hypothalamus controlling adenohypophyseal hormone secretion. The endocrine hypothalamus is also a main target of melatonin action (8) which when depressed could lead to changes in the sensitivity and temporal response of the hypothalamic neurons to peripheral endocrine feedback signals that affect the secretion of the gonadotropins as well GH and TSH. Changes in the diurnal secretion of prolactin in patients with depleted melatonin point to a lowered activity of the dopaminergic neurons that tonically inhibit prolactin secretion.

It may be argued that the presence of endocrine imbalances including the depression of melatonin are of minor importance since they are only transient phenomena confined to the period of primary malignant growth before operation. However, recent studies show that the endocrine balance at the time of surgery is a decisive factor for survival of the patients because a significantly better prognosis exists for those patients with BC who are operated during the luteal phase of the menstrual cycle (9). If the acute endocrine balance at the time of surgery is indeed of great importance then the observed hormonal disturbances in patients with BC and PC could enhance metastatic spread and local recidives leading to a shorter survival. A substitutive therapy with melatonin in patients with depressed pineal secretion may therefore be advisable since melatonin possesses an antineoplastic activity (10,11) and may even alleviate parts of the endocrine disturbances.

It is concluded that the pineal gland and its hormone melatonin play a central role in the chronobiological control of the endocrine system as well as of hormone-dependent tumors. The mechanisms involved in the depression of melatonin secretion require to be elucidated and it is hoped that more extensive future clinical studies using non-invasive urinary aMT6s-determinations (1) may contribute to the solution of this problem.

ACKNOWLEGDEMENT

The authors are research scholars of the Society of Servants of God, New Delhi, India, working in the Cancer Research Programme of the Society.

REFERENCES

1. ALDOUS, M.E. & J. ARENDT (1988): Radioimmunoassay for 6-sulphatoxymelatonin in urine using an iodinated tracer. Ann. Clin. Biochem. 25, 298.
2. BARTSCH, C., H. BARTSCH, U. FUCHS et al. (1989): Stage-dependent depression of melatonin in patients with primary breast cancer: correlation with prolactin, TSH, and steroid receptors. Cancer 64, 426.
3. BARTSCH, C., H. BARTSCH, S.-H. FLÜCHTER et al. (1985): Evidence for modulation of melatonin secretion in men with benign and malignant tumors of the prostate: relationship with the pituitary hormones. J. Pineal Res. 2, 121.
4. BARTSCH, C., H. BARTSCH, A. SCHMIDT et al.: Melatonin and 6-sulfatoxymelatonin circadian rhythms in serum and urine of primary prostate cancer patients: evidence for reduced pineal activity and reliability of urinary determinations. (submitted for publication).
5. BARTSCH, C., H. BARTSCH, O. BELLMANN & T.H. LIPPERT (1991): Depression of serum melatonin in patients with primary breast cancer is not due to an increased peripheral metabolism. Cancer 67, 1681.
6. BARTSCH, C. (1988): Untersuchungen zur Funktion der Zirbeldrüse und ihrer Beziehung zum endokrinen System im Brust- und Prostatacarcinom: tierexperimentelle und humanbiologische Studien. Dissertation, University of Tübingen.
7. BARTSCH, H., C. BARTSCH & T.H. LIPPERT (1991): Melatonin und chronobiologische Aspekte von Krebs. Münch. med. Wschr. 133, 113.
8. BLASK, D.E. (1981): Potential sites of action of pineal hormones within the neuroendocrine-reproductive axis. In: R.J.REITER (ed.), The Pineal Gland, Vol.II, CRC Press, Boca Raton, Florida, p.189.
9. BADWE, R.A., W.M. GREGORY, M.A. CHAUDARY et al. (1991): Timing of surgery during menstrual cycle and survival of premenopausal women with operable breast cancer. Lancet 337, 1261.
10. BARTSCH, H. & C. BARTSCH (1981): Effect of melatonin on experimental tumors under different photoperiods and times of administration. J. Neural Transm. 52, 269.
11. BLASK, D.E. (1984): The pineal: an oncostatic gland? In: R.J.REITER (ed.), The Pineal Gland, Raven Press, New York, p.253.
12. TAMARKIN, L., D. DANFORTH, A. LICHTER et al. (1982): Decreased nocturnal plasma melatonin peak in patients with estrogen receptor positive breast cancer. Science 216, 1003

MECHANISMS UNDERLYING THE PATHOGENESIS OF OSCILLATORY ACTIVITY IN PALATAL MYOCLONUS

H. HEFTER, E. LOGIGIAN, O. WITTE, K. REINERS & H.-J. FREUND

Department of Neurology University of Düsseldorf,
Moorenstraβe 5, D-W 4000 Düsseldorf 1, FRG

INTRODUCTION

In palatal myoclonus (PM) a broad spectrum of complex motor phenomena has been observed which are poorly understood so far. PM is regarded as "the prototype of a hyperkinetic movement disorder depending on a central pacemaker (1), chiefly because the rhythmic jerks were time-locked in different muscles" (2).

The following features are thought to be characteristic for PM: 1) PM may be uni- or bilateral, 2) the contractions are synchronous and rhythmic, 3) they may be associated with contractions of external ocular muscles, and those of larynx, neck, diaphragm, trunk, and limbs, 4) the contractions occur at an average rate of 120 - 130 per minute, with a range of 80 to 180 per minute, 5) they persist during sleep, 6) they disappear after damage to the corticobulbar or spinal motoneuron pathways, and 7) there is hyperthrophic degeneration of the contralateral inferior olive (3).

Only few quantitative measurements of the complex motor phenomena in PM have been reported. We have analyzed in a patient with PM involving eye muscles, the palat, respiratory and extremity muscles some aspects of this movement disorder dealing with the coupling of the oscillatory activity in different motor subsystems and the functional role of feedback.

CASE REPORT

The 65 year old woman had a 14 year history of several brainstem attacks with nausea, oculomotor disturbances, and gait ataxia. After these attacks she fully recovered within a few days. One year before admission to our hospital she suffered from the first severe stroke. We saw the patient 2 days after a second brainstem stroke. She presented with a bilateral palatal myoclonus interfering with speaking and breathing and involving trunk and limb muscles on the left side. The oculomotor system was severely disturbed with a spontaneous horizontal myoclonic jerking of both eyes resulting in oscillopsias, a reduction of optokinetic responses in all directions, an impaired smooth pursuit and hypermetric saccades to the right and a paretic abduction of the left eye. Furthermore, she had a tetraspasticity with a moderate paresis of the left arm and extensor plantar responses on the left side as well as a leftsided hemiataxia in addition to a severe gait and stance ataxia and bradykinesia on both sides. A mild paresis of the right facial muscles and sensory impairment in the area of the right trigeminal nerve were thought to be residual impairment resulting from the first stroke.

METHODS

The patient underwent several recording sessions each lasting not longer than about 30 minutes. For recordings of myoclonic activity in the abductor digiti minimi (ADM) muscle, the flexor digitorum superficialis and the extensor carpi radialis (ECR) muscle, conventional surface-EMG electrodes were placed over the bellies of these muscles. The frequency of involuntary movements of the digiti minimi and the thoracic wall was measured by means of an accelerometer attached to the nail of the finger or the sternum by adhesive tape. For measurement of most rapid voluntary alternating index finger movements the accelerometer was taped to the distal phalanx of the index finger. Both voluntary and involuntary movements were analyzed off-line by spectral analysis.

Most rapid voluntary isometric index finger extensions were recorded by means of a force transducer. In a simple reaction time paradigm the patient had to produce a short force pulse by extending her index finger as fast as possible as soon as possible after hearing a tone signal. (For methodological details of recordings of voluntary and involuntary movements see (4)).

Cutaneous stimulation of the ulnar nerve at the Vth finger of the left hand was carried out with ring electrodes. The electrical stimulus served as a trigger signal for the recording of EMG activity which was digitized over a period of 2 sec with 1kHz. Horizontal eye movements were recorded by means of conventional EOG-electrodes.

RESULTS

Rhythmic myoclonic activity was recorded from the left abductor digiti minimi muscle (ADM). When the wrist was held extended to an horizontal position and the fingers were stretched out completely a mean inter-burst interval of 343 ± 21 ms was observed. Control of interburst intervals during maintenance of finger and wrist position three weeks later yielded similar values (mean 346 ± 22 ms).

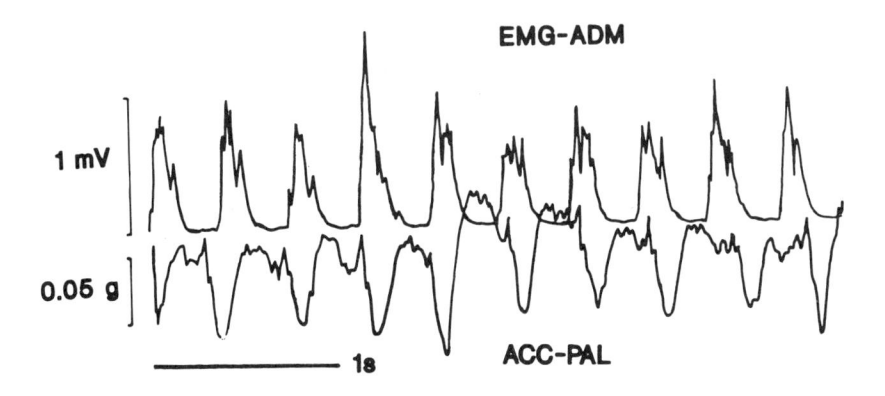

Fig.1: Comparison of smoothed rectified surface-EMG bursts from the abductor digiti minimi muscle with accelerometer recordings of palatal movements. Though an overall 1:1 relationship was found considerable phase shifts occurred.

There was a 1:1 correspondence between finger abductions (Fig.1 upper trace) and palatal myoclonic activity (Fig.1 lower trace). However, no stable phase-locking could be detected. The variation of both the finger abductions and the palatal myoclonus led to considerable phase shifts of up to ± 150 deg. (Fig.1).

Accelerometer recordings of chest movements showed that the amplitude of the myoclonic activity was heavily modulated by breathing. The rhythmic myoclonic jerking of the chest wall was only visible during inspiration. Spectral analysis revealed that the mean dominant frequency was the same for the finger and chest movements.

In the fingers no corresponding variation of frequency or amplitude with breathing was found.

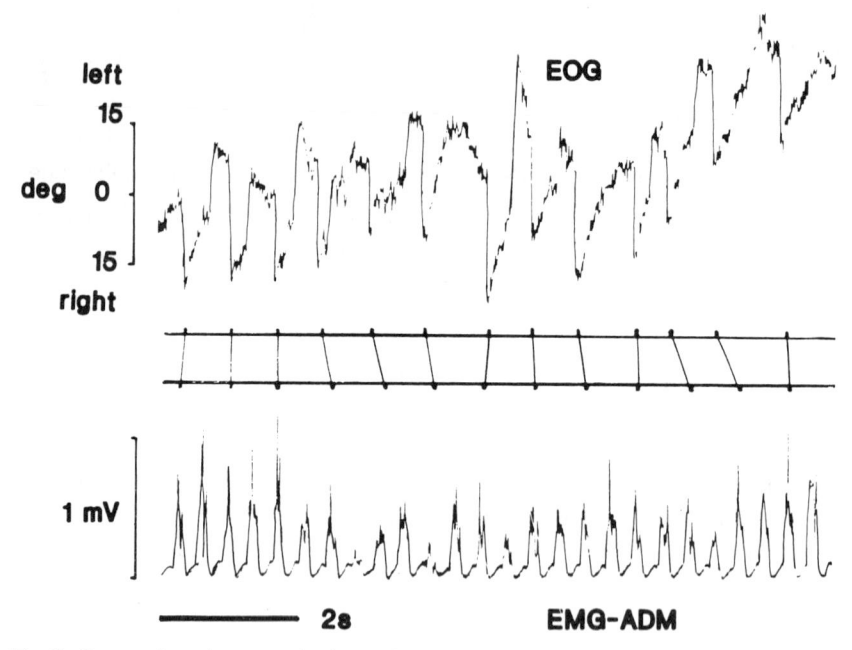

Fig.2: Comparison between horizontal eye and finger movements revealed a 1:2 relationship. Again considerable phase shifts were observed.

When the patient had to perform most rapid voluntary isometric index finger extensions prolonged contraction times were observed. In addition to the slowness of voluntary movements superimposed oscillations were detected with a frequency observed during tonic extension of the entire hand. Voluntary alternating index finger movements were also performed with this frequency. Thus the patient did not succeed in escaping voluntarily the frequency range of the involuntary oscillations. Mean reaction times were prolonged with a considerable variation (283 ± 68 ms). No reaction times were observed in the 320-400 ms range though both shorter and longer reaction times were measured. Thus relative to the start signal, initiation of most rapid index finger extensions was not possible during the time period around the inter-burst interval.

Cutaneous stimulation of the ulnar nerve reset the myoclonic jerking in the abductor digiti minimi muscle (ADM). The best resetting result was achieved when the stimulus was applied 1/4 to 3/4 of a cycle length. When the stimulus was applied at the onset of a myoclonic burst prolongation of this burst was prolonged. This underlines the phase dependent vulnerability of the oscillator.

Comparison of finger and eye movements revealed that the jerking rate in the eyes was half of that observed in the finger movements. Thus a 1:2 relationship was observed. This is indicated in Fig.2 where eye movements and ADM-EMG were compared over a period of a few seconds. However, considerable phase shifts were observed indicating a weak coupling between both oscillations.

DISCUSSION

In the present case of symptomatic palatal myoclonus, rhythmic oscillatory activity in hand and forearm muscles on the left side, in the palat and in trunk as well as external ocular muscles was found. Involvement of the extremities as well as of the eyes is a rare feature in PM (2). In the present case a combination of different unusual clinical features was found. The palatal movements and the myoclonic jerking in the left hand occurred with a rather high frequency but still in the burst range of 60 to 180/ min (3). The rhythmic in- and expiration heavily interfered with the rhythmic movements of the thoracic wall. During expiration the rhythmic movements were nearly abolished. During inspiration the breathing was modulated by the 3 Hz oscillator to such an extent that a loud rhythmic stridor could be heard. Modulatory influence of in- and expiration on amplitude and frequency of PM is well-known (5,6). This has led to the assumption that PM is a human homologue of gill-breathing in invertebrates (5) and that structures close to the inferior olive (IO) as e.g. the nucleus ambiguus and the dorsolateral reticular formation which are related to branchial muscles and respiration are involved in PM (7). A recent PET-study (8) demonstrated that the IO is hyperactive in PM. Thus, inspite of exceptional cases there is convergent evidence that the injured IO influences different motor subsystems as a central pacemaker.

On the other hand the oscillations in the present case could be reset (cf. 9). This indicates that feedback mechanisms may contribute to the generation of PM. Furthermore, the myoclonic jerking in the present case cannot be explained by a single pacemaker leading to phase-locked activity in all affected motor subsystems. Instead there was a 1:2 ratio between eye and finger jerking and a considerable fluctuation of phase relations between different motor subsystems. Thus disinhibition of IO neurons in combination with feedback mechanisms may be the mechanisms for the generation of the complex motor phenomena found in patients with PM.

ACKNOWLEDGEMENT

This study was supplied by grants from the Deutsche Forschungsgemeinschaft (SFB 194, A5)

REFERENCES

1. LLINÁS, R.R. (1984): Rebound excitation as the physiological basis of tremor: a biophysical study of the oscillatory properties of mammalian central neurones in vitro. In: L.J. FINDLEY & R. CAPILDEO (eds.): Movement disorders: Tremor. Macmillan, London, pp. 165-182
2. DEUSCHL, G., G. MISCHKE, E. SCHENCK et al. (1990): Symptomatic and essential palatal myoclonus. Brain 113, 1645-1672
3. MARSDEN, C.D., M. HALLETT & S. FAHN (1982): The nosology and pathophysiology of myoclonus. In: C.D. MARSDEN & S. FAHN (eds.): Movement disorders. Butterworths, London, pp. 196-248
4. HEFTER, H., V. HÖMBERG, H. LANGE & H.-J. FREUND (1987): Impairment of rapid movement in Huntington's disease. Brain 110, 585-612
5. STERN, M. (1949): Rhythmic palatopharyngeal myoclonus; review, case report, and significance. Journal of nervous and mental diseases 109, 48-53
6. MASUCCI, E.F., J.F. KURTZKE & N. SAINI (1984): Myorhythmia: a widespread movement disorder. Clinicopathological correlations. Brain 107, 53-79
7. KANE, S.A. & W.T. THACH (1989): Palatal myoclonus and the function of the inferior olive: are they related? Exp. Brain Res. 17, 427-460
8. DUBINSKY, R.M., M. HALLETT, G. DI CHIRO et al. (1991): Increased glucose metabolism in the medulla of patients with palatal myoclonus. Neurology 41, 557-562
9. SOSO, M.J., V. KAMP NIELSEN & P.J. JANETTA (1984): Palatal Myoclonus. Reflex activation of contractions. Arch. Neurol. 41, 866-869

CIRCADIAN PATTERNS IN THE OCCURRENCE OF PANIC ATTACKS

E. BECKER, J. MARGRAF & S. SCHNEIDER

Department of Psychology, Philipps-University, Gutenbergstr.18, Marburg and Christoph-Dornier-Foundation for Clinical Psychology, D-W-3550 Marburg, FRG

INTRODUCTION

Circadian patterns are prominent in affective disorders. They have been extensively investigated in depressive disorders (2). Surprisingly, there are almost no studies of circadian patterns in anxiety disorders, although the comorbidity with depression is high. During the 1980's panic disorder has been the most frequently investigated mental disorder. It is characterized by panic attacks which are sudden episodes of intense anxiety accompanied by multiple somatic symptoms (primarily palpitations, dizziness, and shortness of breath) although the patient is typically in good physical health. During an attack, the patient is afraid of dying, going crazy or loosing control. The lifetime prevalence of panic disorder is 4-5% and the long-term outcome is unfavorable. There are almost no spontaneous remissions and the suicide rate is about eighteen times higher than in the normal population. Panic disorder also causes significant health care costs because panic disorder patients see their physicians frequently and often undergo expensive medical workups. Studies of circadian patterns in the occurence of panic disorder are sparse and the results published so far vary greatly. CAMERON et al. (1) found more panic attacks to occur in the afternoon or evening, whereas MARGRAF et al. (3,4) found more attacks in the morning. Another study by TAYLOR et al. (5) found a greater number of attacks during daytime, but the attacks were randomly distributed over the day. All these studies can be criticized because data analysis was based on averages across individuals. Thus, the results might reflect an averaging artifact and not resemble any individual pattern. Furhermore, these studies determined circadian patterns on the basis of rather short time intervals, i.e. only one or two days.

METHODS

We studied 25 panic disorder patients (15 female, 10 male) with a mean age of 36 years and a mean duration of panic disorder of seven years. The patients kept a standardized panic attack diary for 10 weeks. Immediately after the occurence of an attack, the patients recorded its onset, duration, severity, symptoms, and setting. In this study we concentrated on the individual patterns. We were interested in intervals across the course of day where the occurrence of attacks was higher than expected by chance. We called these intervals "peaks" and defined them by two sets of criteria: Peaks defined by strict criteria were periods of four hours where three times more attacks occurred than expected by chance. Liberal criteria peaks were periods of eight hours where two times more attacks occurred than expected by chance.

RESULTS

In the total sample of 647 attacks as recorded, there were significantly more attacks during daytime than during nighttime. They were randomly distributed over the day

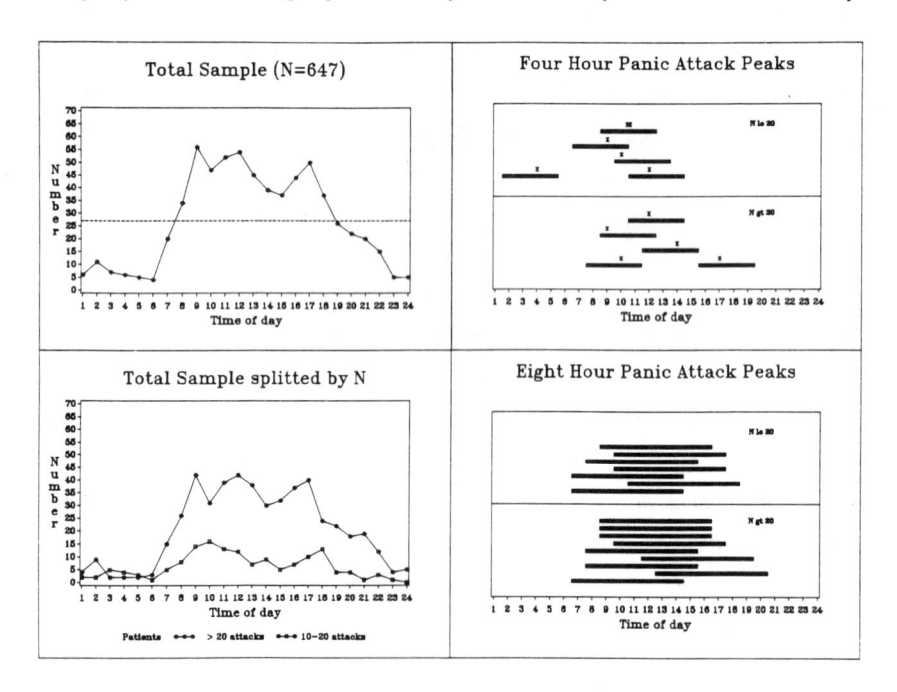

Fig. 1: The left half of the figure shows the time of occurrence of all 647 panic attacks (upper part) and separately for the attacks of patients with more or less than 20 attacks (lower part). The right half of the figure shows the time of occurrence of four- hour peaks (upper part) and of eight hour peaks (lower parts), indicated by black bars. Time of day is given on a 24 hour scale.

with no clear peaks, as can be seen in Fig. 1. This is an average curve, however, and we therefore had to look at individual curves as well.

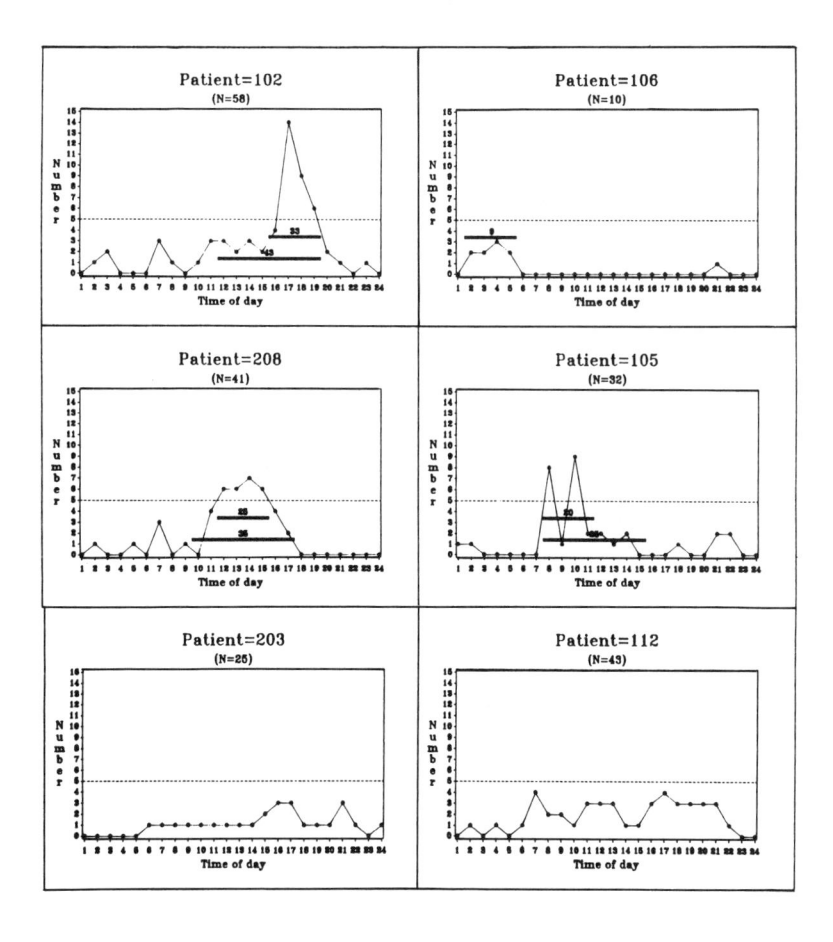

Fig. 2: Time of occurrence of panic attacks in six examples of individual patients. For each patient, the total number of attacks (N) as well as the peaks (if present indicated by black bars) and the number of attacks in those peaks are shown. Time of day is given on a 24 hour scale.

Fig. 2 shows several examples of individual curves. These varied considerably, i.e. all types of patterns occured. Very few of these curves resembled the average pattern. Forty percent of our patients had a four hour peak. Patient 106, for example, was the only subject with a peak at nighttime (and no attacks during the day), whereas patient 105 had a peak in the morning hours and patient 102 a peak in the evening. Sixty-four percent had an eight hour peak (e.g. patient 208), and twenty-eight percent had no peak at all (e.g. 112 and 203). Their attacks were randomly distributed over the day. If patients showed peaks of panic attacks, these were not randomly distributed. Almost all four hour peaks appeared in the morning or around noon and the eight hour peaks consistently appeared during daytime as can be seen in the right part of Fig. 1.

CONCLUSION

The average curve we found in our sample most probably reflects an averaging artifact. It may be suspected that the patterns published earlier were similarly due to artifact. Panic attacks occur primarily during daytime activities. This might be due to confrontation with phobic stimuli at daytime and not to circadian rythms per se. Individual peaks of panic attacks exist in some patients. While such peaks may occur at any time, we observed by far most of them during the morning hours. Preliminary investigations of relationships of circadian patterns with other clinical variables (e.g. level of depression) yielded no significant results. Thus, the clinical significance of our findings remains to be determined.

REFERENCES

1. CAMERON, O., M. LEE, J. KOTUN & K. MCPHEE (1986): Circadian symptom fluctuations in people with anxiety disorders. Journal of Affective Disorders, 11, 213-218.
2. HEALY, D. & J. WILLIAMS (1988): Dysrhythmia, dysphoria, and depression: The interaction of learned helplessness and circadian dysrhythmia in the pathogenesis of depression. Psychological Bulletin 103, 163-178.
3. MARGRAF, J., C.B. WILLIAMS, A. TAYLOR et al. (1987): Panic attacks in the natural environment. Journal of Nervous and Mental Disease 175, 558-565.
4. MARGRAF, J. (1990): Ambulatory psychophysiological monitoring of panic attacks. Journal of Psychophysiology 4, 321-330.
5. TAYLOR, C.B., J. SHEIKH, W.S. AGRAS et al. (1986): Ambulatory heart rate changes in patients with panic attacks. American Journal of Psychiatry 143, 478-482.

CHRONOPHARMACOLOGY AND -TOXICOLOGY

CHRONOPHARMACOKINETICS OF AMITRIPTYLINE IN RATS AFTER I.V. ADMINISTRATION

A. RUTKOWSKA, W. PIEKOSZEWSKI & J. BRANDYS

Department of Toxicology Medical Academy in Cracow, Podchorążych 1, 30-084 Cracow, Poland

INTRODUCTION

Circadian rhythms in the pharmacologic actions of a variety of drugs have been reported (2,3,9). The earlier studies were limited to observations of rhythmicity in drug effects and toxicity and did not explore any mechanism underlying circadian variations. Pharmacological actions of a variety of drugs are determined not only by the drug-receptor sensitivity, but also by kinetic variables. Circadian changes in the kinetics of drugs may be found in each phase of the drug present in the organism (absorption, distribution, elimination) (9). In the case of amitriptyline, few data concerning changes in distribution and elimination according to the time of drug administration may be found. This issue is designed to clarify the effects of the time of day of dosing on distribution and elimination of amitriptyline after a single intravenous dose.

MATERIAL AND METHODS

Animals: male Wistar rats (aged 3 months) were kept in cages in groups of 5, under a light:dark regimen of 12:12 h (light between 6.00 a.m. and 6.00 p.m.), at temperature of 20-22°C, with free access to water and food.

Chemicals: amitriptyline hydrochloride (POLFA). Other chemicals, anal. grade, were delivered by POCh Gliwice.

Methods: Amitriptyline hydrochloride in dose 3 mg/kg (solution in physiological saline) was administered intravenously to rats in the early part of November at the following hours: 6.00 a.m., 10.00 a.m., 2.00 p.m., 6.00 p.m., 10.00 p.m., 2.00 a.m..

From 5 to 180 min after the amitriptyline administration blood and organs (brain, heart, kidney, liver, lungs) were collected. Amitriptyline concentrations were determined by gas-chromatography with AFID detector as described previously (4).

Pharmacokinetic and statistic analysis: pharmacokinetic parameters were determined according PKCALC program (10). Chronopharmacokinetic parameters were calculated using the single cosinor analysis and the Fischer-Snedecor F-test were used to statistical analysis of rhythmicity (5).

RESULTS AND DISCUSSION

Due to its favorable effects in treating depression, amitriptyline is a quite popular medicine. However, this conventional tricyclic antidepressant drug shows a fairly high toxic effect. Quite often, monitoring its level in serum neither gives rise to necessary therapeutical level nor are the toxic symptoms prevented. This might be related to amitriptyline chronopharmacokinetics, which has not been given enough consideration hitherto. Few works in this field described changes in absorption and in the elimination rate of this drug when administered in the morning and evening (7), while a choice of the dosing time was chronobiologically unfounded.

In the presented work, which is aimed at assessing amitriptyline chronodistribution and chronoelimination, the simplest pharmacokinetic model had been provided (single intravenous administration), which, however, enables to get maximum information on the pharmacokinetics.

Tab. 1 shows the values of area under the time-amitriptyline concentration curves (AUC), which measure the drug amount in serum and in the tissues under examination, as well as the mean residence time (MRT) values describing the rate of amitriptyline elimination after the drug is administered at six different times during the day.

Distribution of intravenously administered amitriptyline is largely different than the disposition after intragastric administration (6). After i.v. administration, only ca. 1% of the drug is found in the liver, while a significant portion is accumulated in lungs;

Tab. 1: Pharmacokinetic parameters for amitriptyline after i.v. administration of a single 3 mg/kg dose.

Time of administration [h]	Tissue					
	Serum	Brain	Heart	Kidney	Liver	Lung
AUCtotal [mg/h/l]						
6.00am	1.27	13.47	4.86	9.87	1.33	57.97
	+/-0.08	+/-1.71	+/-0.91	+/-0.74	+/-0.24	+/-11.43
10.00am	0.861	6.15	3.83	5.08	0.815	9.19
	+/-0.21	+/-1.09	+/-0.43	+/-0.69	+/-0.05	+/-8.73
2.00pm	0.85	13.60	4.12	6.86	0.98	63.32
	+/-0.11	+/-1.53	+/-0.84	+/-1.48	+/-0.09	+/-8.08
6.00pm	0.66	13.27	4.00	8.65	1.68	89.59
	+/-0.11	+/-2.73	+/-0.57	+/-1.46	+/-0.24	+/-13.05
10.00pm	0.72	18.37	3.40	6.95	1.48	80.03
	+/-0.22	+/-2.67	+/-0.81	+/-1.68	+/-0.22	+/-15.69
2.00am	0.47	13.87	4.63	7.76	1.08	90.36
	+/-0.18	+/-1.40	+/-0.90	+/-2.72	+/-0.45	+/-13.09
MRT [h]						
6.00am	3.68	1.29	1.15	1.19	2.57	1.46
	+/-0.18	+/-0.01	+/-0.16	+/-0.14	+/-0.18	+/-0.02
10.00am	3.00	1.47	1.70	1.57	4.09	0.86
	+/-0.38	+/-0.19	+/-0.29	+/-0.24	+/-1.41	+/-0.01
2.00pm	1.82	0.95	0.98	1.28	1.23	1.65
	+/-0.10	+/-0.01	+/-0.03	+/-0.19	+/-0.03	+/-0.22
6.00pm	1.36	1.07	1.41	1.60	2.67	1.38
	+/-0.04	+/-0.16	+/-0.05	+/-0.17	+/-0.36	+/-0.08
10.00pm	1.87	1.66	1.15	1.19	3.02	0.99
	+/-0.07	+/-0.27	+/-0.07	+/-0.24	+/-0.01	+/-0.06
2.00am	1.26	0.88	1.43	1.47	4.69	2.00
	+/-0.37	+/-0.17	+/-0.16	+/-0.09	+/-0.75	+/-0.17

on the other hand, p.o. administration results in ca. 40% of the drug being retained in the liver, while the lung concentration in ca. 50% less than after i.v. administration.

Heart, brain, and kidney distribution of amitriptyline after both ways of administration is comparable to each other.

The chronobiological analysis based on "Cosina" program (5) showed circadian changes in the MRT of amitriptyline in serum, brain, liver, and lungs (Tab. 2).

Tab. 2: Cosinor analyses of pharmacokinetic parameters of amitriptyline.

Parameter	AUC					
	Serum	Brain	Heart	Kidney	Liver	Lung
Period [h]	24	12	12	12	24	24
Mesor [mg/h/l]	0.80 +/-0.06	14.79 +/-10.45	4.14 +/-0.18	7.53 +/- 0.41	1.23 +/- 0.07	73.41 +/-3.10
Amplitude [mg/h/l]	0.22 +/-0.08	2.48 +/-10.63	0.54 +/-10.26	1.88 +/- 0.58	0.31 +/- 0.11	16.77 +/-4.38
Acrophase [h]	8.84 +/-1.45	10.17 +/-10.49	4.12 +/-10.91	5.22 +/- 0.59	21.11 +/- 1.31	21.7 +/-1.00
Rhythm [%]	31.6	50.8	22.9	41.5	35.9	49.4
	MRT					
Period [h]	24	12	Circadian rhythm not detected		24	12
Mesor [h]	2.16 +/-0.16	1.19 +/-0.03			3.05 +/- 0.26	1.39 +/-0.04
Amplitude [h]	1.01 +/-0.22	0.32 +/-10.04			1.00 +/- 0.37	0.52 +/-0.05
Acrophase [h]	7.94 +/-0.84	9.07 +/-10.27			3.08 +/- 1.43	3.11 +/-0.20
Rhythm [%]	57.8	77.6			32.1	86.1

Circadian changes in the AUC were observed for serum and for every organ examined (Tab. 2). Maximum AUC value (acrophase) for any organ, except the brain, is for the dark phase. Probably, this is due to increased blood flow through these organ at the time of the rat's elevated activity. Maximum AUC values for the liver and lungs are accompanied by the highest MRT values in these organs. High AUC values of amitriptyline in the liver at the dark phase may be also related to the weight of this organ being reduced (minimum at 10.00 p.m.) and to the increased level of non-enzymatic proteins (maximum at 3.00 a.m.) (8). The acrophase for AUC and MRT of amitriptyline in serum was observed at the light phase (9.00 a.m.), which was probably related to the minimum activity of the drug-metabolizing enzymes (1), shown at 10.00 a.m. Maximum AUC and MRT values are accompanied by minimum clearance value - 2.38 mg/l/h (clearance not tabulated). Minimum AUC value of amitriptyline in serum at 2.00 a.m. may be due to increased binding by the proteins with which ca. 90% of this drug is combined, their concentration at this time showing a peak value (8). Another reason for lower AUC values at the dark phase might be faster amitriptyline elimination observed at 2.00 a.m. (maximum clearance value 6.14 mg/l/h).

Brain is the only organ having AUC and MRT acrophase at the light phase. The differences found cannot be explained on the basis of our experimental results. Further studies would be required on the changes in the blood/brain barrier permeability for amitriptyline and to the changes in the drug binding by the macromolecules of brain cells.

Circadian changes of AUC amitriptyline in brain and heart allow us to presume that with the compound administration at 10.00 a.m. or 10.00 p.m. maximum brain concentration with simultaneous minimum heart levels will be achieved.

A similar correlation had been observed in clinical research; with amitriptyline administered in the evening there was a substantial fall in the cardiotoxic effect, with the therapeutic effect remaining unchanged (7).

Our studies based on the simplest model, i.e. a single intravenous administration, give a fundamental information on amitriptyline chronopharmacokinetics and chronodistribution, and they shall be made-up by a research covering gastric and multiple administration.

REFERENCES

1. FALK, M. (1984): Einfluss von Biorhythmen auf die Arzneimittelwirkung. Pharmazeut. Praxis 39, 59
2. LEMMER, B. (1986): Chronopharmakologie. Wichtiges für die Arzneimitteltherapie. Deutsche Apotheker Zeitung 126, 1030
3. LEMMER, B. & G. LABRECQUE (1987): Chronopharmacology and chronotherapeutics: definitions and concepts. Chronobiology International 4, 319
4. MONICZEWSKI, A., A. RUTKOWSKA, W. PIEKOSZEWSKI et al. (1990): Circadian changes in the disposition of amitriptyline, doxepin and imipramine in rats. Clin. Pharmacol. 6, 28
5. MONK, T.H. & A. FORT (1983): 'Cosina': A cosine curve fitting program suitable for small computers. Inter. J. Chronobiology 8, 193
6. NEGRUSZ, A., J. BRANDYS & W. PIEKOSZEWSKI (1991): The effect of some xenobiotics on amitriptyline and nortriptyline distribution in rats. Pol. J. Pharmacol. Pharm. 43, 27
7. NAKANO, S. & L.E. HOLLISTER (1983): Chronopharmacology of amitriptyline. Clin. Pharmacol. Ther. 33, 453
8. PHILLIPPENS, K.M.H. (1976): The manipulation of circadian rhythms. Arch. Toxicol. 36, 277
9. REINBERG, A. & M.H. SMOLENSKY (1982): Circadian changes of drug disposition in man. Clin. Pharmacokinetics 7, 401
10. SHUMAKER, R.C. (1986): PKCALC: A BASIC interactive computer program for statistical and pharmacokinetic analysis of data. Drug Metabolism Reviews 17, 331

INFLUENCE OF GALANTHAMINE (NIVALIN) ON SYNAPTIC TRANSMISSION IN THE RAT NEOCORTEX IN VITRO

G. HESS

Jagiellonian University, Department of Animal Physiology,
ul.Karasia 6, 30-060 Krakow, Poland

INTRODUCTION

It has been noted previously that the injection of galanthamine (Nivalin), an inhibitor of acetylcholinesterase (9), resulted in a phase shift of the circadian rhythm of the locomotor activity of rodents (1). However, apart from possible effects on central mechanisms regulating circadian rhythmicity, the compound might exert its influence on the locomotor activity pattern by changing the activity of neurons in the motor cortex. Testing this possibility was the aim of the present study.

METHODS

Rat neocortical slices with a nominal thickness of 500 μm were cut in a coronal plane from a region corresponding to the motor cortex. Individual slices were transferred to a submerged-type incubation chamber continuously perfused at a flow rate of 2.8 ml/min with saline of the following composition (in mM): NaCl 124, KCl 5, NaH_2PO_4 1.25, $NaHCO_3$ 26, $CaCl_2$ 2.5, $MgSO_4$ 1.3 and glucose 10, bubbled with 95% O_2 and1990 5% CO2. Conventional techniques for extracellular recording were used. Recording microelectrode was inserted at the distance of 300 - 400 μm from the pial surface. Single stimuli (10 - 20 V, 0.2 Hz) were delivered with a bipolar stimulating electrode situated at the white matter/layer VI border or inserted in the grey matter at the distance of about 1000 μm from the pial surface radial to the site of recording. Data were averaged and digitized on a personal computer. Galanthamine hydrobromate (Nivalin, Pharmachim, Bulgaria) and atropine sulfate (Medexport, USSR) were applied to the incubation fluid.

RESULTS

A typical field potential recorded at the distance of 300 - 400 µm from the surface of the neocortex consists of two major components (Fig. 1B). The early (first) waveform represents the synchronous firing of antidromically activated neurons and/or firing of afferent fibres, the later (second) component is mediated by glutamatergic synaptic transmission (3).

The application of galanthamine to the incubation medium for 10 minutes resulted in a marked reduction of the amplitude of synaptically driven part of the field potential while the early part of the potential remained unaltered (Fig.1 A, B). The clear effect appeared at the concentration of galanthamine of 10 micromol, greater concentration (up to 300 micromol) did not induce in general an additional reduction of the response. On average, galanthamine reduced the amplitude of the later component to 54±10% (±SD) of control (n=12). The peak decrease was reached within 10 minutes after the beginning of galanthamine application and recovery was recorded within 1 h. A tendency to incomplete recovery was observed with concentration of the compound greater than 20 µM. Bath application of atropine (1 µM) before applying galanthamine prevented its suppressing effect on synaptic response (n=2).

DISCUSSION

Galanthamine has been shown in vitro to be a reversible, selective inhibitor of acetylcholinesterase (9). Therefore, observed effects are most likely explained as a result of a prolongation of lifetime of released endogenous acetylcholine acting on muscarinic receptors. The effective concentration of galanthamine found in the present study is consistent with the results of THOMSEN et al. (8) who observed a near-maximal inhibition of cholinesterase activity at galanthamine concentration of 10-100 µM.

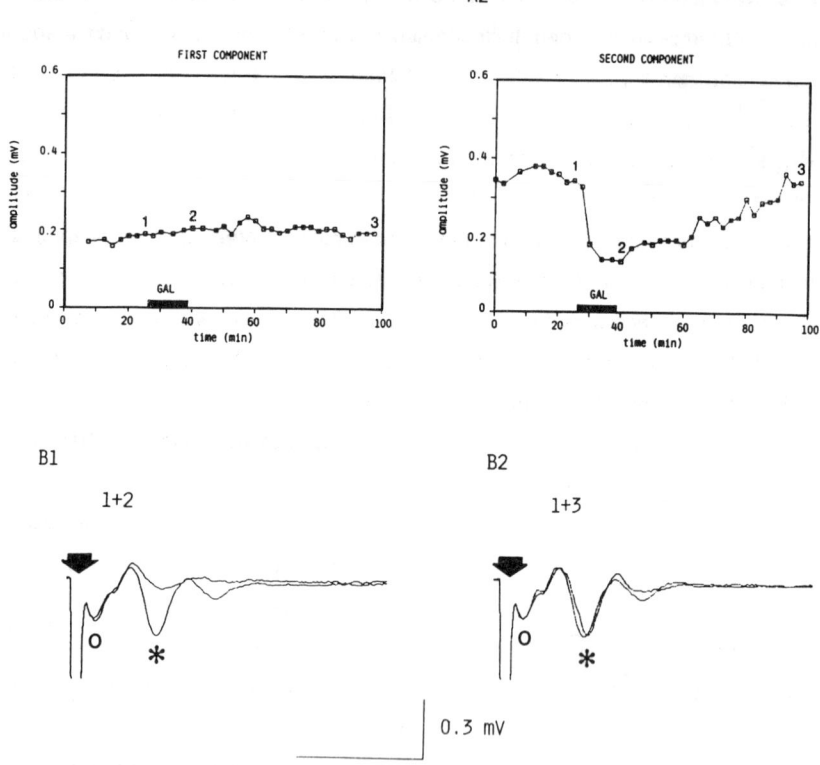

Fig.1: A: During the application of 20 μM galanthamine (GAL - black bar) the first component of the field potential (indicated by a circle in B) remains unchanged (A1) while the second (indicated by an asterisk in B) decreases (A2). Each square represents the average of 8 responses.
B: Superposition of averages of 8 responses taken during control and in the presence of galanthamine (B1) at indicated timepoints (1+2, see A) as well as during control and after recovery (B2; 1+3, see A). Arrows, circles, and asterisks represent stimulus artifact, the first and the second component of the field potential, respectively.

The effects of muscarinic agonists on cortical neurons are complex (see 7 with refs). Galanthamine reduces the amplitude of field excitatory postsynaptic potentials in the area CA1 of the rat hippocampus in vitro (author's unpublished observation). It has been proposed that a presynaptic muscarinic receptors mediate depression of excitatory synaptic transmission in the hippocampus (6,10 with refs). This may also take place in the neocortex. Alternatively, muscarinic activation of inhibitory interneurons (4) can account for the observed reduction of response amplitude, as it was also proposed for the dentate gyrus (2). In any case, a depression of neocortical transmission by galanthamine injection has to be taken into consideration. This may in turn affect the locomotor activity rhythm of experimental animals.

REFERENCES

1. BARBACKA-SUROWIAK, G. & J. SUROWIAK (1991): The influence of the Galanthaminum Hydrobromicum (Nivalin) on the locomotor activity circadian rhytm in mice. In: J. SUROWIAK & M.H. LEWANDOWSKI (eds.): Chronobiology and Chronomedicine, Basic Research and Applications, Peter Lang, Frankfurt
2. BRUNNER, H. & U. MISGELD (1988): Muscarinic inhibitory effect in the guinea pig dentate gyrus in vitro. Neurosci. Lett. 88, 63-68
3. LANGDON, R.B. & M. SUR (1990): Components of field potentials evoked by white matter stimulation in isolated slices of primary visual cortex: spatial distribution and synaptic order. J. Neurophysiol. 64, 1484-1501
4. MUELLER, C.M. & W. SINGER (1989): Acetylcholine-induced inhibition in the cat visual cortex is mediated by a GABAergic mechanism. Brain Res. 487, 335-342
5. SEGAL, M. (1988): Synaptic activation of a cholinergic receptor in rat hippocampus. Brain Res. 452, 79-86
6. SHERIDAN, R.D. & B. SUTOR (1990): Presynaptic M1 muscarinic cholinoceptors mediate inhibition of excitatory synaptic transmission in the hippocampus in vitro. Neurosci. Lett. 108, 273-278
7. SUTOR, B. & J.J. HABLITZ (1989): Cholinergic modulation of epileptiform activity in the developing rat neocortex. Dev. Brain Res. 46, 155-160
8. THOMSEN, T., U. BICKEL, J.P. FISCHER & H. KEWITZ (1990): Stereoselectivity of cholinesterase inhibition by galanthamine and tolerance in humans. Eur. J. Clin. Pharmacol. 39, 603-605
9. THOMSEN, T. & H. KEWITZ (1990): Selective inhibition of human acetylcholinesterase by galanthamine in vitro and in vivo. Life Sci. 46, 1553-1558
10. WILLIAMS, S. & D. JOHNSTON (1990): Muscarinic depression of synaptic transmission at the hippocampal mossy fiber synapse. J.Neurophysiol. 64, 1089-1097

THE EFFECT OF NIVALIN ON THE FACILITATION OF NEOCORTICAL RESPONSIVENESS IN COMPARISON WITH LOCOMOTOR ACTIVITY RHYTHM

M.H. LEWANDOWSKI

Jagiellonian University, Institute of Zoology, Department of Animal Physiology, M. Karasia 6, 30-060 Krakow, Poland

INTRODUCTION

Galanthamine (Nivalin), a phenanthridine alkaloid originally isolated from the snowdrop *Galanthus nivalis*, could offer several advantages compared to the other agents and is currently being investigated in therapeutic trials (9). It has been shown in vitro and in vivo to be a reversible selective inhibitor of AChE (9).

BARBACKA & SUROWIAK (1) reported, that Nivalin induced the phase shift of the circadian free-running rhythm of mice most pronounced in LL and DD conditions.

The aim of the present study was to support the idea that Nivalin produces an effect through direct influence on the mechanism of the reticular formation arousal reaction (i.e. facilitation of neocortical responsiveness).

METHODS

The experiments were performed in 9 male Wistar rats, weighing 320-390 g. Animals were anesthetized with intraperitoneal injection of Urethan (1.5 g/kg). The body temperature was thermostatically held at 36-38°C. One bipolar, concentric stimulation electrode was inserted into optic tract (opt: 4.3mm posterior to the bregma, 4.0mm lateral to the midline and 6.6mm vertical under to the bregma height; B -4.3, L 4.0, V 6.6) and two stimulation electrodes were placed in the ponte reticular formation (PRF: B -9.3, L 1.0, V 7.5). Electrically evoked potentials were recorded by means of silver ball electrodes implanted epidurally over the neocortex.

Electrode positions were adjusted under electrophysiological control to obtain maximum evoked responses (opt) or a maximum facilitation of the cortical afferents (PRF). Electrical stimuli consisted of single pulses of 50 μs duration (opt) or pulse trains of four 50 μs-pulses (PRF) given 60 ms prior to the opt shock. Data acquisition was performed online with a laboratory computer and consisted of compilation of averaged evoked potentials from 8 repetitions at 20 sec intervals of given combination of stimuli. Responses to opt stimulation and to combined opt and PRF stimulation were obtained in successive blocks. As response parameter we measured off-line the evoked potential amplitude between the maximal positive and negative deflection.

Three separate sets of experiments were performed:
1) effect of intramuscularly Nivalin administration (0.25mg)
2) effect of intravenous Nivalin administration (0.25mg)
3) effect of simultaneous intravenous Atropine (0.5mg) and Nivalin (0.25mg) injection.

Facilitation of the opt response by a conditioning stimulus delivered to the PRF were compared before and after the administration of AChE inhibitor (Nivalin) and ACh blocking agents (Atropin).

RESULTS

In any case of experiments electrical stimulation of the PRF resulted in a strong facilitation of the cortical evoked response.

Intramuscularly Nivalin injection produced a slight suppression of the facilitation of cortical evoked potentials. Notice that the potentials evoked by opt stimulation alone do not change. Curves reflect averaged responses of 8 consecutive stimuli (Fig.1).

Both the potentials evoked by electrical stimulation of the opt and strongly facilitated response after a conditioning PRF stimulus were reduced substantially by intravenous injection of Nivalin (Fig. 2).

In a total of 3 experiments administration of choline blocking agent Atropin (muscarinic blocker) before applying Nivalin did not significantly reduce facilitation of the opt response by PRF conditioning stimulation. Typical examples of this effects are shown in Fig. 3.

DISCUSSION

The results of this study show that the administration of Nivalin produced a depression of cortical facilitation by PRF stimulation. It appears that Nivalin influences reticular facilitation of cortical evoked responses.

It may be suggested that also motor cortex activity is reversibly reduced after administration of acetylcholinesterase inhibitor, thus changing the phase of locomotor activity rhythm.

It is well established that the neocortex receives a dense cholinergic innervation (2,6). Increased release of acetylcholine (ACh) in sensory cortex has been observed during behavioral arousal and after stimulating the reticular formation (8).

The fact that the effects of reticular stimulation persist after Atropine injection might indicate that the cholinergic action includes a gain change in cortical inhibitory interneurons. Activation of GABAergic interneurons in the vertebrate neocortex by ACh has been described both in vivo (7) and in vitro (5). The conclusion that cholinergic depression in the neocortex is mediated by activation of inhibitory interneurons is supported also by immunocytochemical studies showing cholinergic transmission in association with presumed GABAergic cells (4).

Recently BICKEL et al. (3) showed rapid and high accumulation of galanthamine in the mouse brain tissue after intravenous injections, into a tail vein. They also measured inhibition of acetylcholinesterase (AChE) in the same sample of brain. These findings and the results of the present study supported our suggestion that Nivalin changes the phase of the locomotor activity rhythm involved cholinergic projection in striate cortex.

435

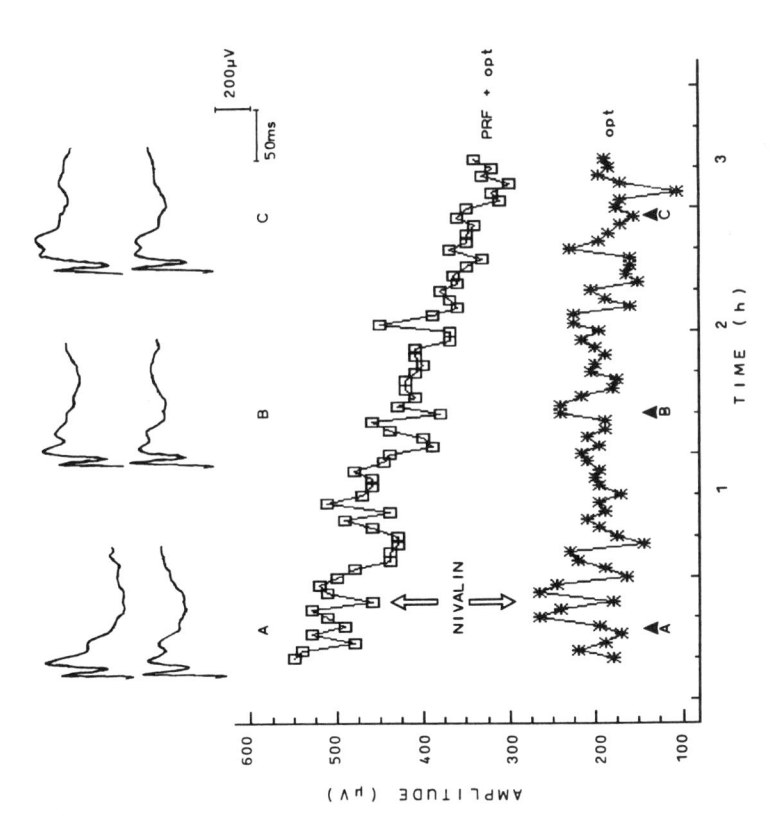

Fig.1: Effect of intramuscularly Nivalin injection on the facilitation of cortical evoked potentials.

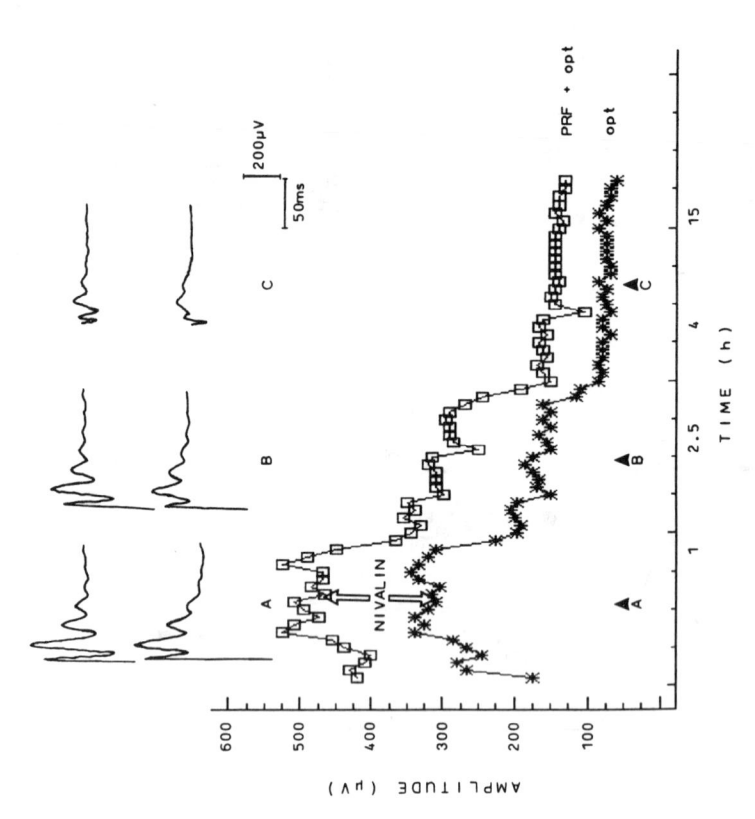

Fig.2: Effect of intravenous Nivalin injection on the facilitation of cortical evoked potentials.

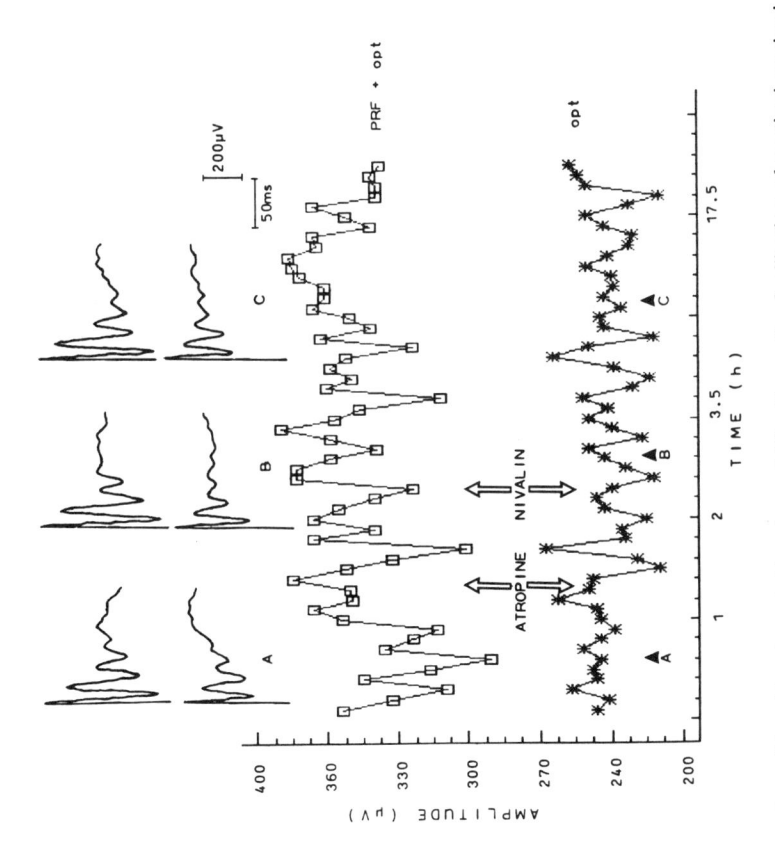

Fig.3: Effect of combined treatment Atropin and Nivalin on the facilitation of cortical evoked potentials.

REFERENCES

1. BARBACKA-SUROWIAK, G. & J. SUROWIAK (1991): The influence of the Galanthaminum Hydrobromicum (Nivalin) on the locomotor activity circadian rhythm in mice. In: J. SUROWIAK & M.H. LEWANDOWSKI (eds.): Chronobiology and Chronomedicine, Basic Research and Applications. Peter Lang, Frankfurt am Main, pp. 51-59
2. BEAR, M.F., K.M. CARNES & F.F. EBNER (1985): An investigation of cholinergic circuitry in cat striate cortex using acetylcholinesterase histochemistry. J. Comp. Neurol. 234, 411-430
3. BICKEL, U., T. THOMSEN, J.P. FISCHER et.al. (1991): Galanthamine: Pharmacokinetics tissue distribution and cholinesterase inhibition in brain mice. Neuropharmacology 30, 447-454
4. DELIMA, A.D. & W. SINGER (1986): Cholinergic innervation of the cat striate cortex: a choline acetyltransferase immunocytochemical analysis. J. Comp. Neurol. 250, 324-338
5. MCCORMICK, D.A. & D.A. PRINCE (1986): Mechanism of action of acetylcholine in the guinea pig cerebral cortex, in vitro. J. Physiol. (London.) 375, 169-194
6. MESULAM, M.M., E.J. MUFSON, A.I. LEVEY & B.H. WAINER (1983): Cholinergic innervation of cortex by the basal forebrain: cytochemistry and cortical connections of the septal area, diagonal band nuclei, nucleus basalis (substantia innominata), and hypothalamus in the Rhesus monkey. J. Comp. Neurol. 214, 170-197
7. MÜLLER, C.M. & W. SINGER (1989): Acetylcholine-induced inhibition in the cat visual cortex is mediated by GABAergic mechanism. Brain Research 487, 335-342
8. PHILLIS, J.W. (1968): Acetylcholine release from the cerebral cortex: its role in cortical arousal. Brain Research 7, 378-389
9. THOMSEN, T., U. BICKEL, J.P. FISHER & H. KEWITZ (1990): Galanthamine hydrobromide in a long-term treatment of Alzheimer's disease. Dementia 1, 46-51
10. THOMSEN, T., U. FISHER, J.P. BICKEL & H. KEWITZ (1990): Stereoselectivity of cholinesterase inhibition by galanthamine and tolerance in humans. Eur. J. Clin. Pharmacol. 39, 603-605

DAILY RHYTHMS OF INSECTICIDE SUSCEPTIBILITY IN NORMAL AND RESISTANT STRAINS OF MUSCA DOMESTICA L. (DIPTERA)

D. DAHLHELM

Zoologisches Institut der Martin-Luther-Universitaet
Domplatz 4, D-O 4020-Halle,FRG

Since HALBERG et al. (3) and BECK (1) many chronotoxicological effects in different species of animals have been described. In insects now the occurrence of resistant strains is one of the most important problems in pest control. Daily rhythms of sensitivity in insects against insecticides are of practical importance in testing, characterization and application of these compounds.

This paper deals with comparative observations between a normal susceptible and a resistant strain of *Musca domestica L.* against the P-organic insecticides Trichlorfon (TCF), Butonat, and Dichlorvos (DDVP). The resistance index R_i of the TCF-resistant strain was 25.

METHODS

Groups of 10-20 5 day old male and female flies (sex ratio 1:1) were exposed for 5 minutes to deposits of the insecticides in the concentrations shown in Tab. 1. (1 ml per 1.150-ml Erlenmeyer)

Tab. 1: Concentrations of insecticides (% in Acetone)

	DDVP	TCF	Butonat
S-strain	0,0005	0,1	0,2
R-strain	0,005	2,0	1,0

After 1 hr the number of flies laying on the back was recorded. After 24 hrs the died animals were counted. There was no significant difference between these values.

The time schedule was 3 hrs. starting with light on at 6 a.m. The photoperiods were LD = 12:12, LD 18:6, and LD 8:16. In each time point the number of tested animals was 150-200. The experiments were carried out in closed boxes with a temperature of 23 °C, 60 % humidity, and 400-500 Lx in L, in D weak read light was used. The experiments with the S-strain and Trichlorfon were repeated in an other series independent from the first study. There are small differences between the two series. But the same conclusions can be done.

RESULTS AND DISCUSSION

Trichlorfon (TCF)

$$\begin{array}{c} CH_3O \diagdown \quad \overset{O}{\underset{}{\|}} \\ \quad\quad P-CH-CCl_3 \\ CH_3O \diagup \quad | \\ \quad\quad\quad OH \end{array}$$

In LD 12:12 (Fig. 1A) the S-strain shows a high sensitivity in L increasing already in D with a maximum of more than 90 % efficacy short after L on. After L off it drops down to 30 % at 9 p.m. The results of the second series of experiments were 63 % at the maximum and 40 % at the minimum respectively.

The resistant (R-) strain (Fig. 1B) shows an inverted pattern. There is a maximum in the middle of D reaching nearly 100 % efficacy and starting to increase with L off. The sensitivity decreases rapidly 3 hrs before L on. In L the efficacy is very low. Only 28-37 % of flies showed on effect.

The normal sensitive strain was tested also in LD 18:6 (Fig. 1C) and LD 8:16 (Fig. 1D) against TCF. There are two maxima under long day conditions (LD 18:6) at 9 a.m. and 9 p.m. with 77.4 % and 76.6 % of died flies. The two minima show 54.4% and 52.5 % killed flies. The mesor is 66.5 %.

Under short day conditions (LD 8:16) the sensitivity was only low. The Mesor is 47.7 %. Whereas in the short L-time the efficacy reached somewhat more than 60 %, in the darkness it is lower than 50 %. Compared with LD 18:6 the pattern shows

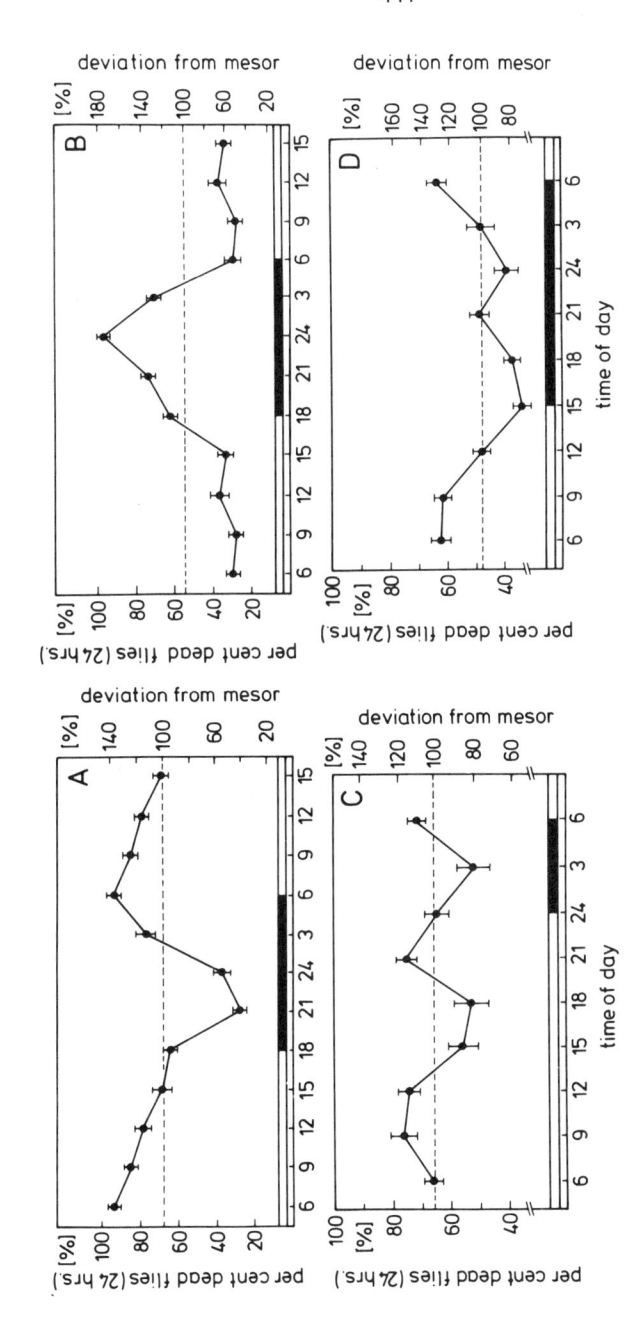

Fig.1: Daily rhythm of sensitivity of:
(A) normal susceptible house flies against Trichlorfon (LD 12:12) (B) Trichlorfon-resitant house flies against Trichlorfon (LD 12:12)
(C) normal susceptible house flies against Trichlorfon (LD 18: 6) (D) Trichlorfon-resitant house flies against Trichlorfon (LD 8:16)

in principle the same position of the maxima and minima of the sensitivity. That may indicate the same endogenerhythmicity.

Butonat (Fig. 2A and 2B)

$$\begin{array}{c} CH_3O \\ \diagdown \\ CH_3O \diagup \end{array} P \begin{array}{c} O \\ \| \\ - CH - O-COC_3H_7 \\ | \\ CCl_3 \end{array}$$

In contrast to the results against Trichlorfon under LD 12:12 the S-strain and the R-strain have not an inverted sensitivity. The mortality is mostly lower than 50 % in L. In D it increases to more than 90 %. The difference between maximum and minimum is 70 %. The mathematical analysis by power spectra suggests a 12-hr periodicity in the S-strain whereas the R-strain is characterized by a 24-hr rhythm.

Dichlorvos (DDVP) (Fig. 2C and 2D)

$$\begin{array}{c} H_3CO \\ \diagdown \\ H_3CO \diagup \end{array} P \begin{array}{c} O \\ \| \\ - O - CH = CCl_2 \end{array}$$

The pattern of the sensitivity in the S-strain and in the R-strain are more ultradian. In short steps the sensitivity alternates between high values near 90 % and low values until 20 % of efficacy. The mathematical analysis shows beneath a weak 23-hr-rhythmicity mainly a strong 6-hr-rhythm.

The different rhythms of susceptibility in normal sensitive and in resistant strains of an insect against different P-organic insecticides show the necessity of chronotoxicological observations. They may help to explain the reasons for resistance and to solve some problems of application of insecticides.

The different time pattern of susceptibility in normal and resistant insects may be caused in the behavior or the metabolism of the animals. Differences between different insecticides should be searched especially in the mechanisms of detoxification. The findings of OTTO and WEBER (5) suggest studies of the general level and the rhythmicity of some detoxificating enzymes as the non specific esterase and the Glutathione-S-transferase in normal susceptible and in resistant strains of insects.

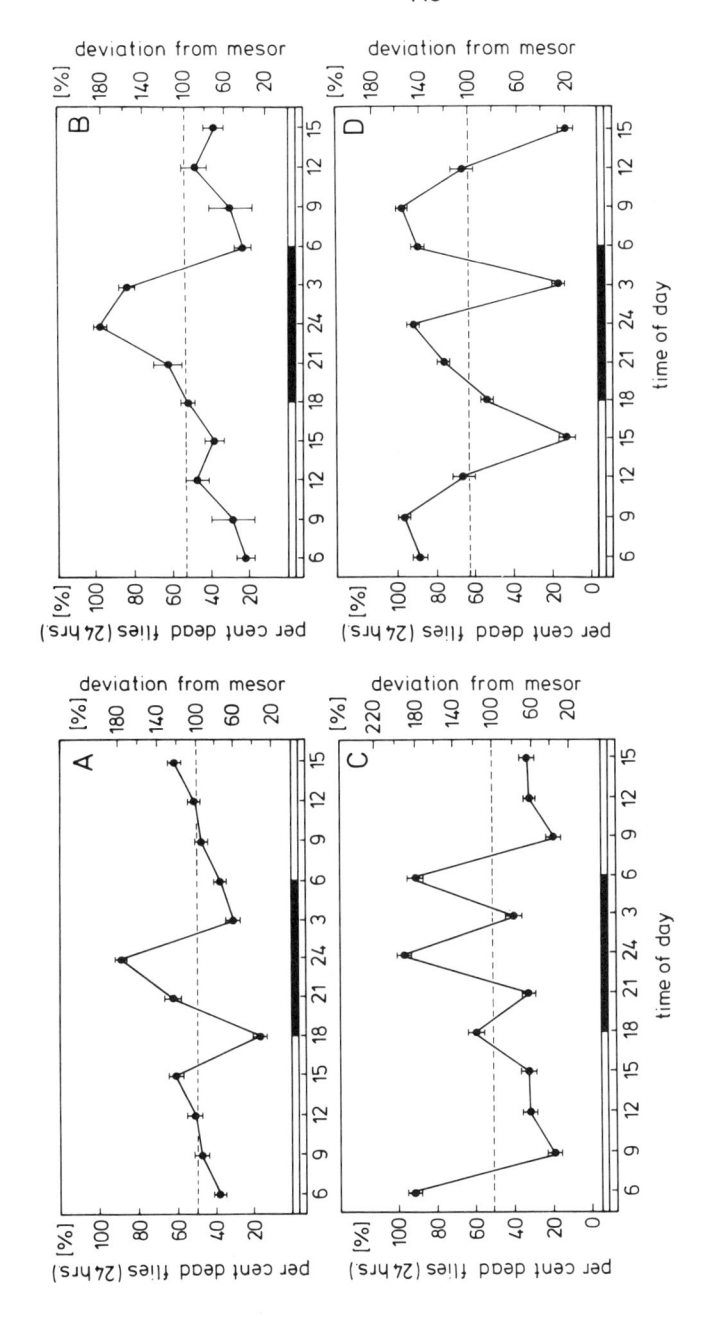

Fig.2: Daily rhythm of sensitivity of:
(A) normal susceptible house flies against Butonat (LD 12:12) (B) Trichlorfon-resitant house flies against Butonat (LD 12:12)
(C) normal susceptible house flies against DDVP (LD 12:12) (D) Trichlorfon-resitant house flies against DDVP (LD 12:12)

REFERENCES

1. BECK, S.D. (1963): Physiology and Ecology of Photoperiodism. Bull. Entomol. Soc. Amer. 9, 8-16
2. BRÜCHER, M. (1986): Untersuchungen zur tageszeitlichen Empfindlichkeitsänderung von Musca domestica L. gegenüber Trichlorfon (Flibol E 40) in Abhängigkeit vom Lichtregime. Diplomarbeit Martin-Luther-Universität Halle-Wittenberg
3. HALBERG, F., E. HAUS, & A.N. STEPHENS (1959): Susceptibility to oubain and physiologic 24-hour periodicity. Fed. Porc. 18, 1
4. HOPPE, I. (1982): Untersuchungen zur tageszeitlichen Insektizidempfindlichkeit gegen drei phosphororganische Verbindungen (Butonat, Flibol E 40, DDVP) an unterschiedlich empfindlichen Stämmen der Stubenfliege Musca domestica L. Diplomarbeit Martin-Luther-Universität Halle-Wittenberg
5. OTTO, D. & B. WEBER (1989): Resistance to trichlorfon in Musca domestica and the effect of synergists. Tag.Ber. Akad. Landwirtsch.-Wiss. DDR, Berlin 274, 253-267

THE INFLUENCE OF SOME NEUROLEPTIC AND NEUROMIMETIC DRUGS ON THE CIRCADIAN WHEEL-RUNNING ACTIVITY RHYTHM IN MICE UNDER CONSTANT DARKNESS CONDITION (DD)

G.BARBACKA-SUROWIAK

Department of Animal Physiology,Institute of Zoology, Jagiellonian University, Krakow, Poland

INTRODUCTION

Our previous investigations (1,2) indicated, that some neuroleptic and neuromimetic drugs, such as Droperidol (dehydrobenzperidol) - adrenergic alpha-receptor blocker and Nivalin (galanthamine) - Cholinesterase blocker and l-adrenaline - sympathomimetic, affect the circadian wheel-running activity rhythm in mice at constant light conditions (LL), however, it does not affect this rhythm under LD 12/12 regime. In the present study we sought to determine whether Nivalin, Neostigmine, both of the cholinesterase blockers, and Droperidol will change the phase of the circadian wheel-running activity rhythm of mice in constant darkness (DD) conditions.

MATERIAL AND METHODS

130 male Swiss mice, 8 weeks old, mean body weight 30 ± 2 g, were used in the investigation. Every time animals were initially grouped housed, and maintained on a 12/12 light/dark cycle (LD 12/12) i.e. light phase lasted from 8.00 a.m. to 8.00 p.m. After two weeks, the animals were transferred to individual cages equipped with running wheels to record the circadian wheel-running activity at LD 12/12 and then at DD conditions. The wheel-running activity was continuously measured by means of a computer. Animals were fed standard diet and water ad libitum. Temperature in the breeding room was 25 ± 2^{0}C, humidity 65 ± 5 %, and light intensity in the room at the LD 12/12 condition 200 lx, and within cages ± 100 lx. After few days or one week of

the free-running locomotor activity rhythm recorded at DD condition, the animals were hypodermically injected with Nivalin (Galanthamine hydrobromide, Pharmachim, Bulgaria) in the doses of 0.0125 to 0.0250 mg per 30g of body weight, either with Neostigmine (Neostigmine methylsulfate - Polstigminum, Polfa, Poland) 0.00025 mg/30g of body weight, or with Droperidol (Dehydrobenzperidol, Richter, Hungary) 0.00015 mg/30g of body weight. The control animals were injected with NaCl solution. The drugs were administered either every 4 hours throughout one or two days, or in early and in late subjective night hours. Phase-shifts in the activity wheel-running rhythm were determined by measuring the phase difference between eye-fitted lines connecting onset of activity for a period of few days (4 - 7) before and few days after experimental drug injections, and by calculation of the acrophases and period by means of the Fourier analysis according to (4).

RESULTS

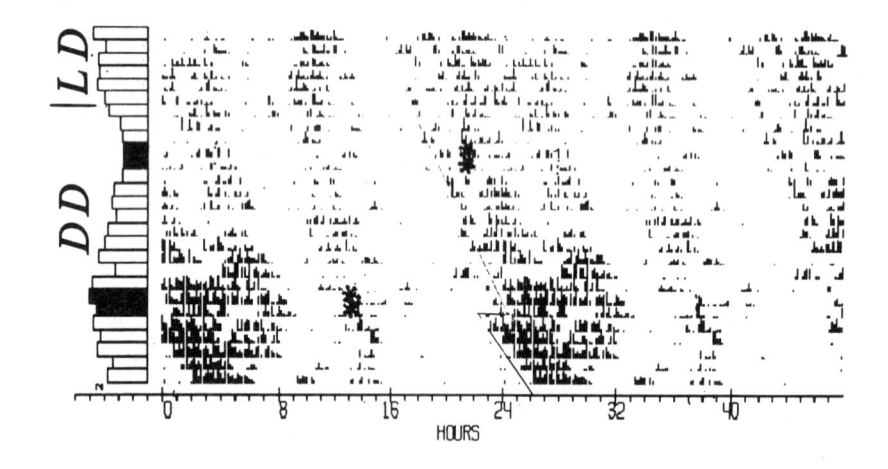

Fig.1: Double-plotted wheel-running activity of the mice in LD and DD conditions. Black bars on the left hand side of actogram : the days of the drugs administrations. Asterisk show the time-point of the Nivalin injections. The actogram demonstrate the phase-advance and free-running rhythm reorganization after twice of drug administration.

Each of the three drugs used changed the phase of the circadian wheel-running activity rhythm of the mice. Dependent on the circadian time-point of the drug's administration, the phase of the free-running rhythm was delayed or advanced (Fig. 1). In the animals with the splitting of the rhythm or with those which were arrhythmic after Droperidol and Nivalin injection, the free-running rhythm was organized and no splitting was observed (Fig.1). The phase response curve showed, that the phase-delay of the rhythm occurred when the drugs were injected in the early subjective night, and the phase-advance took place when the time-point of drug's administration fell in the late subjective night (Fig.2).

PHASE RESPONSE CURVE

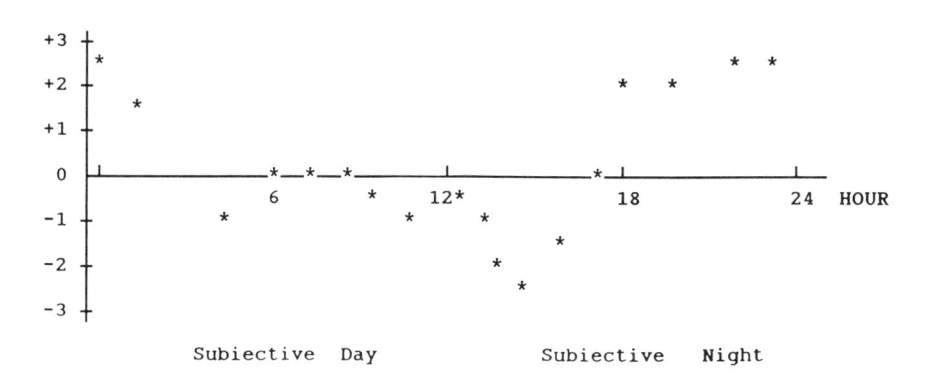

```
+3  
    *                                      *   *
+2                                    *  *
    *
+1

 0  |_____*__*__*_____|_____*_|_____|___
           6         *  12*              18          24  HOUR
-1      *              *   *
                              *
-2                        *
                        *
-3

        Subiective   Day         Subiective   Night
```

Fig.2: Phase response curve for Nivalin, Polstigmine and Droperidol administered in DD condition in mice.

DISCUSSION AND CONCLUSIONS

Our results indicate, that Nivalin, Polstigmine and Droperidol mimic the phase-shift effect of the light pulses on the circadian wheel-running activity rhythm in mice in constant darkness. Acetylcholine is one of the suprachiasmatic nuclei transmitters (6), and Carbachol, muscarinic receptor agonist, mimics the phase shifting effects of light on the circadian wheel-running activity rhythm in mice (10) and in hamster (5,8,9). EARNEST & TUREK (5) concluded, that " may be acetylcholine playes a key role in the mechanism by which light information is transmitted to, or within the biological clock involved, in the generation of the circadian rhythm and photoperiodic time measurement". KEEFE et al (7) demonstrated, that mecamylamine, a cholinergic antagonist, blocked the phase shifting effect of the light on the circadian rhythm of locomotor activity in the golden hamster. Nivalin and Polstigmine as the blockers of the cholinesterase, facilitate conduction of the nerve impulses throughout cholinergic, nicotinic, and muscarinic synapses, and therefore they may mimic the phase-shifting effects of the light on the circadian free-running locomotor activity rhythm of mice in DD conditions. This conclusion is supported by investigations of BICKEL et al (3) which have shown, that Nivalin passes the blood-brain barrier in rats. I cannot interpret the phase shifting effect of the Droperidol, adrenergic alpha-receptor blocker. This problem requires still intensive investigation.

REFERENCES

1. BARBACKA-SUROWIAK, G. & J. SUROWIAK (1990): The influence of the Galanthaminum hydrobromicum (Nivalin) on the locomotor activity circadian rhythm in mice. In: SUROWIAK & LEWANDOWSKI (eds.): Chronobiology & Chronomedicine. Proc. 5th annual Meet. ESC, Cracow 1989. Peter Lang, Frankfurt am Main, pp.51-59
2. BARBACKA-SUROWIAK, G. (1990): The influence of some neuroleptic and neuromimetic drugs on the locomotor activity circadian rhythm in mice (in print).
3. BICKEL, U., T.THOMSEN, J.P. FISCHER et al. (1991): Galanthamine: pharmacokinetics, tissue distribution and cholinesterase inhibition in brain of mice, Neuropharmacol. 30, 447-454
4. DOMOSLAWSKI, J. (1991): All-purpose experimental data processing package for Chronobiology, (in print).
5. EARNEST, D.J. & F.W. TUREK (1985): Neurochemical basis for photic control of circadian rhythms and seasonal reproductive cycles: Role for acetylcholine. Proc. Natl.Acad. Sci. USA. 82, 4277-4281.
6. GROOS, G., R. MASON & J. MEIJER (1983): Electrical & pharmacological properties of the suprachiasmatic nuclei. Feder. Proc. 42, 2790-2795.
7. KEEFE, D.L., D.J. EARNEST, D. NELSON et al. (1987): A cholinergic antagonist, mecamylamine, blocks the phase-shifting effects of light on the circadian rhythm of locomotor activity in the golden hamster. Brain Res. 403, 308-312
8. TUREK, F.W. (1987): Pharmacological probes of the mammalian circadian clock: use of the phase response curve approach. TIPS 8, 212-217
9. TUREK, F.W. (1988): Pharmacological manipulation of the mammalian circadian clock: implications for control of seasonal reproductive cycles. Reprod. Nutr. Develop. 28, 499-513
10. ZATZ, M. & M.A. HERKENHAM (1981): Intraventricular carbachol mimics the phase-shifting effect of light on the circadian wheel-running activity. Brain Res. 212, 234-238

RUBIDIUM AND POTASSIUM:
EFFECTS ON CIRCADIAN RHYTHMS IN HAMSTERS

T.T. BAUER[1,2], H. KLEMFUSS[2], D.F. KRIPKE[2] & B. PFLUG[3]

[1]Universität Bonn, FRG
[2]Dept. of Psychiatry (V-116A), Veterans Affairs Medical Center, San Diego, CA 92161, U.S.A.
[3]Klinikum der Johann Wolfgang Goethe-Universität, Abteilung für Klinische Psychiatrie II, Frankfurt am Main, FRG

INTRODUCTION

Rubidium effects on biological rhythms have been studied in many species including plants, flies, and cockroaches, but results have been inconsistent (1). Lengthening as well as shortening of tau under constant conditions were reported. Little data are available on how rubidium changes rhythms in vertebrates. HALLONQUIST (2) reported that rubidium was able to fuse previously split circadian rhythms in hamsters, implying changes in coupling strength between oscillators. PFLUG et al. (3) monitored a 66 year old female patient who was treated for depression with rubidium chloride. In this study, plasma concentrations of rubidium were related to the improvement of depression, and were accompanied by a decrease in phase angle between sleep-wake and temperature rhythms.

Pharmacological actions of rubidium may be mediated by changes in potassium metabolism. Both rubidium and potassium are alkali metals. Both enter cells through the Na/K ATPase mechanism. Potassium is replaced by an equal number of rubidium ions in the human body. Unlike lithium, which is not conducted through potassium channels, rubidium permeates channels slowly, which can hinder potassium flow.

Potassium itself has also been reported to affect biological rhythms. Potassium advances light-synchronized circadian rhythms in hamsters when administered orally (4), and advances the rhythm of the neuronal firing rate in hypothalamic slices (5). Agents that indirectly affect potassium balance - like potassium ionophores and potassium channel blockers - also affect biological rhythms (1,6). In order to further identify circadian properties of rubidium in vertebrates and the pharmacological interactions between rubidium and potassium, the present study examined changes in activity rhythms caused by oral supplements of rubidium and potassium in hamsters.

MATERIAL AND METHODS

Thirtyeight adult male Syrian Golden hamsters (Charles River: 90-120 g) were housed separately in wheel running cages under short days (8 hr of light; LD 8:16, 30 - 50 lux measured inside the cage). In LD 8:16, onset of darkness precedes onset of activity by several hours, thus minimizing masking effects of light. Hamsters were fed a standard diet (Teklad, Madison WI, TD86299, 171 mmol Potassium/kg) with 7% polydextrose, a water-soluble non-nutritive carbohydrate added. Half of the animals were randomly assigned to receive rubidium (20 mM) in their drinking water. After three weeks, half of the animals in each treatment group were changed to a potassium enriched diet for two additional weeks (TD85191, 1104 mmol Potassium/kg by adding 7% potassium chloride). Control diet was continued for all other animals. Food, water, and access to running wheels were available at all times.

A computer system recorded running wheel activity in five minute bins throughout the experiment. The phase of the activity rhythm was estimated using an automated procedure which identified the time of the greatest increase in locomotor activity. A total of 35 animals were analyzed, of which 27 met criteria for unambiguity of activity onset (mean activity increase $\geq 60\%$). A rater, blind to experimental conditions, estimated the duration of activity (α) by eye.

After two weeks of treatment with high and control potassium diets, hamsters were overdosed with pentobarbital. Blood was drawn by vena cava puncture into a heparinized syringe and centrifuged immediately to separate plasma and blood cells. Concentrations of rubidium and potassium in plasma and hypothalamus were assessed by atomic absorption spectrophotometry. Animals were weighed before and after the experiment. Paired testes were weighed at the end of the experiment. Water intake was measured every week. Four animals died or became ill during the experiment (1 low potassium, 1 high potassium, 2 rubidium plus low potassium). Effects of potassium and rubidium were examined by one and two-way analyses of variance (ANOVA).

RESULTS

During the third week of rubidium administration, activity onset was approximately 40 minutes later in rubidium animals (p < 0.05, ANOVA). The duration of the active phase (α) was shortened by 1.25 hrs on average compared to controls (p < 0.005, ANOVA) (Fig. 1).

Figure 1: Onset and Duration of Activity after 3 Weeks of Rubidium

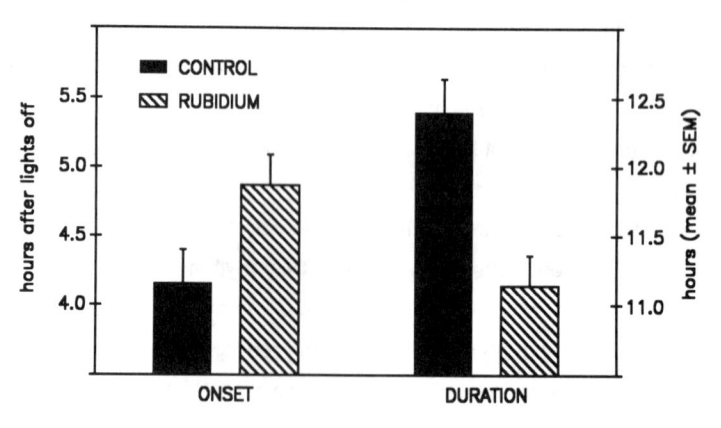

Fig.1: Onset and duration of activity after 3 weeks of rubidium administration.

Supplemental potassium advanced rhythms significantly (high potassium 0.65 ±0.08 hr vs. control diet 0.15 ±0.15 hr, p < 0.005, ANOVA). Animals receiving rubidium plus high potassium diet advanced more than animals receiving high potassium alone (rubidium + high potassium 0.73 ±0.1 hr vs. high potassium 0.43 ±0.1 hr, p < 0.05, ANOVA interaction term between rubidium and potassium effects).

Testicular weight was 25% greater in rubidium-treated animals compared to controls (rubidium 3.1% ±0.12 of body weight vs. control 2.4% ±0.1, p < 0.001, ANOVA). Supplemental potassium had no significant effect on testicular weight, and no interaction was found.

Rubidium treatment decreased plasma potassium levels (p < 0.05 for rubidium plus low potassium vs. low potassium, Newman-Keuls test for multiple comparisons), but potassium content of the hypothalamus was not affected by any treatment (Table1). The high potassium diet increased plasma potassium significantly only in rubidium-treated animals. Rubidium levels in both plasma and hypothalamus were significantly lower in animals receiving supplemental potassium (p < 0.0001, ANOVA, for hypothalamus and plasma) (Table 1).

Rubidium did not cause polydipsia, but when potassium was added, water intake doubled in all animals receiving a high potassium diet (p < 0.0001, ANOVA repeated measures design). Animals treated with rubidium weighed 15 g less than control animals (p < 0.05 for both pairs, Newman-Keuls test for multiple comparisons).

DISCUSSION

Circadian effects of lithium, an alkali metal widely used as an antidepressant, have been well established for many species (7). Rubidium, another alkali metal used as an antidepressant, might be expected to act similarly on the hamster biological clock (8), however, many behavioral aspects are affected differently by rubidium and lithium. For example, spontaneous motor activity and shock-elicited aggressive behavior in rats are decreased by lithium and increased by rubidium.

Table 1:
Rubidium (Rb⁺) and potassium (K⁺) concentrations in plasma (mmol/l) and hypothalamus (µmol/kg protein).

	Rubidium Treatment		Control Treatment	
(N)	Low K⁺ (8)	High K⁺(10)	Low K⁺ (8)	High K⁺ (8)
Plasma K⁺	3.27 ±0.14	5.18 ±0.2*	4.97 ±0.35*	5.75 ±0.33*
Plasma Rb⁺	1.58 ±0.03	1.09 ±0.05**		
Hypothalamus K⁺	629.45 ±50.07	569.92 ±43.47	633.27 ±59.64	641.52 ±48.31
Hypothalamus Rb⁺	162.1 ±14.14	68.7 ±5.74**		

(means ± SEM)
* $p < 0.05$ compared to rubidium plus low potasssium (Newman-Keuls Test).
** $p < 0.001$ compared to rubidium plus low potassium (ANOVA).

These results show that rubidium in hamsters causes a combination of effects that is comparable with other agents that delay circadian rhythms. The combination of phase delays, shortening of activity duration (α) and prevention of testicular regression has also been described for lithium and deuterium (7,9,10) Hamsters show testicular regression and lengthened duration (α) when exposed to short days (less than 12.5 hrs) (11). Phase-delaying agents may cause hamsters to misinterpret short days as long days and therefore prevent testicular regression. EKES & ZUCKER (10) observed complete testicular regression after 13 to 17 weeks of short days, whereas testicles failed to regress in a significant number of deuterium-treated hamsters. Although hamsters in this experiment were kept under short days for only 7 weeks (adaptation period included), a significant preventive effect of rubidium was demonstrated.

Potassium fluxes might be involved in generating biological rhythms (12). Effects of rubidium on potassium balance might therefore be one mechanism for changes in the circadian system. When potassium was added to the diet, running wheel rhythms advanced significantly, as has been described previously (4). Animals pretreated with rubidium, however, advanced more than control animals. This effect might be explained by summation of the primary phase advancing effect of oral potassium plus the relative withdrawal of rubidium, as indicated by lower brain rubidium levels.

Reduction of hypothalamic rubidium levels associated with a high potassium diet might be relevant to the treatment of depression. In order to maintain the efficacy of a rubidium treatment, daily dietary potassium intake should be closely monitored.

A solution of 20 mM rubidium in drinking water was slightly toxic, as indicated by reduced weight gain in rubidium hamsters. Nevertheless, plasma rubidium levels produced with this dosage (1.0 - 1.7 mmol/l) were well above the therapeutic range (0.3 - 0.4 mmol/l) (3). No polydipsia or polyuria was observed in rubidium-treated animals, but therapeutic blood levels of lithium (0.8 - 1.2 mmol/l) may induce severe weight loss and polydipsia. Therefore, rubidium is less toxic than lithium at comparable blood levels in hamsters and might be an alternative treatment in patients where lithium induces severe side effects or is ineffective.

In conclusion, we found that rubidium administered orally for 3 weeks delayed circadian rhythms in hamsters, decreased activity duration, and prevented testicular regression. These effects might contribute to the antidepressant actions of the drug. Rubidium effects on the biological clock might be mediated through changes in potassium balance.

ACKNOWLEDGEMENTS

Supported by the Dept. of Veterans Affairs, Stanley Scholars Program, the Sam and Rose Stein Institute for Research on Aging and MH00117. Bruce Dyke and Dr. Paul Saltman assisted the analyses of rubidium and potassium.

REFERENCES

1. SCHMID, H. & W. ENGELMANN (1987): Effects of Li+, Rb+, and tetraethylammoniumchloride on the locomotor activity rhythm of Musca domestica. J. Interdiscipl. Cycle Res. 18, 83-102
2. HALLONQUIST, J.D. & N. MROSOVSKY (1989): Rubidium fuses split circadian rhythms in hamsters. Soc. Neurosci. Abstr. 15(1), 726 (Abstract)
3. PFLUG, B. (1990): Effect of rubidiumchloride on the circadian system in affective psychosis. In: A. REINBERG, G. LABRECQUE & M. SMOLENSKY (eds.): Chronopharmacology: Biological Rhythms & Medications: Fourth International Conference of Chronopharmacology & Chronotherapeutics, Nice, p. 17
4. KLEMFUSS, H. & D.F. KRIPKE (1990): Light responsiveness of a circadian oscillator during lithium and potassium treatment. Ann. Rev. Chronopharmacol. 7, 5-8.
5. GILLETTE, M.U. (1987): Effects of ionic manipulation on the circadian rhythm of neuronal firing rate in the suprachiasmatic brain slice. Soc. Neurosci. Abstr. 13, 51 (Abstract)
6. KLEMFUSS, H.& D.F. KRIPKE (1989): Potassium channel blockade lengthens the period of the hamster circadian wheel-running rhythm. Soc. Neurosci. Abstr. 15, 492 (Abstract)
7. ENGELMANN, W. (1987): Effects of lithium salts on circadian rhythms. In: A. HALARIS (ed.): Chronobiology and Psychiatric Disorders, Elsevier Science Publishing Co., New York, pp. 263-289
8. WILLIAMS, R.H., A. MATUREN & H.H. SKY-PECK (1987): Pharmacologic role of rubidium in psychiatric research. Compr. Ther. 13, 46-54
9. GIEDKE, H. & H. POHL (1985): Lithium suppresses hibernation in the Turkish hamster. Experientia 41, 1391-1392
10. ESKES, G.A. & I. ZUCKER (1978): Photoperiodic regulation of the hamster testis: dependence on circadian rhythms. Proc. Natl. Acad. Sci. USA 75, 1034-1038
11. ELLIOTT, J.A., M.H. STETSON & M. MENAKER (1972): Regulation of testis function in golden hamsters: a circadian clock measures photoperiodic time. Science 178, 771-773
12. BURGOYNE, R.D. (1978): A model for the molecular basis of circadian rhythms involving monovalent ion-mediated translational control. FEBS Letters 94, 17-19

CHRONOBIOLOGICAL ASPECTS OF THE PRENATALTOXIC ACTION OF CHEMICAL SUBSTANCES

R. SCHMIDT

Institute of Biology, Department of Medicine,
Martin-Luther-University Halle-Wittenberg, Universitätsplatz 7,
D-0-4020 Halle, FRG

INTRODUCTION

The prenataltoxic effect of chemical substances is decisively defined by their physicochemical characteristics, the dose applicated, the phase of pregnancy during which application takes place as well as by the genotype of the animal species and the animal strain respectively. Despite of remarkable increase of knowledge in the field of chronobiological research, in this context too little attention is being given to the question, whether the embryo damaging effect is modified by the circadian phase of application within a definite day of pregnancy (1,2,3).

After we could verify in earlier investigations (4,5) for the cytostaticum Cyclophosphamide a circadian determined teratogenicity and embryolethality pattern, the question arose, whether this also hold for other teratogenes. Searching for a suitable substance showing a different mechanism of action, the glucocorticoid Dexamethasone offered itself. Other than Cyclophosphamide, Dexamethasone acts organotropic and causes exclusively cleft palates. Moreover Dexamethasone acts in a different phenocritical determination phase, is unmetabolized placental permeable and even in the target it releases the teratogenic effect.

MATERIAL AND METHODS

Test animals and keeping:

Males and virginal females, 3-4 months of age, of the outbred Halle:AB and Halle:DBA in 3 mating tendencies:

- females x males Halle:AB
- females x males Halle:DBA
- females Halle:AB x males Halle:DBA (- hybrid embryoes ABD).

Conventional test animals, open system with artificial light condition (L:D= 12:12; on: 6.00 a.m.), standard food and water ad lib.

Mating:

In order to treat possibly identical embryonal stages at different times within the 24-h-period, a modified mating system was used (short time pairing): 3 females : 1 male; four limited times of mating:

- 7.00 - 9.00 a.m. ,
- 1.00 - 3.00 p.m. (no matings happened during this time),
- 7.00 - 9.00 p.m., and
- 1.00 - 3.00 a.m.

Detection of mating at the end of each pairing cycle by vaginal plug techniques.

Substance and application:

Dexamethasone (dissolved in aqua ad inj.); pregnant females of each test group received a single i.p. injection (dose: 7.5 and 15 mg/kg b.w. resp.) at the end of the 13th day of gravidity; application at four different times of day: series 1: 9.00 a.m. ; series 2: 3.00 p.m. ; series 3: 9.00 p.m. ; series 4: 3.00 a.m.

Criteria of prenataltoxic action:
- external malformations;
- cleft palates;
- skeletal malformations;
- postimplantation death-rate (= resorptions and dead fetuses).

For a statistic valuation of the results the U-Test related to each litter was used according to MANN and WHITNEY.

RESULTS

The statistic evaluation of the dissection material resulted in an obvious correlation between frequency of malformation and day-time of treatment. Therein the existence of circadian determined teratogenicity pattern could be verified with respect to cleft palates.

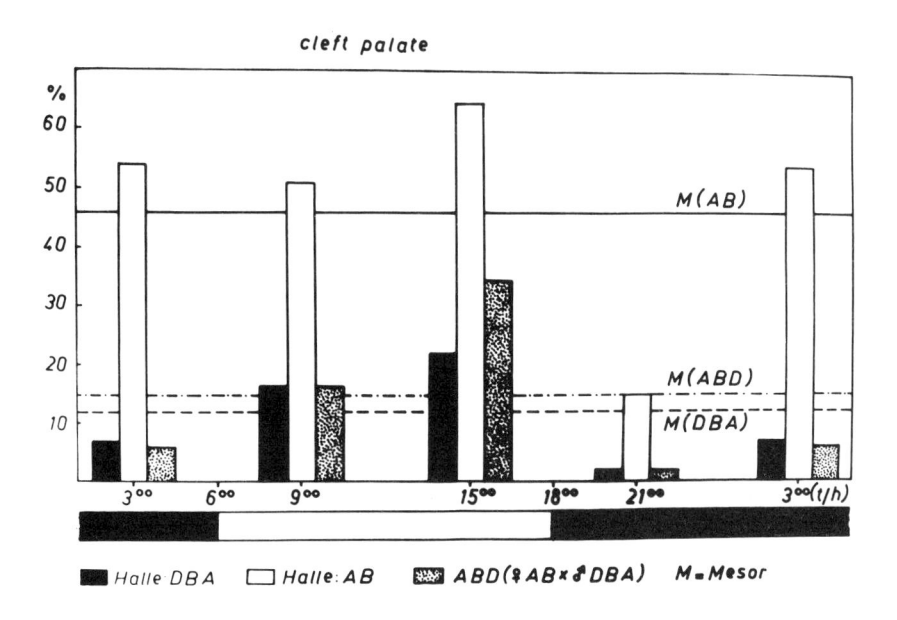

Fig. 1: Incidence of cleft palates due to maternal treatment with 7.5 mg Dexamethasone/kg b.w. at different circadian phases (time post concept.: 13,0 d).

As seen from Fig.1, after treatment of the pregnant females at 3.00 p.m. the rates of malformation correspondingly show a maximum of sensibility in all the 3 strains. After identical Dexamethasone treatment the sensibility minimum was determined for the 3 strains conformably at 9.00 p.m. Compared with the 3.00 p.m. values the cleft rates in this phase of the day are reduced to 15% (Halle:AB) and 2% (Halle:DBA and ABD) and thus differ significantly in all strains from the maximum values.

In our Cyclophosphamide tests it could be stated, that the circadian rhythm within the range of higher dosage is obviously covered, as the malformation rates were, independently of the day-time of treatment, generally at 100%. This phenomenon could not be reproduced for Dexamethasone after doubling the dosage (15 mg/kg b.w.). The cleft incidence is increased in the AB strain, but the basic rhythm described remains preserved. In the DBA strain, which is relatively cleft resistant in comparison to Dexamethasone, no significant changes result after doubling the dosage.

POSTIMPLANTATION DEATH-RATE ⊗

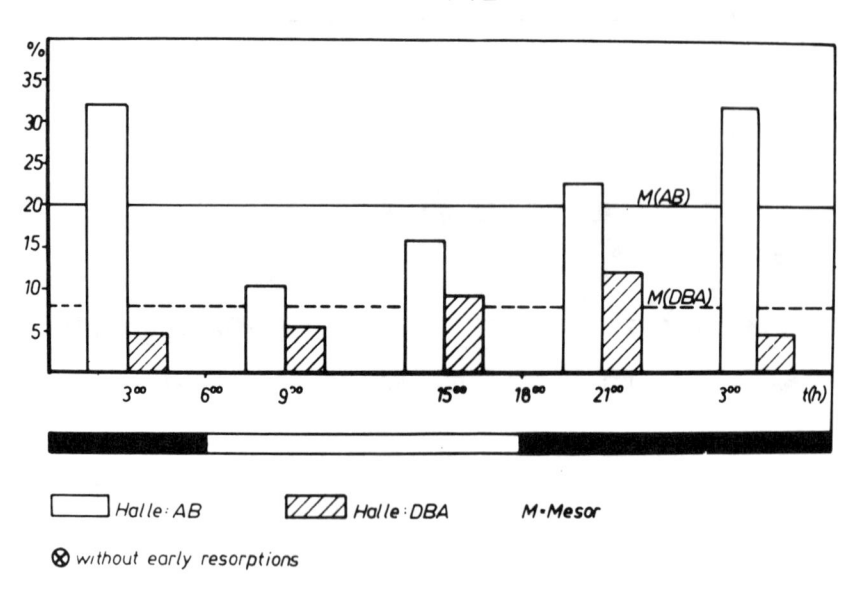

Halle: AB Halle: DBA M•Mesor

⊗ without early resorptions

Fig.2: Postimplantation death-rate due to maternal treatment with 15 mg Dexamethasone/kg b.w. at different circadian phases (time post concept.: 13,0 d).

With respect to the embryolethal effect we could detect biorhythmical pattern for the AB strain under the influence of the increased dosage (Fig.2). Therein the minima and maxima of sensibility are - compared with the teratogenicity pattern - obviously during quite different phases of the 24-h period. This indicates, that Dexamethasone induces its embryolethal effect in a different target with a completely different sensibility rhythm. The results in the DBA strain are not representative, as the values determined are within the standard range of spontaneous postimplantation death-rate.

CONCLUSION

Related to the induceability of cleft palates SCHUBERT (6) could describe the 13th and 14th day p.c. as the period of highest teratogenic sensibility after the Dexamethasone treatment of pregnant mice for the strain Halle:AB. In continuation of earlier investigations with the cytostaticum Cyclophosphamide (4,5) in this context the question arose whether the teratogenic effect is further modified by the circadian phase of Dexamethasone application within a definite day of pregnancy. Using the described short time mating system (7) it could be proved that also the teratogenic effect of Dexamethasone follows a biorhythm. However, the minima and maxima of sensibility for both the pharmaca are in different phases of the day (max. Cyclophosphamide: 9.00 a.m., min. Cyclophosphamide: 3.00 a.m.; max. Dexamethasone: 3.00 p.m., min. Dexamethasone: 9.00 p.m.). The interpretation of the results must at first remain a hypothetical one. Nevertheless one can start from the fact, that the described variations of the prenataltoxic effect in dependence on the time of day can be explained by

- chronopharmacokinetic factors (i.e. circadian fluctuations of the biotransforming enzyme systems);
- circadian changes in the sensibility of the target (i.e. the embryonal tissues and hereditary factors resp.);
- day-time dependent modifications of the placental permeability toward chemical substances.

Although many questions remain unanswered, our investigations ascertain for two teratogenes with different mechanisms of action that besides other factors the reactions of the embryo are decisively determined by the time of day at which the substance is applicated. It must be separately tested for each substance with prenataltoxic properties in which circadian phase the minima and maxima of sensibility will be found.

REFERENCES

1. ISAACSON, R.J. (1962): An investigation of some factors involved in the closure of the secondary palate. Thesis Ph. D. Univ. Minnesota, Minneapolis
2. CLAYTON, D.L., A.W. McMULLEN & C.C. BARNETT (1965): Circadian modification of drug-induced teratogenesis in rat fetuses. Chronobiologia 2, 210-217
3. SAUERBIER, J. (1982): Circadian system and teratogenicity of cytostatic drugs. In: E.A. VIDRIO (ed.): Biological rhythms in structure and function. Alan R.Liss, New York, pp. 143-149
4. SCHMIDT, R. (1977): Zur zirkadianen Modifikation der teratogenen Wirkung von Cyclophosphamid (1.Mitt.). Biol. Rdsch. 15, 314-317
5. SCHMIDT, R. (1978): Zur Tageszeitabhängigkeit der pränataltoxischen Wirkung von Cyclophosphamid. Wiss. Z. Univ. Greifswald, Med. Reihe, XXVIII, 171-174
6. SCHUBERT, J. (1980): Untersuchungen zur medikamentösen Beeinflussung experimenteller Gaumenspalten an der Hausmaus. Med.Prom.B., Univ.Halle
7. SCHMIDT, R. (1985): Ein modifiziertes Verpaarungssystem für die Labormaus. Z. Versuchstierk. 27, 206-208

CAN DOPAMINERGIC AGENTS INDUCE COUPLING OF OSCILLATING MECHANISMS UNDERLYING PSYCHOMOTOR PERFORMANCE ?

T. RAMMSAYER

Department of Psychology, University of Giessen,
Otto-Behaghel-Str. 10F, D-W-6300 Giessen, FRG

INTRODUCTION

Based on studies of temporal numerosity and analyses of the distributions of reaction times several authors have inferred oscillating mechanisms underlying human information processing (e.g. 1,2,3). Periodicities in reaction time are reported to range from a modal period as short as 25 ms to approximately 100 ms. On the other hand, time perception, in particular perception of very short intervals in the range of milliseconds, is interpreted by assuming an internal clock. According to this account, a pacemaker generates pulses and the number of pulses occurring during a given time interval is the internal representation of this interval. Hence, the higher the clock rate the better the temporal resolution of the internal clock, which is equivalent to more accuracy and better performance in time perception tasks. However, as compared to the oscillating mechanisms in reaction time, the periods of the internal clock must be extremely short in order to enable discriminations of time intervals in the range of milliseconds. So far, very little is known about the biochemical basis and the relationship between the oscillating mechanisms underlying time perception and reaction time. Therefore, the present studies were designed to elucidate the effect of various pharmacological substances on both of these psychomotor functions.

METHODS

In a series of placebo-controlled pharmacopsychological experiments, we studied the effects of different pharmacological substances on time perception and reaction time in healthy male volunteers. Reaction time was measured by a Leeds Psychomotor Tester. This apparatus is comprised of an array of six lights and six corresponding response buttons arranged in a semicircle. The time required to lift the hand from the start button and move the index finger to the response button next to the active light was scored as reaction time. For the time perception task, the subjects had to decide which of two auditorily presented intervals was longer. One of the intervals was a constant standard interval with a base duration of 50 msec, and the other interval was a variable comparison interval. As an indicator of performance on time perception, at least in most of our experiments, the 70.7%-difference threshold in relation to the standard interval of 50 msec was estimated for each subject. Thus, better performance is indicated by smaller threshold values. In addition, as control variables, critical flicker fusion frequency (CFF) and self-ratings of "alertness", "drowsiness", and "concentration" were measured.

The pharmacological substances under investigation were: the dopamine precursor L-dopa (100 mg), the D2 receptor blocker haloperidol (3 mg), the tyrosine hydroxylase inhibitor alpha-methyl-p-tyrosine (AMPT, 1750 mg), the 5-HT uptake inhibitor fluoxetine (60 mg), the 5-HT2 receptor blocker ritanserin, ethanol (0.65 g/kg of body weight), caffeine (400 mg), and theophylline (233 mg). All substances were administered orally.

Analyses of variance were computed for comparison of drug and respective placebo groups. For reasons of better comparability, the results of the statistical analyses are presented as percent change from respective placebo groups. In addition, Pearson correlations between performance on time perception and on the other psychomotor tasks within each drug condition were computed.

Table 1: Drug-induced percent changes from respective placebo groups.

Drug	Sample Size	Time Perception	Reaction Time	CFF	"alertness"	"drowsiness"	"concentration"
L-dopa	24	- 2.7	- 1.5	+ 0.6	- 2.0	- 5.2	- 2.1
Haloperidol	24	-19.2**	- 6.4**	- 3.6*	-22.4**	-39.2	-31.7**
Fluoxetine	24	+ 5.8	- 0.1	+ 1.4	-16.3*	-57.0*	- 9.7
Ritanserin	24	+ 5.8	- 2.3	- 4.7*	-22.4**	-65.7**	-16.0*
AMPT	40	- 1.2	+ 0.7	- 0.2	+ 1.1	-11.9	+ 3.5
Ethanol	40	-29.7*	- 4.1	- 3.9	-11.1	0.0	-14.0*
Caffeine	12	+10.0	- 1.4	+ 1.1	+13.0	+18.2	+14.6*
Theophylline	12	-16.7	0.0	+ 3.2	+16.4	+18.6	+20.5**

Increase in performance or vigilance as compared to respective placebo group: +
Decrease in performance or vigilance as compared to respective placebo group: -
*: $p < .05$; **: $p < .01$

RESULTS

Drug-induced changes in psychomotor performance and self-ratings of experienced alertness, drowsiness, and concentration are presented in Tab. 1.

Computed correlations between performance on time perception and the other psychomotor tasks yielded statistically significant results for reaction time only (see Tab. 2). Furthermore, significant correlations indicating a positive linear relationship between performance on time perception and reaction time were found exclusively under dopaminergic treatment, i.e. under L-dopa, haloperidol, and AMPT. No significant correlations between time perception and reaction time were found either with non-dopaminergic drugs or with placebo.

Table 2: Pearson correlations between performance on time perception and performance on reaction time and CFF.

Drug	Reaction Time	CFF
L-dopa	+ .58**	- .06
Haloperidol	+ .52**	+ .23
Fluoxetine	+ .22	+ .22
Ritanserin	- .16	- .02
AMPT	+ .57**	- .13
Ethanol	- .20	- .08
Caffeine	+ .06	- .04
Theophylline	- .17	- .29

**: $p < .01$

DISCUSSION

The pattern of drug-induced changes and the results of the correlational analyses suggest that under natural conditions, i.e. without any pharmacological treatment, the periodicity in reaction time and the oscillations of the internal clock can be viewed to be independent as well as free-running oscillating systems as indicated by non-significant correlations between performance in time perception and reaction time.

The same seems to be true for all the non-dopaminergic drugs studied. Therefore, our results may be interpreted in terms of a dopamine-dependent coupling of oscillating mechanisms underlying both reaction time and time perception. Pharmacologically induced changes in dopaminergic activity might cause a frequency demultiplication suggesting that the oscillating system of the internal clock - i.e. the oscillating system with the shorter natural period - will become an initiating agent for the oscillating mechanism with the longer natural period underlying reaction time. This may account for the significant positive correlations between time perception and reaction time.

One might raise objections to our interpretation, since significant correlations between time perception and reaction time could be caused by either a drug-induced general decrease in vigilance and cortical arousal or drug-induced changes in levels of performance. However, the pattern of drug-induced changes clearly demonstrates that neither decreases in vigilance and cortical arousal as indicated by CFF thresholds and self-ratings on "alertness", "drowsiness", and "concentration" nor changes in levels of performance on reaction time can explain these significant correlations under dopaminergic treatment. Ritanserin, for instance, caused a very pronounced sedative effect that nonetheless did result in a positive correlation between performance on time perception and reaction time. On the other hand, unlike haloperidol, which induced a marked decrease in performance on time perception and reaction time, neither L-dopa nor AMPT influenced the means of both these time-related psychomotor functions or induced changes in the indicators of vigilance. However, all these dopaminergic drugs induced statistically significant correlations between performance on time perception and reaction time. Hence, dopaminergic agents seem to operate by inducing a highly specific functional covariation of oscillator-dependent neural processes underlying time perception and reaction time.

REFERENCES

1. AUGENSTINE, L. G. (1955): Evidence of periodicities in human task performance. In H. QUASTLER (ed.): Information theory in psychology. Chicago, Free Press, pp. 208-226
2. PÖPPEL, E. (1985): Grenzen des Bewußtseins. Stuttgart, Deutsche Verlags-Anstalt
3. WHITE, C. T. (1963): Temporal numerosity and the psychological unit of duration. Psychological Monographs 77, 1-37

ON THE SPECTRUM OF THE REACTIVE PERIODS STUDIED IN PATIENTS TREATED WITH A CYCLIC DESIGN OF PYROGENOUS DRUGS

M. WECKENMANN, J. STEGMAIER & E. RAUCH

Filderklinik, Im Haberschlai 7, D-W 7024 Filderstadt, FRG

In 1990 we reported on the reactive periods of the body temperature of patients treated with a cyclic design of injections with pyrogenous drugs (1). The 7 d period proved to be the basic periodicity, the prominant 10.5, 4.7, and 3.5 d periods being the harmonics 3:2, 2:3, and 1:2. The analysis of 16 longer time series of 9 patients revealed further harmonics (2). Now we can report on 59 time series of 33 patients.

METHODS

25 females and 8 males (aged $\bar{x}=57.03$, ± 11.04 years; range: 26-79 years, suffering from malignant diseases were treated with extracts of *Viscum album* L. (Iscador: Weleda AG, D-7070 Schwäbisch Gmünd, FRG; Helixor: Helixor-Heilmittel GmbH, D-7463 Rosenfeld 1, FRG). The drugs were administered once daily between 08.00-09.00 h in series of either 2 or 3 days; in 8 time series the 3rd dose was given at the 5th or 6th day. The series were repeated cyclically every 7 days. The drugs were applicated subcutaneously; in 6 time series once a week by infusion. The dose was adapted to the reaction observed ($\bar{x}= 206.140$, ± 179.57; range 5-800 mg per week).

After a resting period of 1/2 hr either the rectal or the oral temperature was checked daily at the same time of day, mostly at 18.00 h.

The time series data (48-698 d) were analyzed by periodogram (Fourier analysis). Only the significant periods of 2-35.4 d were recognized ($F_{0.05}$). To optimize the discrimination of the periods the large range of the periods was analyzed fractionatedly with overlapping borderlines (s. Fig. 1).

As phase shifts can occur frequently in infraseptan periods, the time series data were smoothed by a simple moving average with 1 point on either side of the target and the periods of ≥ 10 d evaluated visually too.
Distributions were tested by the Chi-Square Goodness-of-Fit Statistic.

RESULTS

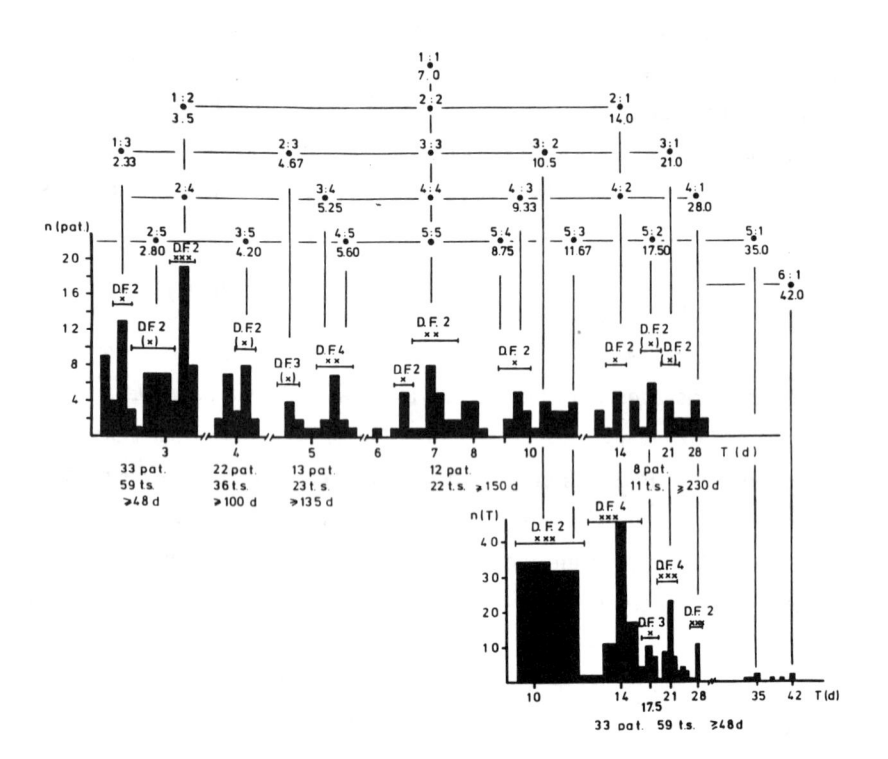

Fig. 1: The distribution of the period lengths analyzed in 59 time series of the body temperature of 33 patients. The scale of the results of the visual analysis (below) was adapted to that of the periodogram (middle). ··· p <0.001; ·· p <0.01; · p < 0.05; (·) p <0.1. The distribution of the harmonics is printed above.

Fig. 1 shows the non-random distribution of the periods. Because of the classification applied the peak of the 7 d period is not representative for the whole group, as 44/59 of the time series exhibited most or very intensive peaks of T = 6-7 ($\bar{x}$= 7.00 ±0.34) d. Many of the quotients of the integer ratios up to the 5th harmonic coincide with maximal peaks, the mean error being -0.10 sx 0.28 d or -0.86 sx 2.76%.

The distribution of the maxima within the 3 classes next to the quotients of the integer ratios shows a maximum in the middle of the 3 classes in 68.4% (Chi-Square 17.42 D.F.2 p <0.001). In these cases even the single distribution can be non-random e.g. T = 3.5 d (1:2), 10.5 d (3:2), 14.0 d (2:1) etc. (s. Fig.1). Exceptions occur if quotients of two integer ratios are similar, e.g. T = 5.25 d (3:4) and 5.60 d (4:5) or T = 8.75 d (5:4) and 9.33 d (4:3). In these incidences the peak can occur in the mean of both of the ratios, e.g. T = 5.35 d or in the class of the more intensive period e.g. T = 9.33 d. Excluding these incidences the peaks coincide with the quotients of the best fitting integer ratios even sometimes in the 5th harmonic (Chi-Square = 6 D.F. 1 p <0.02). The significant peak of T = 6.41 d is not understood. The plots of 2 time series of 637 and 693 d were analyzed visually and yielded two series of T = 12 weeks, the difference between the maximal and minimal points being significant. The mean difference between the period lengths (12:1 and 8:1) of the best fitting ratio and the readings was averg. -1.08 sx 2.45 d and averg. -0.34 sx 3.03 d. One of the 12 week periods was superimposed by a 6 week period.

DISCUSSION

There are good arguments to accept the reactive periods to be a symmetric harmonic system with the basic period of 7 d in the very centre. However, a biological clock may be organized according to this system too.

It is striking that 7 of the harmonics observed were already cited in the Hippocratic Collection as days of crises: 4th, 7th, 11th, 14th, 17th and 20th d (3).

REFERENCES

1. WECKENMANN, M. & J. STEGMAIER (1990): Reactive periods stimulated by a periodic application of mild pyrogenous drugs. Oral presentation, 6th Ann. Meeting of the European Society for Chronobiology, Barcelona, Spain.
2. WECKENMANN, M., J. STEGMAIER & E. RAUCH (1991): On the spectrum of the reactive periods in patients with malignoma. Oral presentation, 7th Ann. Meeting of the European Society for Chronobiology, Marburg, FRG.
3. GÖLL, H. (1876): Die Weisen und Gelehrten des Altertums. O. Stamer Leibzig.

EFFECTIVENESS OF PHYTOADAPTOGENS IN DIFFERENT SEASONS

O.N. DAVYDOVA

Department of Pharmacology of the Moscow Medical Academy,
Moscow, Russia

A study of the effect of seasonal factors on sensitivity and reactivity of living systems to phytoadaptogens, such as bioginseng, eleutherococcus, rhodiola rosea was conducted in experiments in mice. We investigated the effect of these pytoadaptogens on the immune system indices in different seasons, from January to March and in May and June. Adaptogenous action of these preparations was studied during the same seasons on the basis of the values of general physical hardiness, the ability to speed up a recovery process in between repeated physical loads and resistance of an organism to acute hypoxia. The immunotropic effect of bioginseng (preparation from biomass of ginseng tissue culture) and eleutherococcus in case of lowered immunity in different seasons was studied during the first series of experiments. The studied preparations were administered to the animals in a dose range of 0.5 ml/kg to 10 ml/kg orally once in 24 hours during 7 days. The mice were immunized by a test antigen-ram erythrocytes. To reduce immune response cyclophosphamide was administered to the mice in doses of 25 mg/kg and 50 mg/kg subcutaneously the next day after an antigen stimulus had been given to the animals. Evaluation of the immune response was made on the basis of titres of hemagglutinins and hemolysins in series of double dilutions in the volume of 0.5 ml. The highest dilution was recognized as a serum titre when either complete hemolysis or complete agglutination of the standard quantity of erythrocytes could be observed. The results of the experiments through a model of attenuated immune reaction show that in administration of bioginseng and eleutherococcus from January to March immunodepressive effect of cyclophosphamide in the dose of 50 mg/kg is considerably reduced and is completely eliminated in administration of cyclophosphamide in the dose of 25 mg/kg. In this case the titers of antibodies reached the control level. But in summer the immunomodulating effect of the phytoadaptogens was not observed.

In addition to this, we studied their effect on immunity at an organ level. Evaluation of immunomodulating properties of phytoadaptogens was made through a change of mass of the thymus and the spleen. Administration of cyclophosphamide in the dose of 50 mg/kg gave a significant 40 % decrease of the spleen mass and 25 % decrease of the thymus mass. Seven day administration of phytoadaptogens (the dose for bioginseng being 5 ml/kg and 10 mg/kg for eleutherococcus), recovered the spleen and thymus to practically initial value in winter-spring period, but not in summer.

Then we studied adaptogenous effect of bioginseng, eleutherococcus, and preparation from biomass of rhodiola rosea tissue culture on the basis of the values of general physical hardiness and the ability to speed up a recovery period in between repeated physical loads in winter and summer. The preparations were injected in a dose range of 1 ml/kg to 10 ml/kg intraperitoneally one hour before a physical load. The physical load was implemented in two ways: running on a treadmill and swimming. As an assessment criterion of fatigue of animals we used their refusal from proceeding to run on the treadmill or stay for more than 10 sec on an electric site with applied voltage of 25 V. As for swimming, sinking served as an assessment criterion of fatigue. A submaximal load was determined experimentally by the above mentioned techniques and experimental animals were exposed to a load making up 3/4 of the submaximal one. Evaluation of motor activity amount after a dosed physical load was made with the help of the chamber "animex". The experiments showed that the afore named phytoadaptogens significantly increased general physical hardiness and shortened the period of physical working capacity recovery in between repeated physical loads when these adaptogens were administered from January to March. In summer their adaptogenous effect was insignificant. The adaptogenous effect of rhodiola rosea was similar to that of bioginseng and eleutherococcus.

During the third series of experiments we studied antihypoxic effect of bioginseng and rhodiola rosea on the model of hypoxic hypoxia in winter and summer months. Hypoxic hypobaric hypoxia was modelled according to the conventional method. Only animals of medium resistance, that is those who could survive in the conditions of hypoxia not less than 5 minutes were used. Bioginseng in a dose of 5 ml/kg and rhodiola rosea in a dose of 0.5 ml/kg were given interaperitoneally one hour before placing the animals into an altitude chamber. The same amount of distilled water was administered to the control group. At 10.400 meters above sea level we fixed a period of time before appearance of cramps and respiratory stand-still in both groups. Bioginseng and rhodiola rosea have significantly increased longevity of mice in winter months, bioginseng 1.5 times, rhodiola rosea 2.2 times, but a similar effect was not observed in summer months. Thus, the largest chronoesthesia and chronergia to the effect of the phytoadaptogens in winter and spring appear to be connected with reduction of general (nonspecific) organism resistance to different factors of internal and external media. Clinical studies proved seasonal effectiveness of the phytoadaptogens under discussion.

OCCUPATIONAL ASPECTS

CIRCADIAN RHTYMICITY IN SELF-CHOSEN WORK-RATE

G. ATKINSON, A.COLDWELLS, T.REILLY & J.WATERHOUSE*

Centre for Sport and Exercise Sciences, Liverpool Polytechnic, Byrom Street, Liverpool. L3 3AF. UK
*Department of Physiological Sciences, University of Mancester Medical School, Stopford Building, Manchester. M13 9PT. UK.

INTRODUCTION

Conventionally, the heart rate response to submaximal exercise has been employed in predicting maximal aerobic power (1) and physical working capacity (10). In such tests, heart rates at two or more known submaximal workloads are either entered into tables to predict oxygen consumption at an estimated maximum heart rate, or used to extrapolate a work-rate that elicits a certain heart rate (usually 150-170 beats min^{-1}). Significant circadian variation has been found for both maximal heart rate and heart rate during submaximal exercise (6). Therefore, the time of day at which these predictive tests are administered is important. A low heart rate response to exercise results in a high estimate of work capacity in these tests. Since the rhythm in heart rate peaks at 17.00 h (6), the acrophase of predictive work capacity may occur 180° out of phase with acrophase of body temperature and other performance-related measures.

Acceptability scaling has been used in industrial contexts to determine the load-handling capabilities of individuals (9). It derives a work-load based on psychophysical rather than physiological criteria i.e. the subject's feelings of effort, discomfort and fatigue. BAXTER (2) identified a significant circadian rhythm for the rating of acceptable lift with the best performance determined in the early afternoon, in phase with rhythms in both arousal and oral temperature. To date, little research has applied psychophysical acceptability scaling to continuous work such as exercise on a cycle ergometer.

The aims of this study were i) to establish whether self-chosen work-rate (determined on a cycle ergometer) displayed circadian rhythmicity and ii) to examine the relation between circadian rhythms in physiological and subjective measures.

METHODS

Five physically active male subjects, aged 19-24 years, participated in the investigation. A test battery was administered to each subject at different times of the day and night during a period of 3 weeks. Measurements were made at 00.30, 06.30, 10.00, 14.00, 18.00, and 22.00 hours. Each subject was assigned to start at a test time so that no two subjects followed the same sequence. Pre-test diet and physical activity were controlled for each subject.

At the start of the battery, arousal was measured by asking each subject to place a mark on a 10 cm horizontal line between extremes of "Alert" and "Drowsy". Sleepiness was measured using the Stanford sleepiness scale (4). The profile of mood states (POMS) questionnaire (5) was also completed by the subjects at this stage. Rectal and oral temperature were then measured using digital clinical thermometers (Philips) to an accuracy of 0.01 °C for a period until the reading stabilised (<0.02 °C per 8 s). After this, right and left grip strength was assessed using a Takei Kiki Kogyo (Japan) grip dynamometer. The best of three trials was recorded to the nearest 0.5 kg. Resting and submaximal pulse rates were then obtained by palpitation with the index finger from the carotid pulse over a period of one minute. Subjects sat quietly for 5 min before resting pulse rate was recorded. Submaximal pulse rate was measured after 4 min of continuous exercise (2 min at 120 watts and 2 min at 180 watts) on a cycle ergometer. Finally, subjects were required on a cycle ergometer (Monark) for 3 min at a self-chosen work-rate that they believed they were capable of sustaining for 30 min. Mean power production was calculated from the ergometer braking resistance (set at 4% of body mass) and the chosen pedal frequency via a computer (B.B.C.) interface.

Differences between the time of day for all the variables were investigated using cosinor analysis.

RESULTS

The acrophases observed at the six time points for the measured variables are summarised in Fig. 1. Both rectal and oral temperature exhibited significant circadian variation (p<0.05) with acrophases occuring at 19.31 and 20.13 h respectively. Resting pulse rate peaked at 18.13 h. Subjective arousal was found to peak significantly earlier than the temperature measures (p<0.05).

MEASURE

MEASURE	0	4	8	12	16	20	24
RECTAL TEMPERATURE						*	
ORAL TEMPERATURE						*	
LEFT GRIP STRENGTH						*	
RIGHT GRIP STRENGTH						*	
AROUSAL					*		
SLEEPINESS			*				
RESTING HEART RATE						*	
SELF-CHOSEN WORK						*	
SUBMAX. HEART RATE						*	
TOTAL MOOD DIST.			*				
FRIENDLINESS			*				
CONFUSION			*				
ANGER		*					
TENSION			*				

TIME OF DAY (HOURS)

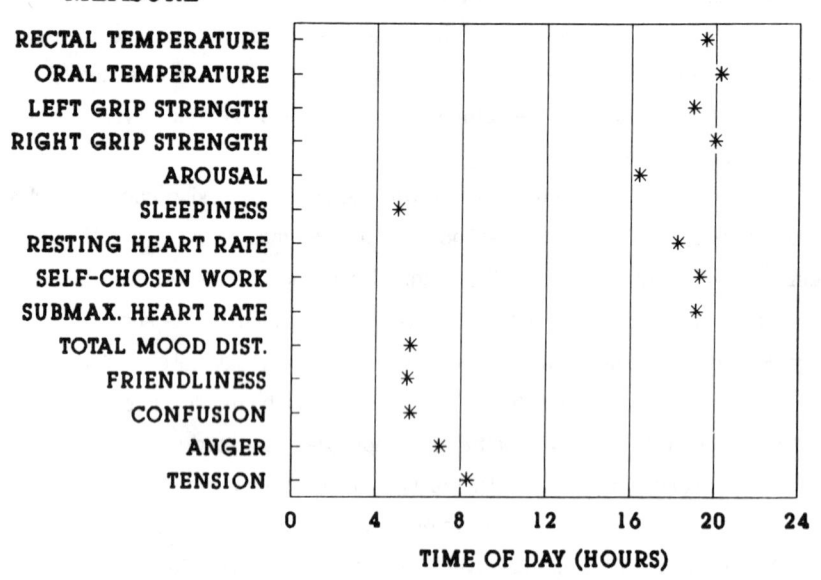

Fig. 1: Rhythm acrophases for all the measures.

Performance rhythms such as right and left grip strength and submaximal pulse rate were all similar in terms of amplitude (5-6% of mean values) and were in phase with the rhythms in body temperature. Self-chosen work-rate evidenced significant circadian variation (p<0.05) peaking in early evening with similar phase and amplitude to other resting and performance measures. The circadian rhythm in self-chosen work-rate is shown in Fig. 2.

WORK-RATE (WATTS)

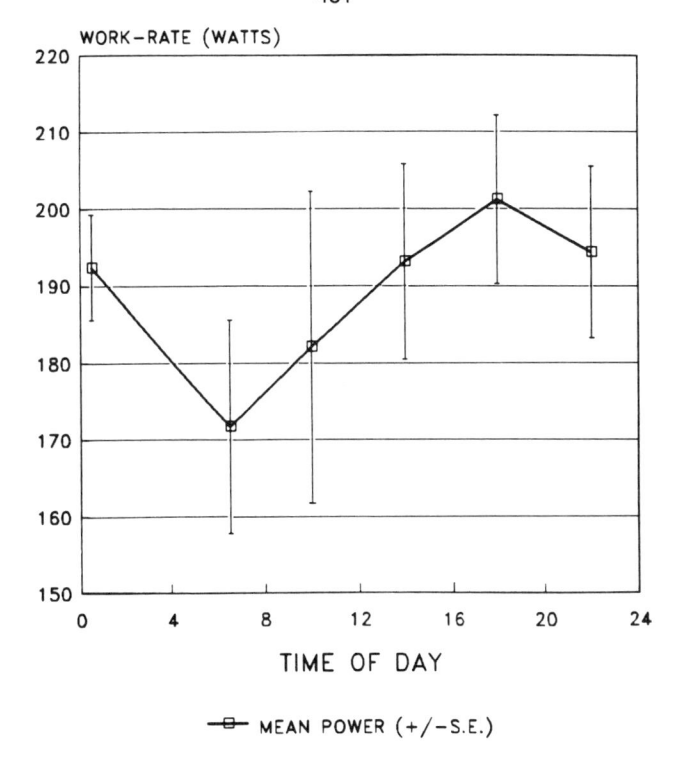

Fig. 2: The circadian rhythm in self-chosen work-rate.

Acrophases in the POMS factors of confusion, anger, tension and total mood disturbance were found to occur at the time of the lowest self-chosen work-rate. Subjects reported higher values for friendliness in the early evening.

DISCUSSION

The major finding of the present study was that self-chosen work-rate exhibited circadian rhythmicity. This psychophysical measure varied in phase with rhythms in body temperature and other performance variables. The amplitudes of both rectal and oral temperature approached the values that have been cited in the literature (5,7). The significance of the body temperature rhythm to human performance may be its relation to rhythms in muscle viscosity, nerve conducion velocity, and joint flexibility (8). The evening rise in core temperature, therefore, has some of the attributes of a 'warm up' which is frequently achieved by athletes prior to competition.

The circadian rhythms of grip strength, resting heart rate, and submaximal heart rate were characteristic of those demonstrated in previous research (7). The amplitude of the heart rate rhythm was 4 beats min^{-1} at rest. This increased marginally to 8 beats min^{-1} under submaximal exercise conditions. This confirms that predictive tests of cardiovascular status, which use resting or submaximal heart rates as criteria, may be comprimised by circadian variation (6).

Subjective arousal, friendliness and vigour exhibited circadian rhythms which showed a sharp increase from 08.00-12.00 h to peak at about 17.00 h. Conversely, negative mood variables (total mood disturbance, confusion, anger and tension) decreased in the afternoon to a trough at 17.00 h. These behavioural factors may be important for human performance since they may alter an individual's predisposition for strenuous physical work. Circadian variations in mood states may also affect the 'team cohesion' of a work-group. For example, the ability to communicate and work together as a team is important for aircraft flight-deck personnel.

Self-chosen work-rate performed on a cycle ergometer was found to exhibit circadian rhythmicity with a trough to peak variation of 27 watts or 15% of the mean work-rate. The rhythm was typical of a performance-related measure, peaking in early evening close to the acrophase of body temperature (6). BAXTER (2) identified circadian rhythmicity for the similar psychophysical measure of acceptable lifting performance. The acrophase of this rhythm occured several hours earlier than the acrophase for self-chosen work-rate. From the results of a longitudinal investigation of one subject, it was hypothesised that the acceptable lift rhythm was better associated with rhythms in behavioural measures (arousal, mood) than with body temperature. In the present study, self-chosen work-rate peaked closer to the acrophase of body temperature than to those of arousal and mood states.

It is possible that self-chosen work-rate is related to the perception of effort (RPE). Minimal RPE peaks in the early evening (3) at a similar time to the acrophase found in the present study for self-chosen work-rate. Significant rhythmicity in RPE has, however, been reported only when work-rates exceed 150 W. The mean work-rate performed by the subjects in the present study was 188 W, an intensity that can be described as moderate to high. The present study has demonstrated that the heart rate response to submaximal exercise is lowest at night. Therefore, more exercise can be performed at a given heart rate at that time (6). As heart rate and RPE were not recorded during the self-chosen work-rate test, it is unclear whether this rhythm reflects the heart rate responses to exercise or an inherent variation in effort perception.

The fact that subjects chose to work harder in the evening than in the morning may have implications for the scheduling of heavy physical work and for the timing of athletic training session. It is possible that greater training and work loads may be tolerated after 18.00 h and, indeed, athletes generally prefer to reserve hard training sessions for this time of day. Factors such as age, chronotype, physical fitness and habitual activity levels may influence an individual's choice of time of day for optimal performance of physical work. The effects of these factors on the circadian rhythm in self-chosen work-rate should now be investigated.

REFERENCES

1. ASTRAND, P.-O. & I. RYHMING (1954): A nomogram for calculation of aerobic capacity (physical fitness) from pulse rate during sub-maximal work. Journal of Applied Physiology 7, 218-221
2. BAXTER, C. (1987): Low back pain and time of day: a study of their effects on psychophysical performance. University of Liverpool: Unpublished doctoral thesis
3. FARIA, I.E. & B.J. DRUMMOND (1982): Circadian changes in resting heart rate and body temperature, maximal oxygen consumption and perceived exertion. Ergonomics 25, 381-386
4. HODDES, E., V. ZARCONE, H.R. SMYTHE & W.C. DEMENT (1973): Quantification of Sleepiness: a new approach. Psychophysiology 10, 431-436
5. MCNAIR, D.M., M. LORR & L.F. DROPPLEMAN (1971): EITS manual for the profile of mood states San Diego: Educational and Industrial Testing Service
6. REILLY, T. (1990): Human circadian rhythms and exercise. Critical Reviews in Biomedical Engineering 18, 165-180
7. REILLY, T. & G.A. BROOKS (1982): Investigation of circadian rhythms in metabolic responses to exercise. Ergonomics 25, 1093-1107
8. SHEPHARD, R.J. (1984): Sleep, Biorhythms and Human Performance. Sports Medicine 1, 11-37
9. SNOOK, S.H. & C.H. IRVINE (1967): Maximal acceptable weight of lift. American Industrial Hygiene Association Journal 29, 531-536
10. WATSON, A.W.S. & D.J. O'DONOVAN (1976): The Reliability of Measurement of PWC 170. Irish Journal of Medical Science 145, 308

SLEEP-WAKE PATTERN TYPE AND OBJECTIVE PHYSIOLOGICAL CHARACTERISTICS DURING NIGHT WAKEFULNESS

V.A. CHEREPANOVA[+*] & A. PUTILOV[*]

[+] Department of Physics, University of Manitoba,
Winnipeg, Manitoba, Canada R3T 2N2
[*] Institute of Physiology, Siberian branch, USSR AMS,
Novosibirsk, RUSSIA 630117

INTRODUCTION

It has become increasingly evident that the individual phase position of the circadian rhythms is one of the major important endogenous factors determining people's ability to adapt to night and shift work (1). The greater tolerance of evening types to night work as compared with morning ones has been demonstrated experimentally (2-4).

One of us has constructed a questionnaire for multidimensional assessment of individual differences in sleep-wake cycle pattern (SWP) and its adaptation to shifts in work-rest schedules (5). The analysis of multivariate structure of the SWP questionnaire revealed 3 scales and 2 subscales which can define as levels of morning (M) and evening wakefulness (E), quality of sleep (S), capacities to be awake (w) and to fall asleep anytime (f). High positive scores mean eveningness on M- and E-scales, good sleep on S-scale and high capacities to change time of sleep and wakefulness on f- and w-subscales.

In the very beginning of process of SWP questionnaire construction M- and E-items were identified as rather independent. So the questionnaire indicates that morningness-eveningness is a combined trait of sleep-wake cycle and differentiates 4 extreme types: "evening sleepers" (ES-type or so-called larks), "morning sleepers" (MS-type or owls), "short sleepers" (SS-type or habitually larks in the morning and owls in the evening), and "long sleepers" (LS-type or owls in the morning and larks in the evening). It was found that w-scores show positive correlation with E-scores and negative correlation with M-scores. It means that not MS-type, but SS-types should have the highest capacity to be awake anytime.

The question arises: which type of sleep-wake pattern - MS- or SS-type - has the greatest tolerance to night wakefulness and why.

MATERIALS AND METHODS

SWP questionnaire structure was studied on a large sample (2000 males and 2126 females aged 9 to 79 years). The great majority of respondents (school children, students, employees, research workers, etc.) lived in Novosibirsk (West Siberia, one and a half million population) and had normal health and work-rest schedule. These 4126 respondents were classified into 9 groups (3x3) on the basis of M and E scores (low, moderate and high values). Means of w-, f-subscale and S-scale scores were calculated for each group and were compared with the overall means.

In order to support this survey study the special experimental investigation was carried out. In October SWP questionnaire was administered to 480 cadets of a high military school well-trained males whose average age was 20.1 ±1.3 years. 18 from them were identified as ES-types, 3 as LS-type, 133 as SS-types, and 40 as MS-types. Then three groups of ES-, MS- and SS-types, nine cadets in each were randomly selected for the sleep deprivation experiments which were carried out at nights between Saturday and Sunday in December.

Starting at 8 a.m. on day of experiment, all subjects were kept continuously awake until 8 a.m. the following day. During an experimental night they were given a low-protein diet after 10 p.m. and before 3 a.m. Two urine samples (from 11 p.m. to 3 a.m. and from 3 to 7 a.m) were collected and analyzed according to the fluorimetric methods (Hytachy). The method for detection of concentration and excretion of adrenaline and noradrenaline (AC, AE, NC, and NE) was developed by MATLINA et al (6). Melatonin metabolites concentration and excretion (MC and ME) were analyzed according to MILLER & MAICKEL (1970) in LEVIN et al.(7) modification.

Subjective vigilance (SV), sensomotor reaction speed (R1 and R2), oxygen and heart cost of physical load (OC and HC) and concentration of sodium in saliva (NA) were measured every 2 hours at fixed times from 10 p.m. to 7 a.m. SV was determined by means of self-assessment WAM-test (Well-being - Activity - Mood)(8). Broken line-test (9) was used for determination of sensomotor reaction speed: number of broken lines written down during 1 min in a small square each line (R1) and without space limitation (R2). OC and HC were calculated from data of a 3-min physical load test (2 W per kg total body weight) on a bicycle ergometer. Oxygen uptake before, during, and till full recovery after exercise was evaluated by using gasoanalyser (SPIROLIT-2). Sodium concentration (NA) was assessed by sodium electrode on pH-METR-5123.

Axillary temperature (AT), heart rate (HR), respiratory rate (RR), their ratio (Hildebrandt index, HI), and time estimation of "individual minute" (IM) were obtained hourly over a 9-hr night span by means of autorhythmometry (10).

A year later in December the same set of parameters plus vascular reactivity to noradrenaline (RN) were measured in these subjects after normal night sleep in the morning (near 7:30) and afternoon (near 15:30). Urine samples were collected from 7:00 till 10:00 and from 15:00 till 8:00.

In order to assess twice more SWP scores and their test-retest reliability the SWP questionnaires (5) has been administered during each experiment. Horne-Östberg (HO) questionaire (11) has been also administered twice to elevate preferred time for go to bed and get up on the items 2 and 1.

The computation was carried out using CSS package. Values of measured variables in ES-, SS-, and MS-types were compared by means of conventional statistics includ Student's t-test for paired data and MANOVA. Differences between types on night and day levels of measured variables are presented in Table 1. Table 2 shows ratios of variable values in the beginning and in the end of night as well as in the morning and in the afternoon.

RESULTS

Results of survey study shows that ES-type had significantly lower capacities both to fall asleep and to be awake anytime. LS-type had the lowest capacity to be awake anytime. MS-type had the highest capacity to fall asleep at anytime. The SS-type had the highest capacity to be awake at anytime.

Table 1: Differences between types on night and day levels of subjective and objective indices

IN-DEX	night			day		
	F	P	pairs	F	P	pairs
SV	11.83	<0.001	ES<MS*<SS*** MS<SS*	4.62	0.014	MS<ES+ MS<SS**
R1	4.37	0.014	SS<ES* MS<ES**	7.48	0.0014	MS<ES** MS<SS***
R2	4.92	0.008	ES<SS** MS<SS*	7.38	0.0016	ES<SS** MS<SS***
OC	0.05	0.95		1.78	0.18	MS<SS+
HC	9.98	<0.001	SS<ES** SS<MS***	0.26	0.77	
NA	0.76	0.47		10.65	<0.001	MS<ES***MS<SS***
AT	1.46	0.23		2.22	0.11	SS<ES+
HR	4.34	0.014	ES<SS**	ES<MS+	1.55	0.22	ES<SS+
RR	1.25	0.29	ES<MS+	0.04	0.84	
HI	2.72	0.06	MS<SS*	0.99	0.38	
IM	3.32	0.03	ES<SS* ES<MS*	1.33	0.27	
RN				3.19	0.05	ES<SS+ MS<SS*
NE	0.44	0.65		1.52	0.22	ES<SS*
AE	3.26	0.04	ES<SS*	1.51	0.23	ES<SS+
ME	1.50	0.23	ES<MS+	1.22	0.30	ES<SS+

Results of one-way MANOVA comparison:
+ p<0.1 * p<0.05 ** p<0.01 p<0.001 (two-tailed differences)

On the results of 3 trials (October and December 1988, December 1989) the correlation coefficients between individual M-scores as well as E-scores in 27 cadets were rather high (0.61-0.75) and were not decreased with increase of the span between surveys. There were no significant differences between SS- and ES-types according to M-scale and between SS- and MS-types according to E-scale. In SS-group sleep time calculated from answers on questions 1 and 2 of HO questionnaire was significantly shorter than in ES- (p<0.01) and MS-group (p<0.05). SS-type had significantly (p<0.05) higher w-scores than other types.

All 3 groups significantly differed one from another according to their night level of SV: The SS-type had the highest level, ES-type had the lowest one. ES- and SS-types had the equal day levels of SV which were higher than MS-type's level (Table 1). During night SS-type had the lowest HC and the highest HI, but differences between types on these variables during daytime were not significant. ES-type had the lowest night levels of HR, IM and AE. Differences between types on R1 and R2 levels were constant in both night and day experiments (Table 1).

SS-type surpasses other types according to the day level of RN (Table 1). This index was higher in the morning than in the afternoon in ES- and MS-types and lower in the morning than in the afternoon in SS-types. In the morning ES-types had the

Table 2: Mean ratios of variables values for different types

IN-DEX	TYPE ES	SS	MS	F	P	DIFFERENCES pairs	TYPE ES	SS	MS	F	P	DIFFERENCES pairs
	midnight/6 a.m.						7:30 a.m./3:30 p.m.					
SV	0.92	1.10	1.13	4.8	0.02	ES<SS* ES<MS*	1.19	0.94	0.90	3.0	0.07	ES>SS+ ES>MS+
R1	1.05	0.91	1.05	4.0	0.03	SS<ES** SS<MS*	0.98	0.84	0.89	6.1	0.007	ES>SS** ES>MS*
R2	1.02	0.99	1.01	0.4	0.71		0.94	0.94	0.94	0	0.99	
OC	1.01	1.00	0.98	0.2	0.86		0.99	1.02	1.14	1.3	0.28	
HC	1.14	1.03	0.87	2.1	0.14	ES<MS+	0.95	1.22	1.15	3.7	0.04	ES<SS* ES<MS*
NA	1.02	1.43	1.05	2.8	0.08	SS>ES* SS>MS*	2.16	1.18	1.15	1.4	0.26	
AT	1.00	1.01	1.01	1.4	0.25		1.00	0.99	1.00	0.3	0.73	
HR	0.99	1.11	1.14	4.9	0.016	ES<SS* ES<MS*	0.93	0.99	0.92	1.4	0.27	
RR	0.98	1.07	1.09	0.8	0.45		0.88	0.95	0.97	0.7	0.51	
HI	1.04	1.05	1.08	0.1	0.91		1.08	1.06	0.97	1.0	0.38	
IM	1.03	0.99	0.99	0.5	0.61		1.03	1.04	1.01	0.1	0.87	
RN							2.65	0.72	1.72	3.7	0.041	SS<ES* SS<MS*
	early/late night						morning/afternoon					
NE	1.37	1.11	1.43	0.3	0.76		2.09	0.93	1.24	4.2	0.028	ES>SS* ES>MS+
AE	1.07	0.82	1.02	1.3	0.28		1.97	0.98	0.87	6.3	0.006	ES>SS* ES>MS*
ME	0.98	0.78	1.13	1.0	0.38		5.89	1.50	2.78	3.2	0.057	ES>SS*

Results of pair t-test comparison:
+ p<0.1 * p<0.05 ** p<0.01 (two-tailed differences)

highest values of different indices of activation and then they were not changed, while in other types they were increased. Morning/afternoon differences between ES- and other types were significant for SV, HC, R1, NE, AE and ME(Table 2).

During night significant interactions between type and time were found for SV, R1, HR, AT (only from 2 to 7 a.m.) and HC (only from 12 to 6 a.m.). SV decreased in MS- and SS-types and increased in ES-types, HR also decreased in MS- and SS-types and was rather stable in ES-types, HC increased in MS-types, it was rather stable in SS-types and it decreased in ES-types. Beginning with 2 a.m. AT in ES-types had been the lowest and then it rose, while in other types it decreased till the morning. Values of R1, NA, AE and NE in MS- and ES-types were almost the same in the early and in the late night, but NA decreased and R1 and NC increased to the end of night in SS-type. AS, AE, ME and MS showed the same tendency, but the correlations between types and the indices during nighttime did not reach statistical significance (Table 2).

DISCUSSION

4126 respondent's samples of survey study showed that people with high M- and E-scores (ES-type) had the highest capacity to fall asleep anytime (f), while not these ones but types with low M- and high E-scores (SS-type) had the highest capacity to be awake anytime (w). The former type can compensate sleep deficits better than others, but the latter types can easier be awaked anytime. Thus one can hypothesize that these types are different in their manner of adaptation to night wakefulness and shift work.

The sleep-wake pattern questionnaire proposes a reason for separating morningness-eveningness trait of sleep-wake cycle into morning and evening components. This result agrees with typology by LEOPOLD-LEVI (1932) who extended WUTH's separation (1931) of "evening sleepers" (early to bed, early to rise) and "morning sleepers" (late to bed, late to rise) by adding two more extreme types: early to bed, late to rise and late to bed, early to rise individuals (cited after 12). Cadets with low M- and high E-scores preferred go to bed (item 2 of HO questionnaire) significantly later than "evening sleepers" (ES-types) and wake up (item 1) significantly earlier than "morning sleepers" (MS-types). It can be assumed that these individuals are "short sleepers" (SS-type).

Comparison of 3 types with respect to different objective indices of physiological activation during night sleep deprivation, and in the beginning and in the end of working day after normal night sleep shows that ES-type has lower level of activation in the first part of night than MS-type's level. However, this level begins to increase in the last part of night and becomes higher in the morning. SS-type fell in between ES- and MS-types, but closer to the last on dynamics of physiological variables. However, they had high and rather stable levels of subjective and objective indices of vigilance both in the first and in the last part of night. Also only SS-type had excretion and concentration of catecholamines in the last part of night higher than in the first part. This peculiarity as well as higher than in other types mean levels of catecholamines and vascular reactivity to noradrenaline might be one explanation of their capacity to keep up high level of alertness during all night.

Our results suggest that SS-subjects can be the intermediate types or the MS-type with respect to AT rhythm as well as to some other vegetative functions rhythms and they can be the special type with respect to sleep-wake cycle and connected with its performance rhythms.

The findings of experimental investigation indicate that SS-type achieves higher than MS-types tolerance to night awaking and that sleep-wake pattern questionnaire may be capable to predict individual differences in circadian rhythms and its adaptation to shift work.

REFERENCES

1. HILDEBRANDT, G. (1980): Survey of current concepts relative to rhythms and shift work. In: L.E. SCHEVING (ed.): Chronobiology: Principles and Applications to Shifts in Schedules. Sijthoff and Noordhoff, Alphen aan den Rijn - Rockville (Md), pp.261-292
2. BREITHAUPT, H., G. HILDEBRANDT, S. DOHRE et al (1978): Tolerance to shift of sleep as related to the individual's circadian phase position. Ergonomics 21(10), 767-774
3. AKERSTEDT, T. & L. TORSVALL (1981): Shift work: Shift-dependent well-being and individual differences. Ergonomics, 24(4), 265-273
4. LAVIE, P. & S.H. SEGAL (1989): Twenty-for-hour structure of sleepiness in morning and evening persons investigated by ultrashort sleep-wake cycle. Sleep 12(6), 522-528
5. PUTILOV, A.A. (1990): A questionnaire for self-assessment of individual peculiarities of sleep-wake cycle. (Russ.) Bull. Siberian Branch, USSR Acad. Med. Sci. 1, 22-25
6. MATLINA, E.SH. & V.N. VASILIEV (1979): Diurnal rhythm of sympatho-adrenal system and its function capacity with young and old persons. VEDA - Publishing House of the Slovak Academy of Sciences: Bratislava, pp. 303-316
7. LEVIN, I.M., I.M. TZVETNOY & N.V. GROMOVA (1988): Improvement of sensitivity and specificity of fluorimetric determination of urine melatonin. (Russ.) Laboratornoe delo 4, 54-58
8. DOSKIN, V.A., N.A. LAVRENTEVA, M.P. MIROSHNIKOV & V.B. SHARAY (1973): Test to differential self-assessment of the functional state. (Russ.) Voprosy Psychologii 19(6), 141-145
9. KULAK, I.A. (1974): Psychic and Physiological Functions of Man and System of NOT (in Russ.). Moscow
10. HALBERG, F., E.A. JOHNSON, W. NELSON et al (1972): Autorhythmometry procedures for physiologic self-measurements and their analysis. Physiology Teacher 1, 1-11
11. HORNE, J.A. & O. ÖSTBERG (1976): A self-assessment questionnaire to determine morningness-eveningness in human circadian rhythms. Int. J. Chronobiology 4(2), 97-110
12. HORNE, J. & O. ÖSTBERG (1977): Individual differences in human circadian rhythms. Biol.Psychol., 5(3), 179-190

A QUESTIONNAIRE FOR SELF-ASSESSMENT OF INDIVIDUAL PROFILE AND ADAPTABILITY OF SLEEP-WAKE CYCLE

A.A. PUTILOV

Institute of Physiology, Siberian branch, USSR AMS Novosibirsk, RUSSIA

INTRODUCTION

Individual differences in human circadian rhythms have become a large interest in chronobiological studies (1,2). The morningness-eveningness preference has received much attention since the publication of an English version of ÖSTBERG's questionnaire (3). This and the majority of other questionnaires (4-6) have concentrated on differences in phase of circadian rhythm and have distinguished between morning and evening types. Only FOLKARD et al. (7) have published a questionnaire to distinguish people whose circadian rhythms still can adjust to night work on the basis of other characteristics associated with flexibility of sleep habits and ability to overcome drowsiness.

The present study is an attempt to construct a questionnaire for multidimensional assessment of individual differences in sleep-wake cycle pattern.

MATERIAL AND METHODS

The questionnaire has been constructed in consequence of selection of affirmative short statements associated with everyday habits. The first variant of the questionnaire included 2 conditionally equivalent versions from 100 statements each. Besides every positive statement each version had similar negative one. There were 5 variants of response: 1) very suited, 2) probably suited, 3) neither suited nor unsuited, 4) probably unsuited, and 5) totally unsuited. Both versions were administered to 117 respondents with normal work-rest schedule. Three or more weeks later they were filled in by 50 of them once more.

The computation was carried out using BMDP package. For each statement was computed the distribution of responses (BMDP2D-program) and the rank correlation coefficient (BMDP1F-program) between the responses obtained in the first and second administration. For the first administration the correlation matrix of all 200 statements' responses were calculated and used to cluster and factor analyses (BMDP1M- and BMDP4M-program).

From each version were rejected all pairs of statements for which the 2nd, 3rd and 4th variants of responses were given more often then 1st or 5th, and/or which hadn't good test-retest reliability, and/or were out of 5 main clusters associated with peculiarities of sleep-wake pattern.

Two examples: All statements about picks of well, being or performance were eliminated because of a bad distribution of responses (not so many persons have evident opinion about their well-being or performance dynamics). All statements about morningness-eveningness of eating habits were excluded because of their isolation in special clusters from corresponding clusters of sleep-wakefulness habits (probably, they associated with different human circadian subsystems).

The next version of the questionnaire had 76 statements and 3 variants of response: 1) yes, 2) neither yes nor no, and 3) no. It has been filled in by 221 respondents. The principles of further selection were the same. The 3rd version had 40 statements and respondents were asked, if they could, to answer only yes or no. After its administration by 306 subjects (40 of them - twice) there were no rejected statements but some of them were slightly improved. Improvement of statements has been ended on the 4th version of questionnaire being administered to 356 people yet.

Stability of questionnaire structure and age- and sex-related differences were studied on 4 samples of more than 4000 subjects (2.000 males and 2.126 females aged 9 to 79 years). The vast majority of respondents (school children, students, employees, research workers, etc.) lived in Novosibirsk (West Siberia, one and a half million population) and had normal health and work-rest schedule.

RESULTS

1. Structure of the questionnaire

The questionnaire consisted of 10 groups of 4-statements each (tetrads). It was suggested that they are connected with the following traits or capacities of individual sleep-wake pattern: 1) to get up at a fixed time, 2) to wake up early in the morning, 3) to be cheerful in the morning, 4) to be cheerful in the evening, 5) to be awake at night, 6) to be awake at anytime, 7) to fall asleep at anytime, 8) to fall asleep at night easily, 9) to sleep deeply at the midnight, and 10) to sleep deeply prior to the morning.

As a rule both cluster (BMDP1M) and factor analyses (BMDP4M) yielded these groups of statements as "simple" clusters or "first-order" factors. However, sometimes sets of statements from a few tetrads jointed in the larger groups which characterized more common peculiarities of sleep-wake cycle, namely: high/low level of morning wakefulness (after night sleep), tetrads 1-3, partly 10, and evening one (before night sleep), tetrads 4-5, partly 8, and quality of night sleep, tetrads 8-10.

The analysis of "high-order" structure in each sample revealed 3 clusters or factors: tetrads 1-3 which can defined as morning wakefulness (M), tetrads 4-5 - evening wakefulness (E) - with rather a isolated 6th tetrad one - awaking at anytime (w) and tetrads 8-10 - quality of sleep (S) - with a rather isolated 7th tetrad - falling asleep at anytime (f).

All correlations between these 5 main characteristics of sleep-wake pattern except one were positive when traits 1-3 had negative marks (the larger mark the lower capacity) and others - positive marks (Tab. 1). For computations of scores each response was expressed as a number -1 or +1 in accordance with these interrelations. So scores for M scale were varied from -12 to +12 (for a total of 4.126 subjects the standard deviation was 6.07 and Cronbach's coefficient alpha was 0.80) , for E - from -8 to +8 (4.75 and 0.77) , for w - from -4 to +4 (2.57 and 0.54), for f - from -4 to +4 (2.47 and 0.54), and for S - from -12 to +12 (6.73 and 0.80). High positive scores claim eveningness on M and E scales, good sleep on S-scale and high capacities to change time of sleep and wakefulness on f- and w-scales, while low negative scores claim morningness, bad sleep and low capacities to change time of sleep and wakefulness.

Tab. 1: Correlations matrix of the 10 tetrads scores (below the diagonal) and of 5 main scales scores (above the diagonal, (n=4.126)

```
            1                       M     E     w     f     S
    1  1.00    2                 1.00   .15  -.20   .13   .18   M
M   2   .41  1.00    3                 1.00   .37   .16   .31   E
    3   .43   .44  1.00    4
                                              1.00   .15   .31   w
    4   .18   .14   .04  1.00    5                  1.00   .33   f
E   5   .13   .13   .01   .51  1.00    6
                                                          1.00   S
w   6  -.06  -.14  -.26   .27   .36  1.00    7

f   7   .12   .16   .02   .12   .16   .15  1.00    8

    8   .21   .10  -.07   .23   .22   .31   .26  1.00
S   9   .23   .10  -.09   .16   .18   .21   .26   .54  1.00   10
   10   .28   .26   .03   .28   .27   .24   .30   .55   .54  1.00
```

Scales of sleep-wake pattern questionnaire:
M - level of morning wakefulness (positive score means more eveningness), E - level of evening wakefulness (positive score means more eveningness), w - capacity to be awake at anytime (positive score means better capacity), f - capacity to fall asleep at anytime (positive score means bettercapacity), S - quality (deep) of night sleep (positive score means better sleep)

2. Age- and sex-related differences

There were no large differences between any sex-age groups in dispersion of scale scores, but differences in mean values were evident (Tab. 2).

All values decreased beginning with the 1st or 2nd age group and crossed zero levels in the 3rd one. It means that sleep difficulties are increasing and ability to change sleep-wake pattern is decreasing with increasing age. Moreover, older individuals have a tendency to awake and go to bed earlier. Thus difference between M- and E-scores remains constant with age.

Practically, in all age groups female M-scores were higher and E-,w-,f- and S-scores were lower than in the correspondent male groups. This means that any adaptive capacity of womens sleep-wake cycle is worse than of males. Women prefer to rise later and go to bed earlier than men of their age group and have more problems with night sleep and unusual sleep-wake schedule adjustment.

Tab. 2: Scores on main scales and subscales (mean and SD) by sex and age

scale	9-20	21-30	AGE 31-40	41-50	51+
MALE n =	598	414	481	268	239
M	-1.6 5.6	-0.2 5.8	-2.0 6.1	-3.4 5.6	-4.5 5.1
à	1.9 3.9	2.5 4.1	1.4 4.4	-0.6 4.6	-1.5 4.4
î-à	-3.5 7.0	-2.7 6.8	-3.5 6.6	-2.7 6.8	-2.9 6.9
w	1.5 2.1	0.9 2.3	0.3 2.5	-0.2 2.7	-0.3 2.6
f	0.4 2.3	0.5 2.3	0.1 2.3	0 2.3	0.3 2.4
s	6.0 5.2	3.7 5.5	0.3 6.0	-1.4 6.0	-2.5 6.1
FEMALE n =	571	424	482	388	261
M	1.5 5.6	1.4 6.0	-0.5 5.8	-1.0 5.6	-3.1 5.6
E	2.2 4.2	0.4 4.8	-0.8 4.8	-1.6 4.7	-2.2 4.4
M-E	-0.7 7.1	0.9 7.0	0.3 6.8	0.6 6.3	-0.9 6.5
w	1.1 2.3	0.1 2.5	-0.2 2.6	-0.6 2.5	-0.6 2.3
f	0.5 2.4	0 2.5	-0.7 2.5	-0.9 2.5	-1.1 2.3
s	4.4 4.9	1.7 5.6	-2.4 6.1	-2.9 5.1	-5.9 5.1

CONCLUSIONS

In contrast with other diurnal type questionnaires the sleep-wake pattern questionnaire was constructed in the long process of items selection from vast first version by using of statistical procedures and rather big samples. The questionnaire measures 5 characteristics of individual sleep-wake pattern: levels of morning (12 statements) and evening wakefulness (8 statements), quality of night sleep (12 statements), capacities to be awake (4 statements) and to fall asleep in anytime (4 statements).

As was shown on a total of 4.126 respondents, all scale scores are near zero in thirtieth and are higher in younger age and are lower in older age, e.g. adaptive capacities of sleep-wake cycle, quality of sleep and eveningness decline to the elderly. Females in all age groups differ from males in direction of lower adaptability of the sleep-wake cycle. These findings correspond with results from instrumental investigations and animal models.

REFERENCES

1. HILDEBRANDT, G. (1980): Survey of current concepts relative to rhythms and shift work. In: L.E. SCHEVING (ed.): Chronobiology: Principles and Applications to Shifts in Schedules. Alphen aan den Rijn - Rockville (Md): Sijthoff and Noordhoff, pp. 261-292
2. SMOLENSKY, M.H., D.T. PAUSTENBACH & L.E. SCHEVING (1985): Biological rhythms shift work and occupational health. In: L. CRALLY (ed.): Industrial Hygiene and Toxicology. 2nd ed. vol. 3b, London-New York: John Wiley & Sons, pp. 156-312
3. HORNE, J.A. & O. ÖSTBERG (1976): A self-assessment questionnaire to determine morningness-eveningness in human circadian rhythms. Int. J. Chronobiology 4 (2), 97-110
4. HAMPP, H. (1961): Die tagesrhythmischen Schwankungen der Stimmung und des Antriebes beim gesunden Menschen. Arch. Psychiatrie und Nervenkrankheiten 201 (4), 355-377
5. PATKAI, P. (1971): The diurnal rhythm of adrenaline secretion in subjects with different working habits. Acta Physiol. Scand. 81, 30-34
6. TORSVALL, L. & T. AKERSTEDT (1980): A diurnal type scale: Construction, consistency, and validation in shift work. Scand. J. Work Environ. Health 6 (4), 283-290
7. FOLKARD, S., T.H. MONK & M.C. LOBBAN (1979): Towards a predictive test of adjustment to shift work. Ergonomics 22 (1), 79-91

CIRCADIAN RHYTHMS OF DIFFERENT SPEED TASKS. INFLUENCE OF SEX AND MORNINGNESS

A. ADAN & M. SÁNCHEZ-TURET

Dep. de Psiquiatria i Psicobiologia Clínica, Unitat de Psicobiologia, Fac. de Psicologia, Univ. de Barcelona, Adolf Florensa s/n, 08028 Barcelona, Spain

INTRODUCTION

Different performance rhythms can have different phases and rates of adjustment, depending on the variable studied and measures taken. Rhythmic patterns of performance tasks differ and depend on speed or memory implication (6,7,11). Tasks implying speed correlate with rhythmic physiological parameters such as body temperature, while memory related tasks show no associated patterns.

Differences between memory tasks have been studied, analyzing the complexity of cognitive demand, temporal interval of memory, etc. Nonetheless, differences between simple or speed tasks have not been subjected to analysis. Studies which include speed tasks - especially those concerning vigilance and reaction times - are not conclusive as far as the similarity of patterns between tasks. But these studies reduce the possibilities to two: on the one hand, that daytime patterns in the tasks covariate with body temperature and are consequently controlled by the "X" pacemaker (4,11). In other words, performance on simple tasks improves over the day to reach a maximum of approximately 20.00. On the other hand, if the result does not show this tendency, then the selected task should be eliminated because of its lack of sensitivity to daytime variations. We think that this attitude is oversimplistic and that simple tasks which demand different abilities or skills can show differential tendencies throughout the day and not be synchronized one with the other.

The study of individual differences, like the variables affecting rhythms of performing simple tasks has given very interesting data. As the most relevant variables we would underline the dimensions of morningness (1,3,8,9) and of extroversion (10,12). The sex of the subjects, however, has not been included as an independent variable and it may have a bearing on performance tendencies. The comparison of daytime and speed tasks also has a practical interest. Many work tasks require the execution of simple, monotonous tasks precisely and at a certain speed.

The aim of the present study is to assess if there are differences throughout the day between speed tasks which require different abilities to carry them out. At the same time, we want to see to what degree these are dependant on individual differences of sex and morningness.

METHOD

Subjects. We selected 20 healthy and right-handed subjects aged 22 to 30 years ($\bar{x}$=24.5, SD=2.76) divided into two groups according to sex, each containing 10 subjects: 3 morning-types (MT), 4 neither-types (NT) and 3 evening-types (ET). The morningness dimension was measured using the reduced scale of the morningness questionnaire (2). The extraversion and neuroticism dimensions were controlled by selecting extroverted and stable subjects respectively (EPQ-A, Spanish Version (5)). In the case of women, the experiment was conducted the week after the onset of menses.

Procedure. The experimental part of the study was performed between May and August 1990. The subjects came to the laboratory twice: the first time for a training session, and the second time to do the experiment. Each subject was tested in 13 sessions throughout the day hourly from 09.00 to 21.00. Each experimental session lasted nearly 10 minutes with a 45 minute break between each, when the subjects were allowed to leave the laboratory, but not the campus.

Three different tasks were studied: auditory reaction time (ART), visual reaction time (VRT), and tapping (T). In each experimental recording a block of 40 trials of ART and VRT, and 40 seconds of T (two 20 seconds blocks) were presented. ART and VRT were recorded with millisecond precision. The median of each individual test block was found, and 5% of the initial tests (2) were eliminated. An RT response greater than 500 msec was considered as a lapse, and responses below 100 msec as anticipated. Later the group averages for RT and lapses were found for each time series. Anticipations were discounted, being practically non-existent. The individual mathematical average was found for each of the two (T) task blocks, as well as the subsequent group average.

The experiment was a three factor factorial design, gender (2) x morningness (3) x time of day (13) repeatedly measured. Data analyses were carried out with the SPSS-X computer package.

RESULTS

AUDITORY REACTION TIME
Sex & Morningness

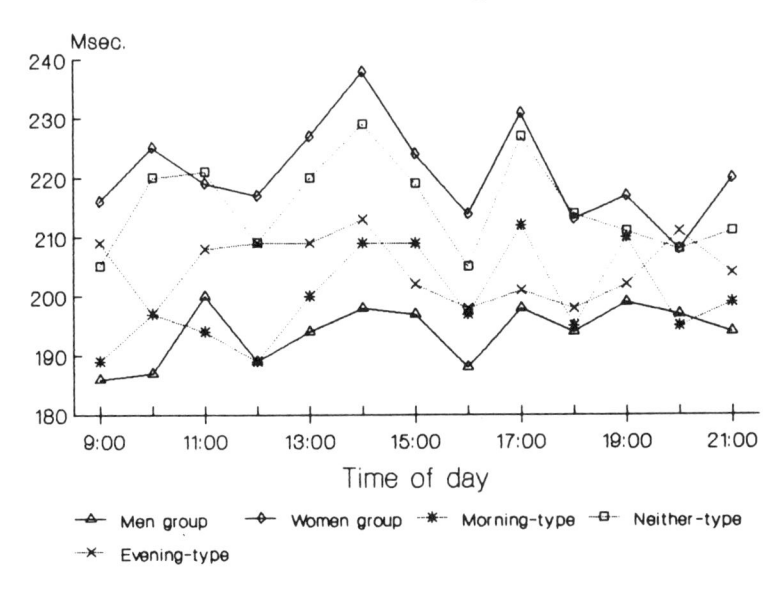

Fig.1: Auditory reaction times according to time of day for the sex and morningness groups.

The group time series (n=20) of ART and VRT, and pulsations in T showed no significant correlation within tasks. The time of day when a task could best be performed does not correspond to maximum performance levels in the others. Best performance values for each of the selected tasks are: ART=201 msec (16.00),

VRT=277 msec (10.00), and T=6.41 number of taps (12.00). Qualitative parameters of reaction time task - lapses - were also correlated. Significant values were found for ART and VRT lapses (r=0.749, p<0.01), and ART latency and VRT lapses (r=0.695, p<0.01). That is, the lapses in both RT tasks covariate according to the hour of the day, and the poorer the ART performance, the greater the VRT lapse. The speed of taps in the T task did not show any significant correlations, either with respect to the TR's, or to the lapses in ART and VRT.

VISUAL REACTION TIME
Sex & Morningness

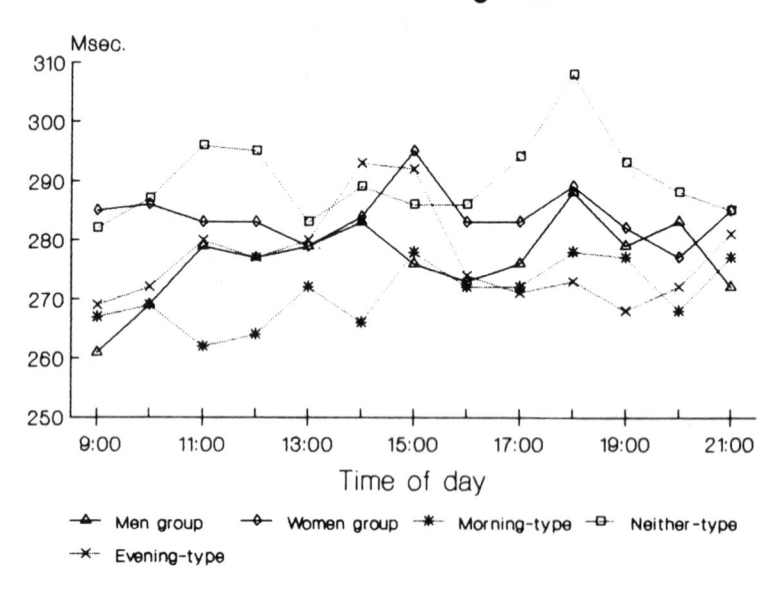

Fig.2: Visual reaction times according to time of day for the sex and morningness groups.

None of the manovas realized showed the hour of the day to have a significant main effect. The three tasks exhibit very small variations between recordings, and high deviation values between the hourly recordings (standard deviation). The ART task shows significant main effects for sex (F=26.31, p<0.0001) and morningness

(F=3.36, p<0.037) as well as for the sex x morningness interaction (F=30.26, p<0.0001). Fig.2 shows ART in relation to time of day for the sex and morningness groups. The best TR's in milliseconds for each of the subgroups are: men =187 (10.00), women =214 (16.00), MT=189 (12.00), NT=205 (16.00), and ET=198 (16.00).

For the VRT task we found significant differences for the main effect of morningness (F=4.87, p<0.009) and sex x morningness interaction (F=22.28, p<0.0001; see Fig.2). The best TR's in msec for each of the subgroups are: men =261 (09.00), women =277 (20.00), MT=262 (11.00), NT=283 (13.00), and ET=268 (19.00).

In the task (T) significant results were found for the main effect of sex (F=7.53, p<0.007) and sex x morningness interaction (F=10.13, p<0.0001). The best tap averages for each of the subgroups are: men =6.55 (12.00), women =6.28 (12.00), MT =6.53 (12.00), NT =6.37 (19.00), and ET =6.41 (12.00).

Such findings stress the influence of morningness on ART and VRT and of sex on ART and T. Interaction of both factors in the three tasks emphasizes the importance of analyzing each factor - morningness and sex - separately, so as not to introduce a bias in results of tasks which were objects of analysis.

DISCUSSION

Our study introduced 3 speed tasks which are rarely chosen in rhythmicity studies, or at least which are normally said to present no clear daytime variations. 2 were reaction time tasks which differed only in the type of sensorial perception involved (auditory and visual) and T task was essentially one psychomotor. There is no doubt that the study of speed tasks requires a more exhaustive selection of tasks, but our results do give us relevant data from various perspectives.

On the one hand, we found evidence to suggest that the tasks analyzed do not exhibit similar patterns throughout the day and that they are influenced to varying degrees by whatever individual difference there may be. We would like to point out the importance of differentiating implied sensory modality from the required skill in a task. Thus, the VRT task shows a global optimum value 5 hours ahead of the ART task and the task T adopts an intermediate peak time. As for individual differences, the men show a phase 5 hours ahead of the women in ART and 11 hours in VRT. The morningness dimension shows an advance phase in MT's of 4 hours in ART compared to the NT's and ET's. For the morningness groups the VRT task represents an ordering in the peak of performance, with an ET phase delay of 8 and 6 hours in MT and NT respectively. Differences between groups in the peak time of performance are few.

The results, especially in the TR tasks, highlight differences between sexes which are equal to or greater than those found in the morningness dimension. More studies are needed which repeat these preliminary findings and their theoretical implications. These differences could possibly reflect an underlying biological control difference between sexes, or perhaps a different type of motivation in or stylistic approach to daytime variations. Whatever the reason for such differences, and considering in addition the significant interactions between morningness and sex in the 3 tasks, we are led to conclude that the study of one of these individual differences necessarily implies the control of the other in order to avoid bias one way or the other.

Finally, we would like to point out the necessity of using qualitative or precision performance indicators in order to evaluate simple or speed tasks. Qualitative monitorisation seems to prove to be more sensitive than the TR's themselves when assessing speed task performance throughout the day. The similarities between the TR tasks throughout the day are particularly noticeable when we evaluate the lapses instead of latent responses.

REFERENCES

1. ADAN, A. (in press): Correlation between body temperature, reaction time and verbal memory: the influence of chronotype. In: A. DIEZ-NOGUERA & T. CAMBRAS (eds.): Chronobiology & Chronomedicine: basic research and applications. Frankfurt, Peter Lang, pp.189-196
2. ADAN, A. & H. ALMIRALL (1991): Horne and Östberg morningness-eveningness questionnaire: a reduced scale. Person. Individ. Diff. 12, 241-253
3. BUELA-CASAL, G., V.E. CABALLO & E. GARCÍA-CUETO (1990): Differences between morning and evening types in performance. Person. Individ. Diff. 11, 447-450
4. CZEISLER, C.A., R.E. KRONAUER, J.J. MOONEY et al. (1988): Biological rhythm disorders, depression and phototherapy. A new hypothesis. In: R.L. WILLIAMS & I. KARACAN (eds.): Sleep Disorders. New York, Wiley, pp. 687-709
5. EYSENCK, H.J. & S.G.B. EYSENCK (1975): Eysenck personality questionnaire for juniors and adults. (Spanish Version, V. Escolar, A. Lobo and A. Seva-Diez. 1984. Madrid, Ediciones Tea)
6. FOLKARD, S. (1990): Circadian performance rhythms: some practical and theoretical implications. Phil. Trans. R. Soc. Lon., B 327, 547-553
7. FOLKARD, S., R.A. WEVER & C.M. WILDGRUBER (1983): Multi-oscillatory control of circadian rhythms in human performance. Nature 305, 223-226
8. HORNE, J.A., C.G. BRASS & A.N. PETTITT (1980): Circadian performance differences between morning and evening "types". Ergonomics 23, 29-36
9. KERKHOF, G.A. (1982): Event-related potentials and auditory signal detection: their diurnal variation for morning-type and evening-type subjects. Psychophysiology 19, 94-103
10. MATTHEWS, G. (1989): Extraversion and levels of control of sustained attention. Acta Psychologica 70, 129-146
11. MONK, T.H., E.D. WEITZMAN, J.E. FOOKSON et al. (1983): Task variables determine which biological clock controls circadian rhythms in human performance. Nature 304, 543-545
12. VIDACEK, S., L. KALITERNA, B. RADOSEVIC-VIDACEK & S. FOLKARD (1988): Personality differences in the phase of circadian rhythms: a comparison of morningness and extraversion. Ergonomics 31, 873-888

CIRCADIAN RHYTHM OF TEACHERS' PHYSICAL ACTIVITY

M. KWILECKA & K. KWILECKI

Institute of Physical Education and Sport,
Academy of Physical Education in Gorzówa, Estkowskiego 13,
66-400 Gorzow, Poland

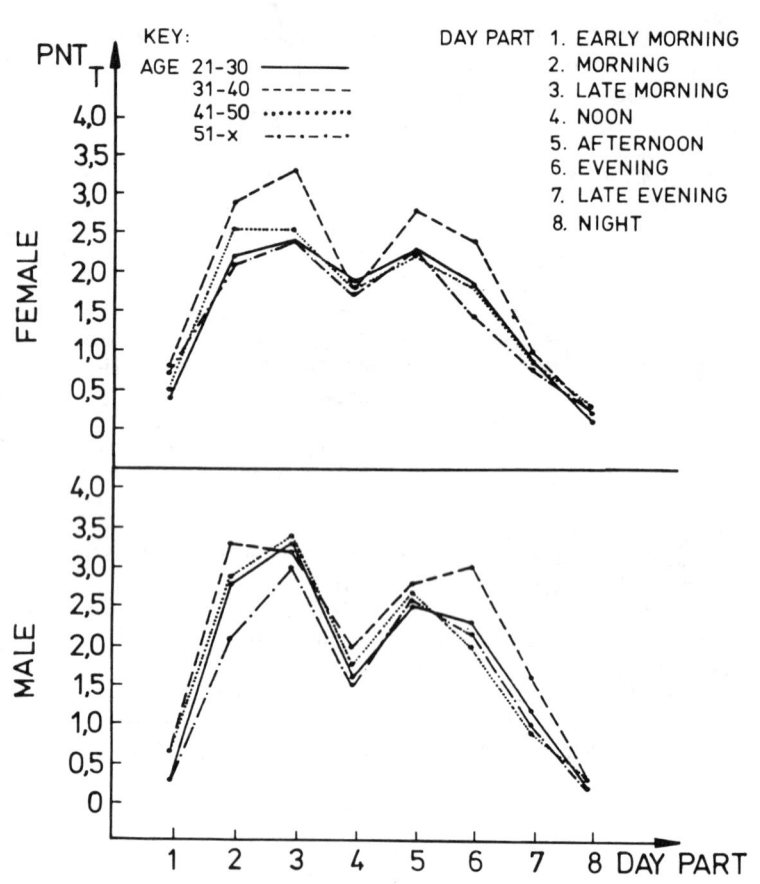

Fig.1: Graphic illustration of circadian state of physical activity according to subjects age

The phenomenon of rhythmicity of the world that surrounds man, as well as various functional properties of man alone, come, beside other issues, within the scope of chronobiology.

In Poland, studies of biological rhythms of the human organism were for the first time undertaken by H. MILICEROWA in the 40's of the present century. At the same time, studies of circadian variability in somatic traits of children and the young were suggested by and then led under the guidance of CWIRKO-GODYCKI (1). A research team directed by DROZDOWSKI (2) carried on their research with rhythms and morphofunctional and motor traits, dealing mainly with data obtained from students and representatives of different sports disciplines. KWILECKA (3) expanded that course by studying professionally active people.

Man's activity, in its broader sense, in rhythms of different (circadian, weekly, monthly, yearly, and of many years') frequency is determined by a number of factors of internal and external nature. It is the subject of studies carried out by specialists from different fields. The subject literature deals most often with information about circadian variability.

REFERENCES

1. CWIRKO-GODYCKI, M. (1956): Outline of anthropometry. Warszawa, PWN
2. DROZDOWSKI, Z. (1965): Methodical remarks reffering to investigations on jumping ability. Scientific Annals. Poznań, WSWF, No. 10
3. KWILECKA, M. (1991): Motorial efficiency of physical workers during about month menstruation cycle. Chronobiology - Chronomedicine, Basic Research and Applications. Peter Lang, Frankfurt/M. - Bern - New York - Paris
4. MILICER, H. (1951): Variability in body building traits as influenced by physical training. Anthropology Review 17, Poznań
5. MILICER, H. (1961): Research into physical development of the youth. Sport and Physical Education 5, Warszawa

A STUDY OF PERFORMANCE RHYTHMS: A FIRST APPROXIMATION OF THE RHYTHMICITY OF DIFFERENT TASKS IN CHILDREN.

D. SÁIZ, A. DIAZ & M. SÁIZ

Department of Educacional Psychology, Edificio B,
Autonomus University of Barcelona, Barcelona, Spain

INTRODUCTION

The study of diurnal variations in different tasks is a matter of continuing debate. Normally, the kinds of task most often studied can be grouped in three categories: a) tasks of visual search or tasks which require speed and precision, generally referred to as "vigilance tasks" in the overall conception; b) tasks which require memorization activity (short and long term memory), and c) tasks which require a more complex cognitive activity (logic reasoning, arithmetic problems, etc.). The results of these works show that the performance rhythms depend to a certain extent on the type of task. Normally, the rhythms are measured with adults, but works with children have a particular interest because they can allow a proper schedulling of the school hours. This matter is at the moment of current interest and the initial emphasis upon "fatigue" has shifted to an emphasis upon "school rhythms". Of course, there are old studies about the variations of performance in children during the school day dating from the beginning of the century (5, 6, 8, 10, 11, 14, 22, 23, 24) but the more scientific studies are from recent years (1, 2, 3, 4, 7, 9, 12, 13, 18, 19, 20, 21). The new research is trying to find the exact variations of psychological and physiological variables in order to make a proper adjustment of schedulling, because it is generally accepted that the best learning takes place when the children have a high level of attention and cognitive capacities. On the other hand, the study with children can also verify if the performance rhythms are regular or stable at different ages.

We have done other works on performance rhythms with university students. The object of the present study consists in verifying whether there are performance differences between the morning and the afternoon in different tasks in children, but also we want to verify if the results with children are similar to university students.

However, as indicated by the title, the present work is only a first approximation to this problem in a sample of Spanish children.

MATERIAL AND METHODS

The sample was taken from Lestonnac School of Mollet del Valles (Barcelona, Spain) and it was of 57 children (31 female and 26 male) between 12 years 1 month and 13 years 11 months old, all of them were of the same school's level (seven of primary). The procedure followed was very similar to that used in another work (16) done with university students. In the present study all of the subjects had a training session. The experimental test consisted of two sessions, one at 09:00 and the other at 15:00. The tasks studied were: visual search, arithmetic problems (addition) and logic reasoning (letters and pictures). The variables studied were: the number of correct answers and the execution time for each task for each time of day.

The material used was different in each session but the level was equivalent. The material for visual search was practically equal to the test Toulousse Pieron, but in our case the child must count the exact number of the items as indicated, the arithmetic problems were additions and for the more cognitive tasks we used logical reasoning with logical sequences of letters (for example, ababababa... - next letter: "b") and pictures.

We controlled the age of the children (12-13 years old), place of experimental sessions (all the time the same place for all subjects) and room temperature (all the sessions were at 19 degrees). Children with learning disabilities or physical illness were omitted.

RESULTS AND DISCUSSION

About results we can say that there were no significant statistical differences in the level of correct answers. When we analysed task to task we can see that visual search and arithmetic problems were best performed at 15.00, but logical reasoning had a difference between letters and pictures. While letters were best performed at 09:00, pictures were at 15:00, but practically performance was more or less similar at the two times of day (Tab. 1).

Tab. 1. Averages and significance level of the comparision between the results at 09:00 and those at 15:00.

Tasks	09:00	15:00.	p
Visual search (errors)	1,56	2,43	0,200
Logical reasoning with letters	9.05	8.61	0,293
Logical reasoning with pictures	12,34	12,45	0,759
Arithmetic problems (additions)	29,12	29,52	0,454
Time of visual search	188,67	175,36	0,090
Time of logical reasoning(letters)	220,26	177,46	0,000*
Time of logical reasoning(pictur.)	259,46	217,41	0,001*
Time of arithmetic problems	610,91	519,47	0,000*

Concerning execution time or the time used to perform the task there were significant statistical differences practically in all the tasks indicating a greater speed in the afternoon.

To summarize, the results show that in general no differences were found in the level of precision of performance, although the tendency is to perform better in the afternoon. The analysis of execution time was more clear, Spanish children perform with more speed in the afternoon without losing their precision. This tendency in Spanish children is similar to adults reaffirming the results found in other studies (15, 16, 17) and shows the tendency in Spanish groups to work better or more quickly in the afternoon.

REFERENCES

1. BAKER, M.A., D.A. HOLDING & M. LOEB (1984): Noise, sex and time of day effects in a mathematics task. Ergonomics, 27(1), 67-80.
2. CANALS, R., E. AÑAÑOS & M. MARTIN (1989): Ritmes circadiaris de l'atenció en una mostra d'alumnes de cicle mitjà. Butlleti Universitari de Psicologia, any IV, 5, juny, pp. 13-15.
3. FOLKARD, S. (1980): A note on "time of day effects in school-children's immediate and delayed recall of meaningful material" the influence of the importance of the information tested. British Journal of Psychology, 71, p.95-97.
4. FOLKARD, S. et al. (1977): Time of the day effects in school children's immediate and delayed recall of meaningful material. British Journal of Psychology, 58, 45-50.
5. GATES, A.I. (1916a): Diurnal variations in memory and association. Univ. Calif. Publ. in Psychology, 1, 323-344.
6. GATES, A.I (1916b): Variations in efficiency during the day, together with practice effects, sex differences and correlations. Univ. Calif. Publ. in Psychology, 2, 1-156.
7. GUERIN, N. et al. (1989): Variations diurnes des résultats de tests psychophysiologiques en milieu scolaire. Approche chronobiologique. Incidences du lieu du déjeuner. L'année Psychologique, 89, 327-344.
8. HECK, W. H. (1914): Studies of mental fatigue in relation to the daily school program. Lynchburg, Va : Bell.
9. KOCH, P. et al. (1987): Variation of Behavioral and Physiological variables in children attending kindergarten and Primary School. Chronobiology International, 4(4), 525-535.
10. LAIRD, D.A. (1925): Relative performance of college students as conditioned by time of day and day of week. Journal of Experimental Psychology, 8, 50-63.
11. MARTIN, G.W. (1911-1912): The evidence of mental fatigue during school hours. Journal of Experimental Pedagogy, 1, 39,45 and 137-147.
12. MONTAGNER, H. (1983): Les rythmes de l'enfant et de l'adolescent ces jeunes en mal de temps et d'espace. Stock/Laurence Pernoud, Paris.
13. POUTHAS, V. et al. (1986): Les conduites temporalles chez le jeune enfant. (Lacunes et perspectives de recherche). L'année psychologique, 86, 103-122.
14. ROBINSON, I.A. (1912): Mental fatigue and school efficiency. Bulletin Winthop Normal and Indus. Coll. of S.C., 5 (2), 1-56.
15. SAIZ, D. (1988): Una aproximación a los ritmos de la memoria. Bellaterra: Universidad Autónoma de Barcelona.
16. SAIZ, D. & E. AÑAÑOS (1990): Communication in the E.S.C. meeting in Barcelona.
17. SAIZ, D. et al. (1990): A preliminary study of the rhythms in different tasks of memory. In E. MORGAN (ed.): Chrobiology and Medicine. Basic research and applications. Peter Lang, Frankfurt, pp. 310-317

18. SOUSSIGNAN, R. (1988a): Relations entre l'activité motrice générale des enfants à l'école et la durée des tâches scolaires. C.R. Acad. Sci. Paris, t. 306, série III, pp. 139-142.
19. SOUSSIGNAN. R. et al. (1988b): Behavioural and cardiovascular changes in children moving from kindergarten to primary school. Journal of Child Psychology and Psychiatry, vol. 29(3), 321-333.
20. TESTU, F. (1983): Variations journalières et hebdomadaires des performances en milieu scolaire et nature de la tâche. In H. MONTAGNER (ed.): Les rythmes de l'enfant et de l'adolescent, Stock, Paris, pp. 175-182
21. TESTU, F. (1988): Apprentissage et variations journalières de performances scolaires. Le Travail Humain, 51(4), 363-375.
22. SCHUYTEN, M.C. (1903): Sur les méthodes de mensuratuon de la fatigue des écoliers. Arch de Psychol. 2, 321-326.
23. WINCH, W.H. (1912a): Mental fatigue in day school children as measured by immediate memory. I. J. Educ. Psychology, 3, 18-28.
24. WINCH, W.H. (1912b): Mental fatigue in day school children as measured by immediate memory. II. J. Educ. Psychology, 3, 75-82.

CHRONOBIOLOGICAL DIFFERENCES IN PERFORMANCE BETWEEN TRAINED AND UNTRAINED CHILDREN

H. OSCHÜTZ

GHK-Universität Fb.Sprachwissenschaft,Heinrich-Plett-Str.40, D-W-3500 Kassel, FRG

INTRODUCTION

During the last years it has become increasingly clear that peak athletic performance can only be obtained if systematic training has begun early in life. Aside from the training contents it is important that the pattern of training over the daily course is constructed on the basis of chronobiological knowledge. Endogenous and exogenous factors influence the formation of the circadian rhythms of bodily functions and physical performance at each age. It was our aim to investigate whether children of the first school years show characteristic daily fluctuations and how far these can be influenced by athletic training.

METHOD

For our study we chose two groups of boys (aged 8). The first group of 22 boys was living under normaly conditiones. The other group of 22 boys started with gymnastic training 2 years before start of experiment. They were trained 2 to 3 times weekly, for 1 1/2 hours per training unit.

Under laboratory conditions we measured between 7 a.m. and 7 p.m. the basal values of skin temperature (forehead), flicker fusion frequency and visual reaction time. For each of the 44 8-years old boys we obtained data of 3 daily curves.

We computed the means and their standard deviations as well as the t-test for independent samples. Tendencies in daily changes with the time-series analysis were evaluated with the empirical regression according to PEIL (4).

RESULTS

The results are presented as the calculated values of the empirical regression of the respective parameters depending on time of day.

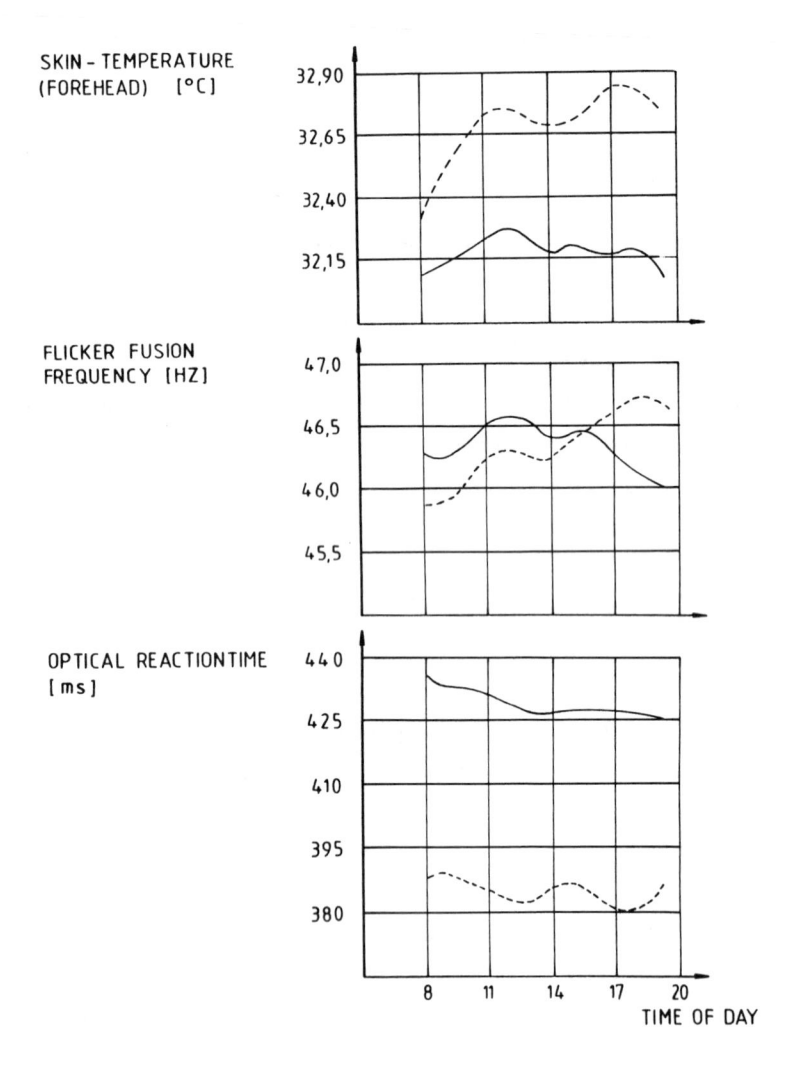

Fig.1: Daily course of skin temperature (forehead), flicker fusion frequency and optical reaction time of trained (- - -) and nontrained (———) 8-years old boys

A daytime dependency was found for skin temperature (forehead), flicker fusion frequency, and visual reaction time. An influence of regular gymnastic training on these parameters could be shown. The curves of trained and not-trained subjects went parallel before noon with a peak around 11 a.m.. After the noon slump between 1 and 3 p.m. the opposite daytime behavior is clearly seen. This is especially obvious with the flicker fusion frequency. While the non-trained subjects stay on their pre-noon level or sink below, the training subjects show an increase with later daytime and reach a second peak around 5 p.m..

In opposite to non-trained, only the trained subjects reach a day peak of psycho-psychological performance in the first half of the day. The influence of training is also seen in a bigger difference between minima and maxima.

SUMMING UP:

(1) There are periods of increased and diminished activity,
(2) continuous athletic training produces changes in the pattern of the daily rhythms.
(3) Training as an exogenous Zeitgeber influences much more the daytime dependency of psychophysiological than of cardiopulmonal and respiratory functions (3) in young school children.

DISCUSSION AND CONCLUSION

Our data can only be discussed with reference to a small body of literature, because most of the research was done with non-trained pupils.

No obvious differences in the daily rhythms of various parameters between 8-year old boys and older children were found. We could confirm the recommendation of OCKEL (2) to utilize the performance peak between 9.00 and 11.30 a.m. for grade school children. Our results show a pre-noon maximum performance of all investigated parameters between 10.00 and 12.00 a.m..

The noon slump lies between 1.00 and 3.00 p.m., in accordance with RUTENFRANZ & HELLBRÜGGE (5) who advise against instructions during this time on the basis of results on the day rhythms of numerical ability (speed of calculation) in children.

All authors concur on the time of appearance of pre-noon peak and noon slump. On the time of appearance of afternoon peak, however, different authors present different results depending on investigated parameters.

It ought to be assumed that the training as exogenous Zeitgeber makes the afternoon maximum to appear during the time of training.

PERFORMANCE

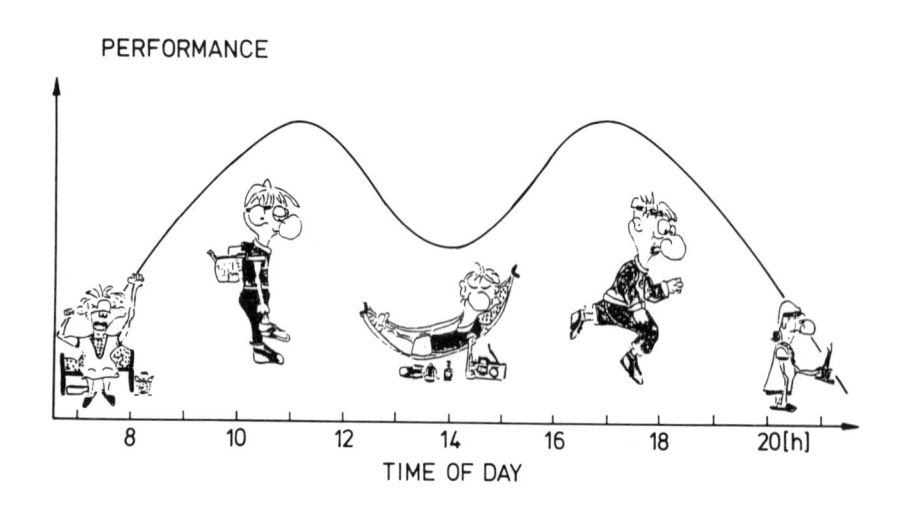

TIME OF DAY

Fig.2: Daytime dependency organization of gymnastic training and recreation in school children

This, however, is not the case. Analyses show that training times were found between 1.30 and 4.00 p.m.. The formation of day rhythms by the exogenous Zeitgeber training seems to lead to a general adaption of performance on a higher level in the afternoon. The more pronounced amplitude is the result of the adaption of the daytime rhythm of the infantile organism to athletic training.

HELLBRÜGGE (1) suggested that the extent of the day rhythm depends on the demands of the organ or the organ system at the respective age.

We like to emphasize that between trained and non-trained subjects no difference in biological age was found which could have suggested a different biological developmental stage of both groups.

On the basis of our results the times between 10.00 and 12.00 a.m. and 4.00 and 6.00 p.m. are preferable for those training units which place high demands on concentration and reaction ability. The time between 1 and 3 p.m. should be used for recreation or compensatory training. It is possible to improve performance in the afternoon with the aid of gymnastic training.

These results imply that persons not in training have the opportunity to improve their sinking psychophysiological performance in the afternoon with gymnastic training.

REFERENCES

1. HELLBRÜGGE, T.H. (1965): Die Entwicklung der Tag-Nacht-Periode im Kindesalter. Wiss. Zeitschr. der Humboldt-Universität zu Berlin. Math.-Naturwiss. Reihe, 2, 263-275
2. OCKEL, E. (1984): Gesundheits- und bildungspolitische Aspekte im Kinder- und Jugendgesundheitsschutz. DDR-Med. Rep. 8, 440-484
3. OSCHÜTZ, H. (1987): Zum Tagesgang ausgewählter physiologischer Parameter im Kindesalter. Wiss. Zeitschrift d. DHfK 28 2, 69-79
4. PEIL, J. & S. SCHMERLING (1977): Ein ALGOL-Programm zur Berechnung von empirischen Regressionen. Gegenbaurs. morph. Jahrb. 4, 538-555
5. RUTENFRANZ, J. & T.H. HELLBRÜGGE (1957): Über Tagesschwankungen der Rechengeschwindigkeit bei 11-jährigen Kindern. Z. Kinderheilkunde 80, 65-82

MONITORING, DATA ANALYSIS

A STUDY OF THE DIURNAL RELATIONSHIP BETWEEN ORAL AND AXILLARY TEMPERATURES

A. ADAN & M. SÁNCHEZ-TURET

Dep. de Psiquiatria i Psicobiologia Clínica, Unitat de
Psicobiologia, Fac. de Psicologia, Univ. de Barcelona, Adolf
Florensa s/n, 08028 Barcelona, Spain

INTRODUCTION

Rectal and oral temperatures are usually considered as excellent rhythm markers in chronophysiological studies. However, in field studies it is necessary to use automatic continuous recordings of body temperature (BT) obtained through an easier and more comfortable placing of the thermistor. MOTOHASHI et al (8), and REINBERG et al (11), using the Chronoterm ambulatory system (9) discovered axillary temperature shows only small and predictable differences in circadian rhythm when compared with rectal temperature. This result is relevant from two perspectives. On the one hand, it allows the continuous insertion of a temperature measuring device which does not restrict the subject's movements, thus allowing their BT to be monitored during a "normal" routine; on the other hand, and from a more theoretical viewpoint, it means that the axillary temperature, conveniently recorded, can be controlled by the "X" pacemaker (4,5,12). If this last point can be corroborated, then the axillary temperature could be considered as a good indicator of a subject's state of activation as rectal and oral temperatures. Nevertheless, in these studies sex differences were not analyzed.

Available literature describes circadian differences and higher thermal readings in women, both in oral and rectal temperatures (2,3,10). These differences, however, do not always have a statistically significant bearing on the results. Such studies do not control the menstrual cycle, although there is clear evidence of circatrigintan variations in BT associated with hormonal changes.

This paper analyses the relationship between oral and axillary temperatures in the waking part of the day, controlling environmental and subject variables. All the subjects were synchronized to natural zeitgebers and to their usual routines. We also intend to verify the existence of sex differences within the parameters studied, controlling the menstrual cycle in the case of women.

METHOD

Subjects: We selected 20 healthy subjects (10 men and 10 women) aged 22 to 30 ($\bar{x}$=24.5, SD=2.76). Extraversion, neuroticism and "morningness" dimensions were controlled. All the subjects were extrovert and stable on the basis of scores achieved on the EPQ-A personality questionnaire of EYSENCK & EYSENCK (6). Each sex group contained 3 morning types, 4 neither types and 3 evening types. The morningness dimension was measured by taking the HORNE & ÖSTBERG (7) reduced questionnaire. The reduced scale contained 5 items obtained through a cluster and correspondence analysis (1), belonging exclusively to the morningness dimension. There were no differences between subgroups with regard to sleep duration nor to the span of time needed to fall asleep.

Procedure: The experimental part of the study was carried out between May and August 1990. Oral temperatures were taken with a clinical thermometer inserted sublingually for a timed 3-minute period, every hour from 09.00 to 21.00. Axillary temperature was measured continuously - from 08.30 to 21.30 - with a Spanish prototype similar to the Chronoterm. A thermistor was placed in the subjects' armpit - with a range of 35-38° C and precision of 0.01° C - with SEMPATAP (10 mm). In order to ensure good contact with the skin it was secured with hypoallergic adhesive tape. The sampling interval was 30 sec, increased later to an average of 10 minutes (20 data), corresponding to simultaneous monitorisation of oral temperature. In the case of women, none of them took oral contraceptives and the experiment was conducted a week after onset of menstruation. All the women had regular menstruation cycles. After each experimental session the subjects had a 45 minute break and were allowed to leave the laboratory, but not the campus. On the day of the experiment both the content and times of meals were controlled. 6 group time series were obtained (from t=1 to t=13) according to the sex and the type of body temperature measuring device used. 2 time series (n=20) were obtained, corresponding to axillary and oral temperatures respectively, as were 2 time series for men (n=10) and 2 time series for women (n=10), 1 for each temperature.

RESULTS

AXILLARY TEMPERATURE

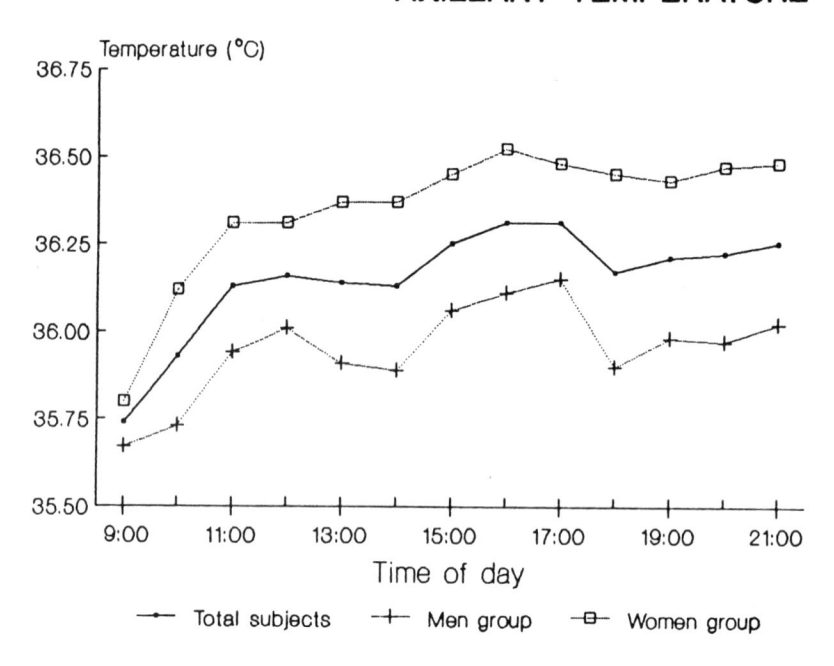

Fig.1: Values for axillary temperature means according to time of day.

Axillary temperature: Fig. 1 shows axillary temperature in relation to time of day for the sex groups and total subjects. Time series obtained for the axillary temperature showed the peak time as follows: total subjects = 36.31° C (16.00-17.00); men group = 36.15° C (17.00), and women group = 36.52° C (16.00). The women's group showed a phase one hour ahead of the men. The averages for the groups were compared for each of the hourly recordings. Significant statistical differences were found in axillary temperature recordings at 13.00 (t=2.15, p=0.045), and at 14.00 (t=2.56, p=0.020). After 16.00 all the differences were significant (0.049 >p< 0.011). In all cases, the thermic value of the women's group is higher than that of the men's. It is worth pointing out, therefore, that the men's group showed greater standard deviations throughout the day than the women's group. The difference between the highest and lowest value of the sample interval measured is larger in the women's group (0.72° C) than it is in the men's (0.48° C).

For axillary temperature there was a highly positive correlation between sex groups (r=0.854, p<0.001). This leads us to conclude that axillary temperature rhythms between sex groups are coupled and any differences are only ones of degree.

ORAL TEMPERATURE

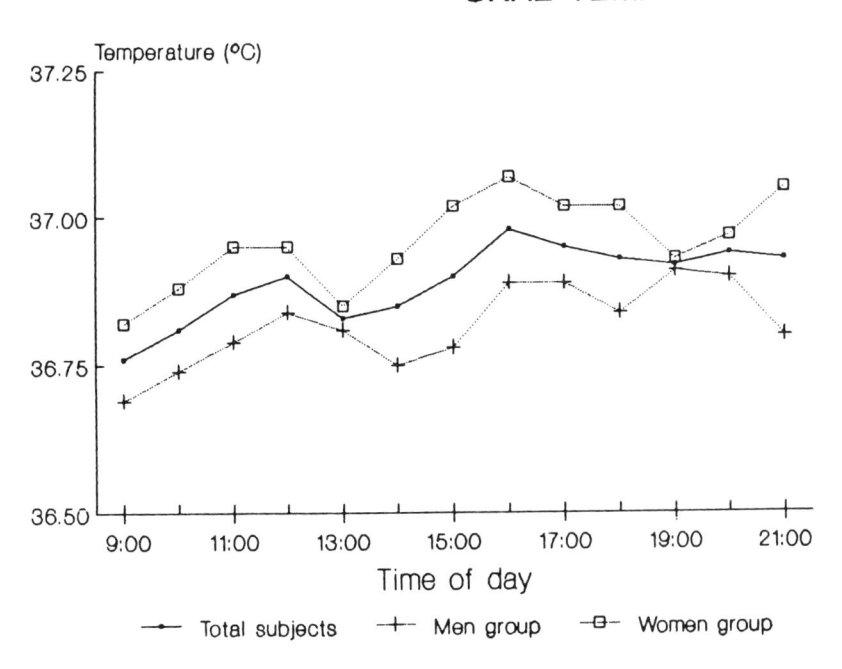

Fig.2: Values for oral temperature means according to time of day.

Oral temperature: Fig. 2 shows oral temperature in relation to time of day by sex group and total subjects. Peak times of the oral temperature were: total subjects

= 36.98° C (16.00); men's group = 36.91° C (19.00), and women's group = 37.07° C (16.00). The women's group showed a phase 3 hours ahead of the men's group. The contrast of averages among time series gave only 3 significant values in the sample interval: at 15.00 (t=-3.19, p=0.005), 18.00 (t=2.85, p=0.011) and at 21.00 (t=3.03, p=0.007). In all cases the women's group showed a higher temperature. The standard deviations for both groups are very similar. The difference between the maximum and minimum values of the sample interval is similar for both groups, although slightly greater in the women (0.25° C) than in the men (0.022° C).

For oral temperature there was no significant correlation between men and women (r= 0.522, p<0.06). These results evidence sex differences in patterns of oral temperature variations over time of day. This result is especially influenced by the differences in peak time between groups and the course of the functions recorded after 18.00.

Relationship between temperatures: Correlations between oral and axillary temperatures were highly positive for the whole sample (r= 0.908, p<0.001), women's sample (r=0.793, p<0.001), and men's sample (r=0.741, p<0.01). These results confirm recent findings and include the sex variable. Both temperatures show similar patterns in the observation period, and exhibit diurnal variations only on a rather small scale. The similarity between them is especially satisfactory in the sample group as a whole and in the women's group. The lesser degree of correlation in the men's group is due to the differential course of both temperatures in the time interval of afternoon and evening.

DISCUSSION

Our study corroborates the high degree of covariation in the daytime period of axillary and oral temperatures, when synchronized with environmental zeitgebers. These results are in line with those obtained by precedent works, using rectal temperature. This conclusion is valid for situations similar to those used in both studies, using subjects displaying moderate activity and without a high level of energy expenditure.

As for the differences in body temperature between men and women, our results ought to be seen in a cautious light, given the reduced number of subjects which made up the groups. Nevertheless, the women's group does tend to show greater thermic values, statistically significant especially for axillary temperature. In the oral temperature measurements the differences are more spurious and are concentrated in the measurements for afternoon and evening. This corroborates the results of previous studies (2,3,10) and extends their validity to include axillary temperature. We believe the fact that daytime variations in temperature functions show more correlation between sex groups with respect to axillary temperature than to oral temperature is significant. This is one result that recommends selecting this parameter if the experimental design includes the sex variable. In later studies, subgroups of women at various stages of the menstrual cycle ought to model the differences in circadian rhythm as a function of the circatrigintan rhythm.

Our study presents empirical data on the validity of using axillary temperature as an alternative measurement of body temperature. Nevertheless, it is possible that the resulting function may be masked by the very synchronization with the environmental zeitgebers. Thus, from a theoretical point of view we cannot hypothesize on the type of control exercised by endogenous pacemakers on axillary temperature. In order to do this, measurements of axillary temperature in isolated conditions would be necessary. Only in this way could we see whether axillary temperature adopts a period similar to that of rectal and oral temperatures or, to the contrary, it adopts a period similar to the sleep/wakefulness cycle. In the first case we would have evidence in favor of maintaining that it is controlled by the "X" pacemaker and in the second that it is controlled by the "Y" pacemaker (4,5,12). The use of axillary temperature in studies on shift work and jet-lag, as well as its later resynchronization will also give very useful results.

REFERENCES

1. ADAN, A. & H. ALMIRALL (1991): Horne and Östberg morningness-eveningness questionnaire: a reduced scale. Person. Individ. Diff. 12, 241-253
2. ASSO, D. (1987): Cyclical variations. In: BAKER, M.A. (ed.): Sex differences in human performance. Wiley, Chichester, pp. 55-80
3. BAKER, M.A. & K. PANGBURN (1982): Female temperature cycles and their relationship to performance. 20th Congress Int. Assoc. Appl. Psychol. Edinburg, U.K.
4. CZEISLER, C.A., R.E. KRONAUER, J.J. MOONEY et al. (1988): Biological rhythm disorders, depression, and phototherapy. A new hypothesis. In: WILLIAMS, R.L. & I. KARACAN (eds.): Sleep Disorders. Wiley, New York, pp. 687-709
5. CZEISLER, C.A., E.D. WEITZMANN, M.C. MOORE-EDE et al. (1980): Human sleep its duration and organization depend on its circadian phase. Science 210, 1264-1267
6. EYSENCK, H.J. & S.B.G. EYSENCK (1975): Eysenck personality questionnaire for juniors and adults. (Spanish Version, Escolar, V., Lobo, A. and Seva-Diez, A. 1984. Madrid, Ediciones Tea)
7. HORNE, J. A. & O. ÖSTBERG (1976): A self-assesment questionnaire to determine morningness-eveningness in human circadian rhythms. International Journal of Chronobiology 4, 97-110
8. MOTOHASHI, Y., A. REINBERG, F. LÉVI et al. (1987): Axillary temperature: a circadian marker rhythm for shift workers. Ergonomics 30, 1235-1247
9. NOUGUIER-SOULÉ, J. & J. NOUGUIER (1984): Microprocessor-based ambulatory temperature and activity recorder. Annual Review of Chronobiology 1, 183-184
10. QUINKERT, K. & M.A. BAKER (1984): Effects of gender, personality and time of day on human performance. In: MITAL, A. (ed.): Trends in Ergonomics / Human Factors I. North Holland, Elsevier, pp. 137-142
11. REINBERG, A., Y. MOTOHASHI, P. BORDELEAU et al. (1989): Desynchronisation interne de rythmes circadiens et tolérance du travail posté. Colloqui L'Home i el Temps. Sant Feliu, Spain
12. WEVER, R.A. (1979): The circadian system of man. Springer-Verlag, New York

AXILLARY TEMPERATURE AND CHRONOTYPE: A PILOT DIURNAL STUDY

H. ALMIRALL

Department de Psiquiatria i Psicobiologia Clínica
Facultat de Psicologia. Universitat de Barcelona.
Adolf Florensa s/n. 08028 Barcelona. Spain

INTRODUCTION

Individual differences in the maximum level of body temperature (BT) throughout the day were described by BLAKE (5). Introverts seemed to present an advanced BT phase with respect to more extraverted subjects.

Lately circadian variation of oral temperature was taken as an external validation of morningness preference (6). However statistical analyses of hourly points revealed no significant differences between Morning-types (M-types), Neither-types (N-types) and Evening-types (E-types). Regarding both extraversion and chronotype ALMIRALL et al. (1) found that E-types showed higher BT at 8.00 p.m. than M- or N-types. However in this work only 6 recordings were made: 9:00, 11:00 a.m., and 1:00, 4:00, 6:00, and 8:00 p.m. throughout the day.

METHODS

Subjects

Nineteen healthy adult (22-28 years old) volunteers were studied. According to their preference subjects were classified as M-types (3 men and 2 women), N-types (3 men and 6 women) or E-types (1 man and 4 women). A reduced scale (2) of HORNE and ÖSTBERG's (6) questionnaire was administered, and also the Spanish version of the Eysenck Personality Questionnaire to select only extraverted and stable people. Sex was not balanced. In a review, ASSO (3) suggested that there are no sex differences in variations in body temperature, with the exception of the work of BAKER & PANGBURN (1982). QUINKERT & BAKER (8) related gender x time of day interaction $F_{(3,72)} = 3.14$, $p < 0.03$, but unfortunately in this study it is not specified how type I error is controlled.

Holter

A specially designed temperature holter was used in thermal registers. This device is now in a development phase and is not commercially available. This prototype is much simpler than the chronoterm described in MOTOHASHI et al. (7). It is not designed as an actographic system and moreover has not software capabilities to program experimental details. It has an autonomy of 24 hours, a range of 35-37.56° C and a precision of 0.01° C.

Temperature was taken continuously every 30 seconds for a period of 12 hours and 20 minutes. Data were stored and read in a PC-compatible microcomputer with RS-232 port after each session. The axillary temperature was registered with a thermistor (S861, Siemens) placed as close as possible to the armpit as described by MOTOHASHI et al. (7). This emplacement was chosen for its obvious advantages in continuous recording.

Procedure

Subjects came to the laboratory at 8.30 a.m. and left it at 9.30 p.m. After the thermistor was attached, it was left for twenty minutes for the temperature to stabilize. Subjects were asked to execute some performance tasks that lasted 10 minutes each hour. During the intervals they were allowed to leave the laboratory but not the flat. All subjects that participated were healthy and had normal sleep-waking routines. Their daily maximum consumption of alcohol and tobacco was less than 40 grams of ethanol and 15 cigarettes, respectively, and they did not take any other psychotropic drugs. On test days, the subjects were asked to cut out stimulant beverages such as coffee, tea and colas. The women were registered the week after the onset of menses.

RESULTS

Tab.1: Mean and standard deviation of axillary temperature, hourly averages according to chronotype.

HOUR	EVENING-TYPE		NEITHER-TYPE		MORNING-TYPE	
	Mean	S.D.	Mean	S.D.	Mean	S.D.
09	35.55	0.37	35.77	0.44	35.70	0.46
10	35.78	0.44	35.99	0.59	35.91	0.58
11	35.97	0.42	36.01	0.58	36.10	0.53
12	35.79	0.46	36.26	0.49	36.12	0.39
13	35.89	0.39	36.30	0.41	36.05	0.48
14	35.99	0.35	36.18	0.47	36.15	0.58
15	35.69	0.33	36.25	0.53	36.32	0.65
16	36.02	0.52	36.39	0.47	36.21	0.81
17	36.13	0.57	36.31	0.28	36.28	0.71
18	36.17	0.54	36.20	0.54	36.31	0.54
19	36.15	0.43	36.12	0.62	36.14	0.72
20	36.27	0.57	36.05	0.63	36.16	0.62
21	36.22	0.49	36.25	0.45	35.96	0.83

Raw series were 1480 data long (from 9.00 a.m. to 9.20 p.m.). Temperature hourly averages were calculated, and also the average of the last 20 minutes. 19 series 13 data long were obtained and studied. Mean and standard deviation of axillary temperature along the day in °C are presented, for each preference group, in Table1.

Univariate analysis

Univariate analysis of peak time and body temperature maximum do not show significant differences between any of the groups (p<0.1267 and p<0.5038, respectively). Peak time according to chronotype was as follows: for E-types, 1 subject presented it at 12 a.m., 2 subjects at 17 p.m., and the other 2 at 20 p.m.; for N-types, 2 subjects presented it at 11 a.m., 2 subjects at 12 a.m., 2 at 13 p.m., 2 at 16 p.m., and 1 at 21 p.m.; for M-types 1 subject at 14 p.m., 2 at 16 p.m., 1 at 17 p.m., and the remaining one at 19 p.m. Acrophases of circadian rhythms could not be obtained as we have not registered several days. In relation to body temperature maximum the values for E-types were: 35.87, 36.53, 36.58, 36.52, and 36.54 being the subjects' order the same as the presented for the peak time. For N-types body temperature maximum were: 36.78, 36.97, 36.62, 36.79, 36.53, 36.62, 37.13, 36.14, and 36.28. Finally, for M-types body temperature maximum were: 36.98, 35.89, 37.12, 36.55, and 35.29.

Multivariat analyses

A Manova approach to repeated measures was carried out with the BMDP4V program. The design was a 3 (between) x 13 (within) factorial design. Preference was not significant ($p > 0.7239$). On the contrary, the within variable was statistically significant ($p > 0.0001$). The Bonferroni procedure was used to control type I errors. Of the 78 possible comparisons only 10 were significant. These correspond to 9.00 a.m. versus all the other hours except 10.00 a.m. and 15.00 p.m. Finally interaction preference x hour was not significant.

CONCLUSIONS

1. There was no significant difference between the peak times of M-, N- or E-types.
2. The axillary temperature maximum did not statistically differ between the experimental groups.
3. Large individual differences in temperature within each group may account for the lack of significant differences between the preference groups along the day.
4. Interaction preference x hour was not significant.
5. The only hour of the day at which temperature was different from the temperature at other hours was 9.00 a.m., although differences between 9/10 and 9/15 were not significant.

REFERENCES

1. ALMIRALL, H., A. ADAN & M. SANCHEZ-TURET (1989): Evolución y relacion de la temperatura corporal y rendimiento durante la vigilia: diferencias individuales. Abstracts III Congreso Nacional de la Sociedad Española de Neurociéncia. Sevilla, 2-5 Octubre 1989, p. 58
2. ALMIRALL, H. & A. ADAN (1990): A reduced scale of morningness: individual differences. Abstracts 6th Meeting European Society of Chronobiology. Barcelona, 6-8 July 1990, p. 3
3. ASSO, D. (1987): Cyclical variations. In: M.A. BAKER (ed.): Sex differences in human performance. Wiley, New York, pp. 55-80
4. BLAKE, M.J.F. (1967): Relationship between circadian rhythm of body temperature and introversion-extraversion. Nature 215, 896-897
5. BLAKE, M.J.F. (1971): Temperament and time of day. In: W.P. COLQUHOUN (ed.): Biological Rhythms and Human Performance. Academic Press, New York, pp. 109-148
6. HORNE, J. A. & O. ÖSTBERG (1976): A self-assessment questionnaire to determine morningness-eveningness in human circadian rhythms. Int. J.of Chronobiology 4, 97-110
7. MOTOHASHI, Y., A. REINBERG, F. LEVI et al. (1987): Axillary temperature: a circadian marker rhythm for shift workers. Ergonomics 30, 9, 1235-1247
8. QUINKERT, K. & M.A. BAKER (1984): Effects of gender, personality, and time of day on human performance. In: H. ANILMITAL (ed.): Trends in ergonomics. Human Factors I. Elsevier, North-Holland, pp. 137-142

ANALYSIS OF INDIVIDUAL DIFFERENCES IN RHYTHMS

J.C. ALDRICH

Department of Zoology, Trinity College, University of Dublin, DUBLIN 2, Ireland

INTRODUCTION

No model has been convincingly proven for many biorhythms, and in numerous cases it would be impossible to do so. Nothing much can be said about the true form of rhythms without relatively frequent sampling. For this reason, few analyses of the actual form of biorhythms have been made, perhaps to the detriment of our understanding of them. On the other hand, the true form of non-biological rhythms is often very well documented as in tides and annual meteorological cycles. These actual forms are accepted because the thousands of data points to ensure the accuracy of such records and the information is used to explain often complicated factors producing the observed patterns. In contrast, the absence of knowledge about the true form of biorhythms could hinder the derivation of the underlying mechanisms (1).

Most contemporary analyses assume an underlying sinusoidal pattern in the data which are fitted either by simple or harmonic Fourier methods (e.g. 5). Methods have recently been developed for the 'unmasking' of biorhythms partially hidden by behavioral and environmental influences; this analysis also is based upon sinusoids or a pattern that can be fitted by harmonics (6). The problem of transformations between the generation and the expression of rhythms is more than one of convenience, herein lies a fundamental question what is the form of the underlying rhythms? If we do not know what the true underlying form of a rhythm is, then we cannot know much about the way in which it is generated. However, if we do know the true underlying form of the rhythm, then its appearance at various stages of 'masking' can tell us something about the mechanisms by which it is actually expressed. A variety of patterns in biorhythms is evident in current literature, ranging from a probable sinusoidal form (4) to a triangular form (9) to a compound pattern (7) and recent work has shown that the sinusoidal model is not essential for analysis; even for overt rhythms having two components, a compound triangular model can sometimes fit the data better (2,3).

In these earlier communications I described ways to analyze data assuming that the underlying cycles were triangular, (described earlier as "relaxation cycles") (BÜNNING, 1973). However, it is unlikely that any single ideal shape could describe even the majority of biorhythms as expressed in current published data. For this reason I believe that data should be analyzed without making any prior assumptions about the form of the expressed rhythm, let the model be derived from the data.

But, how does one proceed when the model is to be derived from the data? The problems are usually not in 'seeing' the overt patterns, but because of the variation between or even within individuals, the problem lies in refining the image sufficiently for comparative and analytical purposes. The material presented here is both the description and test of a method for refining wave forms without having any prior assumptions about a model.

METHODS

Assumptions:
1) Nothing other than the average duration (period) and amplitude is known about the cycles in question
2) The length of each individual cycle varies
3) The amplitude of each individual cycle varies
4) The relative lengths of the rising and falling phases may vary
5) The sampling 'grid' may not coincide with the maximum-minimum peaks (inflexions) in the cycles and the 'grid' or spacing of data collection may even track across successive cycles.
6) Variations in the cycles are random

Important points:
1) The shape of the inflexions (max and min peaks) is a major diagnostic character for the nature of the underlying true cycle
2) Upper and lower inflexions may have different shapes
3) There may be differing numbers of data points in successive cycles.

Analysis:

1) The first task is to divide the data into successive cycles (here a continuous recording in the same individual is assumed, but a number of recordings from different individuals might have to be used). Cycles should be separated at the minimum (or maximum) inflexion points as they appear in the data without any attempt to 'see' the final pattern in advance. Let us assume for simplicity, that the phases do not vary in their relative length within the cycles. In the examples given here, the recordings were divided at the lower inflexions. There will be a common point between every two cycles in a continuous recording and this point must be used twice. In the trials presented here, the division of the cycles at the inflexions resulted in varying numbers of data points per cycle.

2) Scale the time axis (duration) of each individual cycle simply by making the durations equal to 100% (of the length of the cycle) and plotting each data point accordingly. The individual chronological durations are not important here. Plot all data on a grid of 100% equal in chronological length to the average period of the cycles.

3) Scale the amplitude axis for individual cycles as the percentage distance of each point above or below the mean value for that cycle. Thus for a mean amplitude of 6.32: a value of 3.71 would be recorded as 6.32 - 3.71/6.72 = 2.61/6.32 = -41%, a value of 7.98 would be 7.98/6.32 = 1.26 = +26%.

4) Plot all of the data together and derive a preliminary form estimate by drawing the best fit lines by eye. With the general shape of the pattern and the location of the inflexions established, one can proceed to refine the lines. For data that are mainly clumped around similar sampling intervals, averages could be taken at these intervals (this procedure was followed here) and lines drawn again by eye, or using the least squares regression for straight segments. With data randomly distributed on the time

axis the geometric mean regression should be used for straight line segments, or the data could be divided into convenient bands (e.g., 0-5%, >5%-10% etc) and the lines drawn through the averages.

RESULTS AND DISCUSSION

In order to test the fidelity with which this procedure would reproduce true cycles, I made three trials using: a hypothetical complex cycle (Fig.1a), a sinusoidal cycle (Fig.1b), and a triangular cycle (Fig.1c). Each original cycle was sampled using a grid of ten spaces (10%) in duration giving eleven points overall. In order to mimic possible asynchronies between the sampling grid and the cycles, this grid was stepped across the pattern by 1% for each of 10 estimations (Fig.1c) and only in one sampling out of the ten did the sampled points coincide with the inflexions at the maximum and minimum points.

The amplitudes of each individual point in the ten sets of data were then distorted using random numbers (maximum distortion = ± 18% of original value) mimicking random noise in recordings. These distortions occasionally changed the location of the inflexion points giving either ten or twelve data points in a cycle (lower inflexion to next lower inflexion) instead of the original eleven. The cycles were duplicated to facilitate the tracking of the sampling grid. For Fig.1a a further distortion was added by multiplying the overall amplitudes of individual cycles by another random value (range: 67% - 148%).

The resulting forms are given in Fig.2 and individual cycles derived from Fig.1a could now be interpreted as anything from sinusoidal to triangular to square wave. The analysis presented above was then performed, first on data whose amplitudes were not scaled and the result (Fig.1a-ns) was a cycle similar to the original, but of less amplitude. When the varying amplitudes were scaled, the shape in Fig.1a-s resulted - a very close approximation to the original figure.

The same procedure but without scaling the overall amplitudes of the individual cycles, was followed for Figs. 1b and 1c with similarly excellent reproductions of the original forms. Thus, this procedure can refine the image of the underlying wave form where virtually no prior assumptions about models have been made.

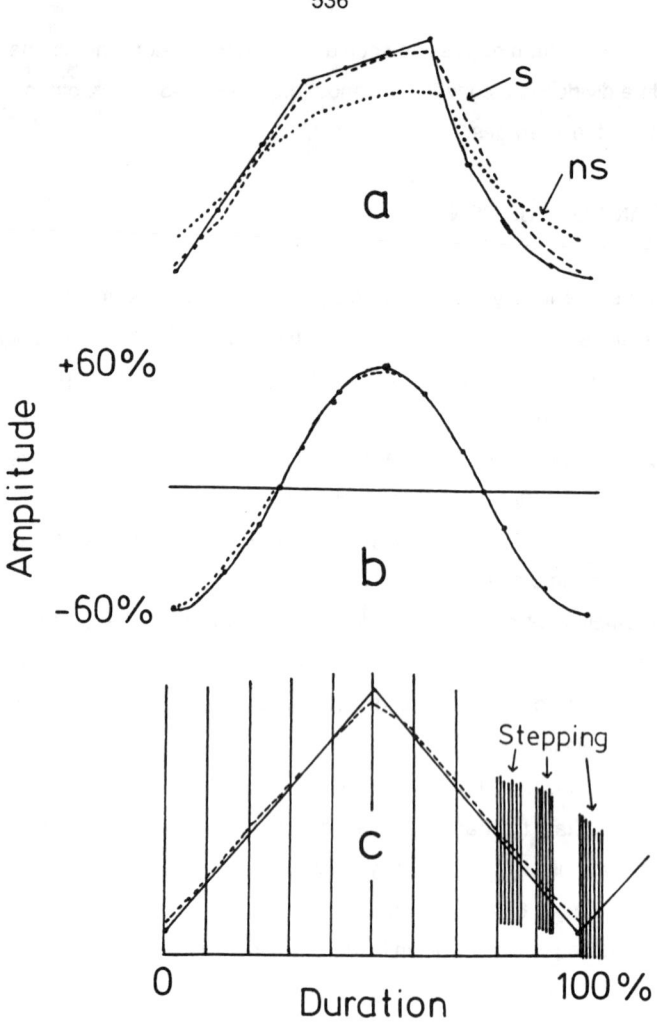

Fig.1: Three underlying forms for comparisons of the effects of random variations and subsequent analysis: a - composite triangular cycle with sinusoidal falling phase, b - sinusoidal cycle, c - regular triangular cycle. Solid lines show the original forms, a-ns= individual cycles not scaled for amplitude, a-s = individual cycles scaled for amplitude. The time grid for sampling is shown in 'c' along with the 'stepping' increments.

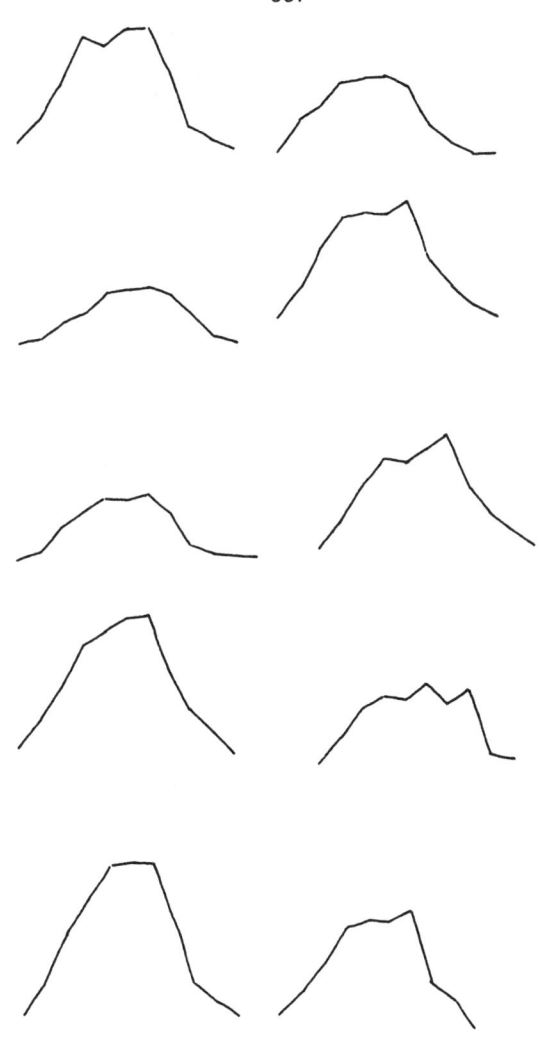

Fig.2: Random variations in Fig.1a including variations in amplitude. Triangular, sinusoidal and square wave cycles are suggested by the random distortions.

However, one must beware of non-coincident inflexions, resulting from differences in relative phase lengths in successive cycles. In a previous paper (1) I discussed the form of the pattern produced when inflexions do not coincide in

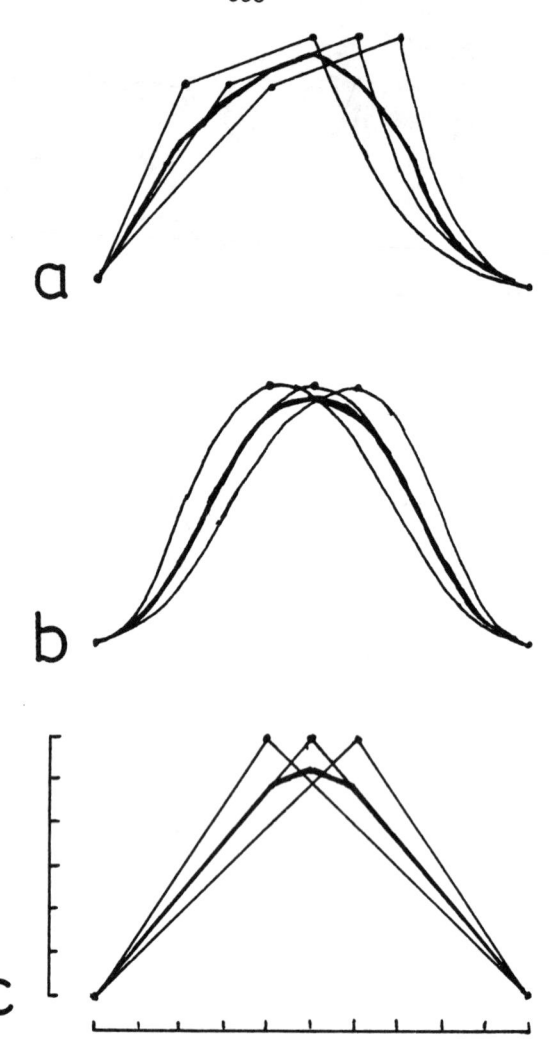

Fig.3: 'Rounded' distortions in the forms of Fig.1 produced by varying the relative durations of the rising and falling phases and simply averaging the data without correcting for relative phase differences, the average figures are shown in heavy lines.

averaged data. Here the severe 'rounding off' of the original forms can be seen when the inflexions do not coincide (Fig.3). The tendency is to produce sinusoidal-like patterns in all forms of cycles and the effect is drastic compared with that produced by variations in overall amplitude (compare Fig.1a-ns to 1a-s and to 3a). Asynchronies in inflexions have worse effects than 'stepping grids' because the entire pattern is distorted. Because the distortion due to varying relative phase lengths could be so great, it could be worth the trouble to divide the data into separate phases, suitably scaled, where necessary.

Another refinement, where the form of the inflexion is in doubt, could be to divide the cycles on the basis of the lower inflexions in order to preserve the form of the upper inflexions, then to repeat the analysis dividing the cycles on the basis of the upper inflexions in order to preserve the form of the lower ones (by preserving the form, I mean keeping the inflexion in the center of the cycle and not at its edges).

There may well be particular amplitudes or phase lengths characterizing the rhythms of individual plants or animals and form estimates based on populations should be used with caution. There may even be different distributions of forms within particular populations, such possibilities will be explored in a future communication.

REFERENCES

1. ALDRICH, J.C. (1990a): The consequences of form in biological rhythms. In: E. MORGAN (ed.): Chronobiology & Chronomedicine. Peter Lang, Frankfurt am Main, pp. 102-114
2. ALDRICH, J.C. (1990b): Explanations of complex chronobiological data using triangles instead of sinusoids as proof that form is more important than theory. J. interdisciplin. Cycle Res. 21, 289-302
3. ALDRICH, J.C. (1991): An analogue to Fourier analysis using triangles, and the consequently different explanations of chronobiological data. In: J. SUROWIAK & M.H. LEWANDOWSKI (eds.): Chronobiology & Chronomedicine. Peter Lang, Frankfurt am Main, pp. 136-151
4. DUFFY, P.H., R. FEURS, K.D. NAKAMURA et al. (1990): Effect of chronic caloric restriction on the synchronization of various physiological measures in old female fisher344 rats. Chronobiol. Int. 7, 113-124
5. HALBERG, H., Y.L. TONG & E.A. JOHNSON (1967): Circadian system phase, an aspect of temporal morphology; procedures and illustrative examples. In: H. v. MAYERSBACH (ed.): The Cellular Aspects of Biorhythms. Proc. International Congress of Anatomists. Springer-Verlag, Berlin, pp. 20-48
6. MINORS, D.S. & J.M. WATERHOUSE (1990): Masking in humans; the problem and some attempts to solve it. Chronobiol. Int. 6, 29-53
7. OKAWA, M., H. SASAKI & K. TAKAHASHI (1991): Disorders of circadian body temperature rhythms in severely brain-damaged patients. Chronobiol. Int. 1, 67-71
8. PALMER, J.D. (1990): The rhythmic lives of crabs. BioScience 40, 352-358
9. ZURAWSKA, E., & J. NOWAK (1991): The melatonin generating system in vertebrate retina. In: J. SUROWIAK & M.H. LEWANDOWSKI (eds.): Chronobiology & Chronomedicine. Peter Lang, Frankfurt am Main, pp. 176-183

ALL-PURPOSE EXPERIMENTAL DATA PROCESSING PACKAGE FOR CHRONOBIOLOGY

J. DOMOSLAWSKI

Institute of Zoology, Jagiellonian University, Karasia 6, 30-050, Krakow, Poland

INTRODUCTION

The development of chronobiology during the last few years indicates the increasing role of mathematical description and analysis used for the estimation of experimental results. There is no doubt that the tendency will deepen. The conviction that more advanced mathematical methods are necessary is more and more common nowadays. The CHRONOS package is an attempt to meet these needs. It is devoted to the analysis of time relationship in living objects. The package performs calculations on numerical data (experimental results) of two kinds i.e. the results of the measurements taken at intervals e.g. every 2-4 hrs during several days, and the results of measurements taken at 1-60' intervals during several days to several months (long runs). The way of treating the first kind of data consists in finding a model that describes the experimental data in a statistically significant manner. In that version a linear model, quasilinear model, and three non-linear models are applied. The second kind of data were processed using actograms and Fourier analysis.

CRONOS OVERVIEW

The package has modular structure which enables both the differentiation of the analysis procedure according to the needs of the user and the extension of the scope of analysis. Some modules have link-up and help functions to ensure certain comfort of user friendliness. For the selection between the particular modules (called CHRONOS options) the program displays the main menu on the screen allowing the user to choose which of the several options will be activated. The highlighted names show the presently available options. At the time of starting CHRONOS, only the following options are available: File, Help, Quit. All other options can be used only after the selection of an appropriate file.

CHRONOS OPTIONS

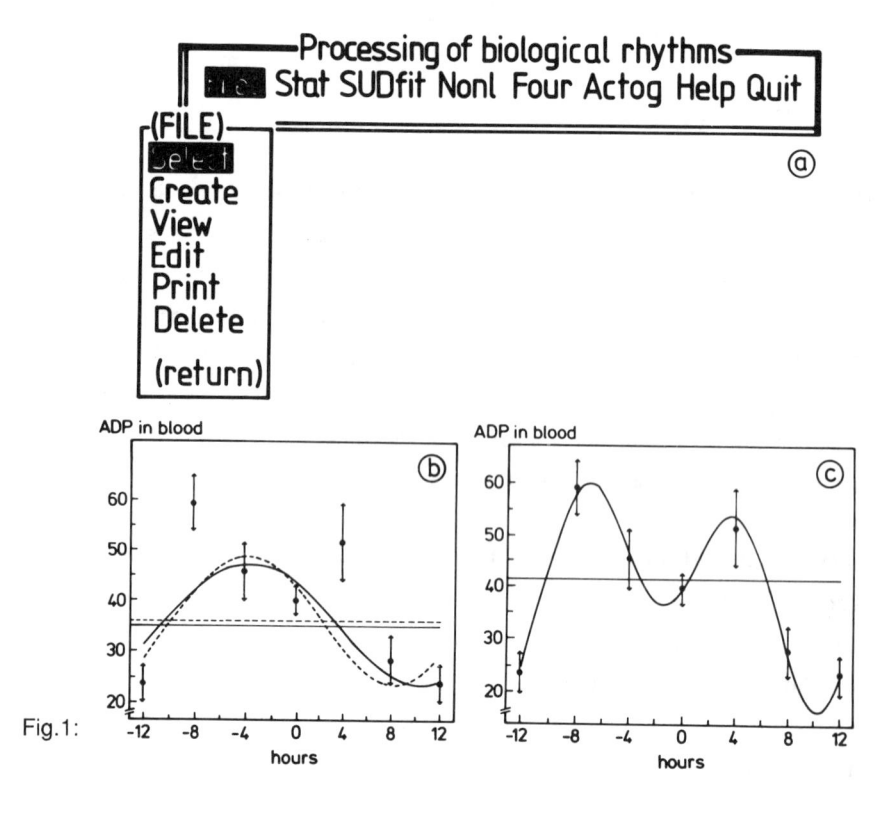

Fig.1:

Processing of biological rhythms
File Stat SUDfit Nonl Four Actog Help Quit

(FILE)
Select
Create
View
Edit
Print
Delete
(return)

(a)

(b) ADP in blood

(c) ADP in blood

HELP - is the description of the program and program options. Always available to the user after returning to the main menu, and selection of the option.

QUIT - selection of that option quits the program.

FILE - selection of that option activates a sub-menu which allows to perform particular operations on the files (Fig. 1a). The most important of them are Create new data file for short runs and Select to chose the previously saved files for both short and long runs. The files for these two kinds of experiments have a different structure. For long runs the files are of a random binary type. The short run files are written in an ASCII code. In this last case two possibilities are taken into account. First, the data which require statistical preprocessing (raw data), and second if our data have been already preprocessed (the mean value and the standard deviation determined). Formally the types of the files are being distinguished by the file-name extension: BIN for long run type, RAW for raw, and DAT for the preprocessed short run data. After the selection of the particular file and return to the main menu the appropriate options can be activated.

STAT - this option performs statistical calculations over the raw short run experimental data. As a result the following data can be found: mean value of every time-point, standard deviation, confidence level of the results for ANOVA variance analysis, and the number of degrees of freedom. As a result of using the STAT option two output files are created. One with the file-name extension DAT is an input file for models fitting options, the second with the extension STO contains the results of the statistical analysis. The CHRONOS package contains two options for the interpretation of short runs experimental data using theoretical models: SVDfit and NONLINEAR.

SVDfit - The option processes the statistically preprocessed data fitting a sinusoid four parameter model to them. The introduction of the additional slope parameter allows to treat the cases in which a linear term resulting from e.g. long term changes superimposes on the oscillations. The fitting procedure is carried out using the concept of the Single Value Decomposition (3). That term means a generalized linearization of the least square problem to rectangular matrices of the order equal to the lower dimension of the matrix. The next processing stage in that option is finding a solution for the classic cosinor model (4). The calculations are completed by displaying several plots on the monitor screen presenting chi-square function versus period, the "error ellipsis" for the amplitude and acrophase, and the fitted function curve against the background of the experimental data (Fig. 1b).

NONLINEAR - By this option statistically preprocessed data can be carried out in three modes (data description according to three nonlinear theoretical models):
1. cosinusoidal oscillation (in principle this is identical with the previous SVDfit model).

2. superposition of two cosinusoidal functions, the period of the first one being fitted while the period of the second is fixed.

3. oscillation of the oscillator with its own period (fitted), operating in the field of an external cyclic force with a constant period. The option uses two minimization methods - Simplex and Gradient (1). The first one is fairly quick far from the minimum region. It does not calculate the covariance matrix but gives the estimation of the order of magnitude of its diagonal values (estimation of the parameters errors). The second one uses the stable Davidon-Flether-Powell (2) variable-metric method. It is extremely fast near a minimum or in any "nearly-quadratic" region but slower if the minimizing function behaves badly. It uses the calculated covariance matrix for the determination of "one-standard-deviation" errors corresponding to a confidence level of 68% for each variable parameter. Later the exact confidence intervals for any parameters may be also calculated. These may differ from the "parabolic" errors derived from the covariance matrix if the model being fitted is highly

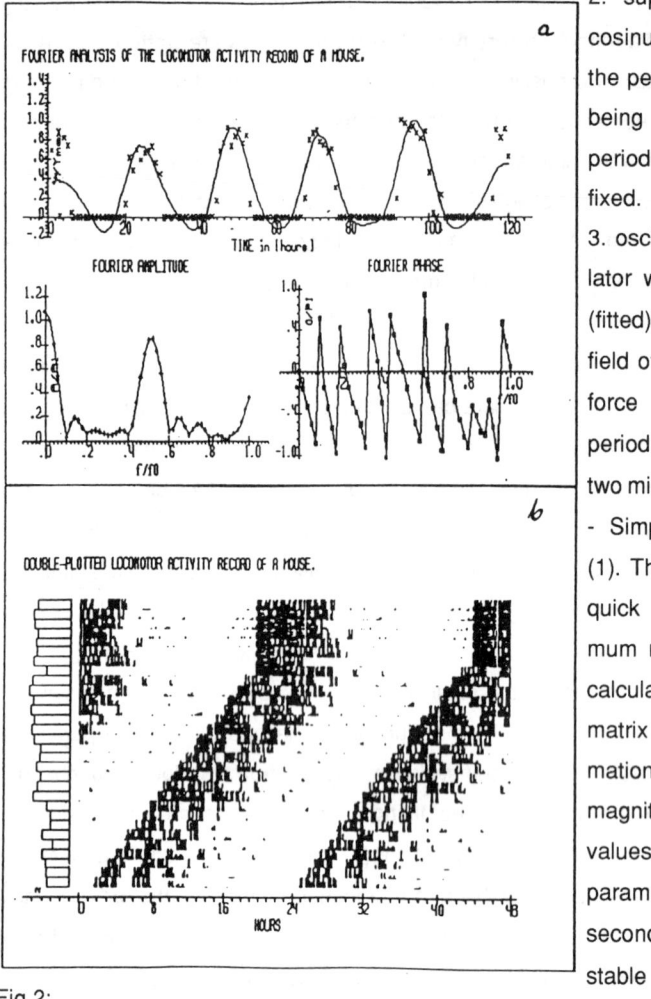

Fig.2:

non-linear. The NONLINEAR option offers some additional possibilities as for example the procedure to improve local minimum by finding a better one. In the work-time the Nonlinear option gives a number of the screen plots which may be printed by print screen command. The final results are both in the graphical (Fig. 1c) and the numerical form.The last two options of the CHRONOS program are connected with the processing of long run data.

FOUR OPTION - It performs the Discrete Fourier Transform (6) on the experimental time series, estimating Fourier amplitude and phase of the data, and smoothes the experimental data by inverting Fourier transformation. That data allows to obtain the dominant periodic components in the investigated data (Fig. 2a).

ACTOG OPTION - It creates the actogram plot for the experimental data. The type of that plot can be chosen from the single plot actogram up to quadruple one (5). The actogram plot allows to estimate a main tendency in the time for the experimental long run data (Fig. 2b).

DISCUSSION

The CHRONOS package can be widely used in chronobiology-oriented analysis. Additional analysis modules (e.g. autocorrelation function and power spectrum) can be added easily. The system requirements include an IBM-PC type computer with hard disc drive and 640 kbyte of RAM memory. It works with all popular graphic adapter cards and can use, by specific interface program, every popular graphic printer. The data processing system that is described in this paper provides an attractive way of analyzing circadian rhythm data. The advantage of this system is that it is reasonably priced, reliable and well supported. Furthermore, it is sufficiently documented so that an individual without training in computer science can get the system up and run it in a few weeks.

REFERENCES

1. EADIE, W.T. et al. (1971): Statistical Methods in Experimental Physics. North-Holland Publ.Co. Amsterdam
2. FLETCHER, R. & M.J.D. POWELL (1963): A rapidly converging descent method for minimization. Comput. J. $\underline{6}$, 163
3. FORSYTHE, G.E. et.al. (1977): Computer Methods for Mathematical Computations. Englewood Cliffs. New York
4. HALBERG, F. (1974): Glossary of Selected Chronobiologic Terms. In: Chronobiology. Georg Thieme Publishers, Stuttgart
5. MOORE-EDE, M.C., F.M. SULZMAN & C.A. FULLER (1982): The Clocks That Time Us. Harvard University Press, Cambridge, MA
6. RAMIREZ, R.W. (1985): The FFT Fundamentals and Concepts. Prentice-Hall. INC, Englewood Cliffs, New Jersey

CLINORHYTHMOMETRY: A PROCEDURE OF PERIODIC-LINEAR REGRESSION ANALYSIS FOR DETECTING OF AGE-RELATED TRENDS IN BIOLOGICAL RHYTHMS

P. CUGINI & L. DI PALMA

Endocrine Pathophysiology (Chronobiology Unit), University of Rome "La Sapienza", II Clinica Medica Policlinico Umberto I, 00161 Roma, Italy

INTRODUCTION

In recent literature, some descriptive models of somatic aging have been derived from the multivariate regression equation and AKAIKE's information criteria (1,2,9,10,14,18,20,22,27).

These methods all propose a linear view of biological functions which decrease with advancing of age. Although the aging process or senescence rises enormous problems of investigation, it seems that the morphology of a linear regression is basically disputable, as it disregards the fundamantal knowledge that the majority of biological functions take place in a periodic fashion (11). The recursive temporality of biological functions is demonstrated by a consistent body of chronobiologic investigations according to which the biologists are now aware that none of the phenomena of living matter escapes the rule of oscillating around the clock along a specific period.

The repetitivity of biological functions should contemplate how the periodic time, which is cyclically zeroed, interacts with the chronological time, which, in turn, is linearly incremented. Basically, the statistical-mathematical approach to aging should elaborate a descriptive equation which considers the cyclical temporality of biological events and the temporal linearity of chronological age.

The present study is aimed to provide such a formula of Periodic-Linear Regression Analysis starting from the concept of combining both the periodic and linear temporalities that time biological events.

IDENTIFYING THE PERIODIC-LINEAR MODEL OF AGING

The task of this study is well represented by Fig. 1. The question is to jump from the view proposed by the left panel to the models provided by the right frames of the plot.

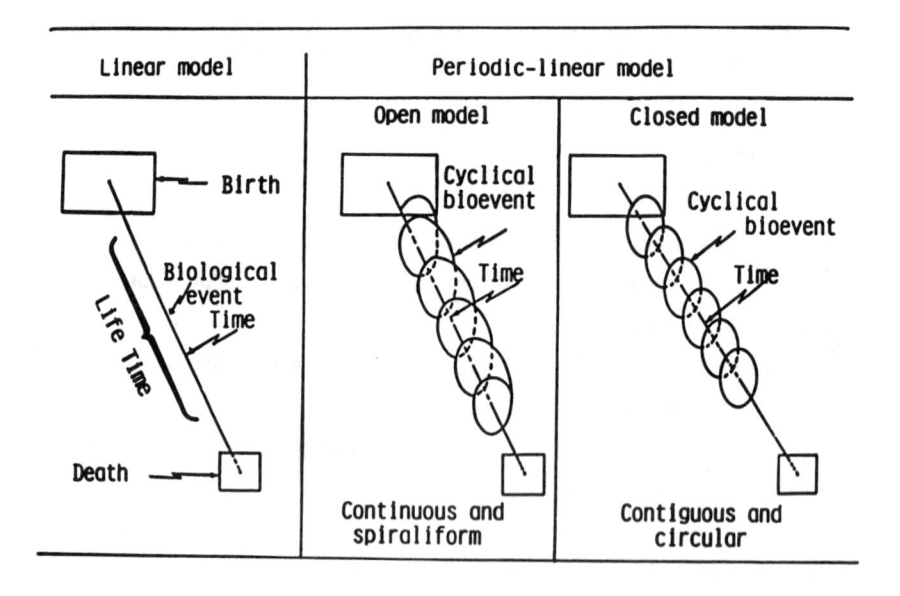

Fig.1: Interpretative models of aging.

According to the non-periodic model, in which the biological phenomena change coaxially to time, aging is seen as a linear relationship between the chronological age, which is the independent variable X, owing to its physical irreversibility, and the biological trend, which is the dependent variate Y, due to its time-dependent decay. For the generation of reference values of aging, the linear model uses the linear regression technique which is schematically represented in Fig. 2.

Discarding the linear interpretation of aging, we are confronted with two possible models of periodic-linear representation. The model on the right assumes the

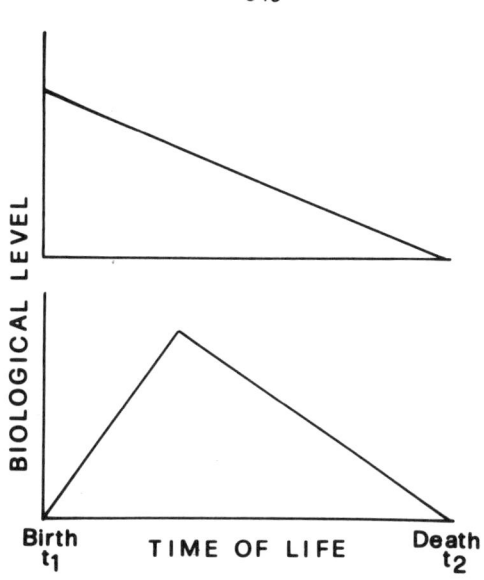

Fig.2: Traditional model of aging seen as a linear decay of biological vitality.

biological rhythm to be a closed event describing a circle in which time is zeroed after a circumvolute revolution. In such a model, the temporal progression of the biological cycle along the axis of chronological time has to be necessarily "discontinuous". The biological rhythm is seen as a "bioperiodic quantum", and its natural course is made by "contiguity" with juxtaposed pauses of inactivity.

Let us remember the ancient aphorysm "Natura non facit saltus" to stress the concept that the biological phenomena take place incessantly in their process of material and energetic metabolism. This motivational reason convinced us that the periodic-linear model of aging could be better expressed by the mid scheme of Fig.1.

This model assumes the biological rhythm to be an open event describing a spiral in which time is zeroed after a progressive revolution. In such a model, the temporal progression of biological cycle along the axis of chronological time has to be necessarily "continuous". The periodic time of biological rhythm is, thus, per se an intrinsic part of the chronological time. Such a counivocity implies that the biological time flows in harmony with the physical time that chronologically dates the aging of living matter.

Assuming the periodic linear model of aging to be spiraliform, one must decide which is the oscillating shape that better represents the cyclicity of biological rhythms. Looking at this question in a bidimensional mode, one can oversimplify the choose among the different waves which all suitably express a periodic signal (Fig.3).

It is well known that biological rhythms are not perfectly sinusoidal in their excursion because of the different proportion of diurnal and nocturnal time. Notwithstanding that, all the mathematicians, who analytically explored the morphologic geometry of biological rhythms, have primarily chosen the harmonic motion as the most representative model. Therefore, all chronobiologists agree that the optimal representation of a biological rhythm is, analytically speaking, a waveform profile, even though they admit that the sinusoidal model does not imply the real rhythm to be sinusoidally arranged. In line with this thought, one must convene that the periodic-linear model of aging may be

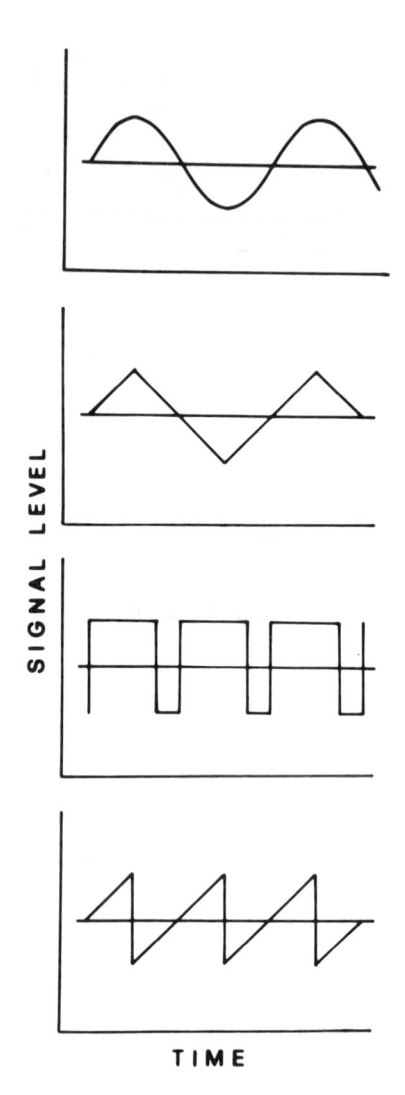

Fig.3: The periodicity of a signal can be given by different waveforms profiles.

schematically represented by the relationship of the sinusoidal decay of a bioperiodic function along the axis of chronological age (Fig.4).

BIOLOGICAL LEVEL

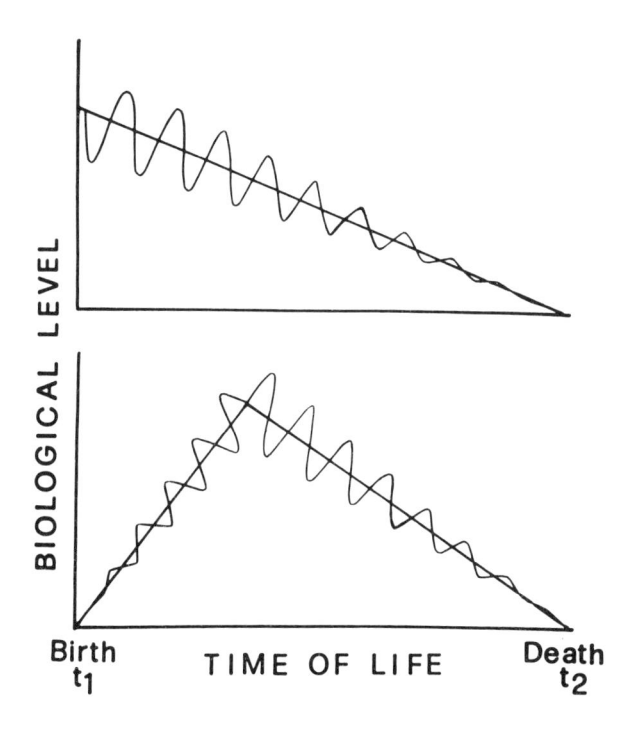

Birth
t_1 TIME OF LIFE   Death
                                   t_2

Fig.4: Chronological model of aging.

This model solves the assumption of combining periodic time and linear time, and suggests that the cause-effect relationship of biological functions with age is, in final analysis, represented by the age-related decline of the biometric parameters which constitute the properties of the oscillating wave.

CONSTRUCTING THE PERIODIC-LINEAR REGRESSION EQUATION OF AGING
No doubt that biological aging is a complex phenomenon. Nevertheless, it simply can be seen as a "variation of biological status along the axis of chronological time resulting in a morpho-functional deterioration of the living organism". According to this definition, aging may be biostatistically seen as a phenomenon whose constituent

events can be concretely regarded as qualitative or quantitative variables that change in their measurable level during the course of life. The binomium "function/variable" allows biostatisticians to use mathematical models to derive cause-effect relationships between biological events and chronological time.

In any casual model, it is fundamental to establish which is the dependent variable that is subject to change. As in living matter any subsequent change is intimately linked to the antecedent status, there is no doubt that time is the independent variable along which the biological aging can take place. Therefore, the chronological time becomes the predictor whose historicity produces an increasing sequence of dates along the X axis of abscissa to which relate the biological values on the Y axis of ordinate. The observed values $Y(i)$, thus, relate to the predictor $X1(i)$, $X2(i)...Xn(i)$ in the following way

$$Y(i) = F(X(i)) \qquad [1]$$

where F is the mathematical function which appropriately describes the relationship between Y and X.

If the biological change has the feature of a trend (T), i.e., a constant variation of negative or positive direction as a function of time (t), the equation will take the form of

$$Y(t) = F(X(t) + T(t)). \qquad [2]$$

If the biological function has a cyclic nature (C) over time (t), the formula can be rewritten as

$$Y(t) = C(t) + T(t). \qquad [3]$$

Mathematically, the analytical expression of a periodic function having a harmonic motion, and, therefore, a waveform profile, is given by the equation

$$Y(t) = M + A \times \cos(wt + \phi) \qquad [4]$$

where M is the mean level of the oscillation, A is the highest amplitude of the fluctuation, w is the angular frequency expressed by the number of cycles, 2π, which occur within the unitary period TAU (w = 2π/TAU), t is a fractional time of TAU.

Importantly, each one of these parameters has a proper term in Chronobiology, as HALBERG and coworkers (12) proposed this formula as the periodic regression method, namely cosinor method, for resolving and quantifying biological rhythms in living matter. M, A and ϕ are, respectively, called Mesor (acronym of Midline Estimating Statistic of Rhythm), Amplitude and Acrophase (Fig.5).

Operationally, the formula 4 may be suitably used to substitute the C(t) term from the equation 4 with the result of having

$Y(t) = M + A \cos(wt+\phi) + T(t).$ [5]

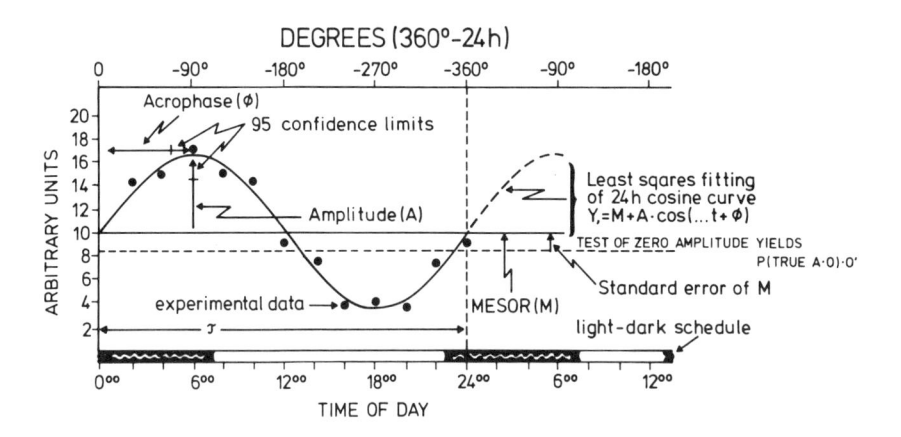

Fig.5 - Cosinor method.

Formally, the equation 5 may be expressed as follows

$Y(t) = (M+T(t)) + (A+T(t)) \times \cos(wt + (\phi+T(t)))$ [6]

suggesting that the time-dependent changes of a bioperiodic function primarily depend on the trend of Mesor, Amplitude, and/or Acrophase.

It is worth to notice that the rhythmometric parameters M, A and ϕ represent the non-periodic part of a regression whose periodicity is dictated by the term $\cos(\Omega)$. Assuming the biological rhythm to maintain unaltered its periodicity during life time, Ω will be the unique term of the equation not to show a chronological change.

This evidence allows us to use the linear regression analysis to assess the trend of M, A and ϕ as a function of chronological age.

In Time Series Analysis (3), a linear trend is expressed by the general equation

$$T = a(\text{constant}) + b(\text{regression}) \times t. \qquad [7]$$

As the linear time t of biologic phenomena corresponds to age, the trend of M, A and ϕ, respectively T(M), T(A) and T(ϕ), can be, thus, computed as

$$T(M) = a(M) + b(M) \times \text{Age} \qquad [8]$$
$$T(A) = a(A) + b(A) \times \text{Age} \qquad [9]$$
$$T(\phi) = a(\phi) + b(\phi) \times \text{Age}. \qquad [10]$$

These equations may be entered into the formula 6 with the effect of having the final expression

$$Y(\text{Age}) = F((a(M) + b(M) \times \text{Age}) + (a(A) + b(A) \times \text{Age}) \times$$
$$\cos((2\pi/\text{TAU} \times t) + (a(\phi) + b(\phi) \times \text{Age}))) \qquad [11]$$

where t still relates to a fraction of time concerning the period of the rhythm under scrutiny.

The equation 11 represents the "Periodic-Linear Regression Function" "PLRF" for investigating aging process in rhythmometric functions as it combines the periodic time of bioperiodic events with the linear time of chronologic age.

The method for exploring aging via the PLRF can be called "Clinorhythmometric method" or "Clinorhythmometry", neologisms which derive from the greek root "clinos" that means trend. Clinorhythmometry is, thus, the formulation proposed to explore the age-related trend in biological rhythms.

CLINORHYTHMOSCOPIC MODELS OF AGING

Basically, the clinorhythmometric equation establishes the concept in virtue of which the biological events necessarily undergo a gerontological change in their cyclic propreties along the systematic, progressive and irreversible dimension of physical time.

This irremediable and underferable changeability may be seen from a gerontologic point of view as a deterioration of the functional properties which are expressed by the parameters of a bioperiodic oscillation.

In a strict sense, the mesor of a rhythm is the biometric expression of the functional level, namely the tonic activity. The amplitude, in turn, is the quantitative measure of the oscillatory level, namely the phasic activity, of the naturally-recurring bioevent. From a chronobiologic viewpoint, aging may be seen as a decay of both the tonic and phasic functions that characterize the biological events in their periodic activity.

One of the most interesting aspects of chronobiologic research in gerontology is the evidence that not all the biological rhythms show the predictable decay in their rhythmometric properties (4-8,13,16,17,19,21,23-25). Some rhythms may show the canonical pattern but some others may show, on the contrary, an increase in mesor and/or amplitude, suggesting that the bioperiodic events can be divided into "causative of aging" and "adaptive to aging" (6).

From the clinorhythmoscopic assessment of biological rhythms, two fundamental rules can be derived, i.e,

1. the age-related changes in mesor and amplitude, namely in tonic and phasic activity, may be concordant or discordant;
2. the chronobiologic changes of aging may not be of negative sign for all the biological rhythms.

From these rules it derives that the clinospectroscopic model of aging may be composite depending on the age-related pattern of the biological rhythm under scrutiny.

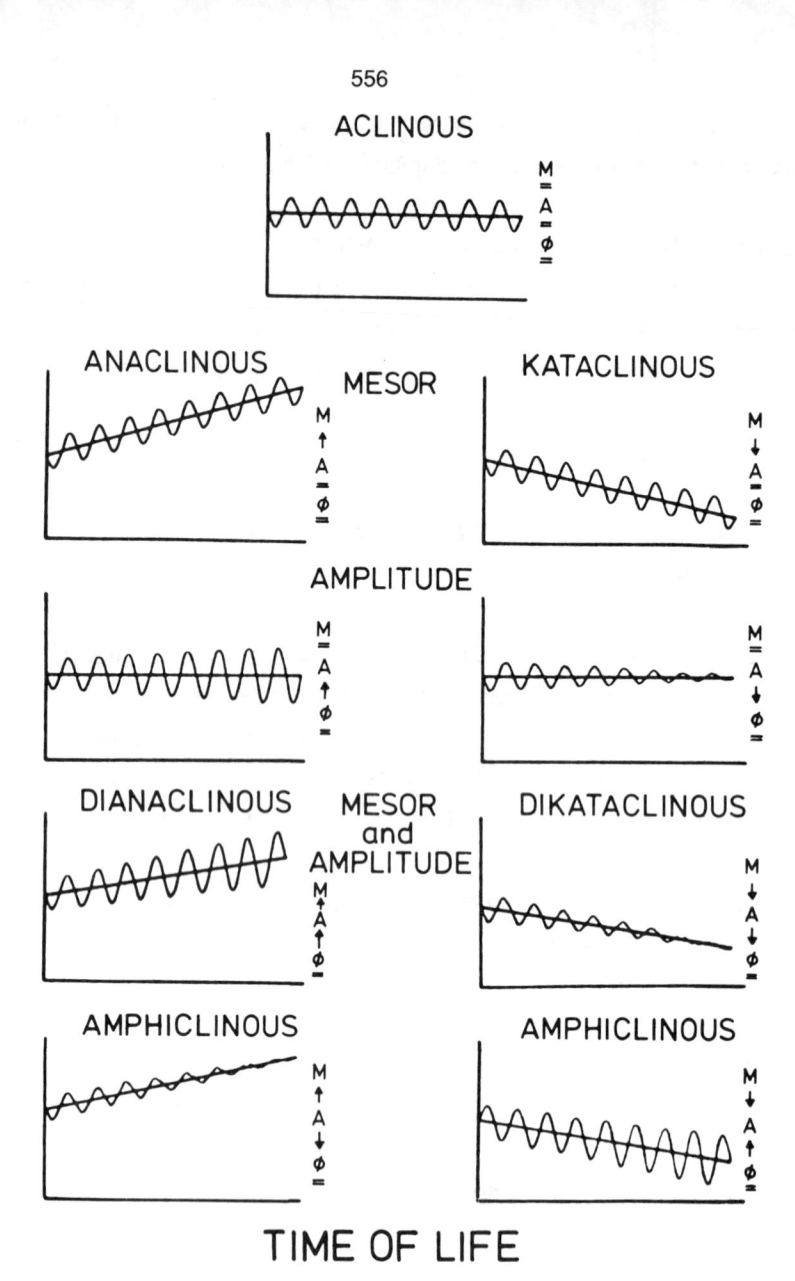

Fig.6: Clinospectroscopic models of the age-related changes in a biological rhythm.

By applying the Clinorhythmometric analysis, several types of clinorhythmoscopic models of aging have been detected. Fig. 6 displays some of these models with a particular reference to the age-related trend in mesor and amplitude, and not in acrophase.

These models derive by plotting the Y(t) value on a bidimensional diagram as a function of chronological age. This diagram has been called "clinorhythmogram".

The continuous line depicts the age-related trend of mesor. The oscillating line displays the gerontologic trend of amplitude with changes every 10 years. Therefore, the clinorhythmogram represents the changes in tonic and phasic activity as a function of age.

TAB. 1: Some examples of clinospectroscopic models in human beings

Model	Biological rhythm	
	Type	Variable
Aclinous	Circadian	Arterial pH
Mesor-Kataclinous	Circadian	Body temperature
Amplitude-Kataclinous	Circadian	Potassium
Mesor-Anaclinous	Circadian	Blood pressure
Amplitude-Anaclinous	Circadian	Cortisol
DiKataclinous	Ciracdian	Renin-Aldosterone
DiAnaclinous	Circadian	Insulin
Amphiclinous: Type I (Mesor-Kataclinous Amplitude-Anaclinous)	Circadian	Heart Rate
Amphiclinous Type II (Mesor-Anaclinous Amplitude-Kataclinous)	Circatrigintan	LH in women

Interestingly, some rhythms show no age-related trend (aclinous models). Some others show negative or positive trends designing respectively "kataclinous" or "anaclinous" models (from greek roots kata = downward, and ana = upward). The trend may occur selectively for mesor (mesor-kataclinous or mesor-anaclinous models) or amplitude (amplitude-kataclinous or amplitude-anaclinous), or, jointly, for both the rhythmometric parameters (dikataclinous or dianaclinous models). Finally, the rhythms may show a discordant trend in mesor and amplitude giving rise to "amphiclinous" models (Type I and Type II).

Tab. 1 lists the biological rhythms which show the above-mentioned clinospectroscopic models.

Importantly, the age-related variations of rhythmometric parameters which describe the clinospectrometric model may be expressed by the symbols displayed in Tab 2.

TAB. 2: Symbols of clinospectrometric changes in rhythmometric properties of biological rhythms

Clino-spectrometric symbols	Type of change	Functional meaning
M ↑	Increment of mesor	Increment of tonic activity
M =	Stationary of mesor	Stationariety of tonic activity
M ↓	Decrement of mesor	Decrement of tonic activity
M X	Disappearance of mesor	Abolition of tonic activity
A ↑	Amplification of wave	Increment of phasic activity
A =	Stationariety of wave	Stationariety of phasic activity
A ↓	Deamplification of wave	Decrement of phasic activity
A X	Disappearance of wave	Abolition of phasic activity
φ →	Posticipation of phase	Delay of phasic temporality
φ =	Stationariety of phase	Stationariety of phasic temporality
φ ←	Anticipation of phase	Anticipation of phasic temporality
φ X	Disappearance of phase	Abolition of phasic temporality
TAU ∗	Shortening of period	Frequency multiplication
TAU =	Stationariety of period	Frequency stationariety
TAU ∗	Lenghtning of period	Frequency demultiplication
TAU X	Abolition of period	Abolition of rhythmicity

CONCLUSIONS

In this paper, we simultaneously applied the linear regression analysis into the periodic regression function with the aim of providing an appropriate equation for a tentative model of aging process.

Obviously, the use of clinospectrometric analysis, briefly the Clinospectror method, does not pretend to exhaustively solve the assumption, as aging is a phenomenon complex enough for being simplistically restricted into a resolutive formulation. Nevertheless, the Periodic-Linear Regression Analysis (PLRA) or Periodic Time Series Analysis (PTSA) has the non-transcurable prerequisite of exploring the age-related changes in biological functions according to the periodically-arranged physiology that characterizes all the phenomena of living matter. Moreover, the PLRA presented here employs more information that did the classical Time Series Analysis. In particular, the clinospectrometric models depend on the actual values of three parameters, Mesor, Amplitude, and Acrophase, for the characterization of each biological function. The results obtained are more extensive than the outcomes of the linear regression models in the sense that the biological function is observed in its tonic and phasic activity during life time. This implies that the model presented here can help to remedy the problem of an inprecise generalization when dealing with the age-related activity in biological functions.

The final comment is to say that the clinospectrometric analysis is relatively easy to apply using microcomputers. Therefore, the application of this technique can lead to newer knowledges on aging process. From this article, it is already patent that aging is characterized by a phenomenon of rhythmic disharmony as the biological rhythms may show opposite trends yet inside their rhythmometric parameters. The effect of the divergent changes in tonic and phasic activity may result in an ever more chaotic coordination, substantiating the concept that aging process tends to a final status of "rhythmic entropy" or "chronoentropy".

REFERENCES

1. AKAIKE, H. (1976): What is Information Criterion AIC? Mathemat. Sci. 153, 5-11.
2. BORKAN, G. & A.H. NORRIS (1980): Biological age in adulthood: comparison of active and inactive U.S. males. Human Biol. 52(4), 787-802.
3. COOK, T.D. & D.T. CAMPBELL (1979): Quasi-experiments: interrupted time-series designs. In: T.D. COOK, D.T. CAMPBELL (eds.): Quasi-experimentation: design and analysis issues for field settings. Rand McNally, Chigago.
4. CUGINI, P., D.SCAVO, F.HALBERG et al. (1981): Ageing and circadian rhythm of plasma renin and aldosterone. Maturitas 3, 173-182.
5. CUGINI, P., D.SCAVO, F.HALBERG et al. (1982): Circadian amplitude of prolactin, cortisol and aldosterone in human blood several decades after menopause. In: P. FIORETTI, L. MARTINI, G.B. MELIS & S.S.C. SEN (eds.): The menopause: Clinical, Endocrinological and Pathophysiological Aspects. Serono Symposium No.39. Academic Press, London and New York, pp. 157-162.
6. CUGINI, P., P.LUCIA, R.TOMASSINI et al. (1985): Temporal correlation of some endocrine circadian rhythms in elderly subjects. Maturitas 7, 175-186.
7. CUGINI, P., P.LUCIA, G.MURANO et al. (1987): The gerontological decline of the renin-aldosterone system: a chronobiologic approach extended to essential hypertension. J. Gerontol. 42, 461-465.
8. DESCOVICH, G., J.KUHL, F.HALBERG et al. (1974): Age and catecholamine rhythms. Chronobiol. 1, 163- 171.
9. FURUKAWA, T., I.MICHITOSHI, K.FUMIHIKO et al. (1975): Assessment of biological age by Multiple Regression Analysis. J. Gerontol. 30, 422-434.
10. FURUKAWA, T. (1976): Model of life span. Mathemat. Sci. 14, 43-55.
11. HALBERG, F. & A.AHLGREN (1978): Chronobiology. A review. In: Biological rhythms. Documenta Geigy pp. 1-3.
12. HALBERG, F., E.A. JOHNSON, W. NELSON et al. (1972): Autorhythmometry-procedures for physiologic self-measurements and their analysis. Physiol. Teacher 1, 1-11.
13. HAUS, E., F.HALBERG, W.NELSON et al. (1979): Age effects upon circadian amplitude in concomitant study of 12 hormones in plasma of women. Chronobiologia 6, 266-276.
14. KIKKAWA, K., T. OGAKI, H. OKABE & J. MATSUMOTO (1988): A statistical approach for physical senility and its control (I). Kyushu J. Phys. Educ. Sports 2, 57-66.
15. MONTALBETTI, N., F. GHIRINGHELLI, P.A. BONIN & L. BONANONI (1982): Adrenal rhythms during human senescence. Acta Endocrinol (Kbh) 119(suppl.), 44 (abstract).
16. MURRI, L., T. BARRECA, G. CERRONE et al. (1980): The 24-h pattern of human prolactin and growth hormone in healthy elderly subjects. Chronobiol. 7, 87-92.

17. NELSON, W., C. BINGHAM, E. HAUS et al. (1980): Rhythm-adjusted age effects in a concomitant study of 12 hormones in blood plasma of women. J. Geront. 35, 512-519.
18. SASAKI, A., Y. TAKASAKI, N. HORIOUCHI & K. OMORI (1970): 'Expected Age' used in evaluation results of multiple health examination. Jpn. J. Geriat. 7(6), 323-332.
19. SERIO, M., P. PIOLANTI, S. ROMANO et al. (1970): The circadian rhythm of plasma cortisol in subjects over 70 years of age. J. Gerontol. 25, 95-97.
20. SHIMOKATA, H., K. SHIBATA & F. KUZUYA (1987): Assessment of biological aging status. Jpn. J. Geriat. 24, 88-92.
21. SILVERBERG, A., F. RIZZO & D.T. KREIGER (1969): Nychtohemeral periodicity of plasma 17-OHCS levels in elderly subjects. J. Clin. Endocrinol. Metab. 28, 1661-1663.
22. SUGIYAMA, K., I. OZAKI, K. USHIZAWA & M.SHIMIZU (1979): Regression analysis of estimating of age in terms of dental defacement. Applied Statistics 5(3), 123-138.
23. TOUITOU, Y., C. TOUITOU, A. BOGDAN et al. (1978): Serum magnesium circadian rhythms in human adults with respect to age, sex and mental status. Clin. Chim. Acta 87, 35-41.
24. TOUITOU, Y., C. TOUITOU & A. BOGDAN (1979): Circadian rhythm in blood variables of elderly subjects. In: A. REINBERG (ed.): Chronopharmacology. Pergamon, Oxford, pp. 283-290.
25. TOUITOU, Y., C. TOUITOU, A. BOGDAN et al. (1979): Circadian rhythms in serum total proteins: observed differences according to age and mental health. Chronobiol. 6, 164.
26. TOUITOU, Y., M. FEVRE, M. LAGOGUEY et al. (1981): Age and mental health-related circadian rhythms of plasma levels of melatonin, prolactin, luteinizing hormone and follicle-stimulating hormone in man. J. Endocrinol. 91, 467-475.
27. WEBSTER, I. & A. LOGIE (1976): A relationship between age and health status in female subjects. J. Geront. 31, 546-550.

MODELFREE EXPLORATORY ANALYSIS OF CHRONOBIOLOGICAL TIME-SERIES BY THE PC-PROGRAM LOADFUAP

J. PEIL, H. HELWIN & S. SCHMERLING

Institute of Anatomy and Department of Mathematics of the Martin-Luther-University Halle-Wittenberg, P.O.B. 302, D-O-4010 Halle

METHOD

In most cases measured data in experimental chronobiological research are given as so-called time-series. The time-series is assumed, as a rule, to be a realization of a continuous stochastic process. In many cases the process is periodical with a known (or assumed) basic periodicity (24 hours, 12 months and the like), (see Fig. 1 at the bottom), or without such a basic periodicity, (see Fig. 1 at the top).

For interpretation of the measured time course of data one has to separate the continuous deterministic component from the pure random component as good as possible. From mathematical point of view there are two basic approaches: the "modelling" way and the modelfree one.

For the modelling approach there must be assumed an analytical expression for describing the deterministic component. The function then will be fitted globally to all measured data within the interval of measurements by the method of Least Squares. Often the sinus function in any form or combination (also in Fourier analysis) is used thougt in the most cases there are not any hints for this function type.

The modelfree approach means that for analyzing of measured time-series data one has not to assume any mathematical model in form of analytical expressions as it is the case, for instance, in periodogram analysis or in the so-called Cosinor analysis. The deterministic component in measured data sequences will be separated from random variations by numerical smoothing procedures based on the method of LOcally ADjusted FUnctional APproximation (1,2). On this way the continuous time course will be calculated in an exploratory manner only on the basis of the measured

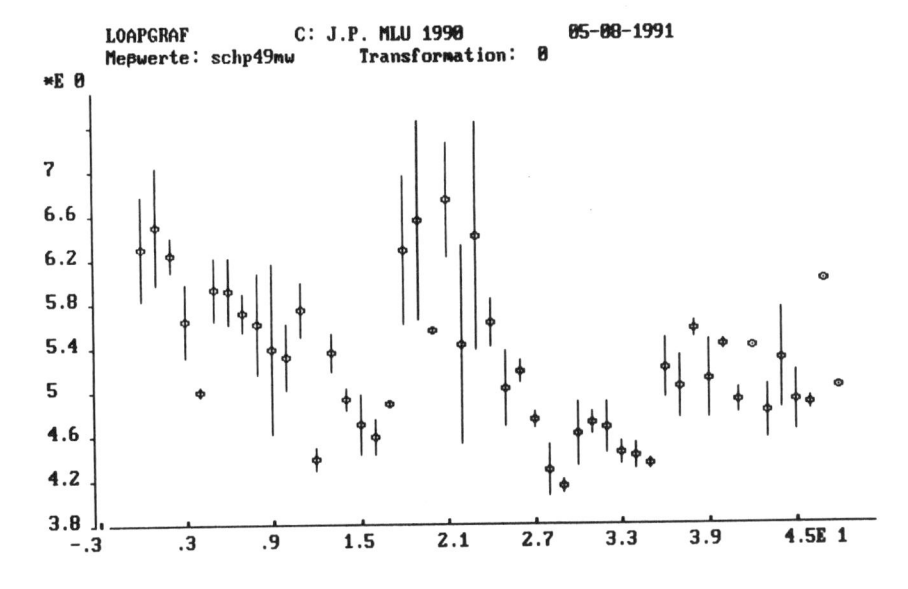

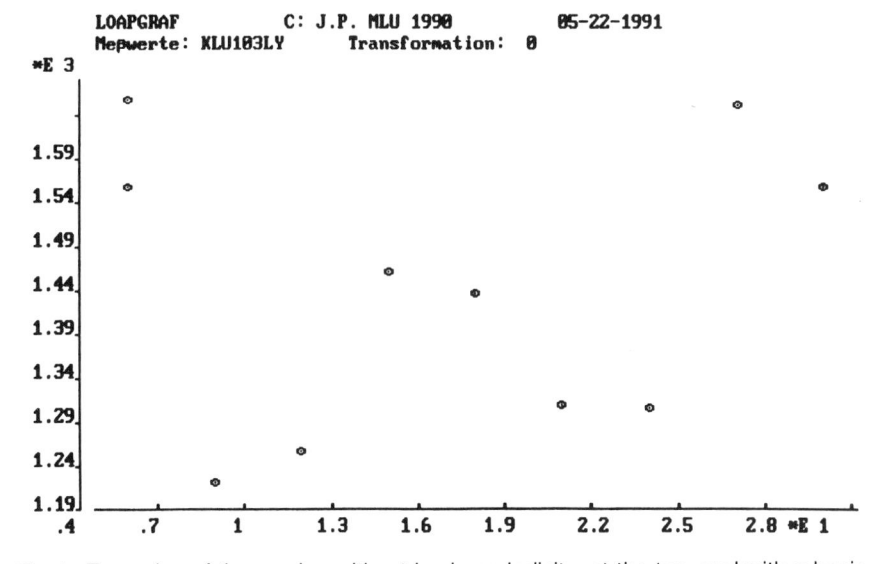

Fig.1: Examples of time-series without basic periodicity, at the top, and with a basic periodicity (of 24 hours), at the bottom.(Hardcopies of monitor outputs of the PC-program LOADFUAP.)

data. The interpretation of the experimental results is not burdened by artifacts as it may be possible by assuming a more or less unfounded analytical expression. "Exploratory manner" means that in the numerical procedures on the basis of locally adjusted functional approximation the user can change the degree of smoothness in a broad range according to his possibly prior knowledge about the biological process in question (3).

EXAMPLES

In Fig. 1 at the top an example of a nonstationary time-series without a natural basic periodicity is shown. Repeated measurements are represented by the mean value and the vertical bar of its standard deviation. Presumedly the deterministic component consists of a globally decreasing trend and of a oscillating portion. What kind of functional expression could describe the whole course of such a deterministic component? (If any could be found it is a very artificial and formal construct).

At the top of Fig. 2 the results are shown calculated by the procedure LOQUAREG (LOcal QUAdratic REGression, code "RV=5") which is implemented in the PC-program LOADFUAP. As mentioned above, by fixing of two different degrees of smoothing (in form of a smoothing parameter in the procedure) one can obtain separately the trend component for a relative large value of the smoothing parameter on the one hand (broken line in Fig. 2) and the oscillating component, still containing the trend component, for a smaller value of the smoothing parameter on the other. If one is more interested in the oscillating part the difference between the two curves may be calculated and further mathematical evaluations may be performed on the basis of these results. Especially in the case of non-equidistant time-series (which often occur in seasonal varying L-D-designed experiments on daily changes of a parameter (4)) the modelfree evaluation is important for classical procedures as Fourier analysis or power spectral analysis which require equidistant time-series.

In Fig. 1 at the bottom an example is shown of a time-series with a naturally assumed periodicity of 24 hours. Measurements carried out over more than one period are projected, as a rule, into the basic periodicity interval. Accordingly, in case of stationarity one has to understand the variety between data at one and the same

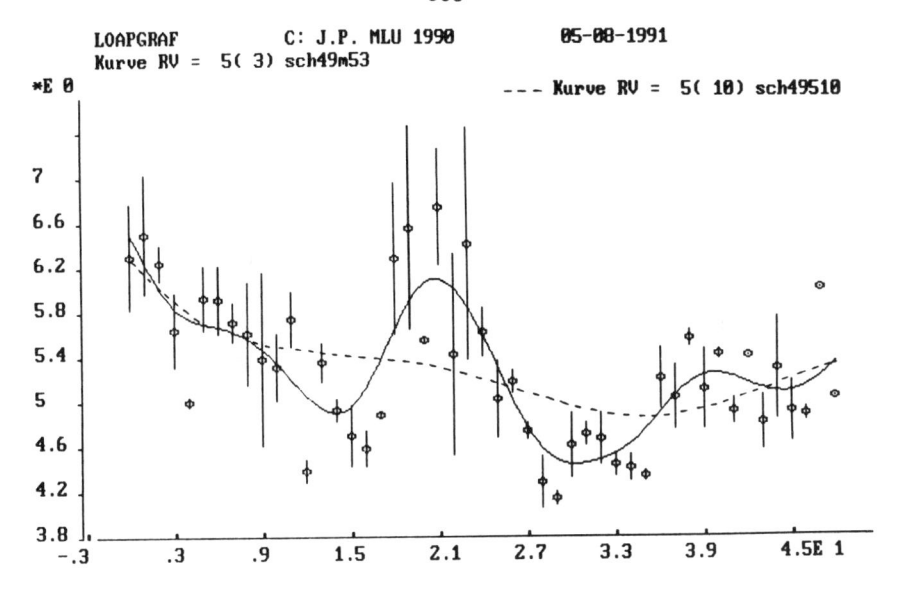

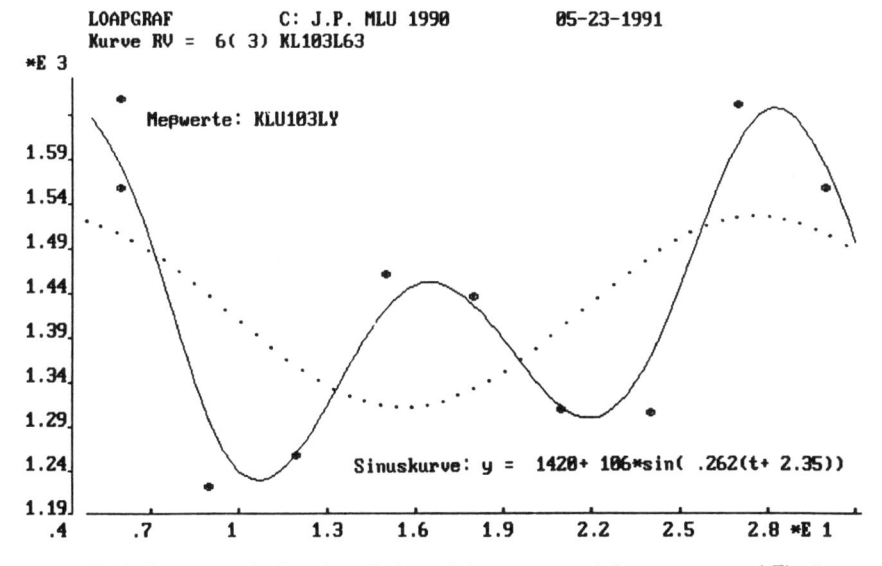

Fig.2: Modelfree quantitative description of the measured time courses of Fig.1 (s. text). (Hardcopies of monitor outputs of the PC-program LOADFUAP.)

time-point of the periodicity interval as random biological variability (in Fig. 1 indicated only for 6.00 a.m.). The attempt to describe the measured course by a globally adjusted cosine function (dotted line in Fig. 2) must fail because of the visible bimodality of the time course of data.

At the bottom of Fig. 2 also the result is shown of local approximation which are calculated by the procedure PERLOCAP (PERiodical LOCal APproximation, code "RV=6") also implemented in the PC-program LOADFUAP. This procedure is a special version of functional approximation using the sinus function for local estimation of the functional relation. As in the other implemented procedures the above mentioned basic periodicity will be taken into account by periodical continuation of the measured data. (This formalism is the algorithmic equivalent to the assumption of basic periodicity. It will be automatically triggered in a program run if the information about periodicity is put in. Numerically this periodical continuation assures the periodicity of the calculated curve as it may be seen in Fig. 2).

PC-PROGRAM

The program LOADFUAP (LOcal ADjusted FUnctional APproximation) is written in (GW-)BASIC running under MS(PC)-DOS (2). It consists of 4 parts (LOAPINIT, LOAPDATA, LOAPCALC, and LOAPGRAF) which will automatically be chained during a run of the program. The user will be guided by dialog.

LOAPINIT installs the pogram on the respective used hardware and presents some program informations.

LOAPDATA is the part for input of actual parameters (length of basic periodicity in the case of periodical time-series, numerical accuracy), for choosing the appropriate version of approximation (locally constant, linear, quadratic, trigonometric), for input of measured data (from keyboard or from diskette), and for handling of these data (correction, omitting, supplementing, calculation of mean values, optional argument transformation).

LOAPCALC tabulates the approximation and the first derivation of the approximation function after settlement of the smoothing parameter and choosing the mode of tabulation.

LOAPGRAF provides graphical representation of the measured data (see Fig. 1), of the approximation results (see Fig. 2), and the curve of the 1st derivative of the approximation on request. (It may also be used for representation of the graph of a function, and possibly of measurements, outside of the program LOADFUAP).

The automatical program return from LOAPGRAF to LOAPCALC offers the possibility of calculation of further approximations (for the same or for changed data file) and for (graphical) comparisons of the approximations (see Fig. 2 at the top).

REFERENCES

1. SCHMERLING, S. & J.PEIL (1985): Verfahren der lokalen Approximation zur nichtparametrischen Schätzung unbekannter stetiger Funktionen aus Meßdaten. Morphol. Jahrb. 131: 367-381.
2. PEIL, J. & S. SCHMERLING (1987): Modellfreie biomathematische Bearbeitung von Meßwerte-Zeitreihen biorhythmischer Untersuchungen. Kongr.- u. Tag.ber. der MLU Halle-Wittenberg, Wissensch. Beitr. 36 (P 30), 330-334.
3. PEIL, J., S. SCHMERLING, D.PESCHKE et al. (1989): Periodische lokale Approximation zur modellfreien Beschreibung von Meßwertverläufen biorhythmischer Vorgänge - Demonstrationsbeispiel für die Wahl des Glättungsparameterwertes. Morphol. Jahrb. 135: 261-269.
4. PESCHKE, E., D. PESCHKE, J. PEIL et al. (1986): Schilddrüsenreaktionen der Wistar-Ratte im Tagesgang nach Gangliektomie unter Normaltemperatur und Kälteexposition unter Berücksichtigung des Einflusses der Epiphysis cerebri. Acta histochem. 80: 63-85.
5. PEIL,J. (1990): Das BASIC-Programm LOANFUAP zur modellfreien Auswertung von Messungen funktionaler Beziehungen y=f(x) auf der Grundlage von N Meßwerten. Inst. für Anatomie der MLU Halle-Wittenberg.

WORKSHOP: COUPLED OSCILLATORS

COUPLED OSCILLATORS IN NEUROSPORA DIFFERENTIATION AND MORPHOGENESIS

L. RENSING

Department of Biology, Institute for Cell Biology, Biochemistry and Biotechnology, University of Bremen, D-W-2800, FRG

INTRODUCTION

Different states of self-organized macroscopic order can develop in nonlinear dynamic systems far from thermodynamic equilibrium, for example in the form of standing or travelling waves, temporal oscillations or steady spatial gradients (11, 18, 26). Periodic patterns represent a large class among these self-organized structures. They are often based on coupled microscopic oscillators. While the interactions of microscopic elements depend on their own properties and internal and external parameters, they do not per se require genetic information. Organisms, however, have "exploited" the self-organizing properties of oscillatory elements in order to achieve certain goals: a mutual coupling of oscillators serves to establish synchronous clock systems in cells and tissues, which is necessary for physiological or morphogenetic functions. The particular design of self-organized structures in living organisms does, in fact, require genetic information, as does the establishment of microscopic oscillators consisting of cellular molecules and structures. Self-organized structures in living organisms thus depend on a) genetic information b) external parameters and c) the self-organizing properties of the interacting elements.

I would like to focus here on the morphogenesis of a fungus - *Neurospora crassa* - because it is less complex compared to most multicellular organisms. *Neurospora* shows, under certain environmental and genetic conditions, periodic differentiation of aerial hyphae and conidia (vegetative spores). In a two dimensional planar mycelial mat this results in a synchronous periodic spatial pattern, such as concentric rings (Fig. 1) of differentiated and non-differentiated hyphae (31).

This pattern is apparently based on a coupling of circadian oscillations between the individual hyphae. Coupling can be enhanced by a diurnal oscillation of light and temperature which forces (entrains) the endogenous oscillators and, in addition, elicits differentiation by itself. The role of yet another periodic process, the cell cycle, in this morphogenetic development is little understood. It is conceivable, in analogy to many known examples that the cell cycle of Neurospora is also coupled to the circadian (and perhaps to the diurnal) oscillations and that cell cycle-dependent processes such as DNA-synthesis, mitosis or the switch to differentiation occur in synchrony.

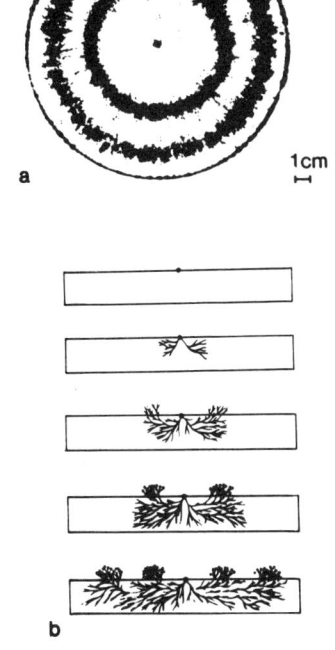

Fig.1:
Morphogenesis of the conidial pattern in mycelium of Neurospora crassa (bd mutant).
a)
A Petri dish in the middle of which a mycelial disc was inoculated. The radially growing hyphae differentiate aerial hyphae and conidia every 21.5 h following an endogenous circadian oscillation (in constant darkness)
b)
Schematic section through a Petri dish with growth medium and agar. It shows the vegatative hyphae growing more or less horizontally and the aerial hyphae growing vertically and above the agar surface. Conidia are formed within the aerial hyphae (4).

As noted by WINFREE (42), one can distinguish coupling between similar and dissimilar oscillators. In the morphogenesis of Neurospora we find interactions of similar (or identical) circadian oscillators within a single hypha and between different hyphae and a - unidirectional - coupling between dissimilar oscillators, between geophysical oscillations and the circadian oscillators and probably between the circadian and the cell cycle oscillator (Fig. 2).

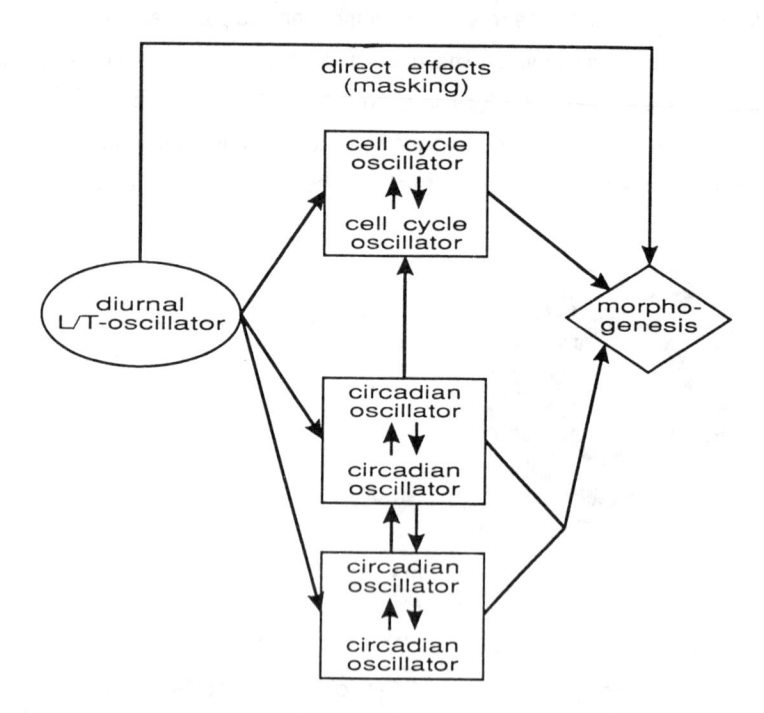

Fig.2: A diagram of the putative coupling of similar and dissimilar oscillators involved in the morphogenesis. The two boxes with circadian oscillators represent two hyphae. Both contain many oscillators which are mutually coupled. They also interact by interhyphal signalling. The box with the cell cycle oscillators represent a hypha; however, it is unclear whether or not the cell cycle oscillators are coupled in *Neurospora*. Unidirectional coupling occurs between the diurnal light and temperature cycle and the circadian oscillator. The effect on the cell cycle is not documented, neither is the coupling between circadian oscillator and cell cycle oscillator.

The morphogenesis of *Neurospora* may result in different structures, such as rings, double rings, radial spikes or constant distribution of conidia, depending on the coupling strength between the various oscillators as well as depending on environmental and genetic factors. *Neurospora* may thus serve as a moderately complex model to analyse the roles of coupled oscillators in morphogenesis. The general topic of oscillations and morphogenesis has been addressed by several workers in the field in a volume edited by RENSING (32). WINFREE (42) presented an extensive overview of the literature on coupled oscillators in general and particularly analysed the theoretical and topological aspects of this problem.

COUPLING BETWEEN SIMILAR OSCILLATORS

Mutual coupling between similar or identical oscillators is a common phenomenon, which may lead to complex dynamic patterns, depending on the coupling strength. Usually, when the oscillating "units" are nearly identical and the coupling is moderately strong, they tend to oscillate in synchrony (25). Very strong coupling may destroy the oscillatory behavior altogether or cause a complex pattern of changes. Weak coupling may result in the independent oscillation of the various oscillatory units with only transient locking between them.

I would first like to present an example of coupling between similar oscillators in an inorganic chemical system, because the parameters of such a system are more easily defined and controlled compared to a living cell.

The heterogeneously catalyzed oxidation of ethanol is a chemical system which oscillates under certain conditions (13). Oscillations appear when the oxidation takes place in a continuous flow calorimeter on an amorphous aluminium-supported Palladium catalyst. When the system is first driven to ignition and the temperature or flow rate is then decreased one finds a continuous oscillation of the products and the temperature. The catalyst consists of numerous units which are strongly coupled by a silver support with high thermal conductivity. A decoupling of the three oscillators was observed at certain temperatures, when the catalyst was mounted on three silver plates which were separated by a Teflon layer with poor thermal conductivity. At an intermediate temperature, a coupling with a doubling of the period length occurred - perhaps due to weaker coupling factors like substrate or product concentrations. Such oscillators can be entrained by an oscillating input (substrate) of the reaction (12).

This rather well defined chemical system already shows a variety of dynamic patterns depending on internal coupling strength by thermal conductivity and external (temperature, flow rate) conditions. It is suggestive to assume an even greater variety of patterns in biological systems.

CIRCADIAN OSCILLATORS

Circadian oscillators reside within a cell and are not (primarily) the result of interactions between cells, as may be seen from the fact that unicellular organisms show circadian rhythms and from dissection experiments with cell populations of pacemaker regions (1). The cellular mechanism of the circadian oscillator is not yet known in any detail. There is indirect evidence that proteins (or perhaps only one protein), membrane-
controlled processes such as Ca^{2+} transport, and protein phosphorylation are involved in this mechanism (reviews: 8, 28, 35). There is no convincing evidence that the circadian oscillator is built up by quantal subcycles as proposed also for the cell cycle by KLEVECZ et al. (15).

Not every part of the cell is required for the function of the circadian oscillator ("clock"), as has been found after cutting the giant algae *Acetabularia* into pieces (34). In this species the nucleus is apparently not necessary for maintaining a rhythm over a long period of time: even rather small parts of the cytoplasmic stalk of the cell were able to oscillate. This observation indicated the existence of many oscillatory "units" in one cell, which are normally strongly coupled, as in the case of the cell cycle oscillator in a plasmodium. Such a completely coupled population of oscillators might be indistinguishable from an oscillating continuum, e.g. a membrane (25).

Recently we observed, however, that under certain conditions different free-running period lengths (s) of two different circadian clock- controlled processes in *Gonyaulax polyedra* may appear: flash activity with a s of 24.2 h and a glow activity with a s of 22.9 h (40). One possible interpretation of this observation is that the various oscillatory units of the circadian clock - perhaps in different compartments of the cell - are uncoupled. This uncoupling may occur when the compartments differ in their respective environment (e.g., in pH or $[Ca^{2+}]$) for the circadian oscillators.

In *Neurospora*, the long, septated hyphae, which are not completely divided into multinuclear compartments, represent a syncytium with, probably, numerous coupled circadian oscillators. Earlier experiments designed to analyze possible differences between circadian rhythms at the growth front of the mycelium compared to the older parts in the center revealed a phase gradient with the older parts lagging behing the growth front (6). However, the oscillatory differences between young and old parts of a single clone of hyphae and its branches are difficult to assay accurately.

Fig.3: Mixing of differently phased mycelia which were first separated by a glass barrier. Left mycelium started at circadian time (ct) 8, right mycelium at ct 16. After mixing, a zone of intermediate phases can be observed (4).

Another problem not yet resolved is that of coupling between different "clones" of hyphae. Attempts to analyse this question are mainly based on experiments in which two differently phased populations of hyphae merge, for example, in a Y-shaped tube (10). We used a similar approach when we separated two differently phased mycelia by a glass wall for some distance in a Petri dish and then observed the interaction in a further stretch without such barrier (Fig.3). Similar experiments were performed by WINFREE & TWADDLE (43) and more recently by NAKASHIMA & HASTINGS (24).

The results of these experiments were not clear and were difficult to interpret. Several cases of intermediate phases within the contact area between differently phased mycelia were observed, like in the experiment presented here. In the experiments with merging populations in a Y-chaped tube, a predominance of one phase over the other was observed. However, this predominance is probably due to the fact that one population which arrived at the merging point slightly earlier inhibits the differentiation (and growth?) of the delayed one (DEUTSCH & RENSING, unpubl.). We did find, on the other hand, further indications of interactions between hyphae like, for example, a synchronous banding pattern in mycelia kept in complete darkness or development of concentric rings from a starting point which contained four differently phased pieces of mycelia (DEUTSCH & RENSING, unpublished).

NAKASHIMA & HASTINGS (24) also observed phase interactions between differently phased mycelia. They fused these mycelia and generated heterocaryotic cultures with the nuclei and cytoplasm of the different strains. The resulting phase was often an average of the parent phases, but sometimes corresponded to one of the initial parental phases. NAKASHIMA (pers. comm.) proposed, furthermore, a "phasing substance", which he isolated from mycelia at a defined circadian phase. This substance was able to influence the phase of another mycelium when added during another phase (much in the same way as was done in the cell cycle experiments with *Physarum*, see below). Again, because the assay of the resulting phase shifts is rather difficult, the problem of how a population of hyphal (or cellular) oscillators interact - either by direct interaction (anastomoses) or by interhyphal signal exchange - is not unambiguously settled.

The *Neurospora* problem is, of course, part of the general problem of how populations of circadian clocks interact, for example, in pacemaker regions like the SCN of vertebrates, the pineal of birds. the optic lobe centers of insects, the eyes of molluscs, etc., but also in tissue cells like the epidermis of insects or parenchymal cells of plants. Again the question of the coupling mechanism, whether it consists of a direct interaction by way of gap junctions or an indirect one by intercellular diffusible substances is not yet resolved.

CELL CYCLE OSCILLATORS

Recent evidence has shown that the main cell cycle oscillator consists of a periodic build-up and break-down of a protein (cyclin), which associates with an enzyme (a protein kinase, also known as the cdc 2 product). This complex, also called "maturation promoting factor" (MPF), controls several processes along the cell cycle, such as DNA synthesis and mitosis and also initiates its own destruction during mitosis by activating a protease that degrades cyclin (23). Even though many details of the oscillating mechanism are not yet known, it appears that the cell cycle oscillation is principally independent of DNA-replication and mitosis, which are normally associated with the cell cycle.

Normally, cell cycle oscillators of individual cells in culture do not couple to each other, except under conditions, like defined temperature cycles, recovery from cell cycle inhibitors or mechanical selection, which externally force such a synchrony (see below). Coupling is observed, on the other hand, in the first cell divisions following fertilization, and, in particular, in syncytia and plasmodia which allow free diffusion of substances relevant to the control of the cell cycle oscillator or elements of the oscillator mechanism itself.

A *Neurospora* hypha represents a syncytium consisting of compartments (segments) which are not completely separated and which contain numerous nuclei. However, when conidia were initiated to germinate synchronously there was no apparent synchrony in nuclear division and DNA synthesis in the growing population of cells. This was interpreted to indicate a different point of cell cycle arrest within the conidia (35). The cell cycle duration is about 1-4 hours, depending on the nutritional conditions (20). Lower temperatures will, of course, prolong this duration as in almost all cases mainly by extending the G1-phase. Again, nuclei of individual hyphae did not appear to traverse the cell cycle phases in synchrony. Thus, in *Neurospora* there is at present no good evidence for a strong coupling of cell cycle oscillators, at least not under the conditions tested.

A much better example for coupling between cell cycle oscillators is the plasmodium of the slime mold *Physarum polycephalum*. It contains millions of nuclei and as many cell cycle oscillators, which seem to be strongly coupled. The nuclei synchronously undergo mitosis with only 4 min deviation at a period length of the whole cell cycle of about 8h.

In order to elucidate the mechanism of cell cycle coupling, crucial experiments were done by LOIDL & SACHSENMAIER (19). They fused plasmodia of different cell cycle phases. Portions of macroplasmodia of different cell cycle phases were fused and the subsequent synchronous mitosis was compared to the nonfused portions of the same plasmodia, which served as controls. Normally, "older" nuclei were delayed by fusion with "younger" plasmodia and younger nuclei were advanced by fusion with older plasmodia. The fused plasmodia, thus, underwent mitosis at a time intermediate between the controls. This was not the case when fusion was induced 30 min before prophase or later, a time when a putative control point at the G2/M border had been crossed.

In order to identify substances involved in this coupling process we (11a) performed the following experiments: Plasmodia of the myxomycete Physarum polycephalum (strain C1) were collected at different times during the cell cycle and extracts were prepared from homogenates using a buffer optimized for microinjection into plasmodial veins. These extracts were injected into plasmodia during the first 3 h of the cell cycle. The time of the following mitosis was monitored and compared with that of the buffer-injected controls. Extracts of plasmodia homogenized 45 min before late telophase accelerated the onset of mitosis in the injected "young" plasmodium up to 70 min, i.e., an advance of 10-14% compared to the 8-10 h cell cycle duration of the controls (Fig. 4). The accelerating activity vanished completely after heating, freezing, or protease digestion, thus indicating the peptide nature of the active agent. Purification of the active compound by means of gel filtration revealed a molecular mass of about 2500 Da (11a).

These results may be interpreted by assuming a mitogenic substance which is synthesized during the whole cycle but bound to (nuclear?) receptors until, at the end of the cycle, the receptors are saturated and the mitogen freely diffuses in the plasmodium. The 2.5-kDa peptide may also represent a protease inhibitor, which prevents the cyclin protease from splitting cyclin during interphase. Shortly before mitosis, higher calcium concentrations, or other events in the cell cycle, may lead to a dissociation between inhibitor and protease. Both assumptions would be compatible with our observation that the injection of protease inhibitors like pepstatin A, leupeptin, and aprotinin advance the cell cycle in *Physarum*, whereas the injection of pronase E delays the cycle.

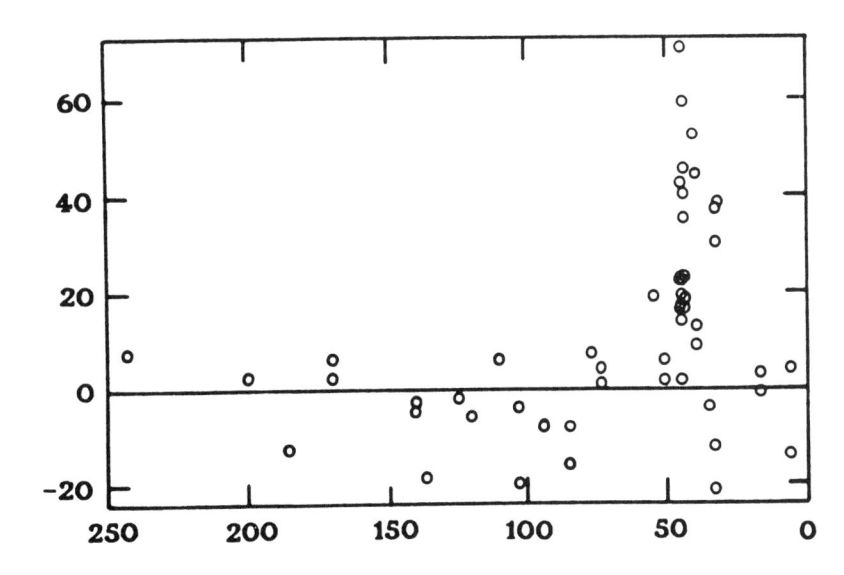

Fig.4: Cellular extracts from different cell cycle phases and their advancing/delaying effect on the third mitosis of the injected plasmodium. Abscissa: Phase of donor plasmodium at the time of harvesting (min before third mitosis). Ordinate: Resulting delay/advance of mitosis in injected plasmodia (min). All injections were administered during the first 3 h after the second synchronous mitosis after the fusion of microplasmodia (11a).

COUPLING BETWEEN DISSIMILAR OSCILLATORS
CIRCADIAN OSCILLATOR - CELL CYCLE OSCILLATOR

Much evidence has been accumulated about the unidirectional coupling between the circadian and cell cycle oscillator (reviewed by 27, 7, 8). Events or processes of the cell cycle such as mitosis and DNA-synthesis are locked to a defined phase of the

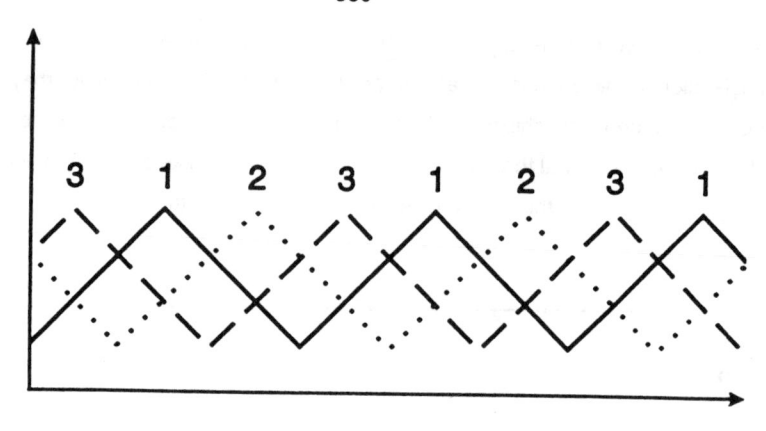

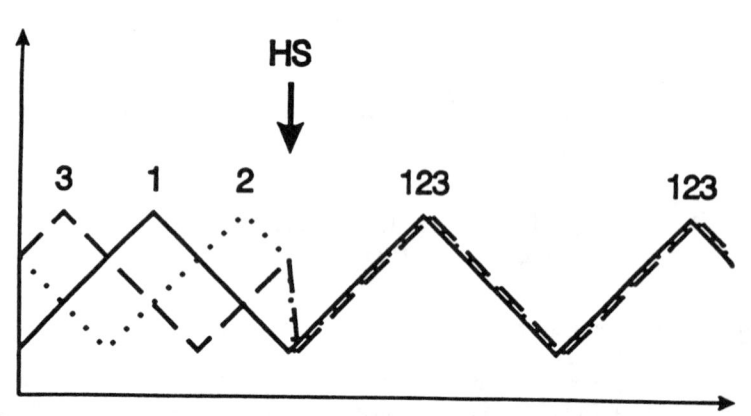

Fig.5: Synchronization of a population of differently phased oscillators (above) by a single heat shock (HS) pulse (below). If the oscillations 1,2,3 represent cell cycle oscillators of individual cells consisting of oscillating protein concentrations, a single heat shock is assumed to lower the concentrations of cells 2 and 3 to the minimum level, whereas cell 3 (at the minimum) is not affected in its phase.

circadian oscillator. If coupling is strong, the event or process takes place only at this phase ("gating"), if the coupling is weaker the frequency of the event or process at any time varies as a function of the circadian phase. There is some evidence that this coupling also occurs in *Neurospora* (21) even though the period length of the cell cycle is shorter (s = 2-6 h) compared to the period length of the circadian oscillator (s (20°C) = 21.5 h). This may lead to a coupling of every tenth or fourth cell cycle to a certain circadian phase. Whether or not the cell cycle duration is kept constant in the course of a circadian cycle is not known.

In most other cells and tissues the reverse is true: the cell cycle needs to be longer than the circadian period length. In general, a shorter cell cycle does not allow the expression of the circadian oscillation as was first observed by EHRET et al. (9). This seems to be true, however, only in those cases in which the cell cycle results in a physical separation of daughter cells (cytokinesis). This is not the case in *Neurospora*.

The signals by which circadian oscillators entrain the cell cycle are unknown. One can only speculate that Ca^{2+} and corresponding kinase activities may be relevant to both oscillators.

(DIURNAL) TEMPERATURE CYCLES - CELL CYCLE OSCILLATOR

In general, all chemical and physical factors which affect an oscillator can be used to "force" this oscillator into a different frequency within a certain range: an oscillating pendulum for example can be forced to adopt another frequency by periodic mechanical striking the pendulum at defined phases. A single mechanical strike will cause a phase-dependent phase shift of the pendulum oscillator, a dependency usually depicted as "phase response curve" (PRC). Because of this phase dependency of the resetting action of a perturbation, a single mechanical strike which hits a series of randomly oscillating pendulum oscillators will cause a synchronization of this population (Fig.5). A complicating problem here is the heterogeneity of period lengths in such a population of cell cycle oscillators in individual cells.

In the case of the cell cycle oscillator, a pulse of elevated temperature (heat shock) induces such a phase dependent phase shift: in *Physarum* a delay of mitosis during the G2-phase (2). In synchronized mammalian V79 cell KLEVECZ et al. (15) reported more complicated PRCs, which they attributed to quantal subcycles. - The

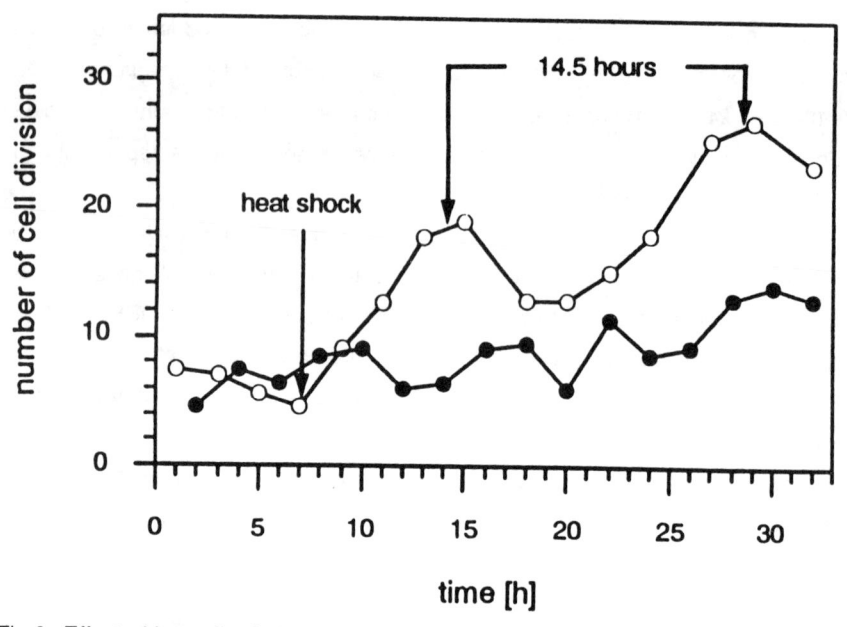

Fig.6: Effect of heat shock treatment on NIE 116 cell growth rate of cytokinesis in a control culture and after a 20 min heat shock (44°C) (open circles) (36).

synchronizing effect of temperature pulses was used extensively, for example, in *Tetrahymena* to achieve synchronous populations (44).

We used mouse *Neuroblastoma* cells (NIE -115) to test the effect of heat shock (20-40 min, 44°C) on proliferation (36). As shown in Fig. 6 a single pulse induces a synchronization of cell division with maxima of mitoses 14.5 h apart. The untreated population of cells, in contrast, showed an almost equal distribution of cell divisions over time. We also measured the variance of the cell cycle times which is typically rather large even among sister cells (3). A temperature pulse also diminished this variation.

In *Neurospora*, I am not aware of any data concerning the effects of elevated temperature on the cell cycle. However, there is good reason to believe that *Neurospora* reacts in the same way as the other species analyzed.

The effects of elevated temperature (heat shock) on the cell cycle may be mediated by a transient increase in the intracellular concentration of Ca^{2+} (36) or of H^+

(41), by the inhibition of "normal" protein synthesis or the induction of heat shock proteins (HSPs). One of the HSPs (HSP 70, for example) is expressed at a high rate at the beginning of the S-phase (review: 22). - It is interesting to note that a pulse of elevated temperature can elicit not only a transient stimulation of cell proliferation but also a later induction of differentiation as observed in *Neuroblastoma* cells (36) and in *Neurospora* (31). This is in contrast to the effect of a constantly elevated temperature which rather inhibits differentiation.

The triggering of differentiation of morphogenesis by a pulsed change, e.g., in the concentration of a stimulant, rather than by a slow change is of general relevance to the discussion of the role of oscillations in morphogenesis (32, 38), because oscillations represent (equally spaced) pulses which are perceived as a different type of signal compared to long term changes (39).

(DIURNAL) TEMPERATURE AND LIGHT CYCLES - CIRCADIAN OSCILLATOR

The unidirectional coupling between diurnal temperature and light changes and the circadian oscillator is so well documented and recently reviewed (42, 29, 8) that I will not discuss this topic here. The mechanism by which this coupling is achieved is, however, still obscure. Again, changes in intracellular Ca^{2+}, protein phosphorylation or in the synthesis rate of specific proteins (light- and/or temperature induced) may be candidates for mediators of the light and temperature forcing. We are particularly interested in the analysis of these mechansism in *Neurospora* (37, 17).

DIRECT EFFECTS OF LIGHT AND TEMPERATURE CHANGES
ON MORPHOGENESIS

In many plant and fungal organisms differentiation and morphogenesis are triggered by light or temperature changes (16). With respect to the circadian oscillator which also triggers these processes endogenously, these direct effects of light and temperature have been termed "masking effects" (see 30). An evolutionary progress may be seen in the development of an endogenous clock program replacing (partly) the exogenous control of differentiation by light and temperature. Such a clock (which needs of course to be tuned to the geophysical periodicity) makes the organism independent of the rather blurred external signals.

TEMPORAL ORDER AND MORPHOGENESIS

Periodic spatial patterns are ubiquitous in nonliving and living dynamic systems far from equilibrium. Their development is based on the same principle: temporal or spatial interaction (inhibition - activation) between temporally or spatially adjacent elements. The periodic pattern can thus develop sequentially - as in the case of *Neurospora*'s rings of conidia - or simultaneously - as in the case of *Drosophila* segments (32). The sequential generation of rings or spirals or linear periodic patterns is a universal order principle, which is often observed in nonliving and in living organisms, for example in *Neurospora* and many other fungi, in the leaf arrangement of plants, in the annual rings of trees, in shells of molluscs, segments of annelids etc. It is based on coupled oscillators of some sort controlling differentiation and morphogenesis or on the imprinting of exogenous periodicities on these processes.

The alternation of vegetative growth of hyphae (including cell division) and differentiation of aerial hyphae and conidia (including cell division and morphogenesis) in *Neurospora* is based on an endogenous circadian oscillation and cell cycles, giving rise to unequal hyphal branches at certain phases. The macroscopic symmetry of - for example - concentric rings of spores seem to be based on coupled oscillators within a mycelium. Both, the vegetative state and the differentiation state probably represent stable networks of interactions between a set of gene products (dynamic attractors, see 14). These attractors, however, may be close to less stable (chaotic) states, such that a single specific perturbation can move hyphae from the vegetative to the differentiated state.

The transition between the vegetative state and the differentiation state can be elicited by an internal (periodic) signal or by a signal (light temperature) from outside. The coupling of the circadian oscillators garantees a synchronous transition and thereby a symmetric form of the whole mycelium.

Whether or not the ultimate differentiated state (conidia) includes a transition to a non circadian state is not clear, it is definitely a non-cycling state concerning the cell cycle. The transitions and pattern are influenced by genes and their mutations, i.e., some mutants are not sensitive to light, some show longer period lengths (frq-mutants). The wild type of *Neurospora* is, furthermore, dependent on the CO_2-concentration for its differentiation. Pattern formation is also sensitive to external conditions, like the viscosity of the agar or the concentration of Ca^{2+} in the medium (4). Thus the expression of a certain periodic phenotype in the mycelium seems to based on a complex chanelling of transitions between different attractors. - We have analyzed this pattern formation by using a cellular automaton approach (4) which derives its rules mainly from experimental results. It is a similar approach to that of KAUFFMANN (14) using a Boolian network.

REFERENCES

1. BLOCK, G.D. & D.G. MCMAHON (1984): Cellular analysis of the Bulla ocular circadian pacemaker system. III. Localization of the circadian pacemaker. J. Comp. Physiol. A 153, 387-395
2. BREWER, E.N. & H.N. RUSCH (1968): Effect of elevated temperatur shocks in mitosis and on the initiation of DNA replication in *Physarum polycelphalum*. Exp. Cell Res. 49, 79
3. BROOKS, R.F., D.C. BEMNETT & J.A. SMITH (1980): Mammalian cell cycles need two random transitions. Cell 19, 493
4. DEUTSCH, A. (1991): Musterbildung bei dem Schlauchpilz *Neurospora crassa*: Mathematische Modellierung und experimentelle Analyse. Dissertation, Universität Bremen
5. DEUTSCH, A., L. RENSING & A. DRESS (1991): Patterns of spore formation in *Neurospora crassa* and their simulation with a cellular automaton. In "Nonlinear Wave Processes in Excitable Media." A.V. HOLDEN et al. (eds.). Plenum Press, New York, pp. 259-267
6. DHARMANANDA, S. & J. FELDMAN (1979): Spatial distribution of circadian rhythm in aging cultures of *Neurospora crassa*. Plant Physiol. 63, 1049
7. EDMUNDS, L.N. Jr. (1984): Cell Cycle Clocks. Marcel Dekker, New York
8. EDMUNGS, L.N. Jr. (1988): Cellular and Molecular Bases of Biological Clocks. Springer-Verlag, Heidelberg
9. EHRET, C.F., J.C. MEINERT, K.R. GROH & G.A. ANTIPA (1977): Circadian regulation: Growth kinetics of the infradian cell. In: B. DREWINKO & R.M. HUMPHREY (eds.), "Growth Kinetics and Biochemical Regulation of Normal and Malignant Cells" , Williams and Wilkins, Baltimore, pp 49-76
10. FELDMAN, J. & J. DUNLAP (1983): *Neurospora crassa*: A unique system for studying circadian rhythms. Photochem. Photobiol. Rev. 7, 319-368
11. HAKEN, J. (1983): Synergetics. An Introduction. Nonequilibrium Phase Transitions and Self-Organization in Physics, Chemistry and Biology. Springer, Heidelberg
11a. HOBOHM, A., G. HILDEBRANDT, & L. RENSING (1990): A purified cellular extract accelerate the cell cycle in *Physarum polycelphalum*. Exp. Cell Res. 191, 332-336
12. JAEGER, N.I., M. LIAUW, P.J. PLATH & P. SVENSON (1990): Communication and synchronization in heterogeneous catalytic reactions . In: Y.S. MATROS (ed.), Unsteady State Processes in Catalysis. VSP, Utrecht, pp. 343-350
13. JAEGER, N.I., R. OTTENSMEYER & P.J. PLATH (1986): Oscillations and coupling phenomena between different areas of the catalyst during the heterogeneous catalytic oxidation of ethanol. Ber. Bunsenges. Phys. Chem. 90, 1075-1079
14. KAUFFMANN, S. (1991): Leben am Rande des Chaos. Spektrum der Wissenschaft, 11/91, 90-99
15. KLEVECZ, R.R., R.A. KAUFFMAN, & R.M. SHYMKO (1984): Cellular clocks and oscillators. Int. Rev. Cytol. 86, 97-128

16. KLOPPSTECK, K. B. OTTO & W. SIERRALTA (1991): Cyclic temperature treatments of dark-grown pea seedlings induce a rise in specific transcript levels of light-regulated genes related to photomorphogenesis. Mol. Gen. Genet. 225, 468-473
17. KOHLER, W. & L. RENSING (1992): Light and recovery from heat shock induce the synthesis of 38 kDa mitochondrial proteins in *Neurospora crassa*. Arch. Microbiol.(in press)
18. KRINSKY, V.I. (ed.) (1984): Self-Organization. Autowaves and Structures Far from Equilibrium. Springer Verlag, Heidelberg
19. LOIDL, P. & W. SACHSENMAIER (1982): Control of mitotic synchrony in *Physarum polycephalum*. Phase shifting by fusion of heterophasic plasmodia contradicts a limit cycle oscillator. Eur. J. Cell Biol. 28, 175
20. MARTEGANI, E., M. LEVI, F. TREZZI & L. ALBERGHINA (1980): Nuclear division cycle in *Neurospora crassa* hyphae under different growth conditions. J. Bacteriol. 142, 268-275
21. MARTENS, C. I. & M.L. SARGENT (1973): Circadian rhythms of nucleic acid metabolism in *Neurospora crassa*. J. Bacteriol. 177, 1210-1215
22. MORIMOTO, R.I., A. TISSIERES & C. GEORGOPOULOS (1990): Stress Proteins in Biology and Medicine. Cold Spring Harbor Laboratory Press
23. MURRAY, A.W. & M. W. KIRSCHNER (1989): Dominoes and clocks: The union of two views of the cell cycle. Science 24, 614-621
24. NAKASHIMA, H. & J.W. HASTINGS (1989): Phase determination of the circadian rhythm of conidiation in heterocaryons between out-of-phase mycelia in *Neurospora crassa*. J. Biol. Rhythms 4, 377-387
25. PAVLIDIS, T. (1973): Biological Oscillators: Their Mathematical Analysis. Academic Press, New York
26. RENSING, L. & JAEGER, N. (eds.) (1985): Temporal Order. Springer Verlag, Heidelberg
27. RENSING, L. & K. GOEDEKE (1976): Circadian rhythm and cell cycle: Possible entraining mechanisms. Chronobiologia 3, 53-65
28. RENSING, L. & R. HARDELAND (1990): The cellular mechanism of circadian rhythms - a view on evidence, hypotheses and problems. Chronobiol. Int. 7, 353-370
29. RENSING, L. & W. SCHILL (1987): Perturbations of cellular circadian rhythms by light and temperature. In: L. RENSING, U. AN DER HEIDEN, & M.C. MACKEY (eds.), Temporal Disorder in Human Oscillatory Systems. Springer Verlag, Heidelberg, pp. 233-245
30. RENSING, L. (1989): Is "masking" an appropriate term? Chronobiol. Intern. 6, 297-300
31. RENSING, L. (1992): Morphogenesis of periodic conidiation pattern in *Neurospora crassa*. In: L. RENSING (ed.), Oscillations and Morphogenesis. Marcel Dekker, New York
32. RENSING, L. (ed.) (1992): Oscillations and Morphogenesis. Marcel Dekker, New York
33. ROSBASH, M. & J. HALL (1989): The molecular biology of circadian rhythms. Neuron. 3, 387-398

34. SCHWEIGER, E., H.G. WALLRAFF & H.G. SCHWEIGER (1964): Über tagesperiodische Schwankungen der Sauerstoffbilanz kernhaltiger und kernloser *Acetabularia mediterreanea*. Z. Naturforsch. 19c, 499-505
35. SERNA, L. & D. STADLER (1978): Nuclear division cycle in germinating conidia of *Neurospora crassa*. J. Bacteriol. 136, 341-351
36. STOKLOSINSKI, A., H. KRUSE, Ch. RICHTER-LANDSBERG & L. RENSING (1992): Effect of heat shock on neuroblastoma (NIE 115) cell proliferation and differentiation. Exp. Cell. Res. (in press)
37. TECHEL, D., G. GEBAUER, W. KOHLER, T. BRAUMANN, B. JASTORFF & L. RENSING (1990): On the role of Ca^{2+}-calmodulin-dependent and cAMP-dependent protein phosphorylation in the circadian rhythm of *Neurospora crassa*. J. Comp. Physol B 159, 695-706
38. VICKER, M. & L. RENSING (1987): Oscillations and the regulation of spatial order in developing systems. In: L. RENSING (ed.), Temporal Disorder in Human Oscillatory Systems. Springer Verlag, Heidelberg
39. VICKER, M. (1992): Cells into slugs: Oscillatory temporal signals, taxis and morphogenesis in *Dictyostelium discoideum*. In: L. RENSING (ed.), Oscillations and Morphogenesis. Marcel Dekker, New York
40. VON DER HEYDE, F., A. WILKENS & L. RENSING (1992): The effects of temperature on the circadian rhythms of flashing and glow in *Gonyaulax polyedra*: Are the two rhythms controlled by two oscillators? J. Biol. Rhythms, in press
41. WEITZEL, G., U. PILATUS & L. RENSING (1987): Cytoplasmic pH, ATP content and total protein synthesis rate during heat shock protein inducing treatments in yeast. Exp. Cell Res. 170, 64-79
42. WINFREE, A. (1980): The Geometry of Biological Time, Springer Verlag, Heidelberg
43. WINFREE, A.T. & G.M. TWADDLE (1981): The *Neurospora* mycelium as a two-dimensional continuum of coupled circadian clocks. In: T.A. BURTON (ed.), Mathematical Biology. Pergamon Press, New York
44. ZEUTHEN, E. (1974): A cellular model for repetitive and free-running synchrony in Tetrahymena and Schizosaccharomyces. In: G.M. PADILLA, I.L. CAMERON & A.M. ZIMMERMANN (eds.). Cell Cycle Controls. Academic Press, New York, pp. 1-30

MODELLING OF CELL CYCLE HETEROGENEITY AND SYNCHRONIZATION WITH CELLULAR AUTOMATA

M.MARKUS & A.SALVADOR

[1] Max-Planck-Institut für Ernährungsphysiologie,
Rheinlanddamm 201, W-4600 Dortmund 1, FRG
[2] Instituto de Investigação Científica Bento da Rocha Cabral,
Cç. Bento da Rocha Cabral, 14, 1200 Lisboa, Portugal

INTRODUCTION

In this contribution we present calculations of the dynamic behaviour of cell aggregates considered as systems of coupled nonlinear oscillators. We will concentrate on two types of collective behaviour:

i) Synchronization of biological cycles, as observed for example in: suspensions of yeast cells (1), neurons during an epileptic seizure (10) and pacemaker cells scattered through the myocard (11). A model of synchronized biochemical oscillations has been proposed as a basis of circadian rythms (24, 25).

ii) Heterogeneity (variability) of periods of individual oscillators in a population.

In particular, we will focus our studies on the direct (cell to cell) coupling of the oscillators that control the cell cycle. Spontaneous cell cycle synchronization has been observed in cultures of *Gonyaulax* (8). Cell cycle synchronization *in vivo* is expected to form the basis of an efficient treatment of malignant tumors (22). Heterogeneity in cell cycle periods is illustrated here in Fig. 1a (mono- and bimodal distribution) and Fig. 1b (polymodal distribution).

It is assumed that the cell cycle is controlled by some endogeneous, biochemical oscillator. Chernavskii et al. (4) (see also (27)) proposed a biochemical mechanism - based on a simple model of lipid peroxidation - for such an oscillator. This mechanism, which is known as the "membrane model", will be assumed in this work.

One way of implementing this model for simulating aggregates of N cells is by integration of $2N$ differential equations (two for each cell), as in the work by Volkov et al. (32, 33). However, this method is enormously time consuming for large N. As a faster alternative, we implement here (see also (21)) the membrane model by using a cellular automaton (CA). The advantages of CAs (good approximations at low computational costs) have been demonstrated by calculations of turbulence with the so-called lattice gas automata (see *e.g.* (28)), of Turing patterns (20, 29) and of waves in excitable media (12, 13, 15, 17).

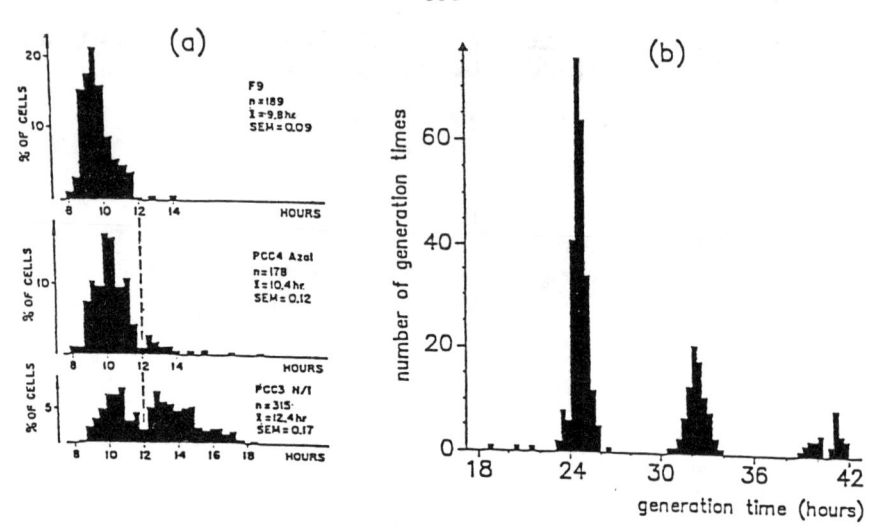

Fig. 1: Measured distributions of cell cycle periods; (a) from (30); (b) adapted from (26).

MODEL AND NUMERICAL METHODS

The differential equations (DE) of the membrane model (4) have the following form (using dimensionless variables) in the version of Volkov and Stolyarov (32):

$$\frac{dL}{dt} = \eta - \kappa - 1.5LR - DL - 0.5RA \tag{1}$$

$$\varepsilon_r \frac{dR}{dt} = \kappa + 0.5LR - 1.5RA - R^2 \tag{2}$$

$$\varepsilon_a \frac{dA}{dt} = \gamma - RA - \delta A \tag{3}$$

Here, L is the fraction of easily-oxidizable lipids, R the concentration of free peroxyl radicals and A the concentration of antioxidants. The parameters η, κ and γ describe the influx of lipids, radicals and antioxidants. The nonlinear terms proportional to RA, R^2 and LR describe the interaction of radicals with antioxidants, the radical recombination, and the lipid-radical oxidation. D and δ characterize the ouflux of lipids and antioxidants. The characteristic evolution time ε_r (resp. ε_a) of R (resp. A) are proportional to the ratio of the average lifetime of radical (resp. antioxidant) molecules and the lifetime of lipid molecules. As in (4) we eliminate A by Tikhonov's theorem, i.e. we

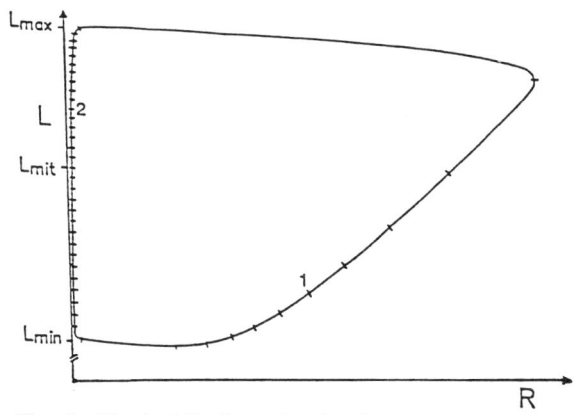

Fig. 2: Typical limit cycle obtained from Eqns (4) and (5) for a single cell. The marks are placed equidistantly in time. Cycling is clockwise.

set $dA/dt = 0$. This reduces the equations to Eq. 1 and

$$\varepsilon_r \frac{dR}{dt} = \kappa + 0.5LR - R^2 - 1.5\gamma \frac{R}{R+\delta} \quad . \quad (4)$$

We introduce intercellular coupling, as in (32, 33) by writing

$$\frac{dL_i}{dt} = \eta - \kappa - 1.5L_iR_i - DL_i - 0.5\gamma \frac{R}{R+\delta} + C\sum_k(L_k - L_i) \quad . \quad (5)$$

This equation corresponds to the ith cell of the aggregate. The index k refers to the neighbours of the ith cell. In this work we fix $\varepsilon_r = 0.01$, $\delta = 0.15$, $\kappa = 0.05$, $D = 0.2$ and $\gamma = 0.5$ (values used in (32)), and we vary the lipid input rate η and the coupling parameter C. Fig. 2 shows a typical limit cycle (LC) obtained by integrating Eqns (4) and (5) for a single cell ($C = 0$). Cycling is highly relaxative since $\varepsilon_r \ll 1$. The LC in Fig. 2 has a slow branch on its left part, where $R \approx$ constant and L increases from $L_{\min}$ to L_{mit} (S and G_2 phase) and from L_{mit} to $L_{\max}$ (G_1 phase). In addition, the LC has a fast branch, where R increases and decreases in a burst-like manner, L decreasing from $L_{\max}$ to $L_{\min}$ (S phase). The marks in the LC in Fig. 2 correspond to equal time intervals Δt. We call them Δt-marks. We choose Δt here as the time between the maximum of L and the maximum of R. We set Δt as our CA time step.

We restrict our calculations to such values of our free parameters η and C that their variation will not change the shape of the LC. We found that this

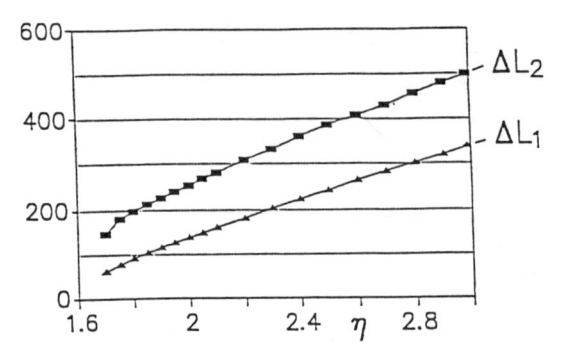

[Graph showing ΔL₂ and ΔL₁ vs η from 1.6 to 2.8, values from 0 to 600]

Fig. 3: Mean length of marks on the limit cycle (see Fig. 2) between L_{min} and L_{mit} (ΔL_2) and between L_{mit} and L_{max} (ΔL_1).

condition is fulfilled for $C \leq 0.68$ and $1.7 \leq \eta \leq 2.8$ (Note: there is no LC for η smaller than the bifurcation value $\eta = 1.7$). By integrating the DEs for two cells we found that in this parameter domain, C does not significantly change the Δt-marks. However, η does change the Δt-marks between L_{min} and L_{mit} in the LC (their length, which is approximately independent of L is given by ΔL_2 in Fig. 3). Between L_{mit} and L_{max} in the LC, η changes the Δt-marks differently (their length, again approximately independent of L, is given by ΔL_1 in Fig 3). η does not significantly change the other Δt-marks, nor L_{min} or L_{max}. We get $L_{min} = 3$ and $L_{max} = 5.6$. We set $L_{mit} = 4.5$. We do integer computation by multiplying all L-values with 5000, thus discretizing the variable L by the set of integers between $L_{min} = 3 \times 5000 = 15000$ and $L_{max} = 5.6 \times 5000 = 28000$ ($L_{mit} = 4.5 \times 5000 = 22500$). Changes of L in the descending part of L after the maximum of R are calculated approximating the Δt-marks by equal intervals defined by $\Delta L_S = (L_{max} - L_{min})/9$. In order to determine ΔL_1 and ΔL_2 for a given η, we linearly interpolated between the two nearest points in Fig. 3.

Fig. 4 shows examples of LCs, as we used in our CA for each cell, for two different values of η. The length of the Δt-marks on the vertical part of these LCs are given by ΔL_1 and ΔL_2 in Fig 3.

Noise is simulated in our CA by adding to η, at each time and in each cell i a quantity $f_i(t) \cdot \nu$, where $f_i(t)$ is a random number, uniformly distributed in the [-1,1] interval, and ν is a parameter. We symbolize $R \in [R_{low}, R_{high}]$ by $\rho = 1$ and $R < R_{low}$ by $\rho = 0$. Thus, the variable R is discretized to two states (0 and 1). We assume a constant number and no movements of cells. The following are the CA rules for the calculation of L and ρ in each cell at time $t + 1$ from their values at time t.

Calculate $\tilde{\eta} = \eta + f_i(t) \cdot \nu$ and compute the effect of the lipid exchange

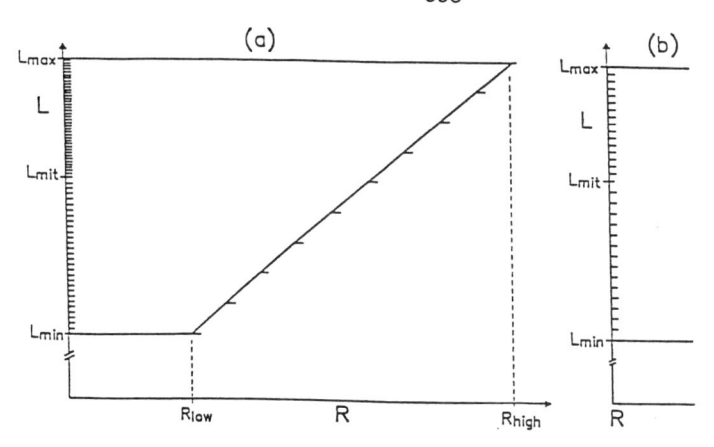

Fig. 4: Limit cycle as used in the automaton for $\eta = 2$ (a) and $\eta = 3$ (b). The marks on the cycles are assumed to be equidistant in time. Only the vertical part of the cycle is shown in (b) because the rest is as in (a).

by $L^*(t) = \left[L(t) + \tilde{C}\sum_k(L_k - L)\right]$, where L_k are the L-values of the neighbours. (For an aggregate of two cells we consider one neighbour, for square cell aggregates four neighbours i.e. a von Neumann neighbourhood). The brackets [ ] indicate rounding to the nearest integer. $\tilde{C} = C \cdot \Delta t$, where Δt is our CA time step (time between maximum of L and maximum of R for a single cell; see Fig. 2). After calculating $\tilde{\eta}$ and $L^*(t)$, proceed as follows:

if $\rho(t) = 0$ and $L^*(t) < L_{\text{mit}}$ then $L(t+1) = L^*(t) + \Delta L_2(\tilde{\eta})$, $\rho(t+1) = 0$;
if $\rho(t) = 0$ and $L_{\text{max}} > L^*(t) \geq L_{\text{mit}}$ then $L(t+1) = L^*(t) + \Delta L_1(\tilde{\eta})$, $\rho(t+1) = 0$;
if $\rho(t) = 0$ and $L^*(t) \geq L_{\text{max}}$ then $L(t+1) = L_{\text{max}}$, $\rho(t+1) = 1$;
if $\rho(t) = 1$ and $L^*(t) > L_{\text{min}}$ then $L(t+1) = L^*(t) - \Delta L_s$, $\rho(t+1) = 1$;
if $\rho(t) = 1$ and $L^*(t) \leq L_{\text{min}}$ then $L(t+1) = L_{\text{min}}$, $\rho(t+1) = 0$;

In an M x M square aggregate, we use periodic boudary conditions:
$X_{i,M+j} = X_{i,j}$, $X_{i,1-j} = X_{i,M+1-j}$, $X_{M+i,j} = X_{M+1-i,j}$, $X_{1-i,j} = X_{i,j}$
($X = \rho, L$ and $i, j = 1, 2, \ldots, M$).

RESULTS FOR TWO CELLS

For a better understanding of coupling of cell cycles, we first consider only two identical cells. Using the DE-method (four DEs), Volkov and Stolyarov (32) obtained two oscillatory modes. In one of these modes the two cells oscillate in phase, thus behaving as if they were uncoupled because $C(L_k - L) = 0$. In the other mode (anti-phase) the cells oscillate with a phase shift of half

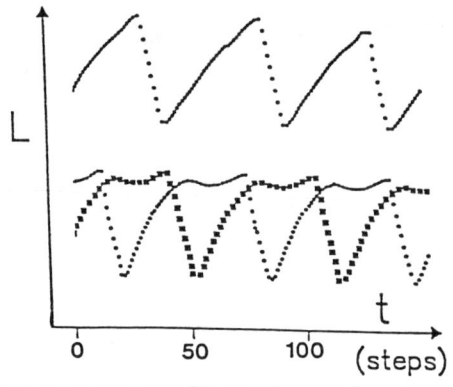

Fig. 5: In-phase ($C = 0.1$ upper) and anti-phase ($C = 0.65$ lower) oscillatory modes obtained with the automaton for two cells, $\eta = 2.7$.

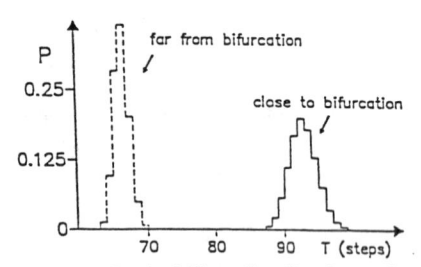

Fig. 6: Probability distribution of cell cycle periods as obtained with the automaton for uncoupled cells ($C = 0$) for η-values close ($\eta = 1.7$) and far ($\eta = 2.2$) from the bifurcation; $\nu = 0.2$.

a period. Both the in-phase and the anti-phase mode are readily obtained with our CA-method, as shown in Fig. 5. The longer period of the anti-phase oscillations is explained as follows. If, for example, cell 1 and 2 are at the points indicated by 1 and 2 in Fig. 2, the lipid exchange will cause a decrease of L in cell 2 and an increase in cell 1, thus retarding the cycling in both cells. The in-phase and the anti-phase oscillations were found to coexist in a large domain of the $\eta - C$ plane; in this domain the occurrence of in- or anti-phase depends only on the initial conditions. We found that for small C and large η (e.g. for $C < 0.2$ and $\eta > 2.3$) as well as for large C and small η (e.g. for $C > 0.57$ and $\eta < 1.9$) only the in-phase oscillations exist. This is also obtained from integration of the DEs (32). However, our CA calculation did not yield the so-called phase death (bifurcation into stable steady states) obtained from DEs at large C and large η ($\eta > 2.9$ or $C > 0.68$) because

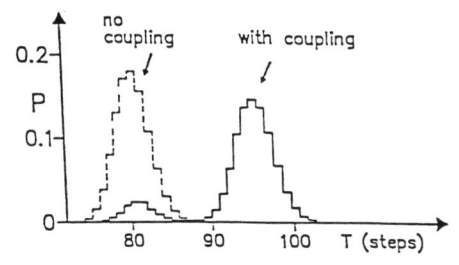

Fig. 7: Probability distribution of cell cycle periods as obtained with the automaton for uncoupled cells (left; $C = 0$) and for two coupled cells (right; $C = 0.35$); $\eta = 2.0$, $\nu = 0.3$

our approximations are not valid in this parameter domain. (Note: In a more general approach (21) it was shown that this automaton does render phase death if Δt-marks are set closer to each other in the descending part of R close to R_{low}).

If noise is added to uncoupled cells (C=0), their cycle period changes in time and distributions such as those in Fig. 6 are obtained. For low η (right peak, $\eta = 1.7$) the distribution is broader than for larger η (left peak, $\eta = 2.0$), the noise level being the same. This result matches well with the DE solutions by Volkov et al. (33). The broadening of the distribution for lower η results from a large sensitivity $dT/d\eta$ of the period T with respect to the noisy η. In fact, $T = \frac{(L_{max}-L_{min})}{\Delta L_1} + \frac{(L_{mit}-L_{min})}{\Delta L_2} + \text{const}$, where ΔL_1 and ΔL_2 change approximately linearly with η. Thus, there is an approximately hyperbolic dependence of T on η, $|dT/d\eta|$ increasing with decreasing η. The broadening of the cell cycle period distribution becomes even larger when coupling is introduced (bimodal distribution in Fig. 7), again in good agreement with the calculation with DEs (33) (Compare also with experiments in the lower part of Fig. 1a). The bimodality comes about by noise-driven 'kicks' between the basins of attraction of the coexisting periodic modes shown in Fig. 5. The resulting distribution can thus be interpreted as strong noise-enhancement, as has also been reported for glycolyzing yeast for the case of coexisting periodic modes (9).

RESULTS FOR AGGREGATES OF M x M CELLS

We now come to take full advantage of our computational shortcut, using our CA for large aggregates (square arrays of cells). A remarkable result in such arrays is that broad distributions of cell cycle periods are obtained even in the absence of noise, as illustrated in Fig. 8, which shows a polymodal distribution for a 3 x 3 array. This distribution is comparable to the measurements shown

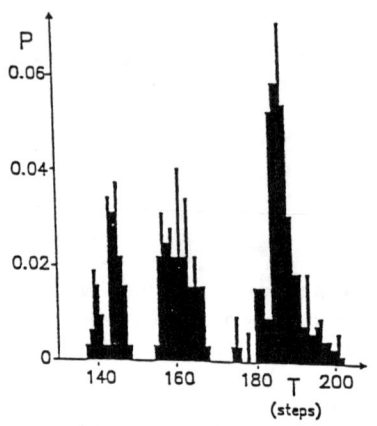

Fig. 8: Polymodal probability distribution of cell cycle periods as obtained with the automaton without noise ($\nu = 0$) for 3 x 3 cells; $C = 0.05$, $\eta = 1.7$.

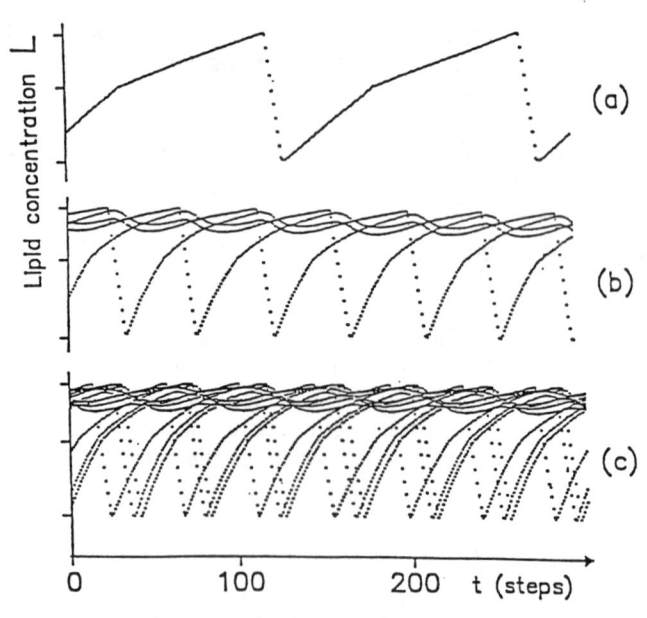

Fig. 9: Coexisting collective modes obtained with the automaton without noise ($\nu = 0$) for 5 x 5 cells; $C = 0.15$, $\eta = 1.7$. The cells are synchronized in (a), have the same period but five different phases in (b), and the same period but fifteen different phases in (c).

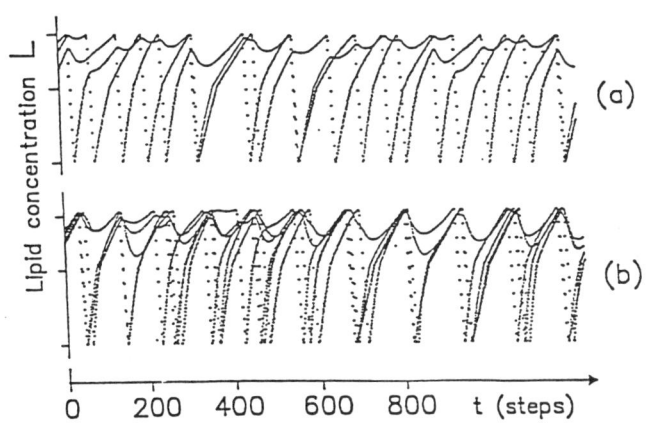

Fig. 10: Collective modes coexisting with those shown in Fig. 9. Three groups of cells (a) and six groups of cells (b) in the cell aggregate differ in period and in phase.

in Fig 1b. Another remarkable result for square arrays is that a large number of attractors (collective modes) may coexist for the same set of parameter values. As an example, Figs 9 and 10 show five different collective modes of an 5 × 5 array for $C = 0.15$, $\eta = 1.7$, and $\nu = 0$ (no noise). In each L vs. t coordinate frame of these figures (and also in Fig. 12) we simultaneously plotted the oscillations of all cells in the array. Cells oscillating in phase yield the same curve. Thus, the number of oscillatory curves in each frame is the number of cell groups within the array differing in their dynamic behaviour. The three small marks in the ordinates of Figs 9,10 and 12 indicate L_{min} (lower), L_{mit} (middle) and L_{max} (upper), Fig. 9a shows synchronization as one of the possible coexisting collective modes. Although in Fig.9 the periods differ from mode to mode, for each single mode (a), (b) or (c) the cells of the aggregate have the same period and differ only in phase. The distribution of the different phases on the cell aggregate are

(a)
$$\begin{array}{ccccc} A & A & A & A & A \\ A & A & A & A & A \\ A & A & A & A & A \\ A & A & A & A & A \\ A & A & A & A & A \end{array}$$
(b)
$$\begin{array}{ccccc} A & B & C & D & E \\ B & C & D & E & A \\ C & D & E & A & B \\ D & E & A & B & C \\ E & A & B & C & D \end{array}$$
(c)
$$\begin{array}{ccccc} A & B & C & D & E \\ A & B & C & D & E \\ F & G & H & I & J \\ K & L & M & N & O \\ F & G & H & I & J \end{array}$$

for the collective modes in Figs 9a, b and c, respectively. (Each letter $A, B, \ldots$ corresponds to one phase within an array, letters in different arrays having different meanings). The solution in Fig. 9b is characterized by

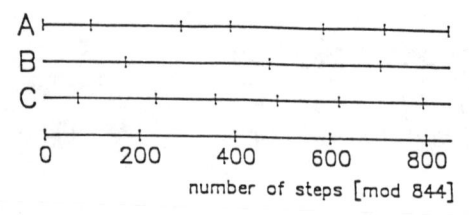

Fig. 11: Times for mitosis (small vertical marks) in the three groups of cells A, B, and C, corresponding to the collective mode of Fig. 10a.

equal phase shifts between five oscillators, so that we are dealing here with a wave running from the upper left to the lower right corner of the cell aggregate. In Figs. 10a and b we show two coexisting modes, in each of which the cell oscillations differ not only in phase but also in period. The distribution of cells in the aggregate are

$$(a) \begin{matrix} A & B & C & C & B \\ A & B & C & C & B \\ A & B & C & C & B \\ A & B & C & C & B \\ A & B & C & C & B \end{matrix} \qquad (b) \begin{matrix} A & D & F & F & D \\ A & D & F & F & D \\ B & E & D & D & E \\ C & B & A & A & B \\ B & E & D & D & E \end{matrix}$$

for the collective modes in Figs 10a and b, respectively. (The letters $A, B, \ldots$ indicate oscillatory behaviour differing in phase and period within an array, letters in different arrays having different meanings). Fig. 11 shows the times at which $L = L_{\mathrm{mit}}$ (given by small vertical marks) for cells A, B and C in the special case of Fig. 10a.

Still another remarkable result is that of "noise induced order". The existence of this phenomenon has been reported for modulated parabolic maps (14,19). For such maps, noise imposed on chaotic oscillations coexisting with periodic ones, drives the system into the latter. In a 5 x 5 cell array, with $C = 0.1$ and $\eta = 1.8$, we found coexistence of a highly disordered collective mode (Fig. 12b) with a more ordered mode (Fig. 12a). As shown in Fig. 12c, a noise amplitude of $\nu = 0.01$ imposed on the disordered mode is not enough to order the system. However, ordering is accomplished for $\nu = 0.02$, as shown in Fig. 12d. In fact, the system is driven into the coexisting mode given in Fig. 12a plus small fluctuations corresponding to the noise.

DISCUSSION

Our calculations have yielded the counterintuitive result that extensive proliferative heterogeneity can be obtained without postulating any physiological differences between the cells. In fact, even without noise, cells may broadly

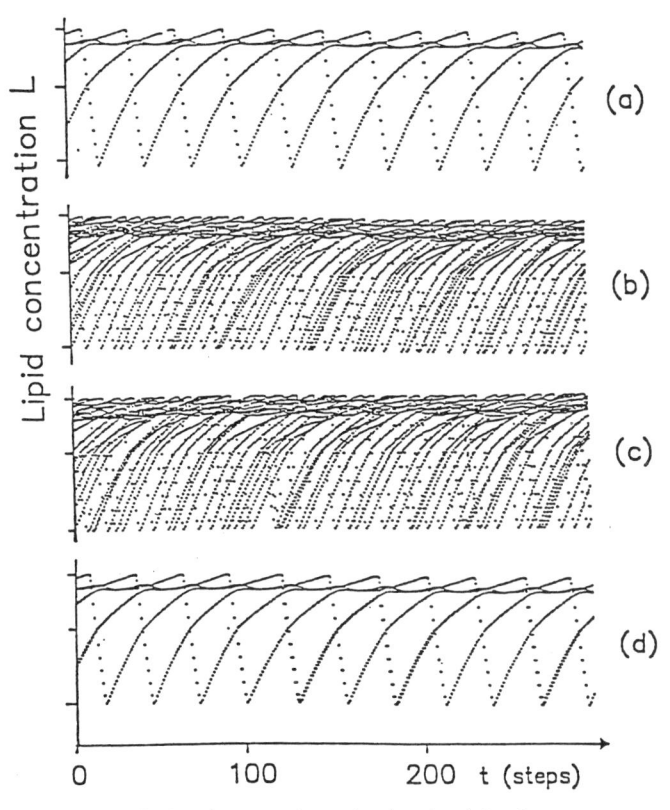

Fig. 12: Collective modes obtained with the automaton for 5 × 5 cells; $C = 0.1$, $\eta = 1.8$. (a,b): coexisting modes without noise ($\nu = 0$); (c): mode (b) with $\nu = 0.01$; (d): mode (b) with $\nu = 0.02$ (noise induced order).

vary in period and phase, solely as a consequence of intercellular transport of chemical reagents. The result that variability increases at low η (Fig. 6) is consistent with the experimental results of Brooks and Riddle (2, 3), in which variability of 3T3 cells increases if the serum concentration is lowered; this consistency holds, of course, if one assumes that serum deprivation is described by low η. We thus obtained agreement with the simulations by Volkov et al. (32, 33), which had been performed using the much slower method of integrating DEs. One must keep in mind however, that the application of the present CA is limited to parameter values such that the following approximations are possible: i) L_{min} and L_{max} and the shape of the LC is constant; and ii) the Δt-marks are equally spaced for $\rho = 1$, for $\rho = 0$

in the L_{mit}-L_{max} interval, and for $\rho = 0$ in the L_{min}-L_{mit} interval.

An interesting result is that coexistence of different oscillatory modes, *i.e.* different modes for the same parameter set and depending only on initial conditions, are not an exception such as in glycolytic oscillations (theoretical: (16), experimental: (18)), but occur in a large domain of parameter space. For M x M arrays we found not only coexistence of in-phase and anti-phase oscillations, as did Volkov et al. (32, 33), but also coexistence of in-phase (synchronization) with modes of high diversity (Figs 9 and 10), including waves. Thus, synchronization, which could be important in tumor therapy (22), may depend on initial conditions.

Finally, we learned that noise may increase variability (Fig. 6) but also that conditions may be such that noise decreases variability (Fig. 12).

OUTLOOK AND CONCLUSION

Future refinements of this model may include: i) more accurate discretization of the LC, *e.g.* by working with coupled map lattices (quasi-continuum of variables); ii) consideration of cell reproduction and cell movement; iii) reproduction stop, *e.g.* by contact inhibition; and iv) biochemically plausible formulation of the noise, especially of its distribution.

A point of criticism to this work could be the choice of the membrane model by Chernavskii et al. (4), in view of the alternative biochemical oscillator models which have been proposed (5–7, 23, 31). However, this will not become a serious point as long as the disputes about these models are not settled; any model is good as any other one as long as their experimental support is poor. The recently proposed model by Norel and Agur (23), although biochemically more plausible than others, does not propose a mechanism for cell coupling.

We conclude by pointing out that the merits of the present work do not lie in the biochemistry but on setting up and testing a useful numerical method. For the time being, this method allows merely to point to the diversity of dynamic behaviour that can be expected in cell aggregates. In the future, more precise predictions could be made by adapting a biochemically correct LC to a CA. The steps of the proposed CA method are: i) determination of the LC of an uncoupled cell by DE integration; ii) discretization of time by some Δt much smaller than the period of the LC, space being discretized by the cells; iii) determination (by DE integration with few cells) of the range in parameter space in which the LC is not significantly distorted (this is the range in which the CA is reliable); iv) inclusion of a coupling algorithm, *e.g.* considering transport of some reagent between neihgbour cells; v) delimitation of zones of the limit cycle in which the steps taken by the phase variables at each Δt can be assumed to be constant; vi) determination of the dependence of these phase variable steps on the free parameters; vii) iteration from $n\Delta t$ to $(n+1)\Delta t$, $n = 0, 1, 2, \ldots$

The rules for a CA of this kind are thus obtained by a discretization of the cycle for a single cell (here obtained by numerical integration of Eqns 4 and 5) and from the assumed coupling. As a result, the computational power needed to run the automaton is not severely restricted by the model for a single cell and by the number of cells. On the other hand, N would indeed restrict the integration of the corresponding $2N$ coupled differential equations. Moreover, data from experimental oscillations of synchronized cells or cell extracts could as well be used for the single cell, instead of a theoretical model.

ACKNOWLEDGEMENTS

This work was supported as Study Project PSS*0303 of the Commision of the European Communities (Directorate-General for Science, Research and Development, Brussels). A.S. acknowledges support by a PhD-grant BD/146/90 - RM from "Programa CIENCIA" (JNICT - Portugal). We thank Evgenii Volkov and Maksim Stolyarov for enlightening discussions and Hans Schepers for careful proofreading of the manuscript.

REFERENCES

1. ALDRIDGE, J. & PYE, E.K., (1976): Nature, 259:670-671
2. BROOKS, R.F. & RIDDLE, P.N., (1988): Exp. Cell. Res, 174:378-387
3. BROOKS, R.F. & RIDDLE, P.N., (1988): J. Cell Sci., 90:601-612
4. CHERNAVSKII, D.S., PALAMARCHUK, E.K., POLEZHAEV, A.A. ET. AL., (1977): Biosystems, 9:187-193
5. DEROCCO, A.G. & WOOLEY, W.H., (1973): Math. Biosci., 18:77-86
6. GILBERT, D.A., (1974): Biosystems, 5:197-206
7. GOODWIN, B., (1970): J. theor. Biol., 28:375-391
8. HASTINGS, J.W. & SWEENEY, B.M., (1959): In R.B. Withrow, editor, Photoperiodism and Related Phenomena in Plants and Animals. American Association for the advancement of Science, Washington
9. HESS, B. & MARKUS, M., (1985): Ber. Bunsenges. Phys. Chem., 89:642-651
10. JEFFREYS, J.G.R. & HAAS, H.L., (1982): Nature, 300:448-450
11. JENSEN, D., (1966): Sci. Am., Feb.:82-90
12. LECHLEITER, J., GIRARD, S., PERALTA, E., ET. AL., (1991): Science, 252:123-126 and cover
13. MARKUS, M., (1990): Biomed. Biochim. Acta, 49:68-69
14. MARKUS, M., (1990): Computers in Physics, Sept/Oct:481-493

15. MARKUS, M., (1991): In O. Arino, D.E. Axelrod, & M. Kimmel, editors, Proc. 2nd Int. Conf. on Math. Population Dynamics, pp. 413–434 Marcel Dekker, New York
16. MARKUS, M. & HESS, B., (1984): Proc. Natl. Acad. Sci. USA, 81:4394–4398
17. MARKUS, M. & HESS, B., (1990): Nature, 347:56–58 and cover
18. MARKUS, M. & HESS, B., (1990): In A. Cornish-Bowden & M.L. Cárdenas, editors, Control of Metabolic Processes, pp. 303–313. Plenum Press, New York
19. MARKUS, M. & HESS, B., (1991): In G. Baier & M. Klein, editors, A Chaotic Hierarchy, pp. 267–283. World Scientific, Singapore
20. MARKUS, M. & SCHEPERS, H.E., (1992, in the press): In Proc. of 1st European Conf. on Mathematics Applied to Biology and Medicine. (Grenoble), Springer-Verlag.
21. MARKUS, M. & STOLYAROV, M.N. & VOLKOV, E.I. (in the press): In Proc. of 3rd Int. Conf. on Math. Population Dynamics
22. NICOLINI, C., (1976): Biochim. Biophys. Acta, 458:243–282
23. NOREL, R. & AGUR, Z., (1991): Science, 251:1076–1078
24. PAVLIDIS, T., (1969): J. theor. Biol., 22:418–436
25. PAVLIDIS, T., (1971): J. theor. Biol., 33:319–338
26. PETROVIC, A., OUDET, B., & STUTZMANN, C., (1984): In L.N. Edmunds, editor, Cell Cycle Clocks, p. 342. Marcel Dekker, New York
27. POLEZHAEV, A. A. & VOLKOV, E.I., (1981): Biol. Cybern., 41:81–89
28. RIVET, J.P., HÉNON, M., FRISCH, V. ET. AL. , (1988): Europhys. Lett., 7:231–236
29. SCHEPERS, H.E. & MARKUS, M., (1992 in the press): Physica A
30. SENNERSTAM, R. & STRÖMBERG, J.O., (1984): Develop. Biol., 103:221
31. TYSON, J.J. & KAUFFMAN, S., (1975): J. Math. Biol., 1:289–310
32. VOLKOV, E.I. & STOLYAROV, M.N., (1991): Phys. Lett., A159:61–66
33. VOLKOV, E.I., STOLYAROV, M.N., & BROOKS, R.F., (submitted for publication)

NONLINEAR OSCILLATIONS AND SOLITONIC EXCITATIONS IN MOLECULAR SYSTEMS

W. EBELING[1] & M. JENSSEN[2]

[1]Fachbereich Physik, Humboldt Universität Berlin, Invalidenstr. 42, D-O-1040 Berlin, BRD
[2]Forschungsanstalt für Forst- und Holzwirtschaft Eberswalde, Alfred Möllerstr., D-O-1300 Eberswalde, BRD

INTRODUCTION

This work is devoted to the study of oscillations and excitations in molecular systems with strong nonlinear interactions

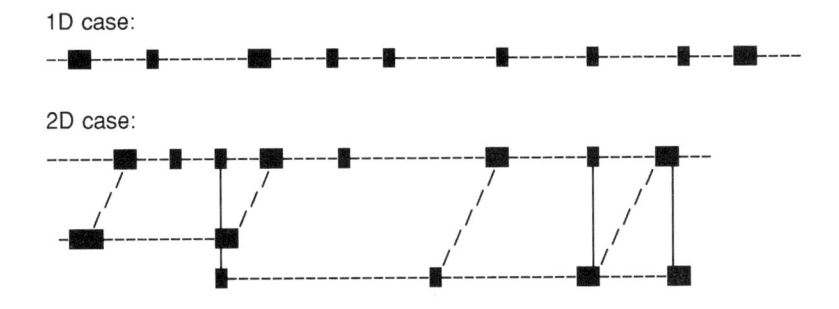

1D case:

2D case:

In equilibrium all of the masses are situated at their rest positions with mutual distances fixed at certain equilibrium values. A very special case of the systems shown above are regular lattices of equal mass molecules connected by bonds of equal strength. Here we will consider the influence of deviations from regularity in system with strong inharmonic forces as e.g. Toda forces:

$$V(r) = \frac{a}{b}\left[\exp(-br) - 1\right] + ar \qquad [1.1]$$

These forces are constant for strong expansions beyond the rest positions and exponentially hard for strong compressions with respect to the equilibrium position. In the limit $b \to \infty$ (ab = const) we get the well known hard core forces:

$$V(r) = \begin{cases} \infty & \text{if } r < a \\ 0 & \text{if } r > a \end{cases} \qquad [1.2]$$

These forces are zero for expansions and infinitely hard for compressions. More realistic molecular forces may be approximated by Morse potentials:

$$V(r) = D\left[\exp(-ar) - 1\right]^2 \qquad [1.3]$$

In the mechanical equilibrium the system possesses only potential energy. By a collision we may accelerate a mass at the border of the system and introduce in this way kinetic energy which will run in form of an excitation through the system. In a thermal regime we may excite even a whole spectrum of excitations. In the case of a purely linear coupling we know all about these excitations: We will observe sinusoidal oscillations and waves, acoustical and optical phonons etc.. Eventually local excitations i.e. wave packets will be observed which however show strong dispersion. In other words local excitations are not stable in linear systems. On the other hand one knows from the theory of infinite chains of molecules with very special interactions as e.g. Toda interactions about the possibility of solitons i.e. absolutely stable local excitations.

The strong interest in local excitation is especially inspired by the theory of reaction rates (1,2). So it may be conceivable, e.g. that catalytic activity in complex reaction systems is supported by nonlinear excitations capable to localize energy at special reaction sites. The questions for existence of solitary waves in biomolecules (3-8) and for their possible role with respect for functional relevant activation processes in enzyme molecules (9,10) or strings of nuclei acids (11,12) were adressed at several places but remain not answered definitely.

Here we try to present a contribution towards a general theory of nonlinear energy localization mechanisms suitable for assistance of local activation processes. We will restrict ourselves to the investigation of simple classical models of molecular systems, one-dimensional chains of masses which are connected to their adjacent neighbours by nonlinear springs of Toda or Morse type.The following part of the paper is decicated to a brief description of a dynamical effect - the soliton fusion - allowing energy localization at a defined site which is part of a nonlinear molecular chain and which can be interpreted as a reaction coordinate to be activated. In thermal equilibrium there is an optimum temperature where energy is mainly partitioned to this reaction coordinate leading to a considerable rate enhancement. In the subsequent section we will turn to an hypothetical example for nonlinear energy localization and recurrence under nonequilibrium conditions. A simple model of global enzyme structure will be used in order to demonstrate the maintenance and efficiency of energy localization for an inhomogeneous chain of molecular masses interacting via Morse potentials with realistic parameters in the presence of frictional forces.

THE EFFECT OF NONUNIFORMITIES ON SOLITONIC EXCITATIONS

A wide class of intermolecular interactions is usually modelled by Morse potentials, Lennard-Jones potentials or other empirical potentials consisting of a steep repulsive and a long-range attractive part. The so-called Toda potential (13) proves to be a special model for this kind of interaction which allows an analytical treatment of the equations of motion. Now we are going to consider the dynamics of a nonuniform chain of masses at position y_n which are connected to their nearest neigbors by Toda springs with the nonlinear spring constant b_n . The Hamiltonian reads

$$H = \sum \left\{ \frac{p_n^2}{2m} + \frac{a}{b_n} \left[\exp\left(-b_n(y_{n+1}-y_n) -1 \right] + a(y_{n+1}-y_n) \right\} \quad [2.1]$$

For an infinite uniform chain ($b = b \; \forall n \; (-\infty + \infty)$)) Toda found the soliton solution (13)

$$\exp\left(-b(y_{n+1}- y_n) \right) - 1 = \sinh \chi^2 \mathrm{sech}^2 \left(\chi n - \sqrt{\frac{ab}{m}} \sinh\chi \; t \right), \quad [2.2]$$

with energy

$$E^s = \frac{2a}{b} (\sinh\chi \cosh\chi - \chi). \quad [2.3]$$

The soliton corresponds to a local compression of the lattice with spatial "width" $c\,\chi^{-1}$. The quantity

$$\tau = \left(\sqrt{\frac{ab}{m}} \sinh\chi \right)^{-1} \quad [2.4]$$

defines a characteristic excitation time of a spring during soliton passage. The energy of a much energy containing and, therefore, extremely localized soliton satisfying the condition

$$\frac{\sinh^2}{\chi} \gg 1 \quad [2.5]$$

reads according to [2.3] and [2.5]:

$$E^s \sim \frac{2a}{b} \cdot \sinh^2\chi \quad [2.6]$$

Now we consider a system consisting of two semiinfinite Toda chains of different spring parameters, $b_n = b \; \forall \; n<0$ and $b_n = b_o \; \forall \; n \geq 0$ with $b < b$. Although this nonuniform chain does not admit exact soliton solutions, the solution [2.2] can be conceived as a right running soliton on the hard part far to the left of the interface where it behaves as in a uniform chain. In the vicinity of the interface however it will be scattered and evolve into reflected and transmitted waves including both solitons and radiation (14,15). In particular we observe sufficiently far to the right on the soft part the formation of a transmitted soliton that was evaluated in a previous paper (15).

$$\exp\left(-b_o(y_{n+1} - y_n)\right) - 1 = \sinh^2\chi_o \operatorname{sech}^2\left(\chi n - \sqrt{\frac{ab_o}{m}} \sinh\chi_o t + \delta\right)$$

$$\sinh^2\chi_o = \frac{b_o}{b} \sinh^2\chi \qquad [2.7]$$

The last expression relates the transmitted soliton to the incident one. δ denotes a constant phase shift that occurs due to the scattering process. In case of strong localization of both incident and transmitted soliton we find from [2.7] and [2.6] for the energy E_o of the latter.

$$E_o^s \sim \frac{2a}{b} \sinh^2\chi \sim E^s$$

Hence the energy of the incident soliton is almost completely transferred to the transmitted one, i.e., scattering losses are less important for energetic solitons. From [2.2] and [2.7] we find according to [2.4] for the characteristic times τ and τ_o of the incident and the transmitted soliton, respectively, the simple relationship

$$\frac{\tau_o}{\tau} = \frac{b}{b_o} \; . \qquad [2.8]$$

The existence of different time scales of soliton motion can be used to generate high energy events by soliton fusion which was demonstrated numerically (15). The energy of two strongly localized solitons of equal magnitudes impinging on the interface will be contained afterwards mainly in one soliton transmitted to the soft spring.

Now we consider a single soft spring embedded in a surrounding, otherwise uniform hard chain ([2.1] with $b_n = b \; \forall \; n \neq 0$ and $b_o < b$) instead of the interface between two extended chains of different stiffness. It turns out that this only soft spring is able to trap and superpose narrow solitons impinging from both directions within characteristic excitation time τ which is demonstrated in Fig. 1. We only note that this kind of soliton fusion leads to considerable concentrations of potential energy at the soft spring (15). Interpreting a compression of this spring up to a certain critical value as activation process we have got a novel mechanism to accumulate the energy of nonequilibrium excitations at a selected degree of freedom to be activated.

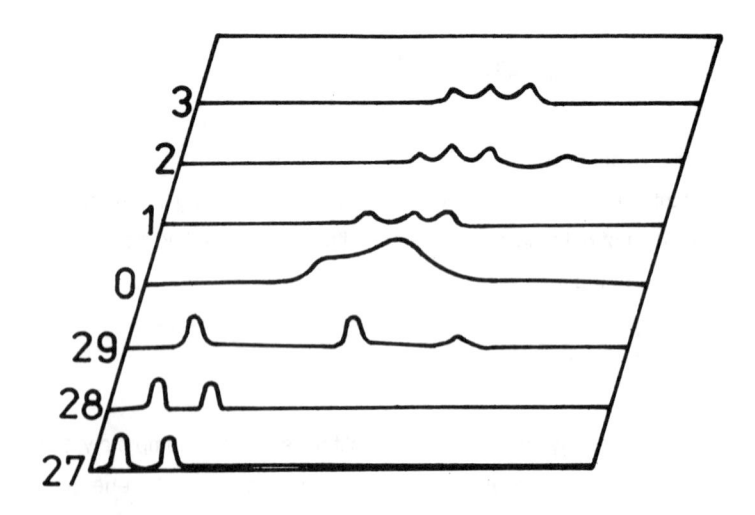

Fig.1: Superposition of two solitons in a single soft spring (n=0) embedded in a hard Toda lattice (n=1,...,29) with parameters m=2, a=10 , b=10 , b/b =10. the solitons (each of energy E=1) are separated from each other by a time 8.4 τ on the hard lattice initially. The potential energy of springs is plotted vs. time t and spring number n.

SOLITON-ASSISTED ACTIVATION PROCESSES

In the last section the fusion of solitons was introduced as a special nonequilibrium effect which is suited to support local activation processes. Now we will give a brief summary of some recent findings (16) proving that an activation enhancement occurs even in thermal equilibrium due to the fusion of thermally generated solitons.

We consider a nonuniform Toda chain [2.1] of N particles which is fixed at the left hand side ($y_0 = 0$) and introduce a pressure P acting on the right end particle (y_0). Among the N springs may be N_o soft springs wiht spring constant b_o. After changing to spring coordinates, $r_n = y_n - y_{n-1}$, the exact classical partition function can be calculated as for the uniform Toda chain (13). Using the notations $\beta = 1/kT$, $\gamma = P/kT$, we obtain

$$Z(\beta, \gamma) = \sum_{n=}^{N} \int_{-\infty}^{+\infty} \int_{-\infty}^{+\infty} dp_n \, dr_n \, \exp\left\{ -\frac{\beta p_n^2}{2m} - \frac{\beta a}{b_n}\left[\exp(-b_n r_n - 1)\right] - (\beta a + \gamma) r_n \right\}$$

$$= \left(\frac{2\pi m}{\beta}\right)^{N/2} \left[\frac{1}{b} \exp\left(\frac{\beta a}{b}\right) \left(\frac{\beta a}{b}\right)^{-(\frac{\beta a + \gamma}{b})} \Gamma\left(\frac{\beta a + \gamma}{b}\right) \right]^{N-N_o} \quad [3.1]$$

$$* \left[\frac{1}{b_o} \exp\left(\frac{\beta a}{b_o}\right) \left(\frac{\beta a}{b_o}\right)^{-(\frac{\beta a + \gamma}{b_o})} \Gamma\left(\frac{\beta a + \gamma}{b_o}\right) \right]^{N_o}$$

The partition function splits into separate factors corresponding to hard and soft springs. The internal energy of the chain reads

$$E = -\frac{\partial}{\partial \beta} \ln Z(\beta, \gamma) = \frac{N}{2\beta} + (N-N_o)\langle u \rangle + N_o \langle u_o \rangle$$

with

$$\langle u \rangle = \frac{a}{b}\left[\ln\left(\beta\frac{a}{b}\right) - \Psi\left(\beta\frac{a+p}{b}\right)\right] + \frac{p}{b} \quad \text{and} \quad [3.2]$$

$$\langle u_a \rangle = \frac{a}{b_o}\left[\ln\left(\beta\frac{a}{b_o}\right) - \Psi\left(\beta\frac{a+p}{b_o}\right)\right] + \frac{p}{b_o}$$

expressing the average potential energies of a hard and a soft spring, respectively. Now we are going to elucidate soliton-induced effects in the thermal behaviour of a nonuniform Toda chain. Because solitons are destroyed at open ends, we are led to fix the total length of the chain, which van be calculated from the partition function [3.1]. Assuming further a vanishing number of soft springs embedded in a chain of hard springs, i.e. a strongly "diluted solution" $\eta = N_o/N = 0$, whereas $N \to \infty$ and $N \to \infty$, one obtains for the pressure (16)

$$p = \frac{b}{\beta}\Psi^{-1}\left[\ln\left(\beta\frac{a}{b}\right)\right] - a. \quad [3.3]$$

By the help of [3.2] and [3.3] we can now calculate the average potential energies of the springs. Especially in the high-temperature limit the ratio of the average potential energies of a soft and a hard spring yields

$$\frac{\langle u_o \rangle}{\langle u \rangle} \xrightarrow[\beta \to 0]{} \frac{b}{b_o} \quad . \quad [3.4]$$

The ratio of spring energies as well as their absolute values in units of thermal energy are presented in Fig. 2 in dependence on temperature. In the high-temperature limit thermal energy is partitioned mainly to the kinetic degrees of freedom. Whereas the ratio of potential energies of soft and hard springs tends up to the maximum value defined by [3.4], the ratio of potential and kinetic energy tends to zero. Hence only a vanishing part of thermal energy is located at soft springs. At low temperatures equipartition theorem is valid and thermal energy is shared equally among all microscopic degrees of freedom. Between these limits there is an optimum temperature where the potential energy of soft springs may reach several k/T2.

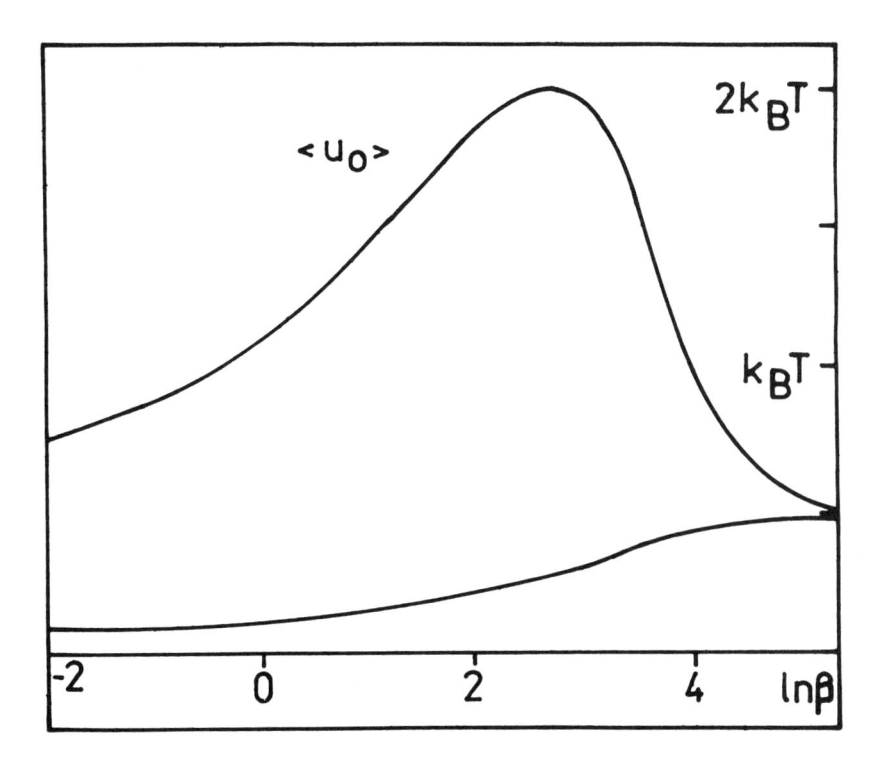

Fig.2: Average spring energies of soft and hard springs in a nonuniform Toda lattice with fixed average length in dependence on b=1/kT.

The pecularities in thermal behaviour of a nonuniform Toda chain can be attributed to the properties of solitons which was outlined previously (16). At high temperatures the dynamcs is completely determined by extreme narrow and hence noninteracting solitons. In the intermediate temperature range that is characterized by a localization of thermal energy at the soft spring thermal solitons become broader and their average distance in time is no more longer than the characteristic time τ of the soft spring. Thus a substantial superposition of incident solitons as presented in Fig. 1 takes place in the soft spring giving rise to the elevation of average potential energy for intemediate temperatures. At low tempera tures however strong interaction is no more confined to the soft spring and individual solitons are destroyed.

Interpreting the soft-spring potential as interaction energy between reactive molecular groups it is possible to develop a simple transition- state theory of chemical reactions based on the obtained results. A considerable rate en hancement comprising several orders of magnitude was estimated for hypothetical transition processes in biomolecular systems (16). The next section will be devoted to the investigation of enzymatic activation processes.

ACTIVATION PROCESSES IN ENZYME MOLECULES

In this chapter we will give a short survey on a model of global enzyme structure which was developed together with ROMANOVSKII (10). We will follow the hypothesis, that solitary wave dynamics could be involved in the primary processes of enzyme catalysis. It is well known that the active site of an enzyme occupies only a few percent of the whole volume of the macromolecule. On the other hand the intra molecular motion of the whole structure is likely to influence the processes at the active site during catalysis. In the following we will treat the enzyme macromolecule in a simplified way as a mechanical system which possesses one selected degree of freedom strongly coupled to catalytic activity. As a crude model for the enzyme α-chymotrypsin we chose (10) a ring of 12 slightly different masses connected by Morse springs:

$$V(r_n) = D_n [\exp(-a_n r_n) - 1]^2 \qquad [4.1]$$

The first term of a Taylor expansion yields the harmonic approximation:

$$V(r_n) \approx D_n a_n^2 r_n^2 = \frac{k_n}{2} r_n^2 \qquad [4.2]$$

The equations of motion read as follows

$$r_n = v_n - v_{n-1},$$

$$v_n = -\gamma_n v_n + m_n^{-1} [V'(r_n)] + F_n(t),$$ [4.3]

$$n = 0, 1, \ldots, 11$$

Here the dots and dashes stand for the derivatives with respect to time t and coordinates, r_n and v_n denote the deviation of the n-th sping from its rest position and the velocity of the n-th mass m_n, respectively. The masses m_n = 2083 a.u. ± 10% model the single domains of α-chymotrypsin (17) which are connected by hxdrogen bonds, D_n = 17 kJ/mole, a_n = 50 nm^{-1} (n=1,...,11), and enclosed by a hydrophobic core, γ =0 n=1,...,10). the spring n=0 with k_o =k/10 and a_o =a/50 stands for a weak elastical interaction between two of the domains between which the active site is located. The motion of the soft spring may be exposed to the solvent, γ_{11} =γ_o =2.86*10^{11} s^{-1}. A compression of this soft or active spring enhances the probability for catalytical reactions significantly (10,18). Now we will discuss the results of numerical integration of [4.1-4.3]. Firstly we considered the action of an initial impact to the active spring modelling the release of vibrational energy due to the adsorption of substrate molecules (10). For this we applied a rectangular pulse F_{11} (t) = - F_o (t) to the soft spring which was adjusted to provide an initial energy of 65 kJ/mole to the system. As can be seen from Fig. 3(a) this adsorption pulse leads to an initial compression of the active spring with a maximum potential energy of more than a half of the total initial energy of the macromolecule. This first compression relaxes thereby inducing the formation of two strongly localized pulses travelling in opposite directions away from the active spring. These pulses are no actual solitons in a strict sense because they possess a finite lifetime which is, however, very large compared to one period around the (undamped) Morse ring. They exhibit to our observation the same qualitative interaction behaviour both in uniform and perturbed chains as the Toda solitons considered above. After a time approximately identical to the natural period of the active spring we observe a first recurrence of the energy due to a fusion of both

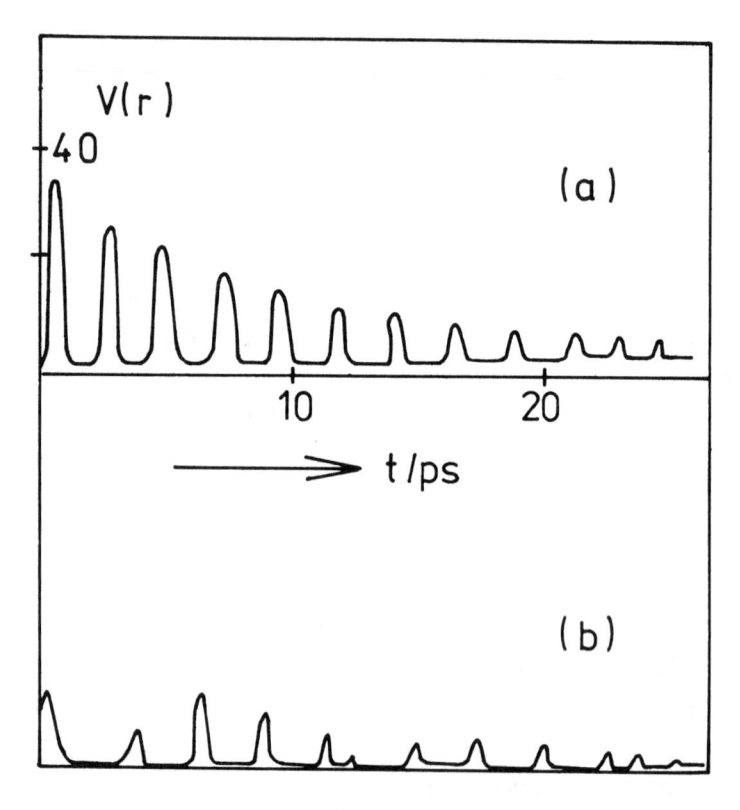

Fig. 3: Relaxation of compressional energy V(r <0) of the active spring vs. time t for (a) the nonlinear and (b) the linear model after an initial compression of the active spring due to an adsorption event (see text).

quasi-solitons after one rotation on the ring. This scenario is repeated until the pulses are damped out due to the solvent friction. It is worth noting that the dissipation of energy proceeds much more slower than for an "isolated" active spring which is not coupled to the rest of the chain. For comparison we integrated [4.3] in the harmonic approximation [4.2] using the same parameters as before, so neglecting the nonlinearity of interaction for demonstration purposes. As can be seen from Fig. 3(b) the initial compression localizes less than a quarter of the total energy at the soft spring. The wave packets made up of the few normal-mode frequencies of the system can be considered moving in opposite directions along the ring as in the nonlinear case. In contradiction to the latter case these wave packets are not stable and we observe an incomplete superposition of the normal waves after a period. The motion of the soft spring is essentially determined by the lowest mode slightly "modulated" by the overtones. Due to the dispersion especially the high-frequency modes do not contribute effectively to a localization of vibrational energy at the active spring in the linear model.

Finally we considered the effect of an initial rectangular pulse put to the fourth spring, i.e. to a hard spring which is positioned asymmetrically with respect to the active one. This pulse was chosen to provide an initial total energy of 65 kJ/mole again. In the linear case the active spring was never excited significantly, whereas in the nonlinear model the energy was always transferred to the active spring where it repeatedly returns also after a certain time when the motion proves to be almost harmonically. This effect of nonlinear energy transfer may be important for the catalytical utilization of external energy sources, collisions with clusters of water molecules.

CONCLUSIONS

Nonlinear oscillation show some special features in comparision to linear oscillations in molecular chains. We have shown that the dynamical effect of soliton fusion provides an efficient mechanism for localization of as well thermal as nonthermal energy at activation sites that are part of a nonlinear molecular chain. This holds true even for an inhomogenous Morse ring under the influence of dissipative forces for parameters applying to biological macromolecules. In thermal equilibrium we proved the existence of an optimum temperature, where energy is preferably partioned to few soft springs embedded into a nonuniform Toda chain consisting mainly of hard springs. Here we restricted ourselves to the investigation of the energetic activation of the soft springs. In the last section we developed some ideas how oscillatory-wave dynamics could support reaction processes under nonequilibrium conditions. Especially we outlined that the nonlinearities occuring in biological macromolecules are sufficient to provide a localization and repeated recurrence of vibrational energy at a functionally relevant part. The effect of thermal energy localization persists also for solutions of soft spheres in hard-sphere solvents what was shown elsewhere (2). A further investigation of thermal energy localization for three-dimensional systems could be of a principle interest to reaction theory.

ACKNOWLEDGEMENT

The authors have to thank Yu. M. Romanovskii for a close cooperation in working on these problems.

REFERENCES

1. P. HÄNGGI, P. TALKNER & M. BORKOVEC, Rev. Mod.Phys. (in press).
2. W. EBELING & M. JENSSEN (1991): Ber. Bunsenges. Phys. Chem. 95, 356
3. S.W. ENGLANDER et al. (1980): Proc Natl. Acad. Sci. USA 77, 7222.
4. A.C. SCOTT (1985): Phys. Rev. A 31, 3518.
5. A.S. DAVYDOV (1985): Solitons in Molecular Systems. Reidel, Dordrecht.
6. A.C. Scott (1987), in: Energy Transfer Dynamics. In: T.W. BARRETT and H.A.POHL (eds.), Springer, Berlin .
7. M. FRANK-KAMENETSKII (1987): Nature 328, 108.
8. V. MUTO, A.C. SCOTT & P.L. CHRISTIANSEN (1989): Phys. Lett. A 136, 33.
9. W. EBELING & M. JENSSEN (1989). In: J. POPIELAWSKI (ed.), The Dynamics of Systems with Chemical Reactions, World Scientific, Singapore.
10. W. EBELING, M. JENSSEN & YU. M. ROMANOVSKII (1989). In: W. EBELING and H. ULBRICHT (eds.), Irreversible Processes and Selforganization, Teubner, Leipzig.
11. M. PEYRARD & A.R. BISHOP (1989): Phys. Rev. Lett 62, 2755.
12. A. FERNANDEZ (1990): J. Phys. A 23, 247.
13. M. TODA (1981): Theory of Nonlinear Lattices, Springer, Berlin.
14. C. CAMACHO & F. LUND (1987): Physica D 26, 379.
15. W. EBELING & M. JENSSEN (1988): Physica D 32, 183.
16. M. JENSSEN (1991): Phys. Lett. A 159, 6.
17. J.J. BIRKTOFT & D.M. BLOW (1972): J. Mol. Biol. 68, 187.
18. YU. M. ROMANOVSKII, A. YU. CHIKISHEV & YU. I. KHURGIN (1988): J. Mol. Cat. 47, 235.

**SATELLITE-SYMPOSIUM:
CHRONOBIOLOGICAL ASPECTS OF PHYSICAL MEDICINE AND
CURE TREATMENT**

INTRODUCTION

G.HILDEBRANDT

Institut für Arbeitsphysiologie und Rehabilitationsforschung, Philipps-Universität, Robert-Koch-Str. 7a, D-W 3550 Marburg/Lahn, FRG
Institut für Kurmedizinische Forschung Bad Wildungen, Langemarckstr. 2, D-W 3590 Bad Wildungen, FRG

A special symposium to deal with "Chronobiological Aspects of Physical Medicine and Cure Treatment" seems to be justified, because in these branches of medicine chronobiological aspects were taken into consideration already since the thirties. In 1935 HAEBERLIN (2) published a book entitled "Lebensrhythmen und Heilkunde" (Biological rhythms and medicine), and VOGT in his Textbook of Balneology and Medical Climatology (4) wrote several pages on chronobiological problems of cure treatment.

Looking at the indications of physical medicine and cure treatment, a prevalence of chronic diseases is to be observed. That means, that the aspect of time must play an important role to understand the therapeutical effects of this mode of therapy.

Cure treatments in Germany not only use natural mineral spring water or mud bathes, but - of course - all other means of physical therapy including hydrotherapy according to KNEIPP. Moreover, dietetic and psychotherapeutical treatment as well as health education play an important role. Not to forget, any cure treatment includes at the same time climatotherapy as well.

Modern aspects of cure treatment have more and more stressed upon the fact, that its various therapeutical procedures utilize quite different mechanisms as compared with pharmacological treatments. Balneotherapy, for instance, offers only very small possibilities to affect the organism by substitution of certain minerals, and particularly by absorption through the skin. However, deposition of minerals within the skin can act as a long lasting stimulus to the autonomic system.

Contrary to the aim of pharmacotherapy to establish a constant plasma level of the respective drug, it is a decisive characteristic of cure treatments, that the applications are performed intermittently, which again points to an aspect of time (Fig.1). The intervals between the single applications are relatively long in order to complete the reactions of the organism for compensation and adaptation. Repeated stimulations lead to the development of long-term adaptive processes, the induction of which is triggered by complex autonomic alterations as evoked by a sympathetic stress response. Cure treatment in spa stations also includes permanent stimulations, e.g. by climatic of psychosocial conditions, and by an unburdening milieu as well. Adaptive processes are therefore combined with processes of deadaptation, including removal of mal-adaptation.

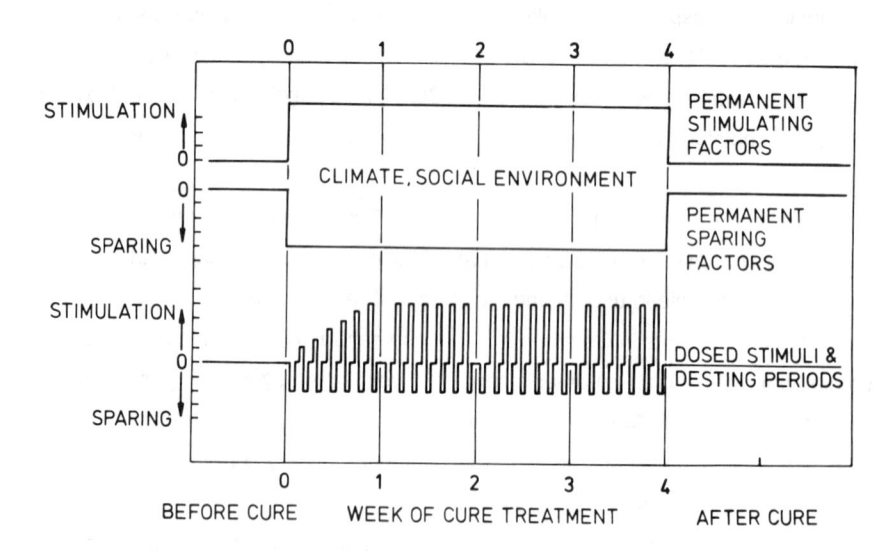

Fig.1: For explanation see text

Consequently, cure treatment utilizes the principles of a so-called natural therapy (Fig.2). There is no attempt to correct or eliminate abnormal processes and deviations, and no attempt to substitute functional or organic deficits, but rather it seeks for to activate the natural endogenous potential of the organism for recovery, regulation, self-defence, regeneration, and adaptation. These effects are not direct ones, they rather represent secondary responses as evoked by certain physical and chemical stressors.

"ARTIFICIAL THERAPY"	"NATURAL THERAPY"
direct effects primary pathogenetic orientation	indirect effects secundary hygiogenetic orientation
1. removal 2. pharmaceut. control 3. substitution replacement	1. sparing, recovery 2. regulation functional adaptation 3. adaptation raising of capacity

Fig.2: For explanation see text

These adaptive processes lead to a characteristic phenomenon, which mirrors the increase of functional economy and regulatory capacity of the autonomic system. That is the so-called normalization. Irrespective of the direction of pathological deviations from the normal range, the functions tend to approach the normal values as a consequence of the increased regulatory capacity. For example: Systolic and diastolic blood pressure values concentrate at the end of cure treatment closer to the normal range. Hyperacidic as well as hypoacidic disorders of gastric secretion are concentrated to the normal range by drinking cures. Even the coordination of different rhythmic functions undergoes the effect of normalization during various modes of cure treatment. Of course, the normalization of the functional order leads to an improvement of the patients subjective condition. (For a review of the literature see (3)).

In the course of functional adaptation and normalization during cure treatment also the state of immunity can be changed. Furthermore, it could be shown that the susceptibility to mental stress reactions is decreased during cure treatment.

From the point of view of chronobiology, another very characteristic mark of this process of functional adaptation is to be observed during cure treatment: That is a periodic structure, which is already pointed to by very old observations, but can - by means of modern findings - substantially contribute to a better understanding of the mechanisms underlying the therapeutical effects of cure treatments.

On the other hand, a mode of stimulating therapy, the effects of which are based on adaptive responses in the autonomous system, demands for a chronotherapeutic time order, taking into consideration the influence of rhythmic changes of responsiveness during the day, the year or the menstrual cycle respectively.

Finally, changes in the autonomic system, as evoked by therapeutical stimulation, will change the internal temporal order of the organism and its synchronization by the outer environment as well.

Hence, there are quite a lot of chronobiological questions and problems in physical therapy and cure treatment (c.f. 1), and we hope, that some of them can be answered or at least can be discussed during this symposium.

It is my pleasant obligation, to thank the speakers, coming from abroad to enrich our program, and to thank our hosts, the burgomaster of Bad Wildungen and the director of the spa stations of Bad Wildungen and Bad Reinhardshausen as well.

REFERENCES

1. AGISHI, Y. & G. HILDEBRANDT (1989): Chronobiological Aspects of Physical Therapy and Cure Treatment. Balneotherapeutic Research Institute, Hokkaido University School of Medicine. Noboribetsu, Japan
2. HAEBERLIN, C. (1935): Lebensrhythmen und Heilkunde. Hippokrates-Verlag, Stuttgart
3. HILDEBRANDT, G. (1985): Therapeutische Physiologie. Grundlagen der Kurortbehandlung. In: W. AMELUNG & G. HILDEBRANDT (eds.): Balneologie und medizinische Klimatologie. Band 1. Springer-Verlag, Berlin-Heidelberg-New York-Tokyo
4. VOGT, H. (1940): Lehrbuch der Bäder- und Klimaheilkunde. J. Springer, Berlin

THE TIME STRUCTURE OF PHYSIOLOGICAL RESPONSES TO CURE TREATMENT

G. HILDEBRANDT

Institut für Arbeitsphysiologie und Rehabilitationsforschung der Philipps-Universität Marburg, Robert-Koch-Strasse 7a, D-W 3550 Marburg/Lahn, FRG

As already pointed out in the introductory remarks, an important chronobiological aspect of cure treatment concerns the fact that the adaptive long-term processes, as evoked by the continuous or repeated stimulation of the autonomous system, exhibit characteristic time structures. In physiology it is wellknown since long time that - in general - the organism tends to arrange its responses in a periodic manner. The period durations depend on the type of response and its time requirements.

Fig.1 gives an overview of different compensatory responses and their period lengths observed. Local compensatory reactions serving local recovery, e.g. in the musculature, need only a few minutes and are structured by a circa 2-min periodicity (6). Recuperation during night sleep needs a time span of several hours and is periodically structured by the ca. 90-min Basic Rest Activity Cycle (REM-NonREM-Periodicity). However, the overcompensating adaptive processes, representing the fundament of those long-term processes leading to therapeutical normalization etc., have a time requirement of several weeks and at the same time are periodically structured by a dominant circa 7-day periodicity, the so-called circaseptan periods. However, circaseptan periodicity represents only one special type of periodic long-term responses which are characterized by adaptive modifications in the organism.

The circaseptan-periodic time structure of the reactive processes, as evoked by cure treatment, was first described in 1959 (7). However, in principle, this time structure of adaptive or self-healing processes must have been well known since the antiquity. For instance, the histograms of critical days in the course of illnesses, as derived from the works of HIPPOKRATES, GALENUS or AVICENNA exhibit a clear cut circaseptan time structure with prominent maxima at day 7, 14, and 21 of the illnesses (Fig.2).

reactive periods

period duration	functional signification
months	trophic-plastic adaptation (overshooting compensation)
weeks	functional adaptation (long-term recreation, normalization)
day—night	general recreation (night sleep)
hours	metabolic recovery (energy stores)
minutes	local recovery (O_2-debt)

Fig.1: Functional significance of reactive periods of different period durations. For further explanation see the text.

During the last centuries the doctors practicing in the spa stations were not aware of this specific time structure during cure treatment but reported the striking experience that patients undergo critical incidents at certain phases of cure treatment, so-called "cure crisis".

Not before the begin of chronobiologically oriented time-series observations during cure treatment we could succeed in an understanding of the real nature of this periodic phenomena.

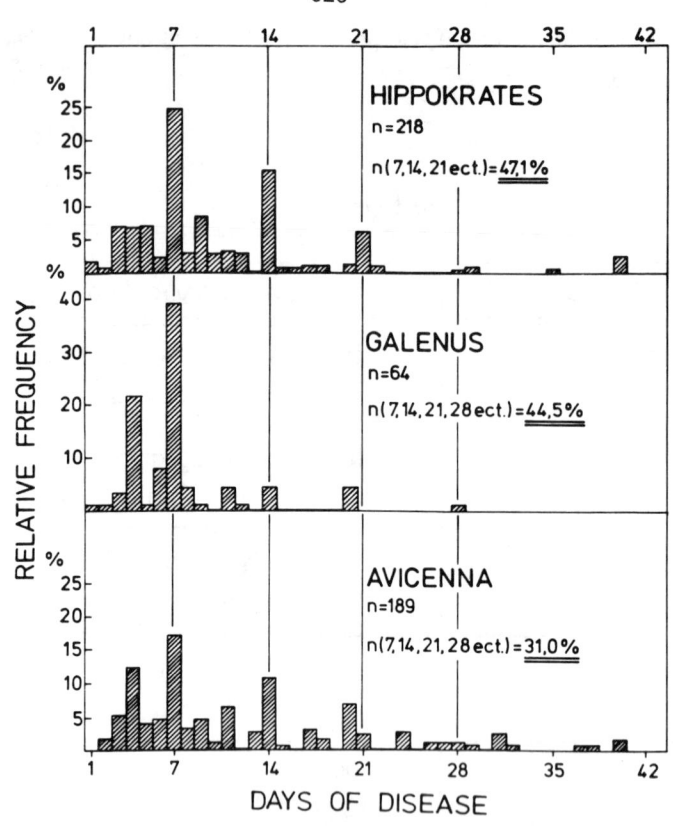

Fig.2: Temporal distribution of the frequency of critical and prognostically important days of illnesses as quoted by Corpus Hippokratikum, Galenus, and Avicenna. (From 12).

Systematic studies could prove, that the circaseptan time structure of cure processes concern all functional levels of the organism and therefore touch also the subjective behavior, feeling, and mental performance. Fig.3 shows a compilation of various findings in groups of patients undergoing different types of European cure treatment. As a prominent result it can be derived that most of the parameters undergo circaseptan periodicities, whereby around the 7th, 14th, and 21st day higher frequencies of disorders, complaints, and deteriorations occur. The critical phases are accompanied e.g. by a lengthening of reaction time and by increases of maximal grip

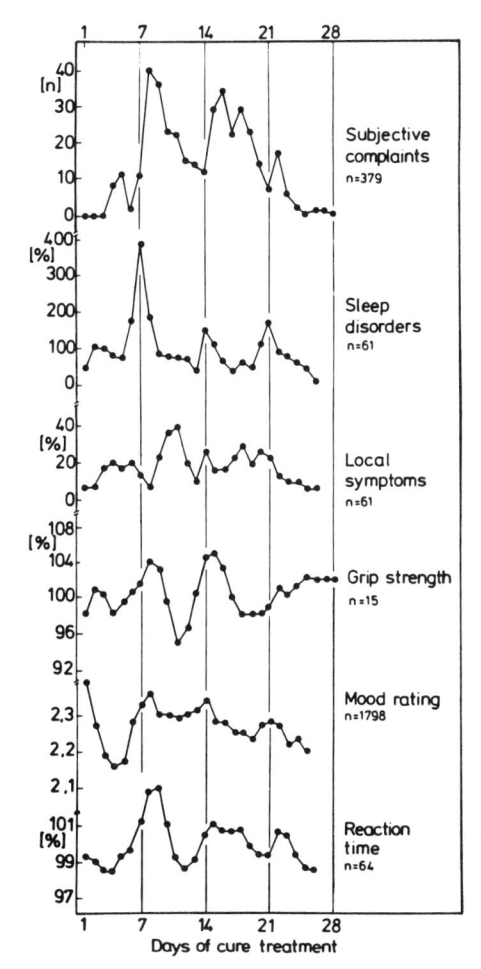

Fig.3:
Courses of various indicators of cure crisis during a 4-week cure treatment at different spa stations. A compilation of results from the literature. (From 3).

strength. These results indicate that the well-known cure reactions (or cure crisis) are located at distinct phases of a periodically structured cure process. However, the course of frequency of local symptoms can deviate from this dominating time structure insofar, as the repeated increase of local symptoms follow more or less an inverse time schedule. This can be understood by the fact that functional disorders precede the organic ones.

According to our experiences, it is possible to control the actual time structure of cure processes already by subjective indicators of the patients, for instance by daily questionnaires.

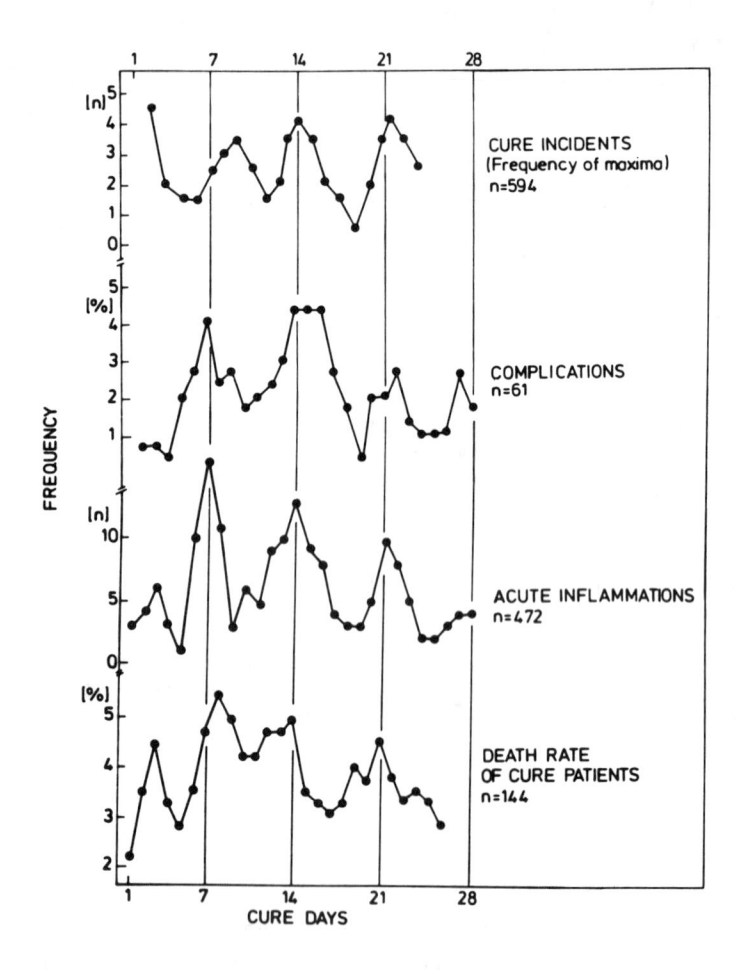

Fig.4: Courses of frequency of cure incidents. (From 11).

From the clinical point of view, it is important to know that even clinical incidents related to cure crisis up to heavy complications or even death rate follow the autonomous time structure of the periodic adaptive reaction of cure treatment. As shown in Fig.4, the courses of frequency of cure incidents at different spa stations are

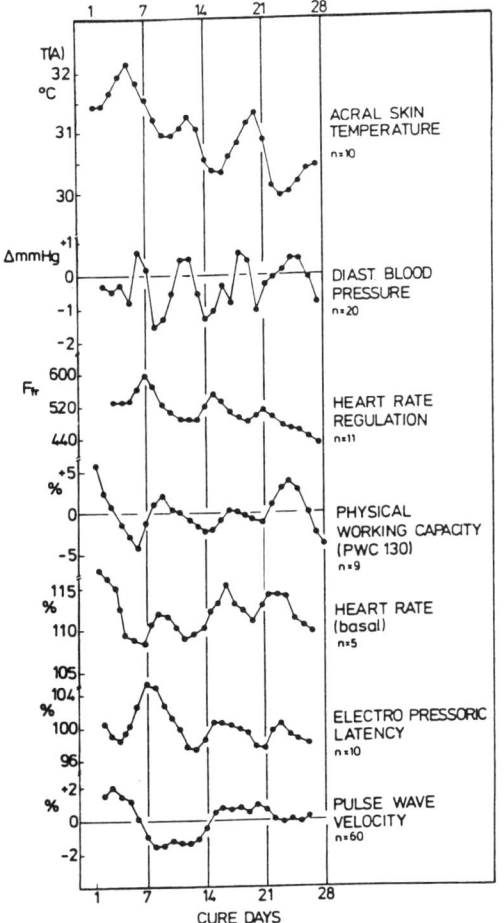

Fig.5:
Mean courses of various cardiovascular functions during a 4-week balneotherapeutical cure treatment. (From 3).

structured by circaseptan periods, only exhibiting slight phase deviations. Particularly inflammatory complications, as collected from various spa stations, vary in frequency exactly following the circaseptan time schedule. Corresponding results from different types of cure treatment point to a great practical importance of these phenomena. Interesting enough that in some curves frequency multiplications of the circaseptan periods can be detected: In the two lower curves additional maxima appear around day 3, indicating a circasemiseptan period.

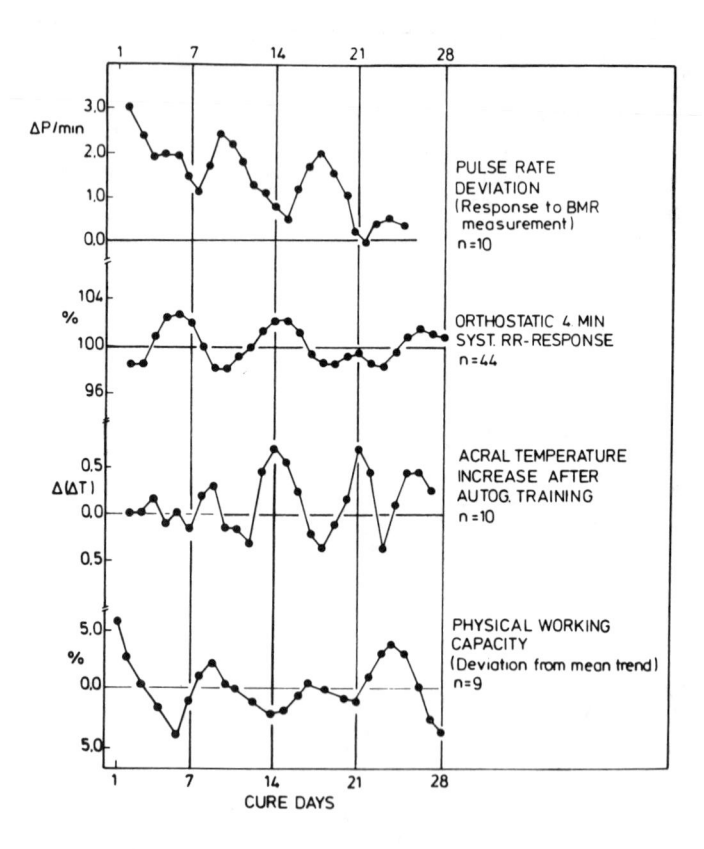

Fig. 6: Mean courses of pulse rate deviation, orthostatic response of systolic blood pressure, acral skin temperature, and physical working capacity during a 4-week cure treatment. (From 10).

Up to now, among the different autonomous functions the time structure of the course of cardiovascular parameters has been studied most extensively, and that during balneotherapeutic, climatic, and hydrotherapeutic treatment as well. Fig.5 shows a compilation of various examples including peripheral blood flow, blood pressure or heart rate as well as cardiodynamic and indicators for the quality of cardiovascular regulation. Most of the average curves exhibit the typical circaseptan periodicity with decreasing amplitudes. The period lengths of the diastolic blood pressure seem to be shorter and that of pulse wave velocity double as compared to the other functions.

Hence, it seems to be evident, that the period duration of the reactive periods can vary and also the phase position of the circaseptan periodicity is not uniform, supposingly depending on the parameters observed as well as on the therapeutical mode.

Circaseptan periodicity includes periodic changes in reagibility. Fig.6 shows a few examples: The pulse rate response to a simple procedure of measurement of the basal metabolic rate or the blood pressure response to orthostasis exhibit circaseptan periodic variations. The same is true for the peripheral vasomotor response to autogenic training, as indicated by the change of acral skin temperature. Pulse rate response to work load also indicates actual reagibility of the cardiovascular system and shows to be influenced by the general autonomous time structure.

As already demonstrated, the well known fact, that symptoms of clinical deteriorations and acute inflammations occur at certain phases of cure treatment, indicates that temporally structured changes of the immunological state are included in the reactive processes. Circaseptan periodic changes of the immunological activity have already been shown in the rejection of kidney allografts or in the periodic systematic controls of immunoglobulin concentration during hot-bath cure treatment in Japan, as performed by AGISHI (1).

All these findings, as taken from different groups of patients under different modes of cure treatment, lead to the conclusion that circaseptan periodicity represents a fundamental process of periodic alternations of the tendencies of the autonomous system.

As already pointed out (11), circaseptan periodicity is mainly controlled by hormonal mechanisms. Unfortunately, our insight into these mechanisms is rather limited up to now. However, recent studies on the plasma concentration of adrenaline (A) and nor-adrenaline (NA), as performed day by day at 6.00 h in the morning during a 4-week cure treatment, made it possible for the first time, to obtain some information on the long-term response of the sympathico-adrenal system. The treatment comprised mainly hot-water bathing and gymnastic therapies in Japanese patients.

The average courses of adrenaline as well as of nor-adrenaline exhibited marked changes at certain phases of the cure treatment whereby the nor-adrenaline concentration showed a significant increasing tendency, and adrenaline concentration tended to decrease. This was true also for the individual curves ($r=-0.53$; $p <0.05$). The ratio NA/A increased markedly during the 4-week treatment and additionally showed a certain periodic time structure. However, there were considerable interindividual differences. Looking at the individual curves, the maximal deviations of the parameters could occur at different phases of cure treatment, which may lead to a differentiation of early and late responding subjects. However, the temporal position of the maxima and minima of adrenaline-plasma concentration and the histogram of the intervals between the peaks show a clear cut prevalence of circaseptan periodicity (Fig.7).

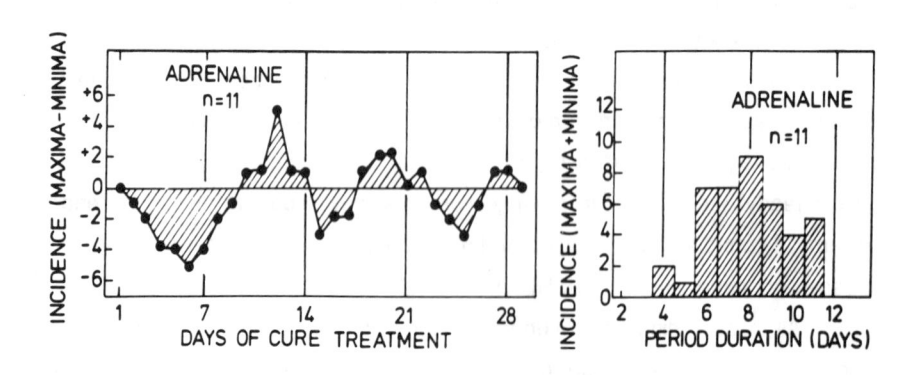

Fig.7: Courses of the incidence of maxima and minima of plasma adrenaline and noradrenaline at 6.00 h during a 4-week balneotherapeutical treatment (left) and the incidence of period durations, intervals between the maxima and minima, of the individual cases (right). (From 2).

In the literature, there are some suggestions, that histamin also is involved in the periodic autonomous response to cure treatment. This is mainly based on the fact, that in animal experiments as well an in humans day by day experimental applications of histamin result in a periodic change of various indicators. It can be presumed that histamin like adrenaline might act as a trigger substance for the induction of adaptive processes, which represent the basic mechanism for the development of long-term cure effects.

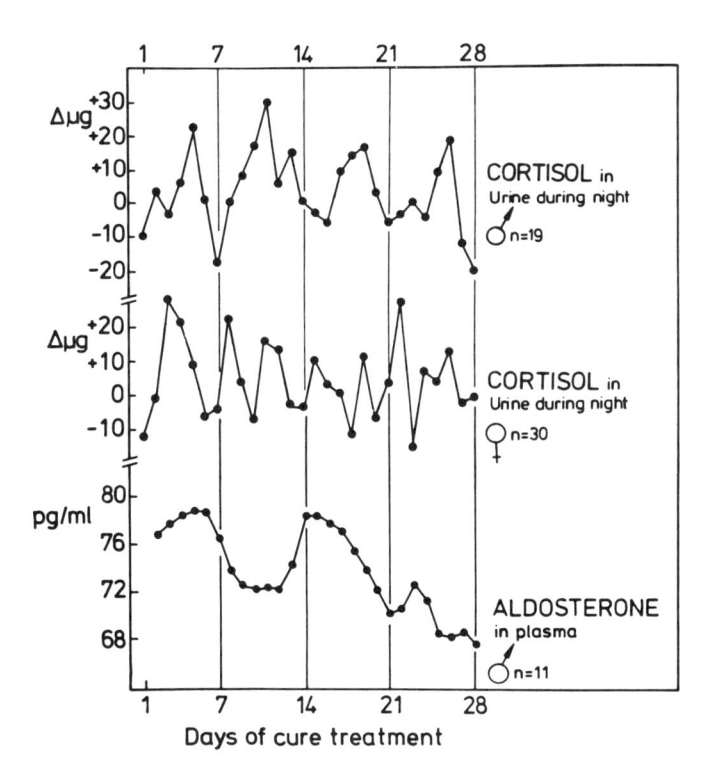

Fig.8: Mean courses of the nightly urinary excretion of cortisol in male and female patients (17) and plasma concentration of aldosterone (2) during 4-week cure treatments.

Concerning the behavior of adaptive hormones, cortisol excretion in the urine as well as cortisol levels in plasma have been studied already during cure treatment by several authors during various types of cures. Fig.8 shows some examples with day by day controls of the cortisol excretion at night time during a hydrotherapeutical cure treatment in males and females (17).

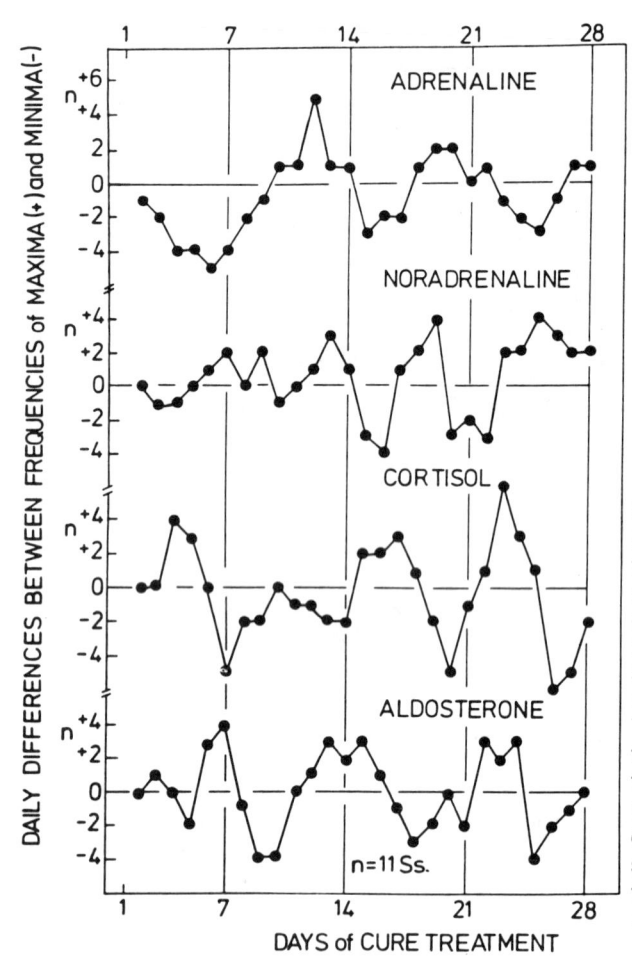

Fig.9:
Average phase positions of plasma adrenaline, noradrenaline, cortisol, and aldosterone at 6.00 h in the same subjects during a 4-week cure treatment. To mark the preferential phase of the location of maxima and minima, the daily frequency of minima was subtracted from the frequency of daily maxima. (From 3).

The mean curve of the male patients exhibits clearly a circaseptan periodicity with troughs located around day 7, 14, and 21, however, the cortisol excretion of the female patients is structured by a circasemiseptan, frequency-multiplicated course during cure treatment.

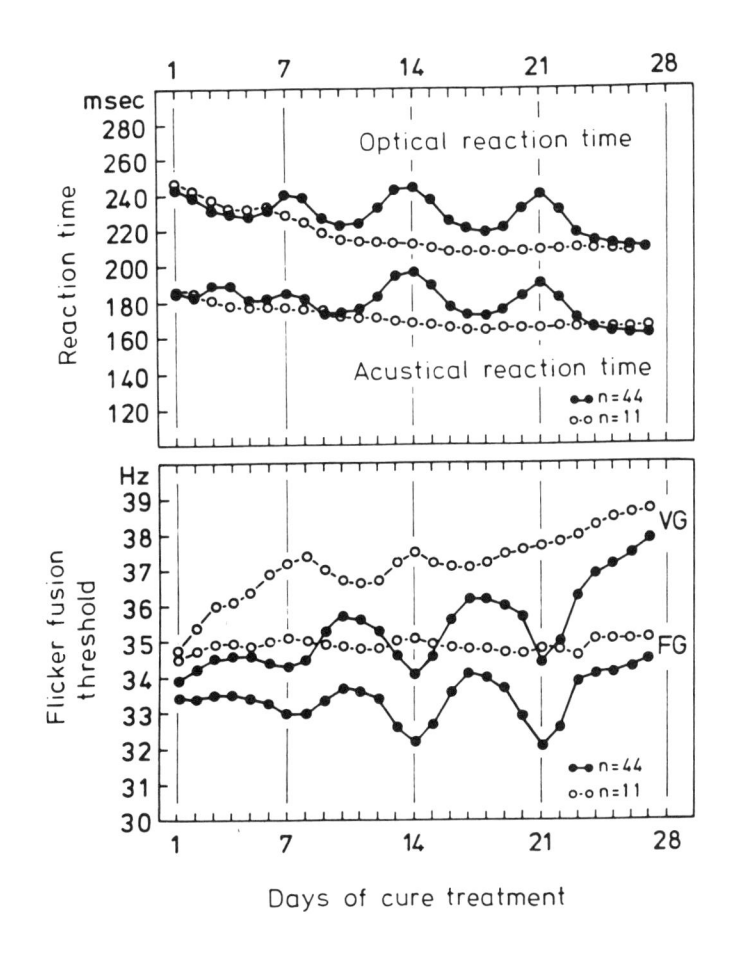

Fig.10: Mean courses of optical and acoustical reaction time as well as of flicker fusion frequencies in patients during a 4-week cure treatment as compared to a group of healthy non-treated inhabitants. (From 15).

As shown in the lowest curve of the figure, the course of plasma aldosteron also in a group of male patients, undergoing hot-bath treatment in a Japanese spa, is structured by a fairly circaseptan periodicity.

In order to compare the time structure of the different hormones during cure treatment, Fig.9 shows the temporal distribution of the maxima and minima of the individual curves in that way, that the daily frequency of minima was subtracted from the frequency of the daily maxima. It can be seen, that in principle all functions undergo a circaseptan periodicity, however, the development of the amplitudes and partially the phase positions do not correspond in all functions.

Till now, we have neglected the important question, whether or not the circaseptan periodicity during cure treatment represents only a consequence of the external week cycle. In fact, therapeutical applications are interrupted at the weekends nowadays.

Fig.10 shows the mean courses of optical and acoustical reaction time as well as of flicker fusion frequencies in patients during a 4-week cure treatment in Bad Wildungen as compared to a group of healthy non-treated inhabitants of this very spa. Whereas the patients develop a circaseptan periodicity of these very sensitive parameters, the control group does not exhibit such a periodicity.

In Fig.11, the group of patients is divided into three subgroups, who started cure treatment at different days of the week. The mean courses of reaction time as well as of flicker fusion frequency do not exhibit any phase difference between the subgroups. These results could be confirmed by other authors using different parameters (For a review see: 11). Hence, we can be sure, that the phase position of the circaseptan periodicity is not synchronized by external periodic events like the lunar cycle or the social week cycle, but is evoked by the onset of cure treatments.

However, in principle, we must concede, that the circaseptan periodicity can be synchronized by the social week cycle, provided that the weekly variations of bodily functions do not represent masking effects only. If we compare the mean amplitudes of the circaseptan periods of both functions as shown in Fig. 11 with the weekly cycles of the control group, the amplitudes of the reactive periodicity in the patients are up to 10 times greater than the weekly variations. Our results confirm, that the circaseptan periodicity, as evoked by cure treatment, represents centrally-coordinated compensatory reactions of the organism like so-called macroperiods, which were first described by DERER in Bratislava 1956 (4). Moreover, there are some indications, that the circaseptan time structure is closely related with the mechanisms of the General Adaptation Syndrome (19).

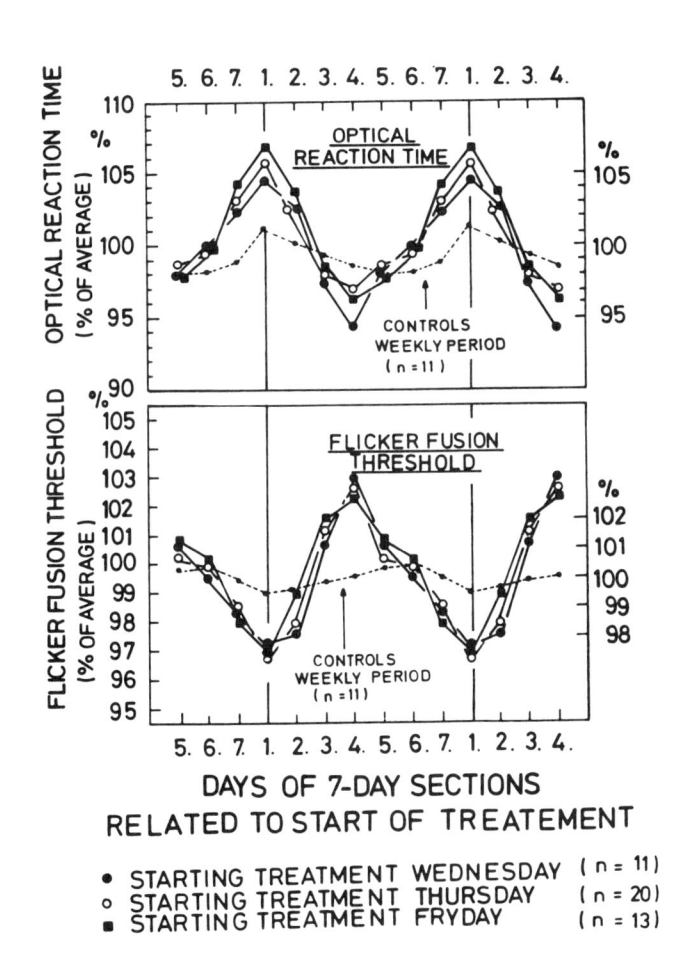

Fig.11: Mean course of acoustical reaction time and flicker threshold of patients starting cure treatment on different days of the week, within 7-day sections of the cure treatment related to the start of treatment. Dotted curves (ordinates on the right margin) represent the mean weekly periods in the control group. (From 16).

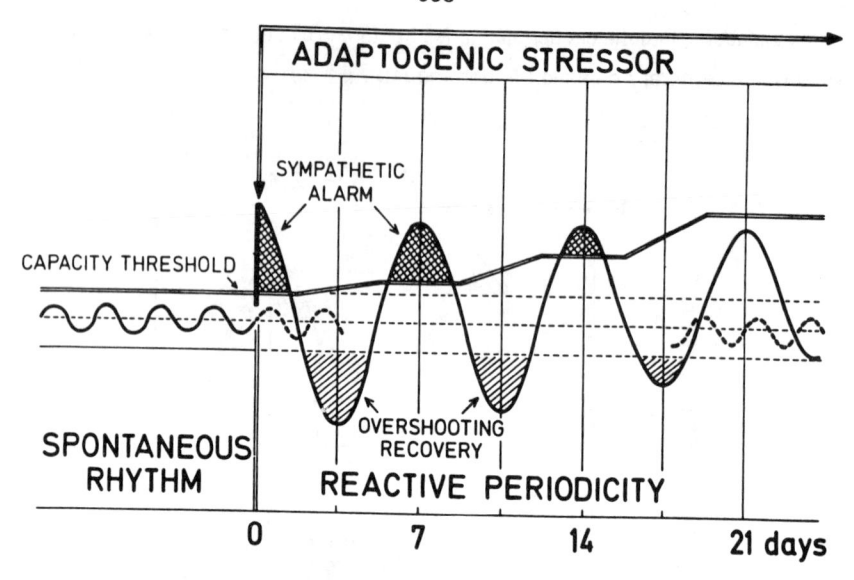

```
                    ADAPTOGENIC STRESSOR
                  ┌─────────────────────────→
                  │    SYMPATHETIC
                  │      ALARM
CAPACITY THRESHOLD│       ↘  ↙
    ～～～～～～～│     ∩    ∩      ∩       ∩
                  │    ↗  ↗
                  │  OVERSHOOTING
   SPONTANEOUS    │   RECOVERY
      RHYTHM      │  REACTIVE PERIODICITY
                  0         7          14         21 days
```

Fig.12: Scheme of the evocation of a reactive periodicity by an adaptogenic stressor (right), starting from the state of complete adaptation with merely spontaneous rhythmicity (left). For further explanation see the text. (From 8).

In order to illustrate the functional significance of this type of reactive periodicity, a scheme is shown in Fig.12. In a state of complete adaptation, the alternation between ergotropic and trophotropic phases in the autonomous system is occurring in the course of spontaneous rhythmicity (left hand part of the figure). An unusual stimulus load, however, causes the autonomous balance to exceed the normal capacity range as far as the limit of the autonomously protected reserves, leading to a sympathetic alarm reaction. This is the starting point of a periodicity with longer waves and higher amplitudes, leading to a recovery overshoot which raises the capacity limit. The following ergotropic half-period, however, again exceeds the actual capacity limit, thus causing renewed sympathetic alarm and introducing the next wave of reactive periodicity. The sequence is repeated until the further elevated capacity limit is no longer exceeded by the ergotropic maxima. This can finally lead to a damped fading of the reactive periodicity, which, again, can be relieved by the spontaneous rhythm, representing the state of complete adaptation.

This scheme is certainly very simplified and rough. However, in principle it might be valid for the periodic structure of compensatory and adaptive reactions of any range. There is no need for adaptational response without an exceeding of the accustomed capacity limit and a triggering of the corresponding sympathetic alarm reaction.

Of course, the question arises, what kind of oscillatory mechanisms could be responsible for the circaseptan periodicity. Is this oscillator permanently active? Or is it in a latent state up to that point, when unusual stimulation occurs in order to evoke compensatory or adaptive reactions of the organism?

If we consider the fact, that the phase position of circaseptan periods is strictly related to the time of stressor action or to the onset of pathological reactions (corresponding with the ancient experiences), the theoretical response curve of the circaseptan oscillator must represent a sort of "Re-set-mechanism", if the oscillator is active all the time and can be hit by the stressor at different phases of the periodicity. On the other hand, in a latent state, the oscillator may be called to activity always in a similar phase position.

In this respect, it is important to remind, that the circaseptan periodicity does not represent the only reactive time structure of the organism. This is true also for the periodic time structure of cure treatments. There exist at least two families of reactive periods, including submultiple as well as multiple period lengths: These are the circaseptan group and the circadecan group. Both groups could be particularly differentiated in patients during 4-week cure treatment, however, several newer findings could confirm this discrimination.

The predominant factor, which determines the period duration of reactive periodicity, seems to be the equilibrium of the autonomous system, as indicated - for instance - by the pulse-respiration frequency ratio. Fig.13 shows the average courses of reaction time in four groups of patients, which are divided according to the ratio. As you can see, patients with high ratios, indicating an ergotropic state of the autonomous system, exhibit a clear cut circaseptan periodicity with an early maximum of amplitudes around day 7. Patients with lower values of the ratio, indicating a trophotropic state, change over to a circadecan shape, exhibiting a later maximum of amplitudes at the end of the 3rd week of cure treatment.

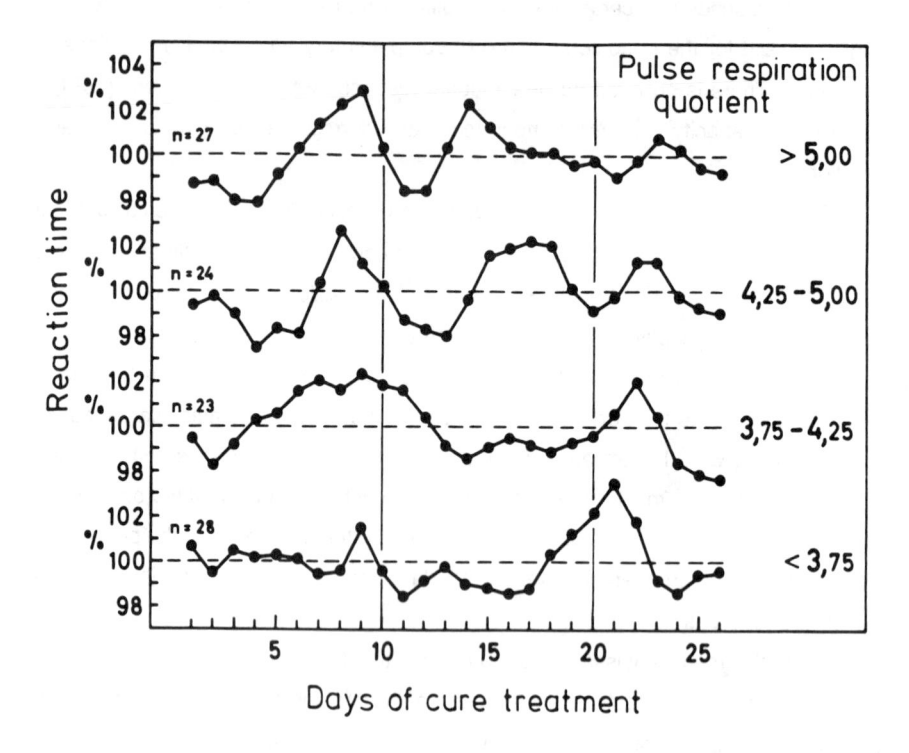

Fig.13: Mean courses of optical reaction time in four groups of patients with different initial values of the pulse-respiration frequency ratio during 4-week cure treatments. (From 5).

Fig.14 shows this difference in the time structure of the course of further parameters during cure treatment between ergotropic and trophotropic patients.

The differentiation between circaseptan and circadecan periodic courses of cure treatment is of practical importance, because both groups of patients need different durations of cure treatment. The later maximal deviation of autonomous functions in the circadecan group demands for a prolongation of cure treatment up to 6 weeks.

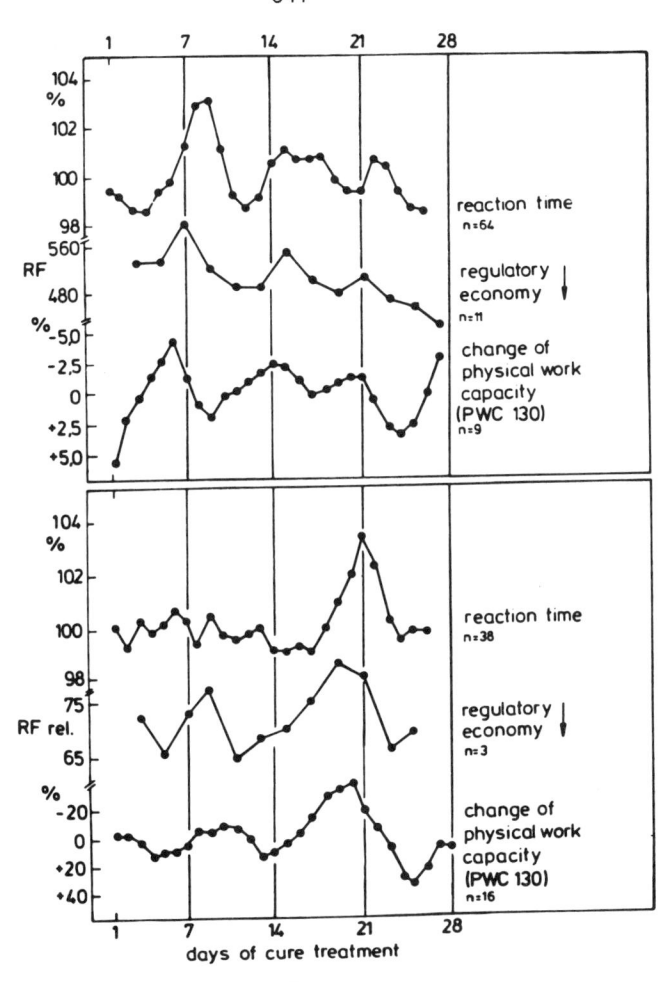

Fig.14: Mean courses of three different parameters during 4-week cure treatment in groups of patients exhibiting ergotropic states (upper part) or trophotropic states of the autonomic system (lower part) at the beginning of the cure. (From 9).

It could be ascertained, that the predominance of the different types of reactive periodicity is influenced by age. As shown in Fig.15, the temporal distributions of subjective complaints in the course of 4-week cure treatments exhibit in the younger age groups a predominance of circaseptan periods. In the older groups, however, a circa 5-day periodicity with increasing amplitudes is preferred, certainly belonging to the circadecan family of reactive periods.

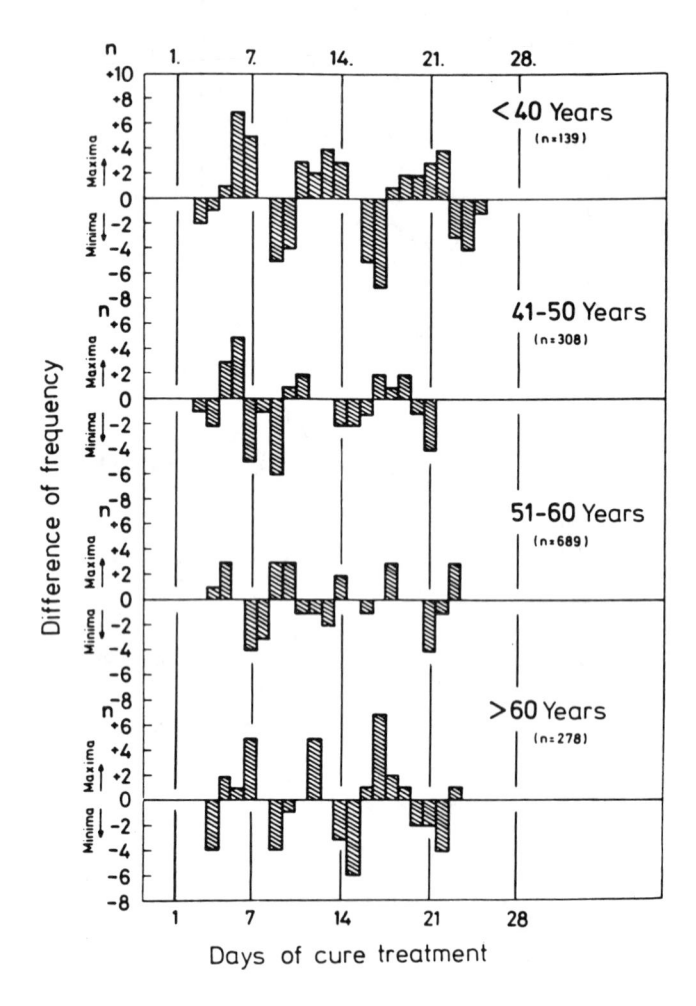

Fig.15: Temporal distribution of the maxima and minima of subjective complaints during 4-week cure treatment in different age groups. (From 13).

Similar results could be demonstrated by PÖLLMANN et al. (18) concerning the temporal distribution of the incidences of odontogenic suppurations in patients undergoing cure treatments. Again, in the younger age groups a prevalence of the circaseptan time structure could be observed, including a decrease of the amplitudes. In contrast, in the older age groups the maximum was shifted to the 3rd week of treatment, and circaseptan structure was diminished.

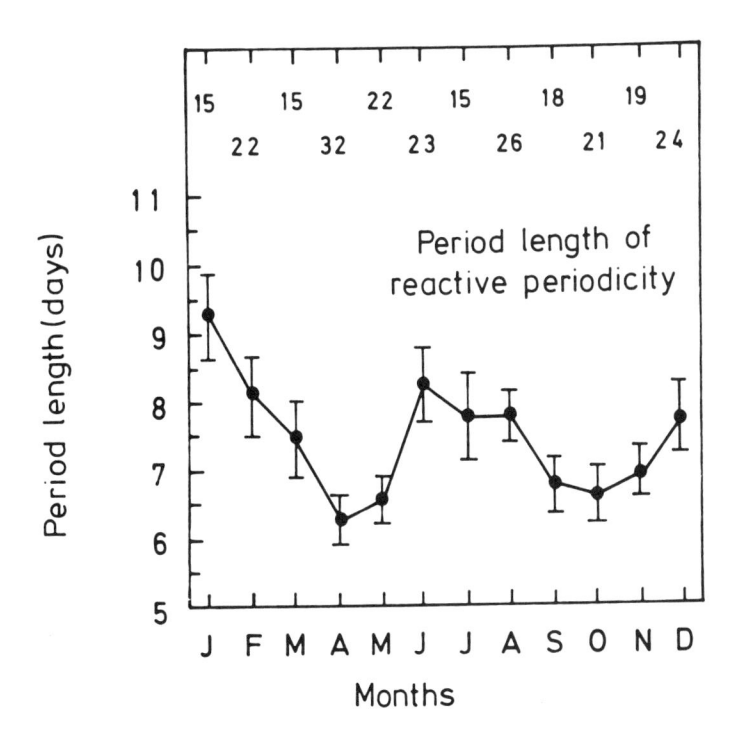

Fig.16: Saisonal changes of the average period length of the variations of the frequency of subjective complaints during 4-week cure treatments. (From 14).

Up to now, the question remains open, whether or not the two families of reactive periods, which structure cure treatment, are based on different oscillators, or the own frequency of the circaseptan oscillator can be shifted by various factors. Our studies on the circannual variations of the average period lengths in the course of subjective complaints of cure patients could show (Fig.16), that the mean period duration in summer time amounted to about 6-7 days, whereas in winter time the mean period lengths were extended up to about 9 days. However, this finding does not prove that the own frequency of a circaseptan oscillator can be modulated. We must consider, that the period lengths of both families of reactive periods represent submultiple periods of the circalunar or circamensual cycle.

To come to an end, our conclusions are as follows:

The periodic time structure of the adaptive processes during cure treatment represents a physiological response of the autonomous system, forming a reactive periodicity mainly of circaseptan period duration. Circaseptan periods are - at least transitorily - free running endogenous oscillations of the autonomous system, which are evoked by external stimulation. However, at least two different families of periodic responses can be differentiated during cure treatment: the circaseptan and the circadecan periods including multiple and submultiple frequencies.

The knowledge of the physiological time structure of cure treatment is an important precondition to understand the adaptive mechanism of cure treatment effects as well as of practical significance in order to avoid critical deviations and to determine the necessary duration of cure treatment.

REFERENCES

1. AGISHI, Y. (1979): Endocrine reactions in hot and cold baths. Z. Phys. Med. 8, 16
2. AGISHI, Y. & G. HILDEBRANDT (1985): Immediat- und Langzeitwirkungen der Balneotherapie auf den Plasma-Katecholamin-Spiegel. Z. Phys. Med. 14, 310
3. AGISHI, Y. & G. HILDEBRANDT (1989): Chronobiological aspects of physical therapy and cure treatment. Balneotherapeutic Research Institute, Hokkaido University, School of Medicine, Noboribetsu, Japan
4. DERER, L. (1956): Concealed macroperiodicity in the reaction of the human organism. Rev. Czechoslovak Med. 2, 4
5. ENGEL, P., G. HILDEBRANDT & H. BERGER (1963): Zur Objektivierung psychophysischer Umstellungen im Kurverlauf. Arch. Phys. Ther. (Leipzig) 15, 335-342
6. GOLENHOFEN, K. (1962): Zur Reaktionsdynamik der menschlichen Muskelstrombahn. Arch. Kreislaufforsch. 38, 202-223

7. HILDEBRANDT, G. (1959): Balneologie und vegetative Regulationen. Therapiewoche 9, 465-475
8. HILDEBRANDT, G. (1982): The time structure of adaptive processes. In: G. HILDEBRANDT & H. HENSEL (eds.): Biological Adaptation. Georg Thieme Verlag, Stuttgart-New York, pp. 24-39
9. HILDEBRANDT, G. (1983): Freizeit und Urlaub. In: W. ROHMERT & J. RUTENFRANZ (eds.): Praktische Arbeitsphysiologie. 3. Aufl., Georg Thieme Verlag, Stuttgart-New York, pp. 381-393
10. HILDEBRANDT, G. (1984): Chronobiological aspects of physical cure treatment. In: A. REINBERG, M. SMOLENSKY & LABREQUE (eds.): Annual Review of Chronopharmacology. Pergamon Press, Oxford-New York-Toronto-Sydney-Paris-Frankfurt, pp. 403-406
11. HILDEBRANDT, G. (1985): Therapeutische Physiologie. Grundlagen der Kurortbehandlung. In: W. AMELUNG & G. HILDEBRANDT (eds.): Balneologie und medizinische Klimatologie. Band 1, Springer Verlag, Berlin-Heidelberg-New York-Tokyo
12. HILDEBRANDT, G. & I. BANDT-REGES (1992): Chronobiologie und Naturheilkunde. Grundlagen der Circaseptanperiodik. Haug-Verlag, Stuttgart (im Druck)
13. HILDEBRANDT, G., L. EMDE, F. GEYER & H. WIEMANN (1980): Zur Frage der periodischen Gliederung adaptiver Prozesse. Z. Phys. Med. 9, 90-92
14. HILDEBRANDT, G. & D. FRANK (1974): Der subjektive Verlauf der aktivierenden Kneipp-Kurbehandlung und seine Abhängigkeit vom biologischen Jahresrhythmus. Z. Phys. Med. 3, 177-194
15. HILDEBRANDT, G. & F. GEYER (1984): Adaptive significance of circaseptan reactive periods. J. interdiscipl. Cycle Res. 15, 109-117
16. HILDEBRANDT, G., F. GEYER & W. BRÜNING (1982): Circaseptan adaptive periodicity and weekly rhythm. In: G. HILDEBRANDT & H. HENSEL (eds.): Biological Adaptation. Georg Thieme Verlag, Stuttgart-New York, pp. 113-116
17. HÜDEPOHL, L. (1985): Untersuchungen der Kurverlaufsperiodik der nächtlichen Cortisolausscheidung mit Hilfe der pergressiven Fourieranalyse. Med. Inaug.-Diss. Marburg/Lahn
18. PÖLLMANN, L., G. HILDEBRANDT & M. HELLER (1987): Circaseptan reactive periodicity during cure treatment. In: G. HILDEBRANDT, R. MOOG & F. RASCHKE (eds.): Chronobiology & Chronomedicine. Peter Lang Verlag, Frankfurt am Main-Bern-New York-Paris, pp. 392-402
19. SELYE, H. (1953): Einführung in die Lehre vom Allgemeinen Adaptationssyndrom. Thieme Verlag, Stuttgart

CIRCADIAN VARIATIONS IN PHYSIOLOGICAL RESPONSES TO THERMAL STIMULI

P. ENGEL

Institut für Arbeitsphysiologie und Rehabilitationsforschung der Philipps-Universität, Robert-Koch-Str. 7a, D-W 3550 Marburg/Lahn, FRG

In Physical Therapy thermal stimuli are applied as local or as whole-body applications to treat acute and chronic diseases of the skin, the muscles, the joints etc. The main effect of a local warm or cold package is a change in the peripheral blood flow. Whole-body exposures, for instance a hot water bath or a Finnish Sauna-bath lead already to more or less hyperthermia of the whole organism. Short lasting exposures in a whole-body cold air chamber of minus 120-150°C are used in the treatment of pains in chronic rheumatism. Moreover, series of thermal stimuli are given in cure treatment as an unspecific stressor to activate the vegetative system and the ability of the organism to react.

We already know from the early findings of chronotherapy by experiments of HILDEBRANDT (10) and also of LAMPERT (15), that physiological responses of the body to thermal stimuli are influenced by the circadian phase of the spontaneous variations of thermoregulation. In practice, it is important to know the reaction of the organism to warm or cold applications at different times of the day, to assist in determining the intensity of treatment. It is known for a long time that diurnal cycle of core temperature is of endogenous nature. This daily course of thermoregulation represents 2 alternating parts of thermoregulatory situations: A warming-up phase in the forenoon, when core temperature is increasing, and a cooling-down phase, when core temperature decreases during the evening. Maximum and minimum of the daily course of the fluctuating functional parameters of the thermoregulation are found around 3.00 h at night and 15.00 h in the afternoon (10). ASCHOFF (2) has shown that circadian variations of thermoregulation are mainly changes in heat loss from the body by circadian changes in skin blood flow. HEISER & COHEN (7) have first proved

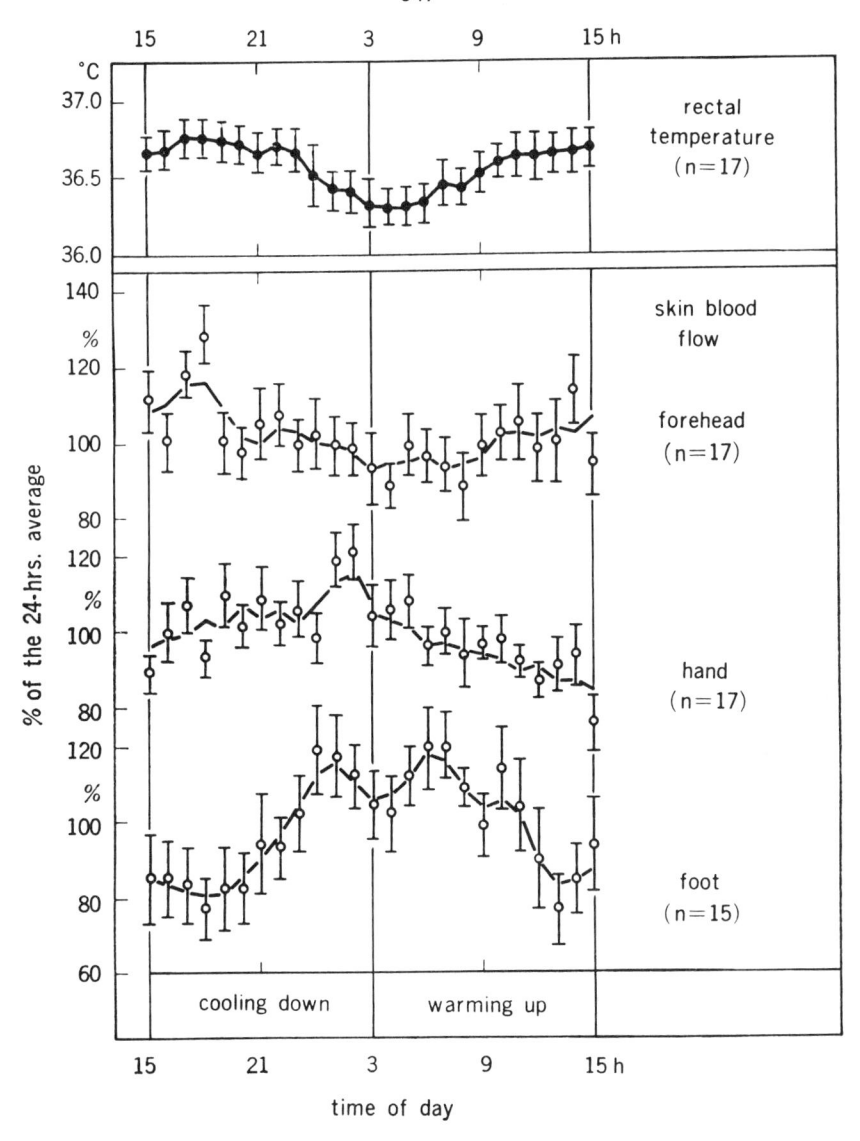

Fig.1: Mean daily variations of rectal temperature (above) and skin blood flow of forehead, hand, and foot (below), as measured by means of heat clearance technique under strict resting conditions in a climatic chamber. Brackets indicate standard errors (From 4).

under constant ambient climatic conditions, that skin temperature in the extremities shows a controversy daily course to that of the core temperature. Other investigators like HILDEBRANDT et al.(9) and later also KOE et al. (13) have confirmed this finding. But the skin temperature at the trunk has a parallel circadian course to that of the core temperature.

It also must be mentioned that the circadian variations in muscle blood flow in the extremities runs contrary to that of skin blood flow, as KANEKO et al. (12) have shown. The different diurnal variations in central and peripheral vegetative functions of thermoregulation are summarized in Fig.1: DAMM et al.(4) have investigated under resting conditions and constant climate the diurnal variations of skin blood flow at different regions of the trunk and extremities. At the top of the figure the mean daily course of rectal temperature of the 17 investigated subjects during the warming-up and cooling-down phase is represented. Skin blood flow in the forehead shows the same circadian phase position than core temperature. In contrast, skin blood flow of the hands and the feet have a controversy circadian course with higher blood flow during the cooling-down period than during the warming-up phase. Vasoconstriction of the skin vessels in the extremities during the warming-up phase hinders heat loss. Vice versa, during the cooling-down phase heat loss is supported by vasodilation of the skin. Core temperature falls.

It must be expected, that thermal stimuli can support the circadian changes of thermoregulation, so that the stimuli response may increase the predominant phase direction of thermoregulation. In other words: A warm application during the cooling-down phase in the evening will lead to a stronger and longer lasting response than during the warming-up phase. A warm stimulus is supporting vasodilation of skin vessels at this time of the day. Contrary to this, a cold application will strengthen the effect of vasoconstriction during the morning warming-up phase of thermoregulation.

Many investigations of circadian responses to different thermal applications have confirmed, that the average sensitivity to cold stimuli is lower during the circadian warming-up phase, and higher to heat during the cooling-down phase. Some examples, namely of our own investigations, shall demonstrate this in the following figures. Already HILDEBRANDT et al.(9) had found out by circadian studies of sweating response at the forehead after drinking of 55 ^{0}C hot diaphoretic tea that no

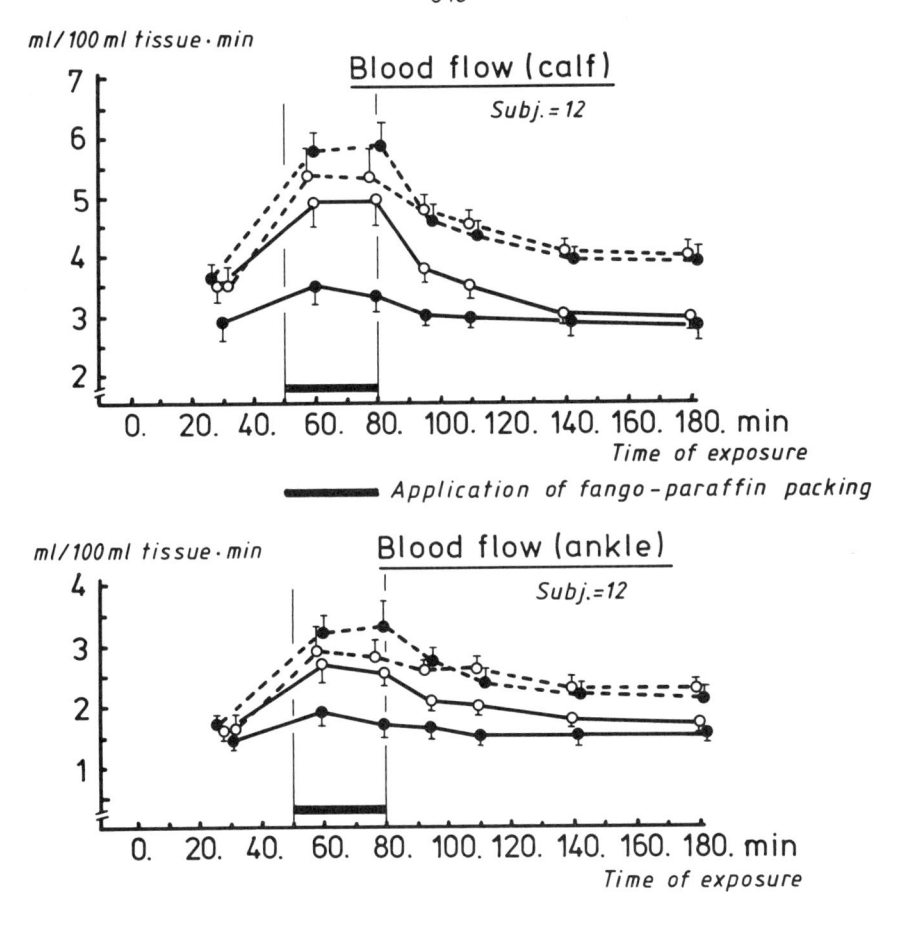

Fig.2: Mean (±SE) course of blood flow in the leg (calf and ankle) from 12 healthy subjects at different times of day (00:30, 08:00, 12:30 and 20:30) before, during and until to 100 min after the end of local heat application (fango-paraffin packing) on the arm (5)

sweating reaction could be evoked in the morning at 9.00 h. But the same stimuli, applied during the afternoon and evening, led to an intensive evaporative sweat production. HILDEBRANDT (10) also demonstrated, that the duration of vasoconstrictor reaction in the finger after a cold hand-bath is considerably longer in the morning than in the evening.

In the practical field of physical therapy ENGEL & HELLER (5) investigated the circadian variations in thermal and circulatory response to an fango-paraffin packing of constant 47.5°C. The high temperature of this earthy composition, heated in an oven, and than wrapped around the arm, can be tolerated without danger for the skin. The consensual response of muscle and skin vessels of the lower leg was measured by means of venous strange-gauge plethysmography. In Fig.2 the result of the legs blood flow response to a 20 min hot packing on the arm can be seen: The curves demonstrate the mean values of measurements on the calf and the ankle for 12 subjects at four separate times of the day. Greater divergences in the increase rate occurring during the fango-paraffin application are to be observed at 4 different times of the day: During the circadian warming-up phase in the morning hours, only a small and brief increase in blood flow is to be seen, whereas at noon, already a greater, but equally short lasting increase is to be noted. In contrast to this, a higher increase in blood blow is found during the cooling-down phase, recorded at 20.00 h in the evening and 3.00 h at night, respectively.

Fig.3 shows the circadian variations of the mean differences between rest values and those obtained during stimulus application, both for circulatory response and subjective thermal parameters. As can be seen for all parameters, the rate of response to a local warm stimulus is greater, when the experiments are conducted at test times between 15.00 h in the afternoon and 3.00 h at night, and smaller at test times between 3.00 h in the early morning and 15.00 h in the afternoon. The differences in mean values of the increased blood flow at 9.00 h in the forenoon and 21.00 h in the evening are statistically significant.

In principle, physiological response to a thermal whole-body exposure at different times of day has to expect similar circadian variations to those of a local thermal stimulus. Fig.4 demonstrates the circadian variations of the mean differences between rest values and those obtained after a 15 min exposure to extreme hot-dry heat in a Finnish Sauna-bath, both for circulatory and thermoregulatory responses. For clarity, the curves have been smoothed and plotted twice by sliding averages. The

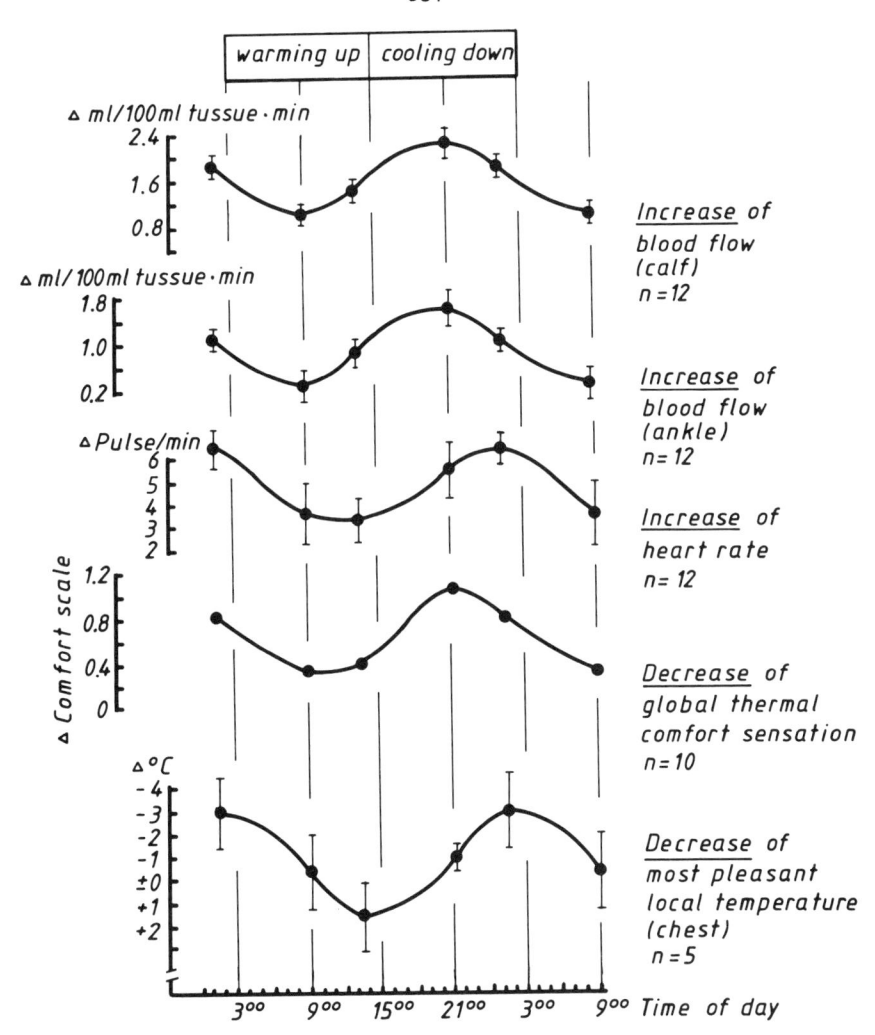

Fig.3: Daily course of mean (±SE) increase of blood flow in the leg and of heart rate as well as mean decrease of overall thermal comfort sensation and preferred local temperature on the chest of groups of healthy subjects during local heat application of a 30 min fango-paraffin packing on the arm (5).

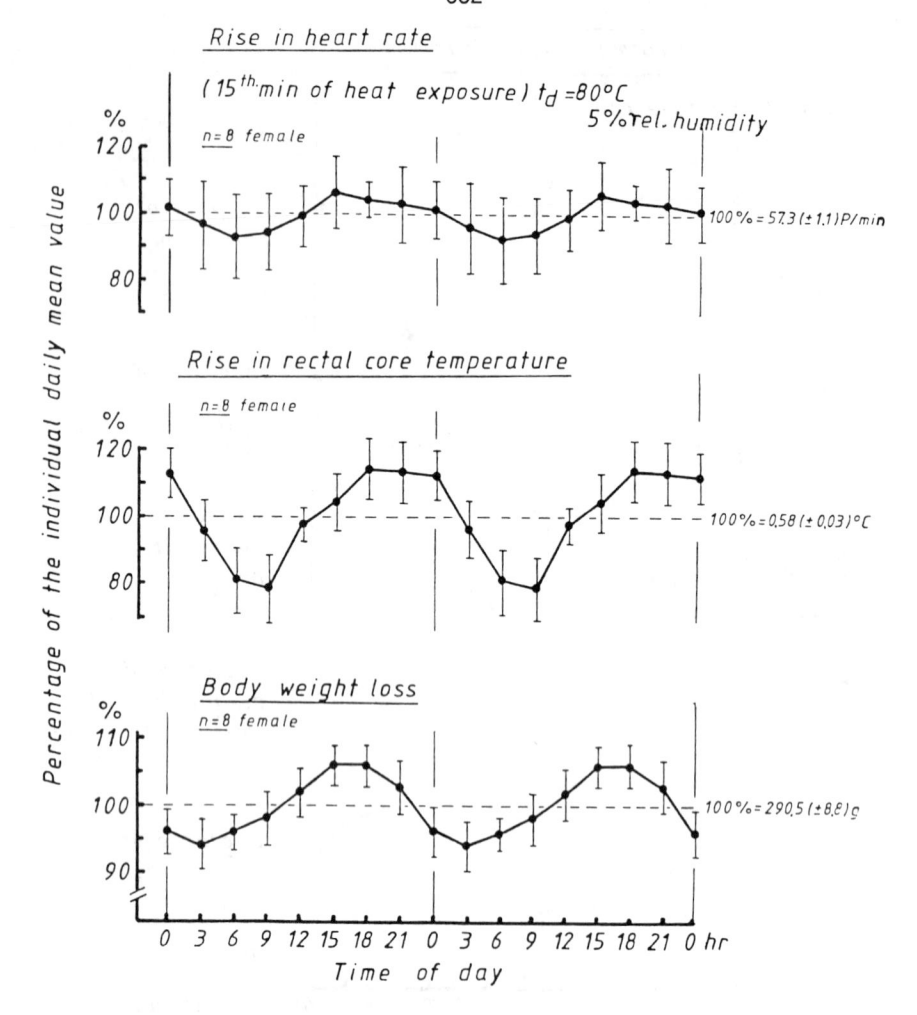

Fig.4: Daily course (in percentage) of mean (±SE) rise in rectal temperature and in heart rate as well as of mean loss of body weight of a group of healthy female subjects after a 15-min Sauna bathing. (For greater clarity the curves are smoothed and plotted twice by sliding average) (From 6).

tests were performed in summer. It can be seen in all curves, that the physiological responses to that extreme heat stress are greater when the Sauna-bathes were taken during the late afternoon and evening, and less during the second part of the night. The amplitude of the circadian variation of the rise in rectal temperature was on average 36%, and for loss of body weight by sweating 13%. These differences were statistically significant. Increase of heart rate was on average 13%, but the amplitude of the circadian variation was statistically not significant.

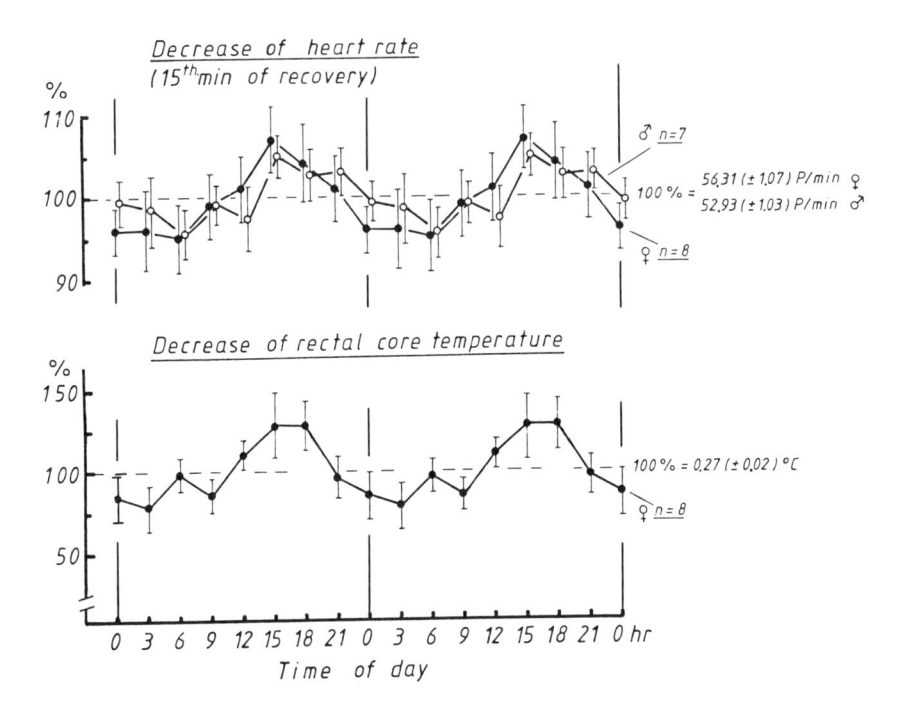

Fig.5: Circadian variations (in percentage) of mean (±SE) decrease of heart rate and of rectal temperature of the test groups at the 15th min of recovery after Sauna bathing (From 6).

A circadian variation can also be seen in the speed of body recovery processes after the Sauna-bath without usual cold water shower. Fig.5 shows the daily course (in percentage) of the mean decrease of heart rate of 7 male and 8 female subjects and of the rectal temperature (for the female group only) at test time 15 minutes of spontaneous cooling-down after Sauna-bath. The fastest speed of heat loss after Sauna is seen during the late afternoon and the early evening during the begin of the circadian cooling-down phase.

°C Rectal Temperature
n = 10

[graph with x-axis hours 6, 12, 18, 24 h; y-axis °C 36.5, 37.0, 37.5, 38.0; curves labeled HEAT and REST]

Zahorska-Markiewicz et al. 1990

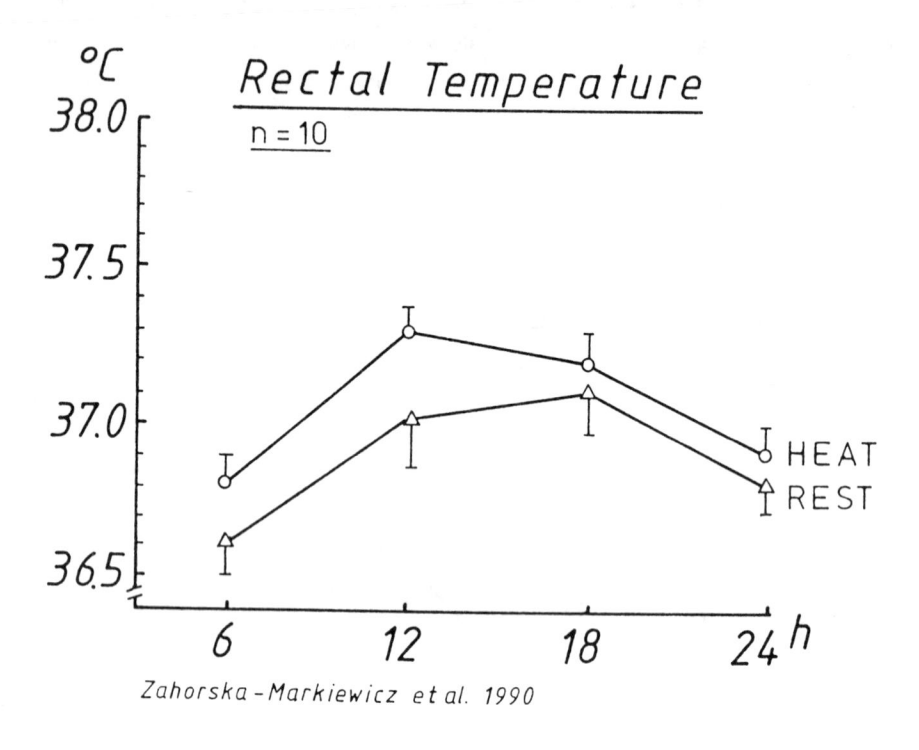

Fig.6: Circadian variations of mean (±SE) rectal temperature of 10 male subjects at rest and after a 50 min heat exposure at 40 °C and 56 % rel. humidity (From 18).

In contrast to our findings, ZAHORSKA-MARKIEWICZ et al.(19) received a different diurnal course of increase in rectal temperature after a 50 min heat exposure at rest in a climatic chamber at 42°C and 60% rel. humidity. These tests with 10 healthy men were conducted in (European) winter at 4 different times of the day between 9.00 h in the morning and midnight. As can be seen in Fig.6, increase of

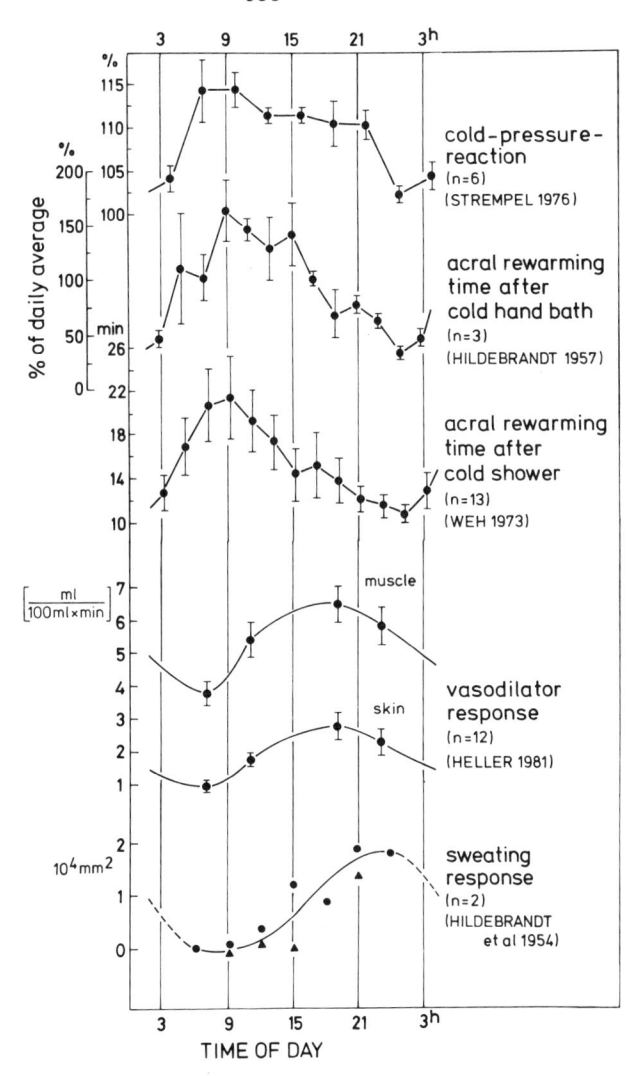

Fig.7: From top to bottom: Mean daily course of the cold-pressure reaction of systolic blood pressure, of the acral rewarming time after a cold hand bath (15°C, 5 min), of the acral rewarming time after cold shower according to KNEIPP (1821-1897) of the vasodilator responses of muscle and skin vessels of the lower leg after a hot pack, and of the sweating response of the forehead to a standardized hot stimulus. Brackets indicate standard errors (From 11).

core temperature is on an average higher in the morning than in the evening. Maximum heat storage was found at noon (12.00 h). The authors concluded, that heat stress has its greatest physiological effect during the early part of the day, when body temperature is lowest. They also found the highest increases of core temperature and heart rate in circadian tests they combined with additional workload. Also KOLLER et al. (14) found similar diurnal variations in heat stress response to those of ZAHORSKA-MARKIEWICZ et al. (19), but the respective increases of core temperature were very small and have not been proved to be statisticaly significant.

The next figures give some examples of circadian variations of physiological response to cold stimuli. WEH (18) investigated the circadian variations of peripheral thermal response to a 40 sec cold water shower of 12°C, applied over both shoulders and arms as an so-called "Oberguss", according to KNEIPP (1821-1897). The acral rewarming time of skin temperature of the middle finger after that cold stimulus was recorded. Circadian variations of the duration of vasoconstrictor reaction after the cold shower is shown in Fig.7. Rewarming time in the finger, as shown in the middle of the figure, is considerably prolonged during the forenoon warming-up phase. At 9.00 h in the morning, rewarming time was on average maximal, during the night minimal. The upper curve in the figure shows the mean daily course of the cold-pressure reaction, as investigated by STREMPEL (17). Blood pressure reaction after 1 minute ice-cold hand bath is much stronger in the morning than in the evening. The lower curves of the figure are a survey of the already discussed circadian variations of responses to warm stimuli.

Further investigations of AGISHI et al. (1) have shown, that the circadian variation of responsiveness to thermal stimuli also influences other regulatory systems like hormonal and metabolic reactions. A comparison of the hormonal response of plasma cortisol to hot and cold water bathing at 9.00 h in the morning and 21.00 h in the evening showed a stronger hormonal reaction to cold in the morning and that to hot bathing in the evening.

These findings on circadian variations of responses to thermal stimuli have shown that the circadian maximum of sensitivity to thermal stimuli does not occur at the time of the maximum or the minimum of the body temperature, but rather approximately in the middle of the two circadian phases, namely at the time, at which body temperature is changing most rapidly during its diurnal cycle.

To summarize, the organism is more sensitive to cold stimuli during the circadian warming-up phase in the morning, and more sensitive to heat stimuli during the cooling-down phase in the evening. Hence, in each case there is a particularly stronger response to those stimuli, which cause reactions, that increase the predominant phase direction of thermoregulation.

Of interest in this context are also measurements of ATTIA et al.(3) of thermal comfort as a drive for behavioral thermoregulation at different times of the day, resulting in higher cold defense in the forenoon.

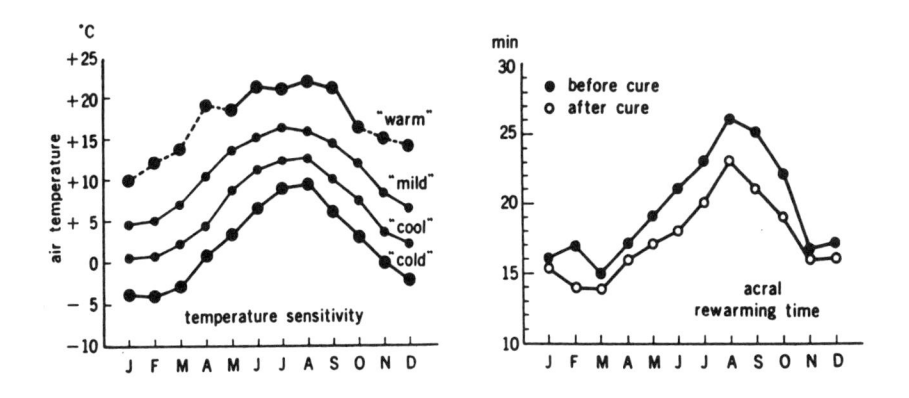

Fig.8: Left: Annual change in mean air temperature for different scaled subjective sensitivity to temperature at the time of going to outdoor; humid cooling grade of 70-80 mg cal/cm^2/sec. (From 16). Right: Annual change in mean acral rewarming time in patients with cure treatment in an station on the Baltic Sea, as measured at the beginning and at the end of cure treatment. Presented as smoothed monthly mean values (From 8, 11).

The lecture may be completed by looking at some remarkable fluctuations in thermal sensitivity also during the annual biological rhythm. After findings of SCHULZ (16) there is an annual variation of subjective sensitivity threshold for warm and cold. According to Fig.8 (left) the annual change in mean air temperature for different scaled subjective sensitivity to temperature at the time of going to out-door. The in February. This is in covariation with the annual extremes of the acral rewarming time after cold hand baths, as HENTSCHEL & SCHIRGEL (8) have found (Fig.8;).

In general, it is of practical importance to consider the various aspects of the spontaneous rhythmic variations in order to evoke optimal responses to thermal stimuli in Physical Therapy.

REFERENCES

1. AGISHI, Y. (1982): Circadian differences in the excretion responses of electrolytes and their related hormones to thermal stimuli by head-out water immersion in man. Int. J. Biometeorol. 26: 191
2. ASCHOFF, M.: Der Tagesgang der Körpertemperatur beim Menschen. Klin. Wschr. 33: 545-551
3. ATTIA, M., P. ENGEL & G. HILDEBRANDT: Thermal comfort during work. A function of time of day. Int. Arch. Occup. Environ. Health 45: 205-215
4. DAMM, F., G. DÖRING & G. HILDEBRANDT (1974): Untersuchungen über den Tagesgang der Hautdurchblutung und Hauttemperatur unter besonderer Berücksichtigung der physikalischen Temperaturregulation. Z. Phys. Med. u. Rehab. 15: 1-5
5. ENGEL, P. & M. HELLER (1984): Circadian variations in circulatory response to thermal stimuli. In: E.HAUS & H.F.KABAT (eds), Chronobiology 1982-83., S.Karger, Basel, pp 268-270
6. ENGEL, P. & W. HENZE (1989): Circadian variations of physiological response to a brief extreme heat exposure (Sauna bath). In: J.B.MERCER (eds.), Thermal Phyiology, Excerpta Medica, Amsterdam, pp. 311-314
7. HEISER, F. & L.H. COHEN (1933): Diurnal variations of skin temperature. J. Industrial Hyg.: 15: 243-256
8. HENTSCHEL, G & L. SCHIRGEL (1960): Beobachtungen über Funktionsänderungen der akralen Durchblutung als klimatherapeutischer Effekt. Arch.phys.Thr. (Leipzig): 12: 235-240
9. HILDEBRANDT, G., P. ENGELBERTZ & G. HILDEBRANDT-EVERS (1954): Physiologische Grundlagen für eine tageszeitliche Ordnung der Schwitzprozeduren. Z.klin.Med 152: 446-468
10. HILDEBRANDT, G. (1974): Circadian variations of thermregulatory response in man. In: L.E. SCHEVING, F.HALBERG (eds), Chronobiology. Thieme Stuttgart, 234-240
11. HILDEBRANDT, G. (1986): Chronobiologische Grundlagen der Ordnungstherapie. In: W.BRÜGGEMANN (ed.), Kneipptherapie (2.Auflage), Springer, Berlin-Heidelberg-New York, 170-221
12. KANEKO, M., N. ZECHMANN & R.E. SMITH (1968): Circadian variation in human peripheral blood flow levels and exercise responses. J. appl. Physiol. 25: 109-114

12. KANEKO, M., N. ZECHMANN & R.E. SMITH (1968): Circadian variation in human peripheral blood flow levels and exercise responses. J. appl. Physiol. 25: 109-114
13. KOE, F.K., W. HÖFER & K. LÜDERS (1968): Mittlere Hauttemperatur und periphere Extremitätentemperatur bei den tagesperiodischen Änderungen der Wärmeabgabe. Arch.phys.Ther.(Leipzig) 20: 221-226
14. KOLLER, M., M. KUNDI, F. KORENJAK & M. HAIDER: An experimental study on day and night exposure to combined work loads. In: OGINSKI, A., J. POKORSKI & J. RUTENFRANZ (Eds.), Contempory advances in shiftwork research; theoretical and practical aspects in the late eighties, Med. Acad. Krakow, 53-64
15. LAMPERT, H. (1953): Rhythmische Reizbarkeitsänderung des Organismus und ihre Bedeutung für die Krankenbehandlung. Verh. III. Konf. Int. Ges. f. Biologische Rhythmusforschung 1949. Acta med. scand. Suppl. 278: 141-144
16. SCHULZ, L. (1960): Der jahreszeitliche Gang der Temperaturempfindung des Menschen anhand einer zehnjährigen Beobachtungsreihe. Arch.phys. Ther. (Leipzig): 12: 245-255
17. STREMPEL, H. (1976): Der Tagesgang der Cold-Pressure-Reaktion unter Ausschluß von Kälte-Habituation. Z.Phys.Med. 5: 37-41
18. WEH, W. (1973): Tageszeitliche Wirkungsunterschiede des Obergusses nach Kneipp. Ein Beitrag zur Tagesrhythmik der Thermoregulation. Med. Inaug. Diss. Marburg-Lahn
19. ZAHORSKA-MARKIEWICZ, B., M. DEBOWSKI, F. SPIOCH et.al. (1989): Circadian variations in psychophysiological responses to heat exposure and exercise. Eur. J. Appl. Physiol. 59: 29-33

INFLUENCE OF BALNEOTHERAPY ON SKIN VASOMOTION

W. SCHNIZER

Privatklinik St.Raphael, Praxis Parkhotel, Am Kurwald 10,
D-W-8394 Bad Griesbach, FRG

To stimulate and to train organs and their functions is common practice in balneotherapy and an established basis of medical treatment. There exists a large amount of information about reactions which are induced by physical stimuli in different organs. Primarily various aspects of the circulatory system have been subject of investigation.

The Laser-Doppler Flowmetry is a relatively new method for registering the microcirculation of the skin in a sensitive and noninvasive manner (4,5). This is of special interest in the research of balneotherapeutic and physical medicine since many therapeutic treatments involve applications to the skin and are considered to have an effect on the circulation. Therefore the effect of thermic and chemical stimuli on the vasomotor reactions of the skin have been studied in various ways.

With the Laser-Doppler method it is possible to register the so called vasomotion of the skin in an objective manner. It is a spontaneous rhythmic phenomenon, which can be observed in various organs and tissues in both humans and animals. It is generated in small arterial vessels which open and close periodically. These fluctuations in blood flow are of quasisinusoidal character, found in about 70% of cases, and have also been described by means of other parameters, i.e. oscillation in capillary pressure, rhythmical change in vessel diameter and capillary erythrocyte velocity (3,7). The sinusoidal curves that are obtained by Laser-Doppler Flowmetry can be analyzed with parameters such as frequency and amplitude. In general the frequency ranges from 3 to 10 per minute. At the same locus it remains constant, whereas the amplitudes can be more or less variable.

The phenomenon seems to be a form of autoregulation of vascular smooth muscle tone, but the various cellular mechanisms underlying vasomotion are not yet fully understood. Vasomotion is considered to be a clue phenomenon of microcirculation. It seems to be a primary quality of a locally well functioning circulatory system and to influence all reactions at the level of microcirculation, such as for instance blood distribution, capillary fluidand metabolic exchange, as well as lymphatic fluid production. Hydrotherapy as well as CO_2 baths belong to the traditional balneotherapeutic methods. We therefore mainly studied thermic influences and the effect of CO_2 on the microcirculation of the skin (1,2,6). The Laser-Doppler flowmeter Periflux (Perimed K B, Stockholm, Sweden) was used to record findings on healthy volunteers ranging from 23 to 46 years of age. The room temperature was kept at 22°C ±0.5, with a relative humidity of 25-35%.

Measurements were conducted on the skin by means of a specially designed chamber which also served as a laser probeholder. This was attached to the skin by double adhesive tapes. The chamber served for local application of small aquaeous solutions of CO_2. In this way the laser beam penetrated the sample before entering the skin. The area of the skin was kept at a constant temperature by means of a water thermostat. To study the isolated effects of thermal stimuli, the chamber was used as a thermode, perfused by thermoregulated water.

At first some results about these thermal effect, especially concerning vasomotion. With the laser method it can easily be demonstrated that an increase in temperature leads to an increase in circulatory flow whereas a decrease in temperature leads to a smaller flow (Fig.1). A strong local cold stimulus leads immediately to a vasoconstriction. In this case the vasomotion is interrupted. Sometimes a decrease of the vasomotion during the cooling-down phase can be observed with a very low-frequency blood flow fluctuation. After a rewarming period the vasomotion recovers very rapidly. The frequency of vasomotion shows a clear variability with the temperature and a positive correlation with a temperature increase. Under extreme local temperature conditions the vasomotive process becomes disturbed. This is to be observed at a temperature above 39-40°C and below 12-15°C.

We also carried out experiments to test the consensual reaction of vasomotion in respect to thermic stimuli. In this situation the locus of the thermic exposition and the locus of measurement are separate from each other. In connection with this we

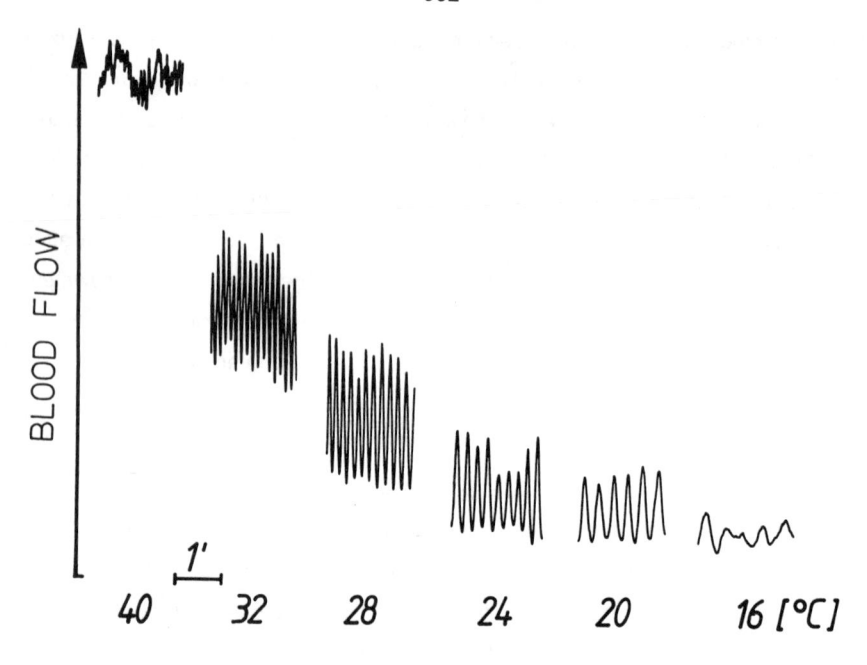

Fig.1: Skin blood flow recorded by Laser-Doppler-Flowmetry, effects of temperature on microcirculatory flow and vasomotion

also took measurements of the nasal mucous membranes. A consensual vasomotoric reaction to cooling follows always in the same direction as the local reaction. The consensual vasoconstriction is of a comparatively smaller degree and does not demonstrate a disturbance of vasomotion. Sometimes, however, one can observe an accentuation of vasomotion during a consensual reaction. This is demonstrated by an increase in amplitude without a change in frequency.

These were some examples of the thermic influences on vasomotor activity and vasomotion of the skin. This subject has not been studied systematically and in all details, as for instance the adaptive changes during thermic training are concerned.

Another aspect of our balneotherapeutic research with the Laser-Doppler flowmetry was the study of the effect of the CO_2 baths. The induction of skin hyperemia by means of the CO_2-bath is a wellknown phenomenon in balneotherapy. The external application of CO_2 leads to local hypercapnia and dilates the cutaneous blood vessels. We investigated these effects by using the probe-chamber mentioned above.

After CO_2 equilibrated water was filled into the chamber, blood flow increased within less than one minute, reaching a plateau after 2 to 3 minutes. Concerning vasomotion there was a marked increase in vasomotion amplitude (Fig.2). This figure shows CO_2-effects which were obtained by the application of varying concentrations of CO_2 on the same area of skin. Blood flow and also vasomotion amplitude showed dependence on CO_2-concentration and both increased markedly. After cessation of exposure to CO_2, when the chamber was emptied and refilled with water, levels reverted to those of the control period within 1 to 2 minutes. Higher temperature and CO_2 concentration resulted in higher values of blood flow and partially also of vasomotion amplitude, while vasomotion frequency was influenced only by temperature. The lowest (minimal) effective dose was 400-600 mg CO_2/l H_2O.

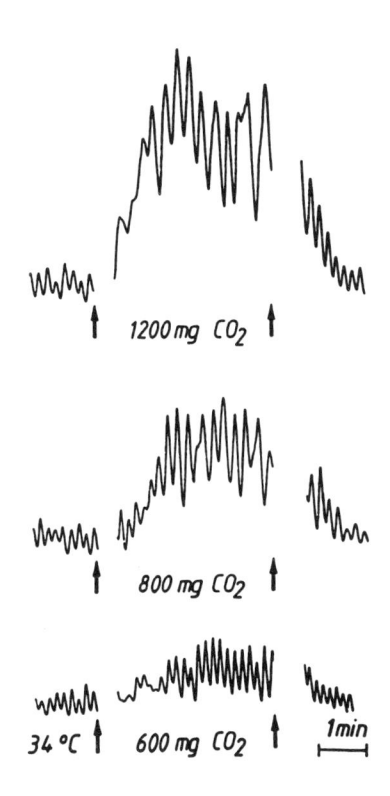

Fig.2:
Effects of external locally applied CO_2 on skin blood flow as measured with Laser-Doppler-Flowmetry; increase in microcirculatory flow and stimulation of vasomotion amplitudes.

Carbon dioxide as a physiological product of metabolism is one of the local chemical factors in the vasculature environment that contributes to the regulation of blood flow. The effects of external CO_2 application on vasomotion of skin microcirculation which was presented in this study constitute a new finding. Our findings suggest that it is possible to differentiate between two microcirculatory effects of CO_2 on the skin, i.e. an increase in total perfusion and an activation of rhythmical vasomotion, both depending on the concentration. It might be possible to deduce whether both these effects can be obtained independently from one another.

The application of hydrotherapy of CO_2 baths for the treatment of peripheral circulatory disturbances and hypertension is an established tradition in balneotherapy. However, the success of this form of treatment is largely gauged empirically and therefore it is useful to have an objective quantitative assessment of the vascular effect of temperature and of CO_2 on the skin. By using Laser-Doppler Flowmetry it is possible to study blood flow as well as vasomotion.

Vasomotion is a special modality of microcirculation and it could be that as a parameter of balneotherapeutic and physiotherapeutic effects it warrants more attention.

REFERENCES

1. ERDL, R., W. SCHNIZER, P. SCHÖPS et al. (1986): Experimenteller Beitrag zur Wirkung hydrotherapeutischer Teilanwendungen auf das konsensuelle Durchblutungsverhalten an Haut und Schleimhäuten. Z. Phys. Med. Baln. Med. Klim. 15, 411-414
2. ERDL, R., W. SCHNIZER, R. GRÖTSCH et al. (1987): Kältevasokkonstriktion, Kältevasodilatation und reaktive Hyperämie der Haut, dargestellt anhand der Laser-Doppler-Flußmessung. Z. Phys. Med. Baln. Med. Klim. 16, 94-98
3. FUNK, W., B. ENDRICH, K. MESSMER & M. INTAGLIETTA (1983): Spontaneous arteriolar vasomotion as a determinant of peripheral vascular resistance. Int. J. Microcirc.: Clin. Exp. 2, 11-25
4. NILSSON, G.E., T. TENLAND & P.A. ÖBERG (1980): A new instrument for continuous measurement of tissue blood flow by light beating spectroscopy. IEEE Trans. Biomed. Eng. BME 27, 12-19
5. NILSSON, G.E., T. TENLAND & P.A. ÖBERG (1980): Evaluation of a laser-Doppler flowmeter for measurement of tissue blood flow. IEEE Trans. Biomed. Eng. BME 27, 597-604
6. SCHNIZER, W., R. ERDL, P. SCHÖPS & N. SEICHERT (1985): The effects of external CO_2 application on human skin microcirculation investigated by laser Doppler flowmetry. Int. J. Microcirc.: Clin. Exp. 4, 343-350
7. TENLAND, T., E.G. SALERUD, G.E. NILSSON & P.A. ÖBERG (1983): Spatial and temporal variations on human skin blood flow. Int. J. Microcirc.: Clin. Exp. 2, 81-90

CIRCADIAN VARIATIONS OF PHYSICAL TRAINING.

CHR. GUTENBRUNNER

Institut für Kurmedizinische Forschung Bad Wildungen, Langemarckstr. 2,
D-W-3590 Bad Wildungen, FRG

INTRODUCTION

Training is one of the most important therapeutical priciples in physical medicine and cure treatment. Rehabilitative training on the one hand aims at an increase of functional capacity of the muscular and cardio-respiratory system, on the other hand at an improved coordination of motoric functions. Influences of the circadian system have to be considered mainly under two aspects, which are the actual state of fitness and motivation and possible variations of long term training effects. However, the interpretation is difficult because of the complexity of factors which influence physical fitness and training processes (see also 13, 22).

ACTUAL PHYSICAL FITNESS

The locomotor system is controlled by central nervous functions, of which first of all vigilance shows a marked circadian variation (18). In Fig. 1 the circadian variation of vigilance as measured by an acoustical reaction test is shown in the upper curve (23). The other curves represent different coordinative functions. All curves show maxima during daytime and minima during night around 3.00 h (1, 7, 17, see also 13). Congruent circadian variations are known in sensoric functions e.g. optical and tactile performance. The well known depression of vigilance about noon does not play an important role in these registrations in recovered test persons which were carried out under constant resting conditions. It occurs only in states of insufficient recovery, e.g. after night shift work or in the state of jet lag (14).

The circadian course of voluntary muscle strength is congruent with the vigilance functions (Fig. 2). In accordance with numoreus studies the circadian

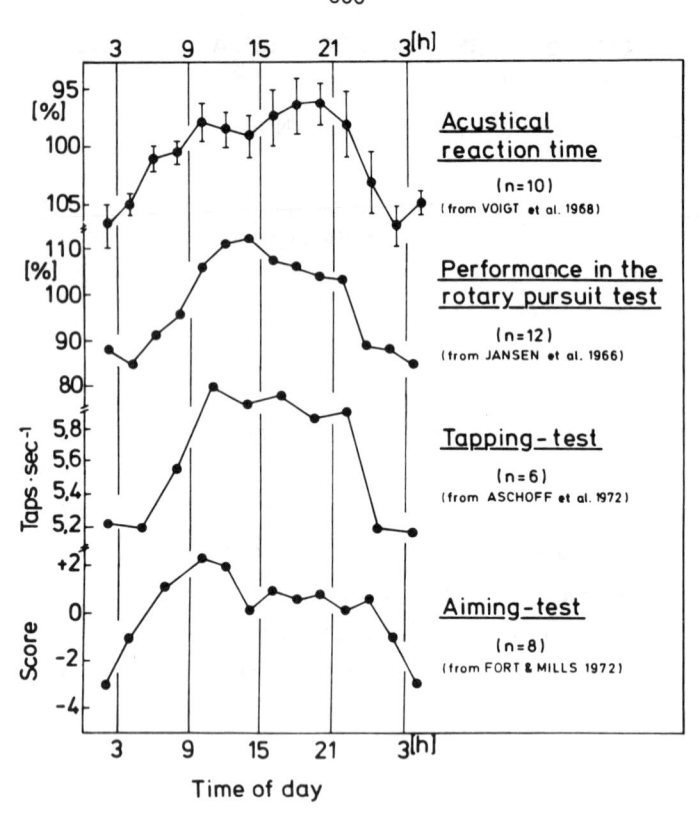

Fig.1: Circadian variations of acoustical reaction time and performance in different sensomotoric tests. Data from the literature (13).

amplitude amounts to around 10 % (8). However, vigilance plays an important role in voluntary muscle strength. It is influenced by variations in the peripheral neuro-muscular junction. As BERG et al. (4) could show, the electrical energy needed to stimulate a skeletal muscle in man is minimal during daytime and exhibits maximum values at 3.00 h (Fig.3).

However, the working capacity of the skeletal muscle exhibits an inverse circadian variation. As Fig.4 shows the ergographic muscular performance is maximal at 3.00 h (5). This variability correlates with the cardio-respiratory capacity, which

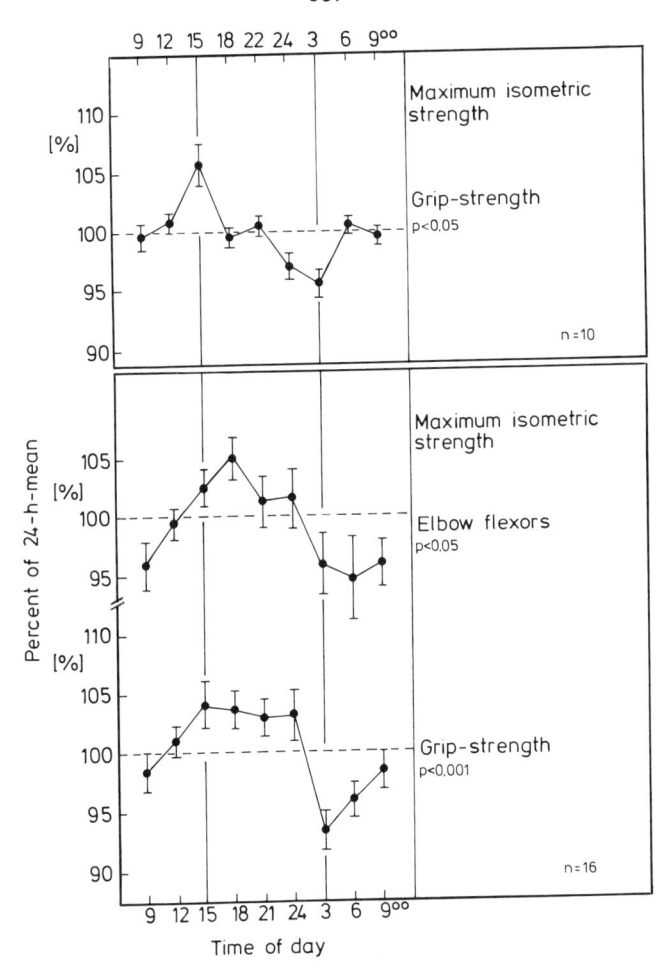

Fig.2: Circadian variations of the mean voluntary strength of different muscle groups in healthy subjects. Brackets indicate standard errors (8).

is demonstrated in the second curve of the figure. These data from VOIGT et al. (23) show that the physical working capacity increases during night time synchronous with a minimal oxygen consumption for a defined work load (24). However, this characteristic is only detectable in the range of submaximal performance and does not occur

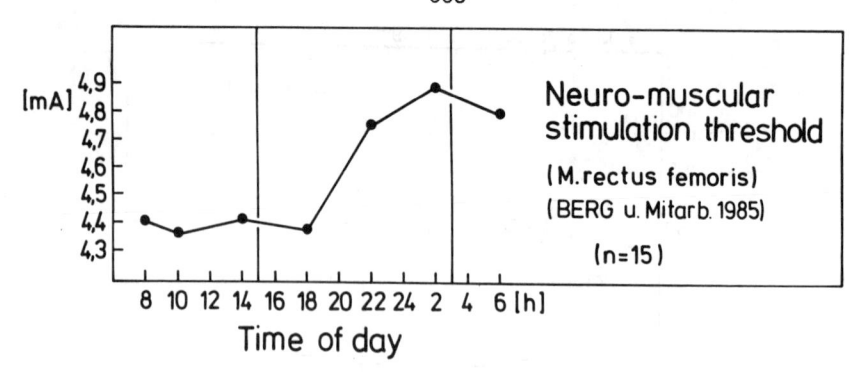

Fig. 3: Circadian variations of the neuromuscular stimulation threshold of the M. rectus femoris in healthy test persons. Data from (4).

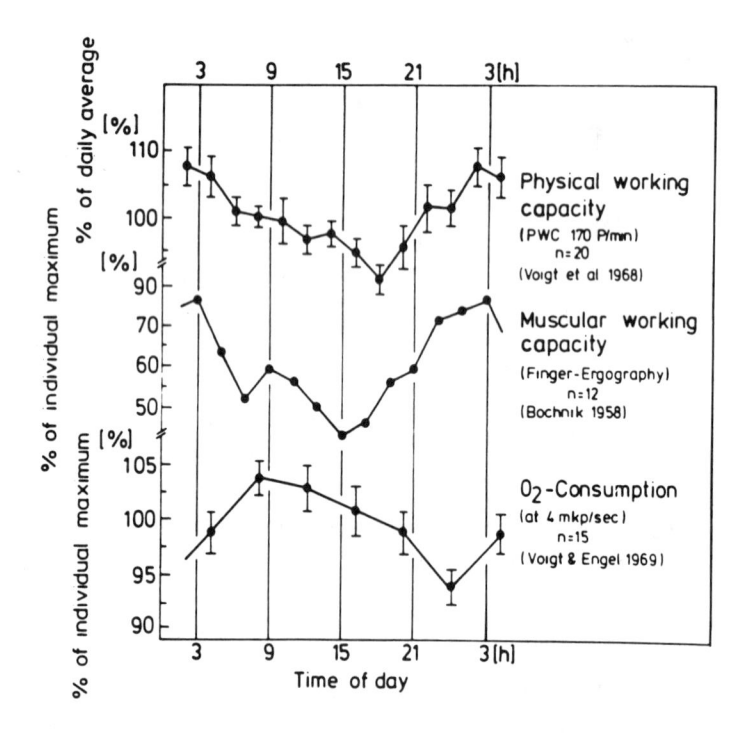

Fig.4: Circadian variations of the physical and muscular working capacity and the O_2-Consumption at a defined work load. Data from the literature (13).

in the range of maximal oxygen consumption (25). The nightly maximum of physical working capacity can be explained with the increased trophotropy of the autonomous nervous system combined with a higher degree of coordination of rhythmical functions of the circulatory and respiratory system (13). However, other factors, e.g. changes of blood viscosity, red blood cell count, activity of oxydative enzymes or the electrolyte balance, have to be considered.

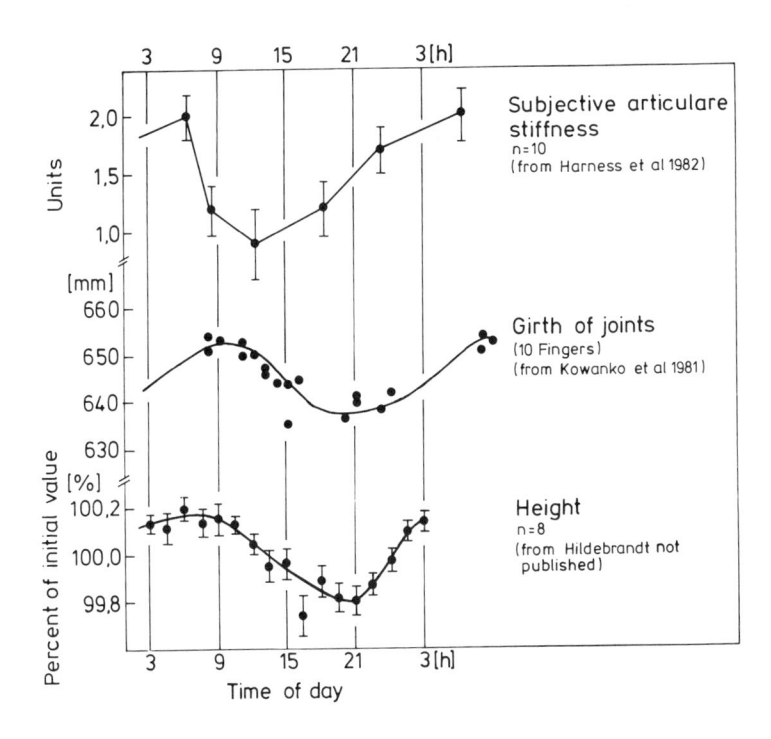

Fig.5: Circadian variations of the subjective articulare stiffness in patients suffering from rheumatoid arthritis, the girth of the finger joints, and the heigth of healthy subjects. Data from the literature (13)

In kinesitherapy often an optimal movability of the joints is the predominant aim of the exercises. Therefore the circadian variations of flexibility and cartilage water content are of some importance. In healthy test persones in the literature only exist reports on flexibility of the spinal cord, which is maximal during the early afternoon

(3). In patients suffering from rheumatism the so-called morning-stiffness is a well known symptom. Systematical ratings of the subjective stiffness of the joints investigated by HARKNESS et al. (10) are shown in Fig.5. Maximum stiffness occurs in the early morning, decreasing rapidly till noon. Almost inversely run the curves of the articular swelling of the fingers (19) and the height of healthy test persons (13). Obviousely the state of hydration of the articular cartilage varies around the day exhibiting a higher water content in the early morning.

TRAINING PROCESSES AND OVERSTRAIN-REACTIONS

Besides the actual state of physical fitness, the load capacity and training ability influence the therapeutic results. Interestingly the circadian variations of the training

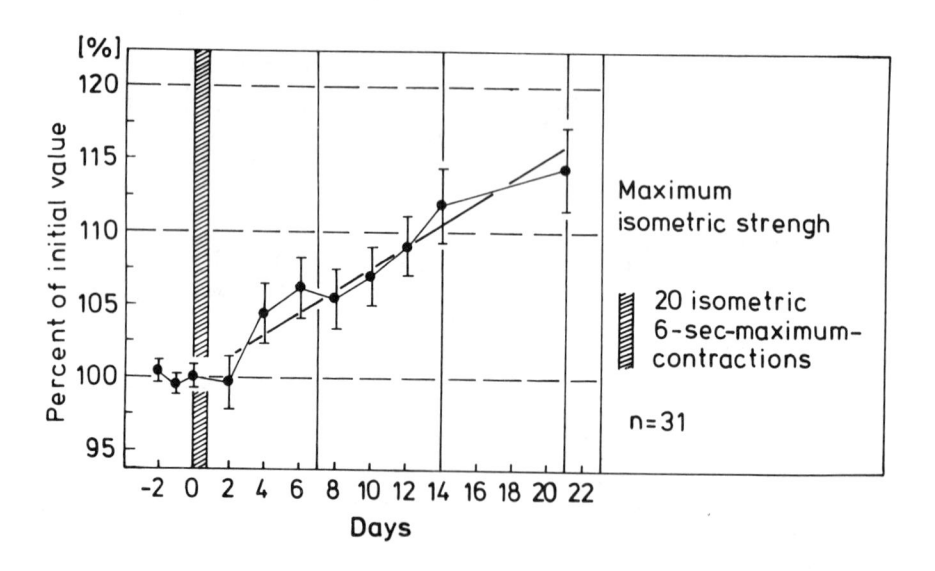

Fig.6: Mean course of the maximum voluntary strength of the elbow flexors in healthy test persons after a one-day isometric training consisting of 20 voluntary maximum contractions, expressed in percent of the mean initial values. Brackets indicate standard errors (8).

ability of muscular strength and cardio-respiratory capacity show similar circadian rhythms. The circadian variations of the response of skeletal muscles to isometric strength training was investigated in our group in several experiments. First of all we trained 31 healthy test persons with a one-day isometric training using 20 maximum voluntary contractions of 6-sec duration each. The training concerned only the elbow flexors. Subgroups of 4 subjects carried out the training at different times around 24 hours (for details see 8).

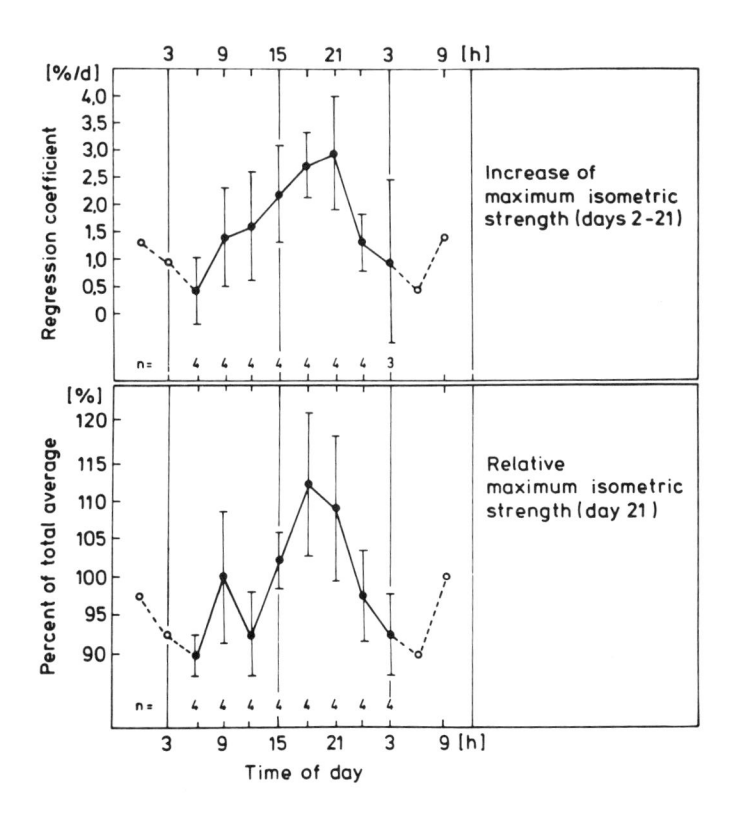

Fig.7: Mean daily increase of maximum isometric strength and mean maximum strength 21 days after a one-day isometric strength training of the elbow flexors in healthy subjects, expressed in percent of the total average. Brackets indicate standard errors (15).

In order to prove the effect of the relatively low training load, in Fig. 6 the mean course of the maximum voluntary strength of the respective muscles is shown during three weeks after the training period on day zero. The increase of maximum isometric strength amounts to five percent per week, which is a normal value for muscular training experiments (12). The mean daily increase of maximum isometric strength as well as the maximum strength which was reached after three weeks show a clear circadian variation (Fig.7) (8,15). Maximum trainig ability obviously occurs during the early evening between 18.00 and 21.00 h. These results coincide with an earlier study of RIECK & KASPAREIT (20).

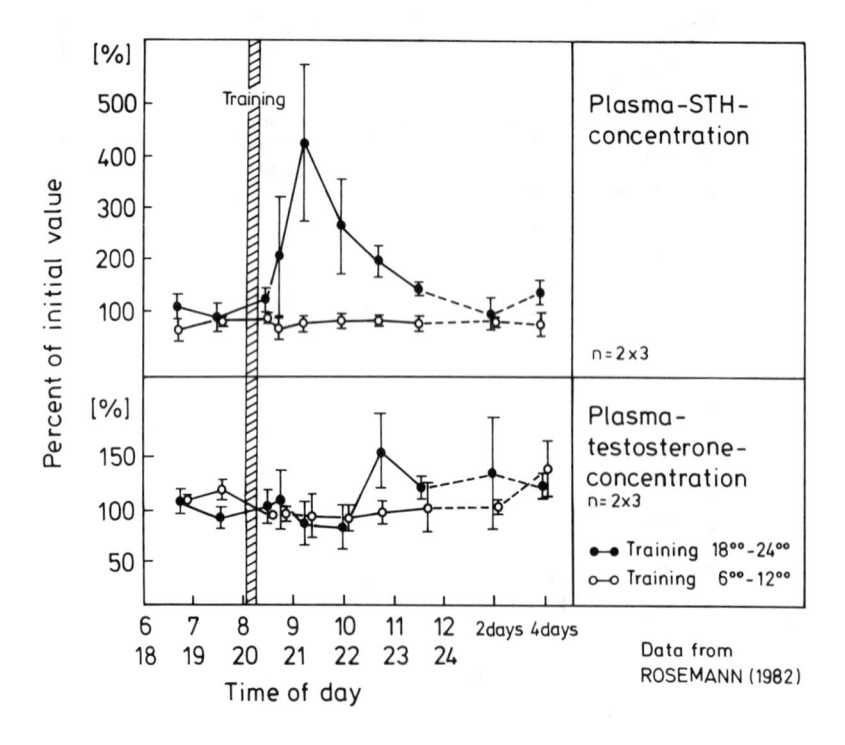

Fig.8: Mean plasma STH- and testosterone-concentrations before and after an isometric muscle training of the quadriceps muscle, consisting of 10 maximum voluntary contractions. The training was carried out either between 6.00 and 12.00 (open symbols) or between 18.00 and 24.00. Brackets indicate standard errors (21).

Looking for the mechanism of this phenomenon, ROSEMANN (21) followed up the hormonal responses to a one-day isometric training of the quadriceps muscle. The training consisted of 10 maximum voluntary contractions. Fig. 8 shows the mean responses of the somatotropic hormone and testosterone in two groups of test persons of which the one carried out the training in the early morning, the other in the afternoon. Whereas the morning-training group showed almost no increase of the mean plasma concentration of the somatotropic hormone, the afternoon group exhibited a marked increase with maximum values one hour after the training. The same tendency was found in testosterone, however, in this case the difference was not significant. It can be concluded, that the circadian variation of muscular training ability is caused by differences in the hormonal responses to the training. This fact can be understood from the well known interactions between the optic nerve, the supraoptic nuleus and the pituitary gland.

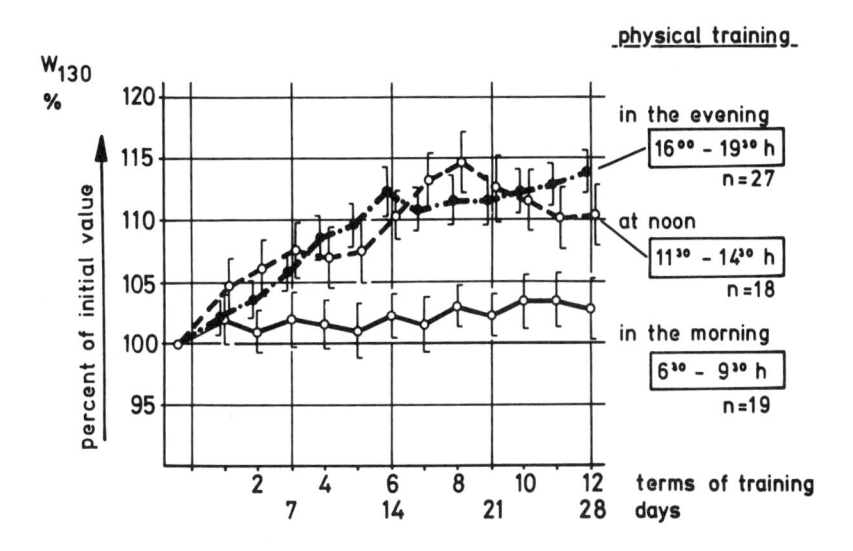

Fig.9: Mean physical working capacity (W_{130}) during a 28-days ergometer training at different times of day. Brackets indicate standard errors (2).

The effect of a dynamic ergometer training at different times of day was followed up by BAIER & ROMPEL (2). They trained three groups of test persons for four weeks on a treadmill ergometer. The increase of physical performance was followed up by means of the physical working capacity for 130 pulses per minute. As Fig.9 shows, the training in the morning had almost no effect as compared to the other subgroups with training at noon or in the evening.

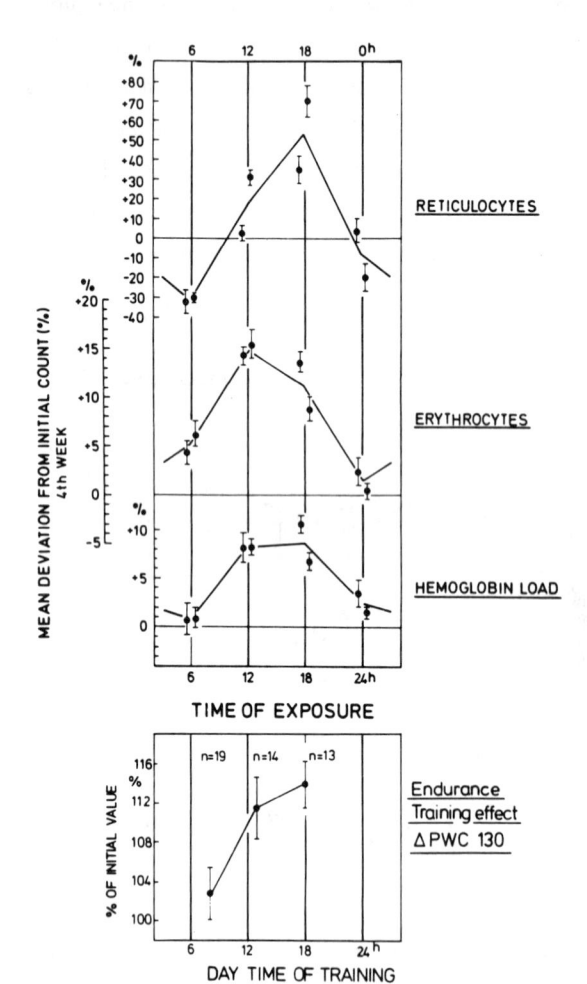

Fig.10: Circadian courses of the mean changes of the indicated hematologic parameters of healthy subjects 4-weeks after a intermittent experimental low oxygen stress in a low pressure chamber corresponding to an altitude of 2000 m (above)(11) and mean increase of the physical working capacity after a 4-week ergometer training (2). Brackets indicate standard errors (13).

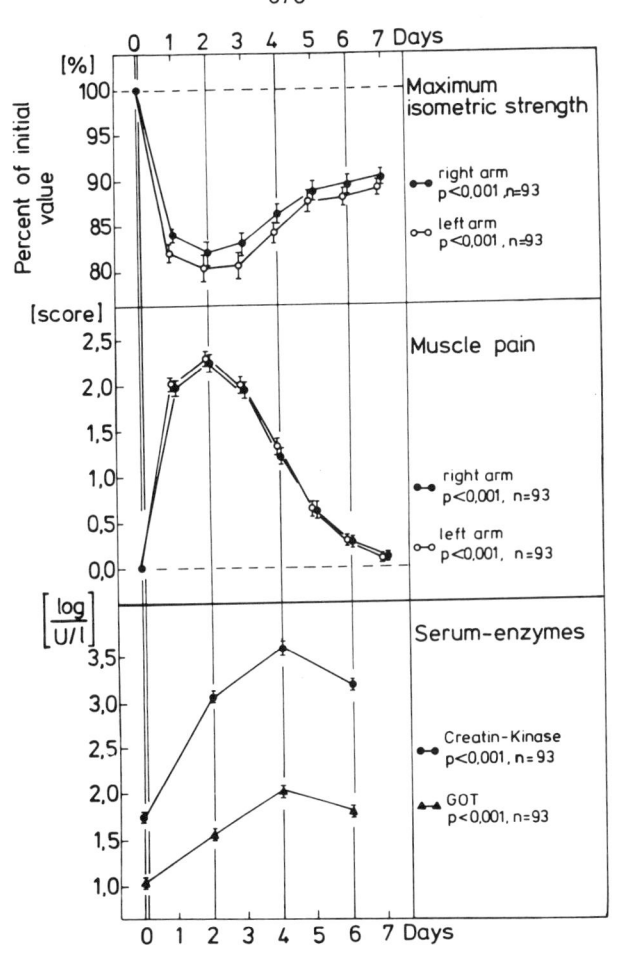

Fig.11: Mean courses of maximum isometric strength, muscle pain and serum-enzymes after a one-day heavy dynamic muscle strain in healthy test persons. Brackets indicate standard errors (8).

The specific stressor for the increase of the cardio-respiratory capacity is the lack of oxygen in the peripheral tissues. Therefore the circadian variability of the reaction to mild hypoxia is of interest. In Fig. 10 three parameters of the erythropoetic reaction after intermittent experimental low oxygen stresses, corresponding to an altitude of 2000 m, are demonstrated. These results, which were published by HECKMANN et al. (11) show the same phase dependency of the adaptive reactions as we saw in the study cited above: The maximum reaction occurs in the afternoon around 18.00 h. This circadian phase-position corresponds to the circadian minimum of actual physical working capacity which obviously is linked to an higher necessity of adaptation.

Besides of therapeutical adaptation physical therapy in the case of overdosage may lead to overstrain reactions (ref. see 8). In the case of muscular training overstrain causes muscle pain and destruction of myofibrils. Fig.11 shows a one-week follow-up of different overstrain parameters after a one-day heavy dynamic muscle strain, with a high amount of negative work. Besides of the pain which is most intensive on the second day after the strain a marked loss of maximum isometric strength occurs which persists even after the disappearance of the pain. This so-called delayed onset muscle soreness includes the occurence of cellular enzymes in the plasma, which is shown in the lower part of the figure.

In cooperation with THEIL (9) we performed a study in which delayed muscle soreness was evoked at different times of the day. The work load was standardized and dosed dependent on the individual muscle strength. The intensity of the mentioned overstrain symptoms is demonstrated in Fig.12. All parameters show significant circadian variations with the highest intensity of overstrain symptoms in the group which was trained around noon. This maximum corresponds with the phase of maximum voluntary muscle strength what can be understood to a certain degree taking into account the fact that the symptoms of the delayed onset muscle soreness are evoked by a mechanical damage of the muscle cells (see 8).

In sensomotoric adaptation the circadien variability of effectiveness exhibits a different phase position. Fig.13 shows the results of a study of HILDEBRANDT & STREMPEL (16) who measured the increase of sensomotor performance using the pursuit rotor test. As the figure shows the increase of contact time is maximal in the forenoon. This characteristic coincides with the maximum dampening of the cold pressure reaction and the best function of the short time memory (6). Sensomotor performance and motoric coordination is highly correlated with central nervous functions especially short time memory.

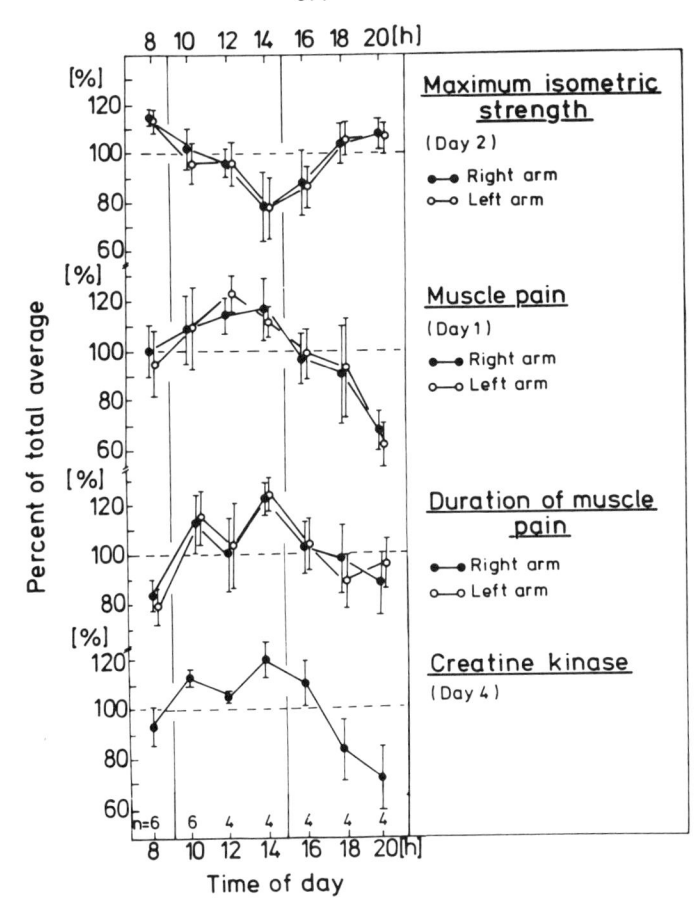

Fig.12: Mean intensity of the indicated overstrain symptoms after a one-day heavy dynamic muscle strain, which was carried out at different times of day. Data in percent of the respective total mean. Brackets indicate standard errors (9).

All in all, the results demonstrated show that a general recommendation for an optimal time of day for physical training therapy cannot be given. Such recommendations have to be referred to the specific aim of therapy, e.g. increase of muscle strength, physical capacity or sensomotor performance.

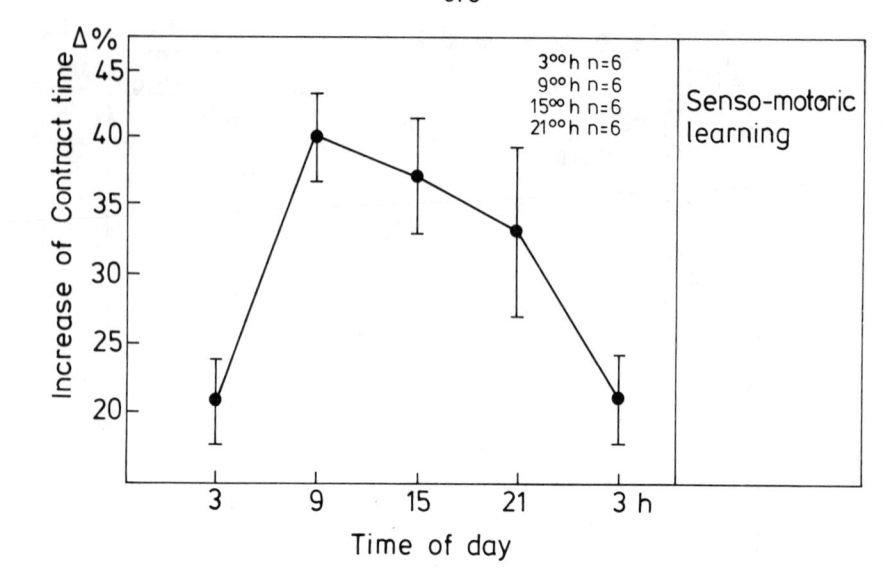

Fig.13: Mean circadian variation of the increase of contact time in the pursuit rotor test in healthy subjects. Brackets indicate standard errors (16).

REFERENCES

1. ASCHOFF, J., H. GIEDKE, E. PÖPPEL & R. WEVER (1972): The influence of sleep interruption on circadian rhythms in human performance. In: W.P. COLQUHOUN (Ed.): Aspects of human efficiency - Diurnal rhythm and loss of sleep. The Engl. Universities Press, London, pp. 128-152.
2. BAIER, H. & Chr. ROMPEL (1977): Der Einfluss thermischer Umgebungsbedingungen auf den Trainingserfolg beim Ausdauertraining. Arbeitsberichte des Sonderforschungsbereiches "Adaptation und Rehabilitation" (SFB 122), Band IV: 547-582.
3. BAXTER, C. & R. REILLY (1983): Influence of time of day on all-out swimming. Brit. J. Sports Med. 17: 122-127.
4. BERG, A., D. GÜNTHER & J. KEUL (1985): Neuromuskuläre Erregbarkeit und körperliche Aktivität. Dt. Z. Sportmed., Sonderheft 1985: 4-8.
5. BOCHNIK, H.J. (1958): Tagesschwankungen der muskulären Leistungsfähigkeit. Dtsch. Z. Nervenheilk. 178: 270-275.
6. FOLKHARD, S. & T.H. MONK (1985): Circadian rhythms in human memory. Brit. J. Psychol. 71: 295-307.
7. FORT, A. & J.N. MILLS (1972): Influence of sleep, lack of sleep and circadian rhythm on short psychometric tests. In: W.P. COLQUHOUN (Ed.): Aspects of human efficiency - Diurnal rhythm and loss of sleep. The Engl. Universities Press, London, pp. 155-172.
8. GUTENBRUNNER, Chr. (1990): Muskeltraining und Muskelüberlastung. O. Schmidt-Verlag, Köln.
9. GUTENBRUNNER, Chr. u. P. THEIL (1988): Tagesrhythmische Schwankungen der Überlastungsempfindlichkeit der Skelettmuskulatur bei dynamischer Arbeit. In: E. BAUMGARTNER, W. BRENNER, P. DIETRICH & J. RUTENFRANZ (Hrsg.): Verhandlungen der Dt. Ges. f. Arbeitsmed. Gentner-Verlag, Stuttgart, S. 375-380.
10. HARKNESS, J.A.L., M.B. RICHTER, G. PANAYI et al. (1982): Circadian variation in disease activity in reumatoid arthritis. Brit. Med. J. 284: 551-554.
11. HECKMANN, Chr., G. HILDEBRANDT, E. HOHMANN et al.(1979): Über den Einfluss der Tagesrhythmik auf die erythropoetische Reaktion. Untersuchungen nach intermittierender Unterdruckbelastung. Z. Phys. Med. 8: 135-144.
12. HETTINGER, Th. (1983): Isometrisches Muskeltraining. Thieme-Verlag, Stuttgart.
13. HILDEBRANDT, G. (1988): Die Bedeutung circadianer Rhythmen für die Bewegungstherapie. Z. Phys. Med. 17: 126-141.
14. HILDEBRANDT, G. (1990): Allgemeine Grundlagen. In: H. DREXEL, G. HILDEBRANDT, K.F. SCHLEGEL u. G. WEIMANN (Hrsg.): Physikalische Medizin. Band I. Hippokrates-Verlag, Stuttgart, S. 13-80.
15. HILDEBRANDT, G., Chr. GUTENBRUNNER, Chr. REINERT & A. RIECK (1990): Circadian variation of isometric strength training in man. In: E. MORGAN (Ed.): Chronobiology and Chronomedicine. Vol. II. P. Lang, Frankfurt-Bern-New York-Paris, pp. 322-329.

16. HILDEBRANDT, G. & H. STREMPEL (1977): Chronobiological problems of performance and adaptational capacity. Chronobiologia IV: 103-115.
17. JANSEN, G., J. RUTENFRANZ u. R. SINGER (1966): Über die circadiane Rhythmik sensomotorischer Leistungen. Int. Z. angew. Physiol. 22: 65-83.
18. KLEIN, K.E., R. HERMAN, P. KUKLINSKI & H.M. WEGMANN (1977): Circadian performance rhythms: Experimental studies in air operations. In: R. MACKIE (Ed.): Vigilance: Theory, Operational Performance, and Physiological Correlates. Plenum Publ. Corp. New York, pp. 111-132.
19. KOWANKO et al. (1981), cited by HILDEBRANDT 1988.
20. RIECK, A. & A. KASPAREIT (1976): Zur Frage tagesrhythmischer Änderungen von maximaler Muskelkraft und Extremitätendurchblutung nach isometrischen Kontraktionen. In: G. HILDEBRANDT (Hrsg.): Biologische Rhythmen und Arbeit. Springer- Verlag, Wien-New York, S. 21-29.
21. ROSEMANN, J.P. (1982): Untersuchungen über endokrine Reaktionen nach isometrischem Krafttraining zu verschiedenen Tageszeiten. Med. Inaug.-Diss., Marburg/Lahn.
22. WIGNET, C.M., C.W. DeROSHIA & D.C. HOLLEY (1985): Circadian rhythms and athletic performance. Med. Sci. Sports Exerc. 17: 498-516.
23. VOIGT, E., P. ENGEL u. H. KLEIN (1968): Über den Tagesgang der körperlichen Leistungsfähigkeit. Int. Z. angew. Physiol. 25: 1-12.
24. VOIGT, E. & P. ENGEL (1969): Tagesrhythmische Schwankungen des Energieverbrauchs bei Arbeitsbelastung. Pflügers Arch. ges. Physiol. 307: 89.
25. WAHLBERG, I. & I. ASTRAND (1973): Physical working capacity during the day and at night. Work Evironm. Health 10: 65-68.

CHRONOBIOLOGICAL ASPECTS OF REHABILITATION OF UROLOGICAL DISEASES.

H.M. SCHULTHEIS & CHR. GUTENBRUNNER

Klinik am Kurpark, Ziergartenstr. 2,
D-3590 Bad Wildungen-Reinhardshausen, FRG

INTRODUCTION

Chronobiological phenomena up to now don't play an important role in therapy of urological diseases including physical therapy and balneotherapy, however, the chronobiological research in this field made some progress in the last few years. Exept of research in the field of formation of renal calculi there exist no scientific research on chronobiological urology.

Rehabilitation in the field of urology mainly concerns renal stone formation, urinary tract infections, motoric disturbances of the bladder and postoperative states of urological tumors as well as states after radiation or chemotherapy. The treatment in the field of urological rehabilitation includes - besides of pharmacological treatment - physical therapy and balneotherapy as well as accompanying psychological treatment. This therapy mainly consists in electrotherapy, ultrasonic therapy, massage, and physiotherapy directet to specific objectives e.g. muscle training of the pelvic floor. The balneotherapy of urological diseases includes hot bathes, hot packs, mineral water cures and others. The normal duration of treatment is 4 to 6 weeks. The cooperating psychologists are specialized in the field of psycho-oncology.

CIRCADIAN RHYTHMS

The most evident chronobiological phenomenon with respect to the urinary tract is the circadian variation of diuresis (1, 16, and others). Because of the multiple factors who determine urine excretion and composition the explanation of the mechanism is very complex (15). Hormonal, circulatory, and neural factors influence renal urine production. These factors underlie a marked circadian variability.

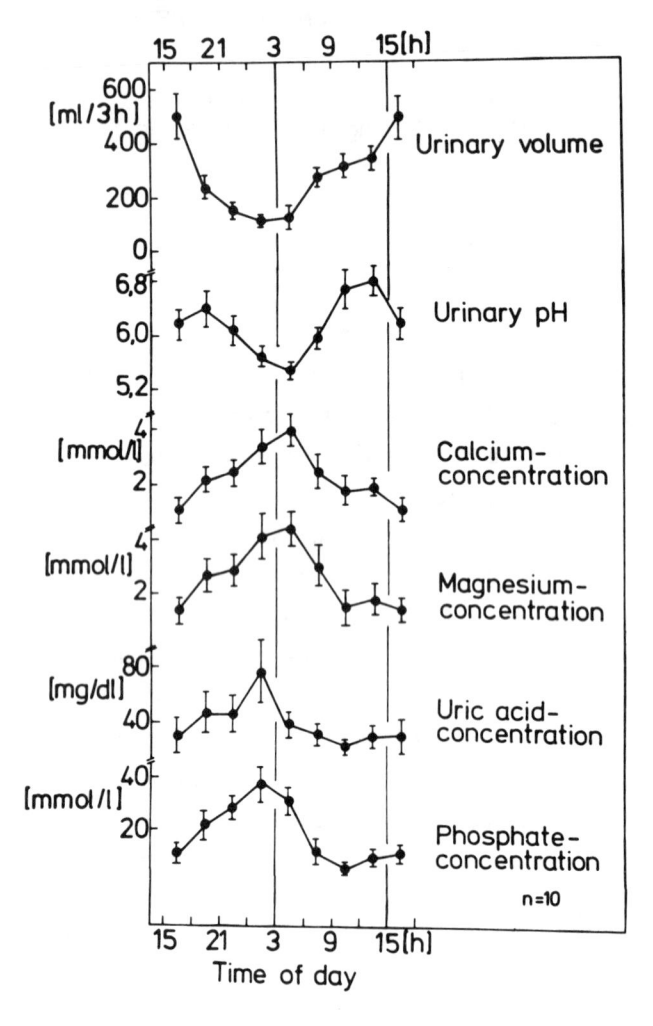

Fig.1: Mean circadian courses of urinary volume, urinary pH, and electrolyte concentrations in healthy subjects under constant bed rest and equally distributed food and fluid intake. Brackets indicate standard errors (8).

The circadian variation of urinary volume, pH, and some electrolyte concentrations are shown in Fig. 1 (8). These results are based on a study, in which nutritional and behavioral influences were excluded. The test persones stayed during the 24-hour experiment in a climatic chamber under constant bed rest. Food and fluid intake was equally distributed and given in 4-hour portions (see 6). About 250 ml of tap water were given every 3 hours. As the figure shows, diuresis is reduced during night time and increases during the day. Parallel with the antidiuresis urinary pH decreases and the electrolyte concentrations rise. This nightly minimum of urine production exhibits an increased risk of calculus formation and the development of infections (14, 21). Correspondingly most renal calculi have a characteristic laminated structure (18), of which each layer corresponds to a night of additional growth (2, 20).

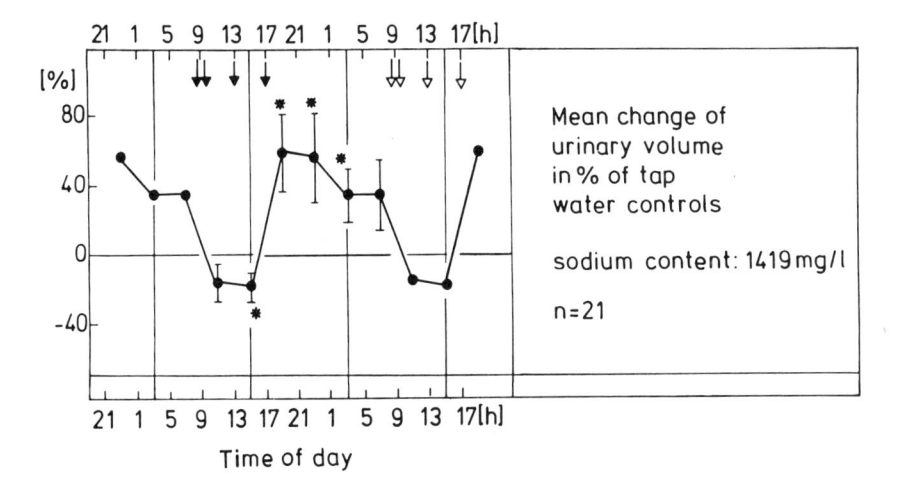

Fig.2: Mean circadian courses of the change of urinary volume after intake of a sodium containing (1.419 mg Na/l) natural mineral water intra-individually compared to tap water controls. Brackets indicate standard errors. For better overview, the circadian curves are plotted twice (4).

Using drinking cures with sodium containing natural mineral waters the circadian variation of diuresis can be influenced (14). In order to prove the effect of a sodium-containing medical mineral water on the circadian variation of diuresis, we measured the mean urinary volume after intake of the mineral water in comparison to tap water controls. The comparison was made in the same test persons, who came two times for 24 hours into our laboratory (4). All measurements were carried out under equally distributed food intake. In Fig. 2 the mean change of urinary volume is

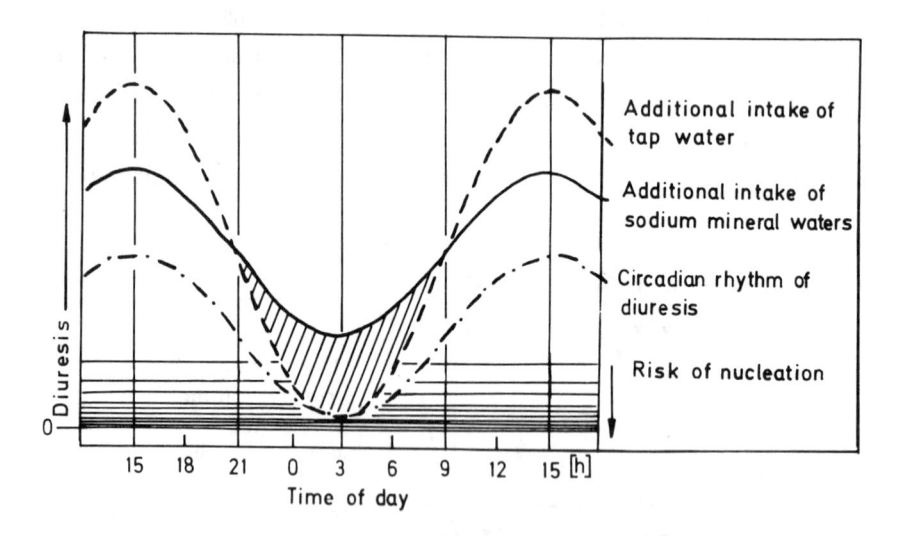

Fig.3: Scheme of the circadian variation of diuresis under normal conditions, additional intake of tap water, and sodium containing mineral waters respectively (3).

plotted. The negative values during forenoon correspond to the well known water retention after sodium intake (19). Starting 8 hours after the first mineral water portion, the urinary volume is increased compared to the tap water controls. The increase amounts in this case to about 50 % in the night. Testing different mineral waters we found, that the amount of the increase of the nightly urinary volume is correlated with the sodium concentration of the drunken mineral water (5).

The therapeutical effect of the described phenomenon is demonstrated in Fig. 3 (3). The lowest curve represents the normal circadian rhythm of diuresis. The dotted line above shows the increase of urine excretion after additional intake of tap

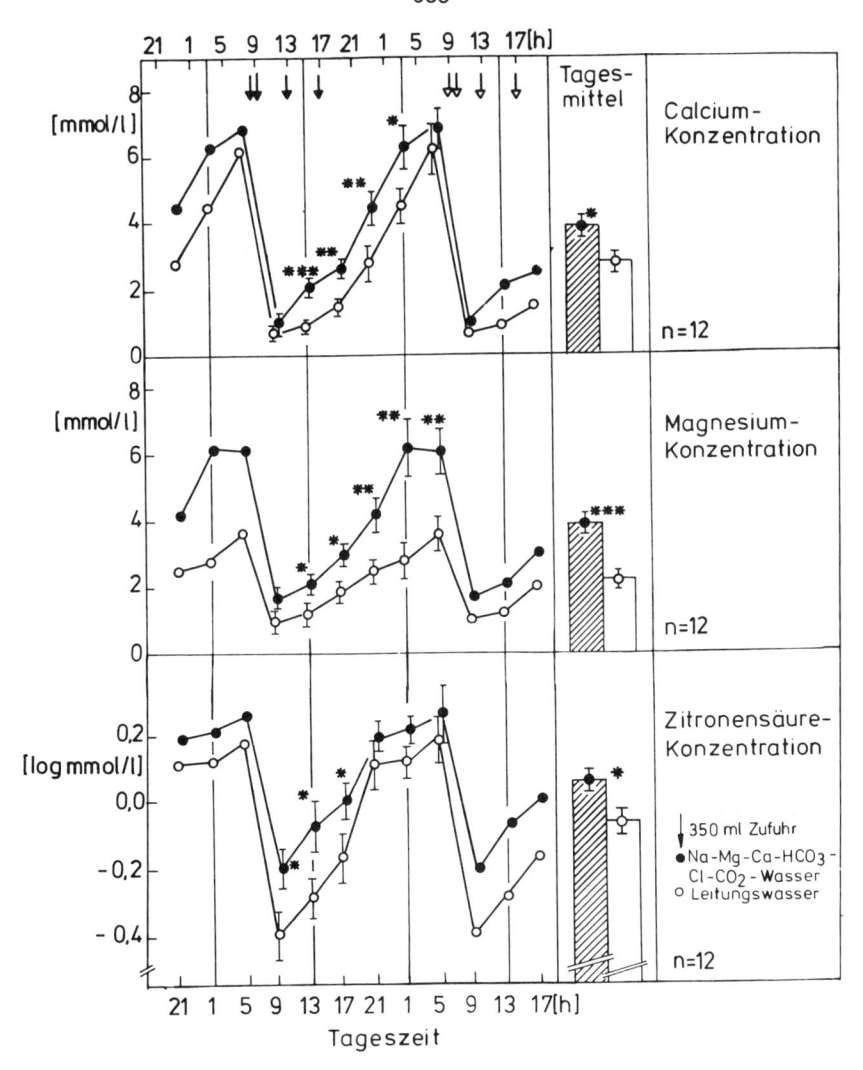

Fig.4: Mean circadian courses of urinary calcium, magnesium and citrate concentrations of healthy test persons under equally distributed food intake and additional intake of tap water or a medicinal mineral water (724 mg Na^+/l, 245 mg Mg^{++}/l, 312 mg Ca^{++}/l, 608 mg Cl^-/l and 3068 mg HCO_3^-/l) respectively. Brackets indicate standard errors. For better overview, the circadian curves are plotted twice (4).

water during daytime. No change of the nocturnal minimum was found. The immediat diuresis after the intake of a sodium mineral water is reduced compared to the tap water controls, however, the nightly diuresis is increased leading to a reduction of stone formation risk.

Because of the described circadian variation of urine composition, the effects of medicinal mineral waters must be evaluated with respect to chronobiological phenomena. So we examined the circadian variation of electrolyte excretion after intake of an alcaline calcium-magnesium water in healthy test persons compared to tap water (4). As Fig. 4 indicates, the calcium concentration is slightly increased during the whole 24-hour period. The magnesium concentration immediately after drinking the mineral water during forenoon exhibits only a slight increase, however, the maximum of the magnesium concentration could be seen during the night. This is of great importance, because of the stone protective effect of magnesium and the increased nocturnal risk of stone formation. Additionally the citrate concentration was increased after mineral water intake caused by urinary alcalization by hydrogen carbonate (11). Numerous studies verify the protective effect of high urinary magnesium and citrate concentrations on the risk of calcium-oxalate nucleation in urine (17).

CIRCASEPTAN RHYTHMS

The characteristic time structure of functional adaptative processes is the circaseptan periodicity (12, 13). This periodicity we found during 4-week balneotherapy in our patients suffering from relapsing renal calculi. We measured urinary composition day by day in the night urine (8). Fig. 5 shows for example in the upper curve the mean course of the uric acid concentration. The lower curve represents the frequency of peaks with uric-acid concentrations over 90 mg%. This value means a significant risk of uric acid precipitation in respect to the actual urinary pH values of these patients. Both curves demonstrate significant decreasing tendencies. Especially the lower curve has a periodic time structure with marked peaks around day 8 and day 17.

In order to prove the dominating time structure, we calculated the period durations of the individual curves of different urine parameters in the patients

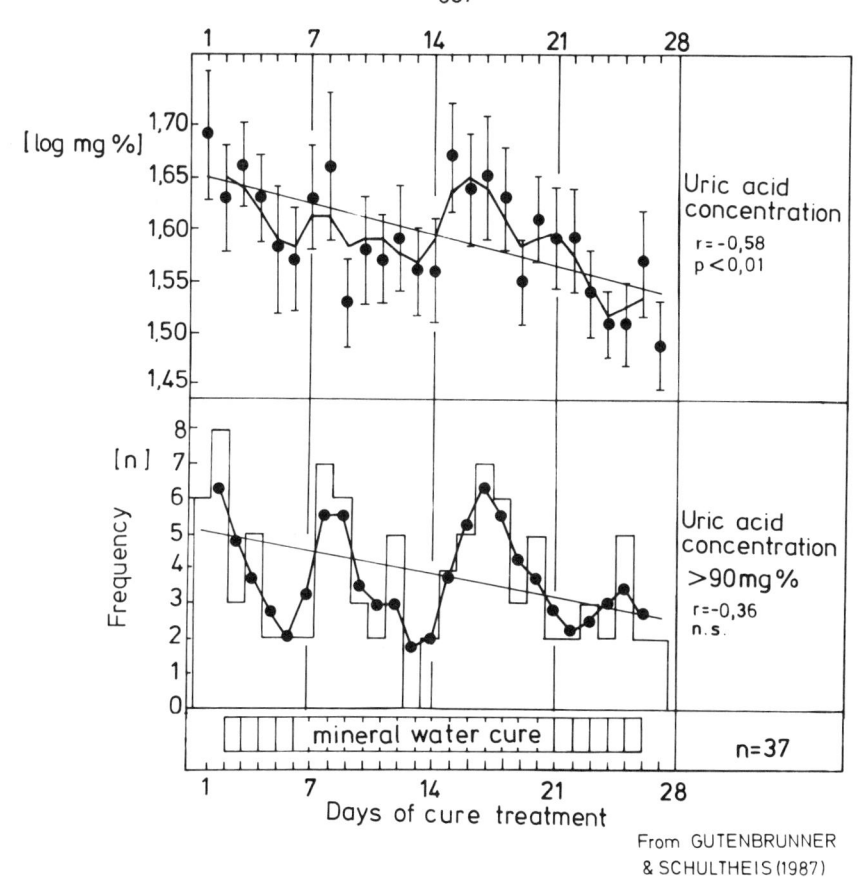

Fig.5: Mean uric acid concentration (upper curve) of the night urine (23.00 to 8.00) and frequency of concentration peaks over 90 mg% (lower curve) of 37 patients suffering from relapsing stone formation during a 4-week balneotherapeutic treatment. Besides the original data smoothed curves are plotted. Brackets indicate standard errors (8).

concerned (Fig. 6). The figure shows the frequency distribution of the maxima- and minima intervals of all electrolyte excretions as well as of the urinary volume curves (9). The frequencies are calculated as deviation from the expected values. Period duration from 4 to 8 days exceed the expected values, and the maximum occurs at a period duration of 7 days, showing that a circaseptan periodicity is dominating. This time structure is characteristic for functional adaptative processes.

In order to illustrate the therapeutic relevance of the changes in urine composition during mineral water cures in patients with renal calculi, we put the results in a diagram of uric acid precipitation risk (Fig. 7) (7). During the first days of treatment, the mean value was in a range of high precipitation risk. After treatment the mean solution-product of these patients was shifted into the range of uric acid solubility.

Functional adaptive processes not only concern metabolic functions but also other functions controlled by the autonomous nervous system. In the field of urology e.g. the motoric bladder function is involved (10). In a long-term study urinary flow values in urological patients during complex balneotherapy including mineral water cures were studied (Fig. 8). As figure shows, the average and maximum flow rate, the mean urinary volume, and the mean time of maximum flow rate significantly increased in the course of treatment. Consequently the residual volume decreased. All curves of the figure exhibit a marked circaseptan periodicity indicating that functional adaptive changes are involved. This hypothesis is supported by the fact that the urinary flow values underwent a significant normalization tendency (Fig. 9): The maximum flow rate (lower part of the figure) increased in patients with low initial values and decreased in patients with high initial values. The so-called cross-over point (= no change) was congruent with the normal value. Similar changes were found in the total voiding time (upper part of the figure).

The treatment of carcinoma in the last few years has been practiced with increasing radical surgery. Therefore a growing necessity of an adequate post-surgical treatment occured. Therefore in Germany the institution of so-called Anschluss-heilbehandlung (AHB) - translated as follow up curative treatment - was established. In a pilot study we examined subjective and objective parameters in 206 patients with different tumors of the urinary tract during a four week treatment period.

Fig. 10 shows for example the mean courses of different parameters of subjective well being, which were rated daily. The upper curve of general well-being shows an initial decrease in the first week. During the following three weeks of treatment the well-beeing ameliorates continously. The other parameters exhibit

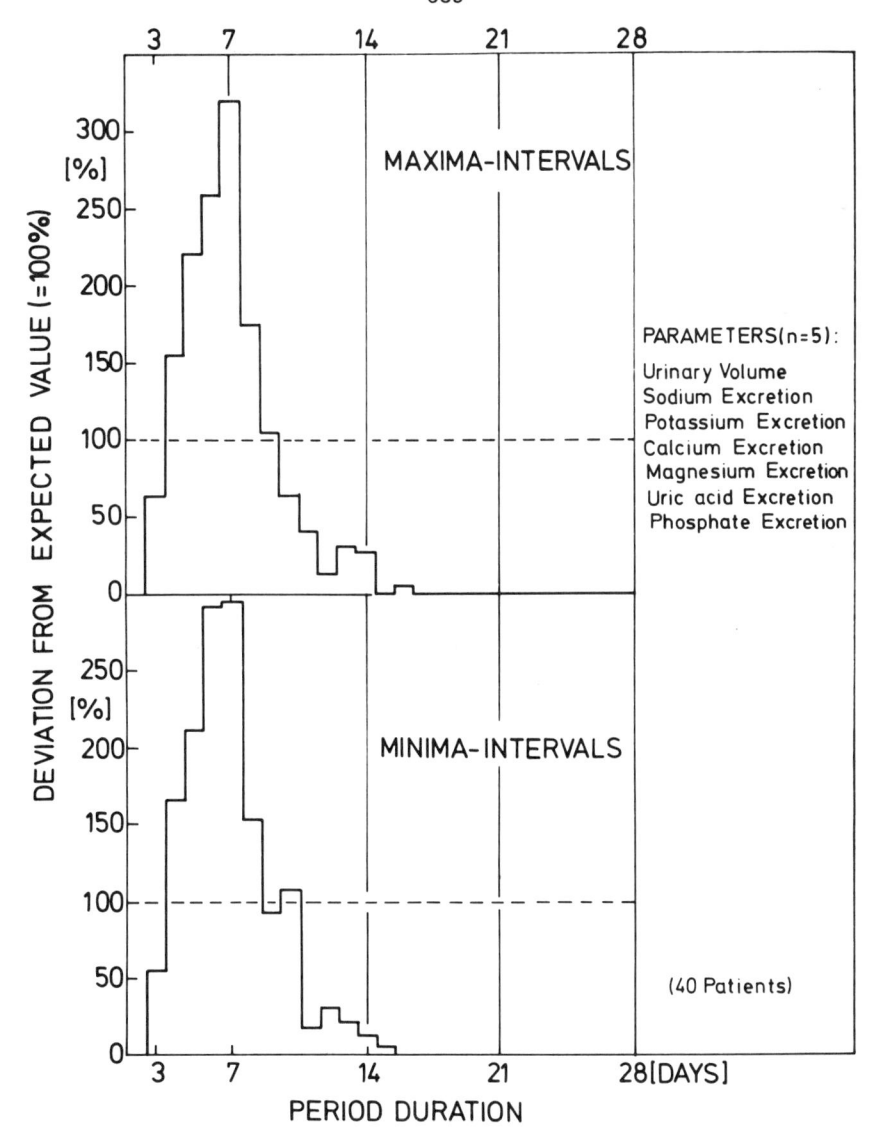

Fig. 6: Frequency distribution of the maximum and minimum intervals of the individual curves of the indicated parameters in daily sampled night urine probes (23.00 to 8.00) of 37 patients suffering from relapsing stone formation during a 4-week balneotherapeutic treatment (9).

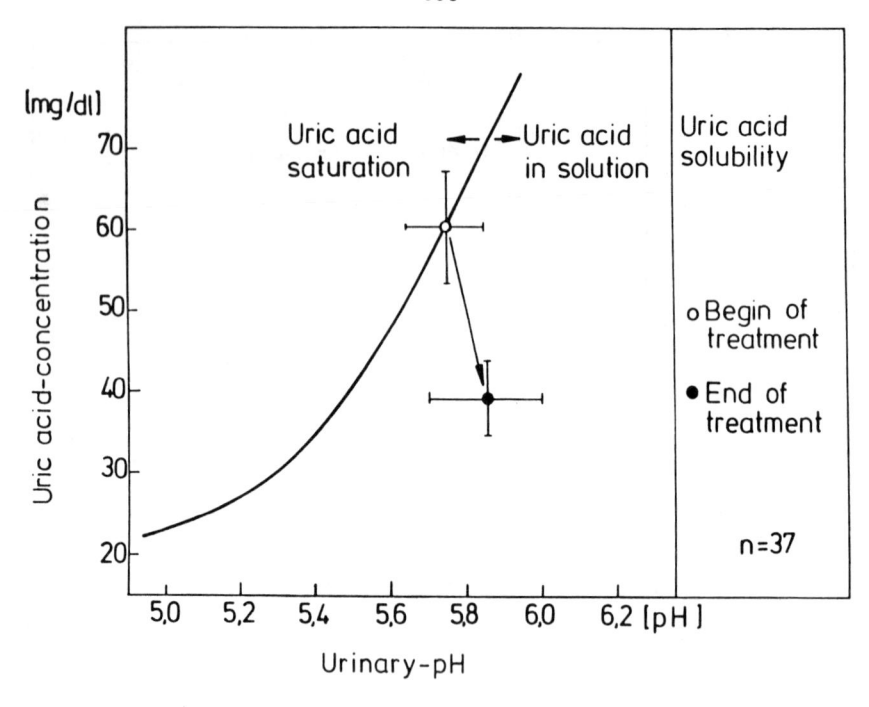

Fig.7: Mean uric acid concentration and pH-values of the night urine (23.00 to 8.00) of 37 patients suffering from relapsing stone formation before and after a 4-week balneotherapeutic treatment including a mineral water cure. Brackets indicate standard errors (7).

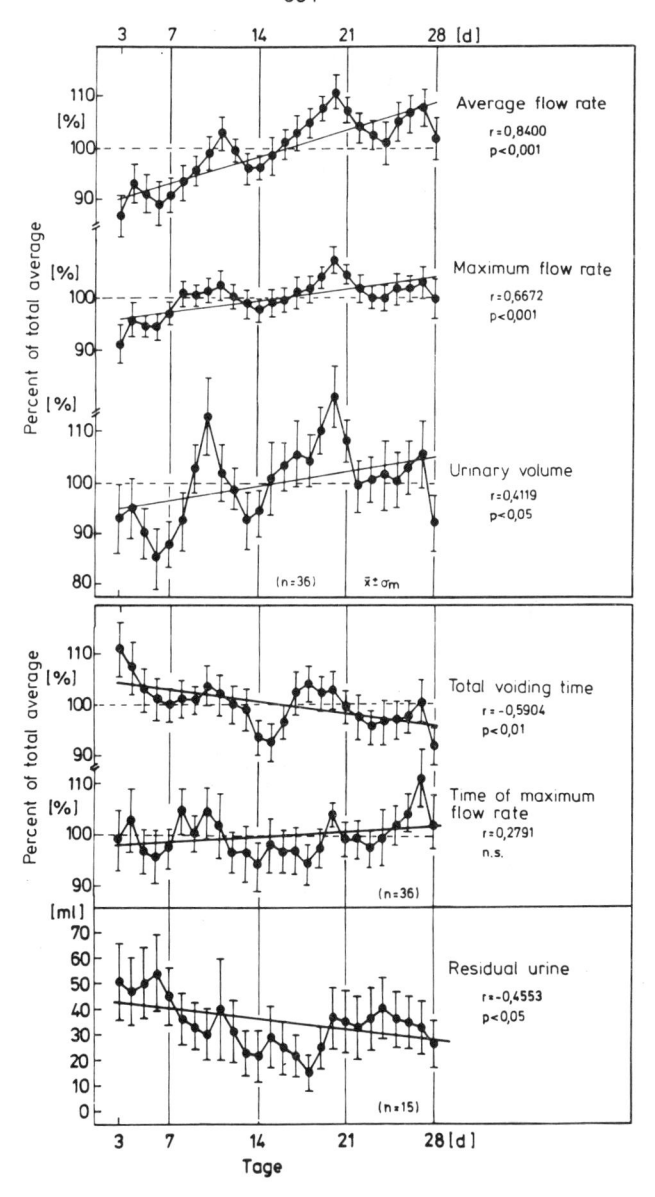

Fig.8: Mean courses of the indicated parameters of daily uroflow measurements of 36 patients during a 4-week balneotherapeutic treatment including mineral water cures. Brackets indicate standard errors (10).

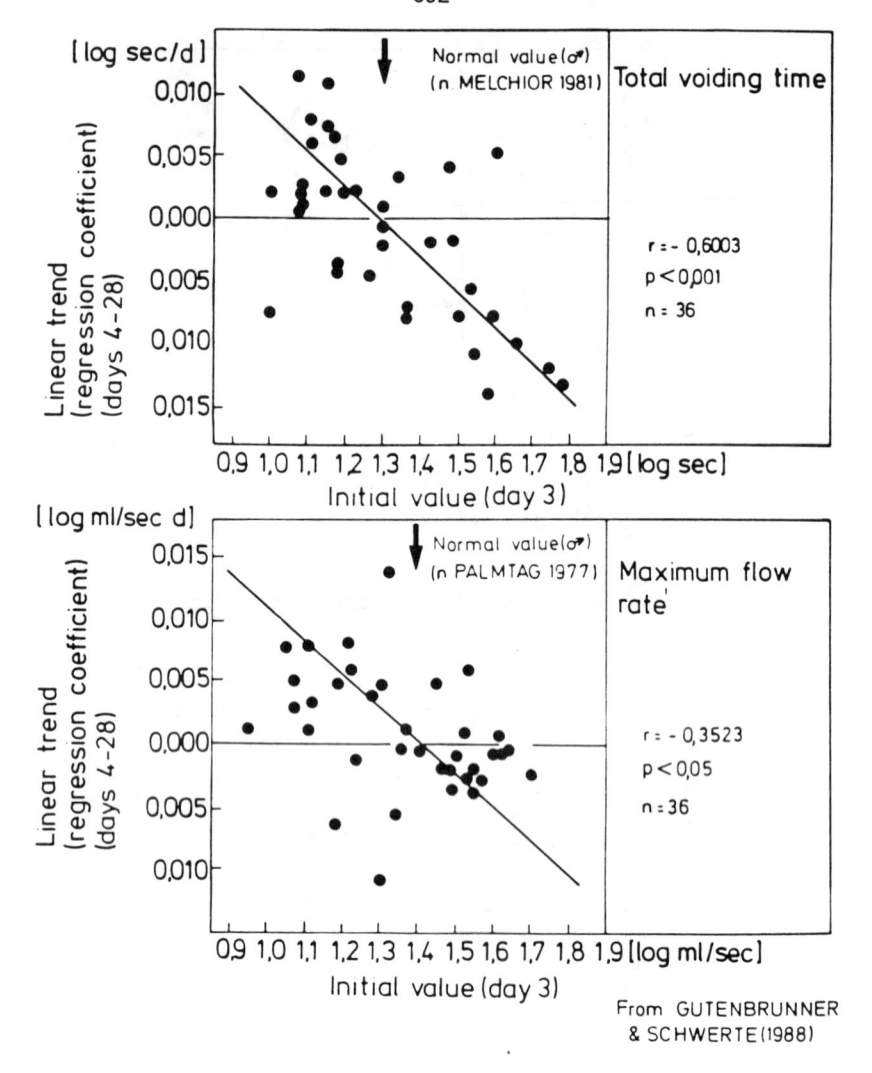

Fig.9: Correlation between the initial value and the respective linear trends of the indicated parameters of daily uroflow measurements of 36 patients during a 4-week balneotherapeutic treatment including mineral water cures. Normal values from the literature are indicated on top of the figure (10).

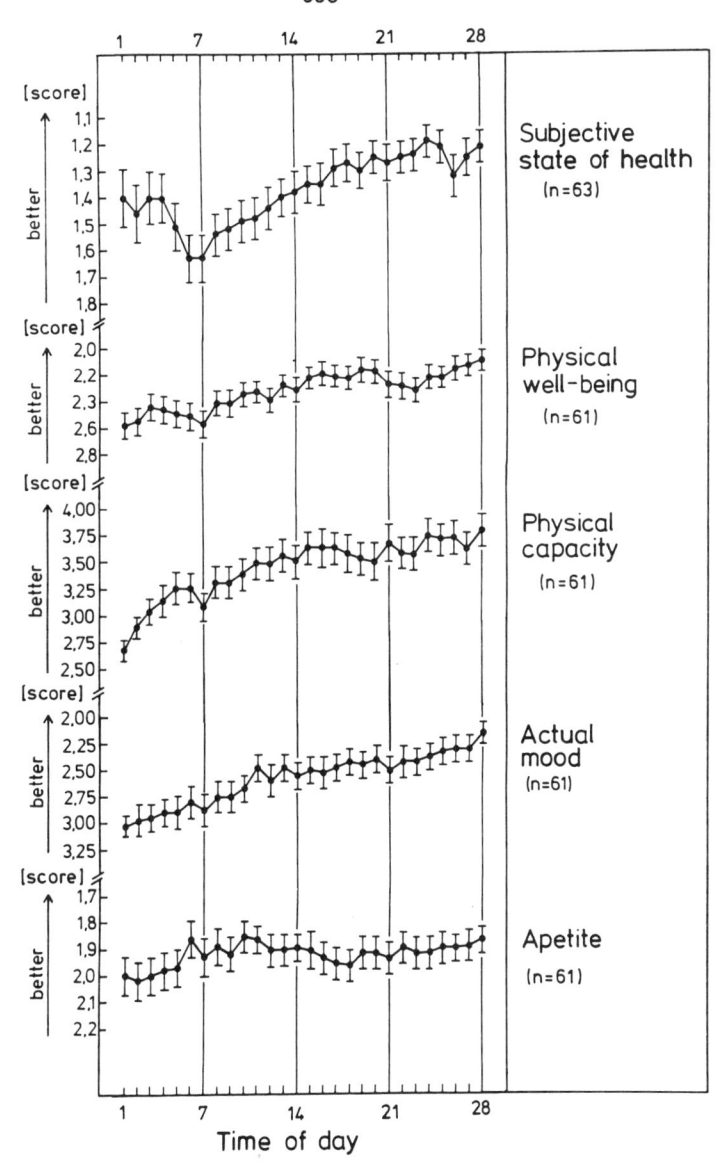

Fig.10: Mean courses of the indicated parameters of subjective well being of 63 patients undergoing a follow-up curative treatment. Brackets indicate standard errors.

increasing tendencies as well. The decrease till day seven of treatment is a hint for phasic processes, however, final conclusions are not yet possible. The mean number of erythrocytes and the mean hemoglobin concentration of these patients increased during the treatment and the mean kreatinine concentration decreased refering to non specific functional improvements. It can be supposed that functional adaptive processes are necessary, to be causal for these effects, however, further investigations are necessary, especially in order to investigate the time structure of these processes.

CONCLUSIONS

The consideration of circadian variations of renal functions is important in the therapy of urological disorders especially because of the circadian maximum of the stone forming risk at night. The chronobiological investigations show new aspects of the effect of mineral water cures, e.g. the possible flattening of the circadian rhythm of diureses caused by sodium containing mineral waters. However, our knowlege is still small and further investigation is needed. Evidently balneotherapy can cause adaptive processes which are also relevant in urological rehabilitation. The time structure of the courses of different functions show a dominating circaseptan periodicity which is characteristic for functional adaptive processes.

REFERENCES

1. CONROI, R.T.W.L. & J.N. MILLS (1970): Human circadian rhythms. Churchill, London
2. GASSER, G. (1987): Personal Message.
3. GUTENBRUNNER, Chr. (1986): Erkrankungen der Nieren und ableitenden Harnwege. In: W. AMELUNG u. G. HILDEBRANDT (Hrsg.): Balneologie und medizinische Klimatologie. Band III. Springer-Verlag, Berlin-Heidelberg-New York-Tokyo, S. 120-129.
4. GUTENBRUNNER, Chr. (1988): Untersuchungen über den Einfluss calciumhaltiger Mineralwässer auf die Harnzusammensetzung. In: G. GASSER u. W. VAHLENSIECK (Hrsg.): Pathogenese und Klinik der Harnsteine XIII. Steinkopff-Verlag, Darmstadt, S. 247-253.
5. GUTENBRUNNER, Chr., K.F. HOLTZ, B. MÜLLER & M. PETRI (1984): Influence of sodium mineral waters on the circadian variation of urine excretions. In: A. REINBERG, L.E. SCHEVING & G. LABERCKE (Eds.): Annual Review of Chronopharmacology. Pergamon Press, Oxford, pp. 339-402.
6. GUTENBRUNNER, Chr. & U. SCHREIBER (1987): Circadian variations of urine excretion and adrenocortical function in different states of hydration in man. In: J.E. PAULY & L.E. SCHEVING (Eds.): Advances in Chronobiology, Part A. Alan R. Liss, New York, pp. 309-316.
7. GUTENBRUNNER, Chr. u. H.M. SCHULTHEIS (1986): Untersuchungen über die Wirkungen kurörtlicher Trinkkuren auf die Wasser- und Elektrolytausscheidungen bei Harnsteinträgern. Niere Blase Prostata aktuell 11/2: 8 (1986).
8. GUTENBRUNNER, Chr. u. H.M. SCHULTHEIS (1987): Untersuchungen über die Wirkung kurörtlicher Trinkkuren auf die Wasser- und Elektrolytausscheidungen bei Harnsteinbildnern. In: W. VAHLENSIECK u. G. GASSER (Hrsg.): Pathogenese und Klinik der Harnsteine XII. Steinkopff-Verlag, Darmstadt, S. 233- 238.
9. GUTENBRUNNER, Chr. & H.M. SCHULTHEIS (1987): Circaseptan reactive periodicity of renal functions during 4-week balneotherapeutical cure treatment. In: G. HILDEBRANDT, R. MOOG & F. RASCHKE (Eds): Chronobiology and Chronomedicine. P. Lang- Verlag, Frankfurt-Bern-New York-Paris, pp. 398-402.
10. GUTENBRUNNER, Chr u. U. SCHWERTE (1989): Klinische Längsschnittuntersuchungen des Uroflow im Verlauf vierwöchiger komplexer Kuren. Z. Phys. Med., 153-162 (1989).
11. HESSE, A., I. BÖHMER, R.M. SCHÄFER u. W. VAHLENSIECK (1987): Zur Wirkung eines Bicarbonat-, Magnesium- und Calcium-haltigen Mineralwassers auf die Harnzusammensetzung. VitaMinSpur $\underline{2}$: 73-77.
12. HILDEBRANDT, G. (1985): Therapeutische Physiologie - Grundlagen der Bäder- und Klimaheilkunde. In: W. AMELUNG u. G. HILDEBRANDT (Hrsg.): Balneologie und medizinische Klimatologie, Band 1. Springer-Verlag, Berlin-Heidelberg-New York- Tokyo 1985.

13. HILDEBRANDT, G. (1991): The time structure of physiological responses to cure treatment. This volume.
14. HILDEBRANDT, G., CHR. GUTENBRUNNER u. CHR. HECKMANN (1983): Trinkkuren - neue Forschungsergebnisse. Heilbad u. Kurort 35: 34-54
15. KOOPMAN, M.G., D.S. MINORS & J.M. WATERHOUSE (1989): Urinary and renal circadian rhythms. In: J. ARENDT, D.S. MINORS & J.M. WATERHOUSE (Eds.): Biological rhythms in clinical practice. Wright, London-Boston-Singapore-Sydney-Toronto- Wellington, pp. 83-98.
16. MINORS, D.S., J.N. MILLS & J.M. WATERHOUSE (1978): The circadian variation of urinary electrolytes and the deep body temperature. Int. J. Chronobiology 4: 1-28.
17. ROBERTSON, W.J. & M. PEACOCK (1985): Pathogenesis of Urolithiasis. In: H.J. SCHNEIDER (Hrsg.): Urolithiasis: Etiology - Diagnosis. Springer-Verlag, Berlin-Heidelberg-New York-Tokyo, P. 185-334.
18. SCHNEIDER, H.J. (1985): Morphology of urinary tract concretions. In: H.J. SCHNEIDER (Ed.): Urolithiasis: Etiology - Diagnosis. Springer, Berlin-Heidelberg-New York-Tokyo, pp. 1-136.
19. STARKENSTEIN, E. (1924): Über die Abhängigkeit der Diurese vom Salzgehalt des getrunkenen Wassers. Arch. exper. Pathol. Pharmakol., 6-22.
20. ÜBELHÖR, R. (1957): Trinkkuren bei Nierensteinleiden. Wien. med. Wschr., 669-704.
21. VAHLENSIECK, W., D. BACH & A. HESSE (1982): Circadian Rhythm of Lithogenic Substances in the Urine. Urol. Res., 195-203.

INDICES

AUTHOR INDEX

ADAN, A. 499, 520
ADLER, N.T. 88
ALDRICH, J.C. 533
ALMIRALL, H. 527
AMBROSI, A.D' 288
ARNDT, H. 8
ARRÓYAVE, R.J. 396
ASLANIAN, N.L. 240
ATKINSON, G. 478
AYALA, D.E. 103, 263, 361, 370, 396
BALAZOVJECH, I. 271
BALSCHUN, D. 151
BALZER, I. 113, 128
BARTSCH, C. 390, 405
BARTSCH, H. 390, 405
BAUER, T.T. 450
BECKER, E. 417
BEHRMANN, G. 113
BENNETT, M.F. 53
BENZI, G. 254
BÖHM, H. 8
BOLDT, W. 3
BÖSE, B. 229
BOTHMANN, M. 247
BRANDYS, J. 422
BRODBECK, B. 254
BURNS, J.T. 172
CARANDENTE, F. 35
CHARVATOVA, I. 63
CHEREPANOVA, V.A. 485
CHIBISOV, S.M. 47
CORDINER, S. 177
COURT, L. 41
CRONIBUS, J. 168
CUGINI, P. 547
CYMBOROWSKI, B. 12
DAHLHELM, D. 439
DAVYDOVA, O.N. 473
DETTI, L. 254
DIAZ, A. 508
DOLCI, C. 35
DOMOSLAWSKI, J. 541
DUKAT, A. 271
EBELING, W. 603

EBNER, J.	216
EIMERT, H.	145
ENGEL, P.	646
ENGELKE, P.	305
FALKENBACH, A.	168
FERLINI, M.	288
FERNÁNDEZ-LORENZO, J.R.	103
FERNÁNDEZ, J.R.	103, 370
FIŠER, B.	98
FRAGA, J.M.	103
FREUND, H.-J.	229, 411
FRÖLICH, M.	378
FROLOV, V.A.	47
GARCIA, L.	370
GOURLET, V.	41
GÜNTHER, R.	339
GUTENBRUNNER, CHR.	345, 665, 681
HAJEK, D.	386
HALBERG, F.	70
HARBACH, B.	30
HARDELAND, R.	113, 128
HECKMANN, C.	305
HEFTER, H.	229, 411
HEINDRICHS, E.	345
HELFRICH-FÖRSTER, C.	134
HELWIN, H.	562
HERMIDA, R.C.	103, 263, 361, 370, 396
HEROLD, M.	339
HESS, G.	428
HIEKEL, H.-G.	8
HILBIG, H.	141
HILDEBRANDT, G.	194, 247, 320, 333, 624
HOFFMANN, P.	26
HONZÍKOVÁ, N.	98
HUDSON, R.	157
IGLESIAS, T.	370
JANKE, S.	151
JENSSEN, M.	603
JOHL, C.	216
KAWASAKI, T.	247
KAYSER, M.	299
KESTING, G.	305
KLEMFUSS, H.	450
KLINKER, L.	163
KONVIČKOVÁ, E.	98
KRIPKE, D.F.	450
KWILECKA, M.	506
KWILECKI, K.	506

LEEUWEN, P.van, ... 305
LEWANDOWSKI, M.H. 432
LIETAVA, J. 271
LINKE, W. 3
LITYŃSKA, A. 16
LODEIRO, C. 370
LOGIGIAN, E. 411
LÓPEZ-FRANCO, J.J. 396
MAGGIONI, C. 254
MARKUS, M. 589
MARGRAF, J. 417
MARX, U. 325
MATYEV, E.S. 47
McEACHRON, D.L. 88
MELLO, G. 254
MERAT, P. 41
MIANI, A. 35
MIKES, Z. 271
MIKULECKY, M. 356
MLETZKO, H.-G. 93
MLETZKO, I. 30, 93
MOGYLEVSKY, V. 47
MOJÓN, A. 103, 370
MONTANARI, L. 288
MONTARULI, A. 35
MOOG, R. 247, 333
MORGAN, E. 177
MOSHKIN, M.P. 57
OSCHÜTZ, H. 513
ONDREJKA, P. 356
PANKOW, D. 26
PATZAK, A. 216
PEIL, J. 562
PENZEL, T. 325
PERRAMON, A. 41
PETER, J.H. 325, 333
PFLUG, B. 450
PIAZENA, H. 163
PIEKOSZEWSKI, W. 422
PIOTRKIEWCZ, M. 229
PÖGGELER, B. 113
PÖLLMANN, L. 351
PORTALUPPI, F. 288
PUTET, G. 41
PUTILOV, A. 485
PUTILOV, A.A. 493
RAMMSAYER, T. 463
RAUCH, E. 469

REINERS, K.	411
RENSING, L.	570
REUSS, S.	21, 121
REY, A.	103
RICHTER, R.	299
RIEMANN, R.	21
RODRÍGUEZ-CERVILLA, J.	103
ROELFSEMA, F.	378
ROTHENBERGER, K.A.	172
RUDOLPH, H.-J.	310
RUDORFF, KH.	305
RUMLER, W.	70
RUTKOWSKA, A.	422
SÁIZ, M.	508
SÁIZ, D.	508
SALVADOR, A.	589
SAMMECK, R.	77
SÁNCHEZ-TURET, M.	499, 520
SCHMERLING, S.	562
SCHMIDT, R.	457
SCHNEIDER, H.	325
SCHNEIDER, S.	417
SCHNIZER, W.	660
SCHUH, J.	3, 70, 82, 145, 151
SCHULTHEIS, H.M.	681
SEIFFERT, U.	168
SITAR, J.	315
SITKA, U.	70
STEBEL, J.	223
STEGMAIER, J.	469
STEPHANI, U.	77
STRESTIK, J.	63
STUBBE, A.	185
STUPFEL, M.	41
SÜSSMUTH, T.	182
TAFIL-KLAWE, M.	320
TSCHUCH, G.	182
UBERTI, E.degli	288
UEZONO, K.	247
UNKELBACH, U.	168
VERGNANI, L.	288
VIZZOTTO, L.	35
VOGEL, M.	333
WECKENMANN, M.	469
WEINERT, D.	70, 82, 145
WITTE, O.	411
WICHERT, P.von	325
WOLNA, P.	26

WUSSLING, M. 3
ZASLAVSKAYA, R.M. 277
ZHIGULINA, E.I. 57
ZÖPHEL, U. 185

INDEX

β-glucuronidase	168
5-htp	172
5-methoxylated indoleamines	128
acrophases	82
ACTH	378
activity rhythms	185
adaptability	493
adults	145
age	82, 88, 547
adults	145
infants	41, 77
children	370, 508, 513
development	64, 70
infants	41, 77
juvenile	145
neonatal	3, 70, 103
newborns	77, 98
ontogeny	93
premature	41, 98
rhythmogenesis	77
animals	
chicks	41
drosophila	134
fish	177
gonyaulax polyedra	128
guinea-pigs	41
hamsters	450
insects	12
mammals	121
mexican cave fish	177
mice	41, 141, 145, 445
microtus brandti	185
monkeys	41
musca domestica l	439
newts	53
pigs	41
quail	41
rabbit	47, 157
rat	3, 16, 21, 30, 35, 41, 422, 428
squirrel	172
amitriptyline	422
amplitude	3, 216, 41, 82
analysis	
all-purpose data processing package	541
chronobiometry	77
clinorhythmometry	547

modelfree	548, 562
multiple component	370
periodic-linear regression	547
spectral analysis	98
rhythmometry	361
apnea	325
arthritis	339
atrial natriuretic peptide	288
balneotherapy	660
blood flow	320
blood pressure	103, 254, 263
breast cancer	405
breathing rate	216, 320
brightness	163
C-reactive protein	339
cancer	396, 405
carbon dioxide	41
cardiac arrhythmia	299
cardiovascular	70, 103, 240, 247, 315
cell	3, 8, 134, 345, 589
children	370, 508, 513
chlorella vulgaris	8
chronobiometry	77
chronopharmacokinetics	422
chronotype	527
chicks	41
circadian	12, 26, 30, 35, 82, 93, 113, 121, 134, 141, 145, 163, 177, 185, 240, 254, 271, 288, 294, 333, 334, 339, 345, 351, 378, 417, 439, 445, 450, 478, 499, 506, 520, 527, 646, 665,
circadian phase type	485, 493, 499, 527
circannual	185, 361, 390, 396
circaseptan	351
clinorhythmometry	547
coherence of rhythm	88
cold water	310
congestive heart failure	288
coordination	186, 194
coronary artery disease	271
cortisol	378
coupling	229, 463, 569, 570
cure treatment	617, 624
cyst induction	128
cytochrome P-450 IIE1	26
darkness	445
development	64, 70
diarrhea	356
disease	271, 305, 681
dopaminergic	463

drosophila 134
drugs, pyrogenous 469
ecdysteroid 12
ecological 57
endocrine 405
entrainment 111
erythrocyte 339
evolution 113
feeding 151
fish 177
flights 163
galanthamine 428
geomagnetic 63
gonyaulax polyedra 128
graves' disease 305
guinea-pigs 41
hamsters 450
hand movements 229
heart   3, 98, 103, 216, 277, 288, 294, 277, 288
    cardiac arrhythmia 299
    cardiovascular 70, 103, 240, 247, 315
    congestive heart failure 288
    heart cells 3
    heart rate 98, 103, 216, 294
    heart rate variability 98
hemi-parkinson 229
hemodynamic 271
hemolymph 12
hibernation 172
hypothalamic 121
immune 345
immunoelectrophoresis 16
immunogenic hyperthyroidism 305
individual differences 533
    chronotype 527
    circadian phase type 485, 493, 499
    sex 378, 499
infants 41, 77
infarction 294
infections 356
insect 12
insecticide 439
insulin 370
interleukin-2 339
intestine 30
juvenile 145
kidney 151
l-dopa 172

L/D-proportion	163
laser	3
laterality rhythm	320
light exposure	168
liver	35
locomotor activity	134, 432
low temperature	128
lunar	315
mammal	121
man	163, 194, 229, 263, 271, 299, 305, 325, 339, 345, 378, 405, 469, 506
melatonin	113, 157, 113, 157
mexican cave fish	177
mice	41, 141, 145, 445
microphthalmus	141
microtus brandti	185
microvascular	320
molecular	603
monkeys	41
moon	356
morningness	499
morphogenesis	570
morphometric	35
mortality	315
motilar activity	30
mucosa	320
musca domestica l	439
myocardial	47
myocardial infarction	294
myoclonus	411
na/k	386
nasal	320
ncpap	333
neocortex	428, 432
neonatal	3, 70, 103
neuroleptic drug	445
neuromimetic drugs	445
neuropeptides	121
neurosecretory	134
neurospora	570
newborns	77, 98
newts	53
night wakefulness	485
nivalin	428, 432
nonapeptides	21
notophthalmus viridescens	53
ontogeny	93
oscillations	603

oscillator 121, 569, 570
palatal myoclonus 411
panic attacks 417
pathogenesis 411
patients 229, 271, 299, 305, 325, 339, 405, 469
peptide 151, 288, 151, 288
perception of time 223
performance 463, 508, 513
phase 145, 182, 194
photoperiod 111, 151
photoperiodism 113, 128
phylogenesis 113
physical activity 506
physical medicine 617
physical training 665
phytoadaptogens 473
pigs 41
pineal gland 21, 390
potassium 450
power spectra 216
pregnancy 254, 263, 277
premature 41, 98
prenataltoxic 457
prostate cancer 405
protein 339
proton 128
psychomotor performance 463
pulse-respiration ratio 305
pyrogenous drugs 469
quail 41
questionnaire 493
rabbit 47, 157
rat 3, 16, 21, 30, 35, 41, 422, 428
reactive periods 469
receptor 339
regulation 21
rehabilitation 681
respiration 310, 325
responsiveness 432
rheumatoid arthritis 339
rhythms 194
    circadian 12, 26, 30, 35, 82, 93, 113, 121, 134, 141, 145, 163, 177,
                      185, 240, 254, 271, 288, 294, 333, 334, 339, 345, 351, 378,
                      417, 439, 445, 450, 478, 499, 506, 520, 527, 646, 665
    circannual 185, 361, 390, 396
    circaseptan 351
    ultradian 41
    seasonal 47, 53, 57, 121, 473

semilunar	315
spectrum	469
rhythmogenesis	77
rhythmometry	361
rubidium	450
ß-glucuronidase	168
ß-n-acetyl-d-glucosaminidase	16
salmonella	361
seasonal	47, 53, 57, 121, 473
sedimentation rate	339
self-assessment	493
semilunar	315
sex	378, 499
sinus arrhythmia	310
skin	351, 660
sleep	325, 485, 493
sleep apnea	325
sleep-wake	485, 493
solar	63
solitonic excitations	603
spectral analysis	98
spectrum	469
speed tasks	499
spermophilus tridecemlineatus	172
spontaneous	247
suprachiasmatic nuclei	121
swimming activity	177
synaptic transmission	428
synchronization	589
tasks	499, 508
temperature	70, 520, 527, 646
thermoregulation	57
thyroid	88
training	665
tuberculin	351
tumor	390, 396, 405
ultradian	41
urine	386, 390
urological	681
uterine	396
valvular diseases	277
vasomotion	660
vegetative rhythms	223
weather	163
wheel-running	445
work-rate	478
zeitgeber	145, 151

Antoni Díez-Noguera / Trinitat Cambras (Eds.)
Chronobiology & Chronomedicine
Basic Research and applications

Frankfurt/M., Berlin, Bern, New York, Paris, Wien, 1992. VIII, 415 pp., numerous fig.
ISBN 3-631-44960-7   pb. DM 98.--

The development of Chronobiology is illustrated by the diversity of fields, both fundamental and applied, in which the study of rhythmicities is increasingly becoming an essential part of the research. This book covers the Proceedings of the VI Annual Meeting of the European Society for Chronobiology, which took place in Barcelona in July 1990. It describes the recent advances in Chronobiology by scientists from Europe working on basic research as well as on medical applications.

Contents: Cellular mechanisms – Circadian rhythm mechanisms – Regulation by endogenous and exogenous factors – Human rhythms in health and disease – Rhythms in social and occupational health

Verlag Peter Lang Frankfurt a.M. · Berlin · Bern · New York · Paris · Wien
Auslieferung: Verlag Peter Lang AG, Jupiterstr. 15, CH-3000 Bern 15
Telefon (004131) 9411122, Telefax (004131) 9411131

− Preisänderungen vorbehalten −